DIFFUSION

MASS TRANSFER IN FLUID SYSTEMS

THIRD EDITION

Diffusion: Mass Transfer in Fluid Systems brings unsurpassed, engaging clarity to a complex topic. Diffusion is a key part of the undergraduate chemical engineering curriculum and at the core of understanding chemical purification and reaction engineering. This spontaneous mixing process is central to our daily lives, important in phenomena as diverse as the dispersal of pollutants to digestion in the small intestine. For students, this new edition goes to the basics of mass transfer and diffusion, illustrating the theory with worked examples and stimulating discussion questions. For professional scientists and engineers, it explores emerging topics and explains where new challenges are expected. Retaining its trademark enthusiastic style, the book's broad coverage now extends to biology and medicine.

This accessible introduction to diffusion and separation processes gives chemical and biochemical engineering students what they need to understand these important concepts.

New to this Edition

- **Diffusion:** Enhanced treatment of topics such as Brownian motion, composite materials, and barrier membranes.
- **Mass transfer:** Fundamentals supplemented by material on when theories work and why they fail.
- **Absorption:** Extensions include sections on blood oxygenators, artificial kidneys, and respiratory systems.
- **Distillation:** Split into two focused chapters on staged distillation and on differential distillation with structured packing.
- **Advanced Topics:** Including electrolyte transport, spinodal decomposition, and diffusion through cavities.
- **New Problems:** Topics are broad, supported by password-protected solutions found at www.cambridge.org/cussler.

Professor Cussler teaches chemical engineering at the University of Minnesota. His research, which centers on membrane separations, has led to over 200 papers and 4 books. A member of the National Academy of Engineering, he has received the Colburn and Lewis awards from the American Institute of Chemical Engineers, the Separations Science Award from the American Chemical Society, the Merryfield Design Award from the American Society for Engineering Education, and honorary doctorates from the Universities of Lund and Nancy.

DIFFUSION
MASS TRANSFER IN FLUID SYSTEMS

THIRD EDITION

E. L. CUSSLER
University of Minnesota

CAMBRIDGE
UNIVERSITY PRESS

CAMBRIDGE
UNIVERSITY PRESS

University Printing House, Cambridge CB2 8BS, United Kingdom

Cambridge University Press is part of the University of Cambridge.

It furthers the University's mission by disseminating knowledge in the pursuit of
education, learning and research at the highest international levels of excellence.

www.cambridge.org
Information on this title: www.cambridge.org/9780521871211

First published 1984
Second edition 1997
Third edition 2009
12th printing 2018

Printed in the United Kingdom by TJ International Ltd. Padstow Cornwall

A catalog record for this publication is available from the British Library

Library of Congress Cataloging in Publication Data

Cussler, E. L.
Diffusion : mass transfer in fluid systems / E.L. Cussler. – 3rd ed.
p. cm.
Includes index.
ISBN 978-0-521-87121-1 (hardback)
1. Diffusion. 2. Mass transfer. 3. Fluids. I. Title.

TP156.D47C878 2008
660'.294–dc22   2008018927

ISBN 978-0-521-87121-1 Hardback

For Jason, Liz, Sarah, and Varick
who wonder what I do all day

Contents

List of Symbols

a	surface area per volume
a	major axis of ellipsoid (Section 5.2)
a, a_i	constant
A	area
A	absorption factor (Chapters 13 and 14)
b	constant
b	minor axis of ellipsoid (Section 5.2)
B	bottoms (Chapters 10, 12 and 13)
B, b	boundary positions (Section 7.3)
c	total molar concentration
c_1	concentration of species 1, in either moles per volume or mass per volume
c_{CMC}	critical micelle concentration (Section 6.2)
c_T	total concentration (Chapter 6)
$\bar{c}_1$	concentration of species 1 averaged over time (Sections 4.3 and 17.4)
c_1'	concentration fluctuation of species 1 (Sections 4.3, 17.3, and 17.4)
$\underline{c}$	vector of concentrations (Section 7.3)
c_{1i}	concentration of species 1 at an interface i
C	capacity factor (Section 13.1)
$\tilde{C}_p, \hat{C}_p$	molar and specific heat capacities respectively, at constant pressure
$\tilde{C}_v, \hat{C}_v$	molar and specific heat capacities respectively at constant volume
d	diameter or other characteristic length
D	binary diffusion coefficient
D	distillate (Chapters 12 and 13)
D_{eff}	effective diffusion coefficient, for example, in a porous solid
D_i	binary diffusion coefficient of species i
D_0	binary diffusion coefficient corrected for activity effects
D_{ij}	multicomponent diffusion coefficient (Chapter 7)
D_{Kn}	Knudsen diffusion coefficient of a gas in a small pore
D_m	micelle diffusion coefficient (Section 6.2)
D^*	intradiffusion coefficient (Section 7.5)
E	dispersion coefficient
E	extraction factor (Chapter 14)
$E(t)$	residence-time distribution (Section 9.2)
f	friction coefficient for a diffusing solute (Section 5.2)
f	friction factor for fluid flow (Chapter 21)
F	packing factor (Section 10.2)
F	feed (Chapters 12 and 13)
F	Faraday's constant (Section 6.1)
$F(D)$	solution to a binary diffusion problem (Section 7.3)
g	acceleration due to gravity

G	molar flux of gas
G''	mass flux of gas (Sections 10.2 and 13.1)
G'	molar flux of gas in stripping section (Chapters 12 and 13)
h	reduced plate height (Section 15.5)
h, h_i	heat transfer coefficients (Chapters 20 and 21)
H	partition coefficient
$\tilde{H}, \hat{H}$	molar and specific enthalpies (Chapters 20–21 and Chapter 7, respectively)
$\bar{H}_i$	partial specific enthalpy (Chapter 7)
HTU	height of transfer unit
i	current density (Section 6.1)
j_v	volume flux across a membrane (Section 18.3)
j_T	total electrolyte flux (Section 6.1)
j_i	diffusion flux of solute i relative to the volume average velocity
j_i^m	diffusion flux of solute i relative to the mass average velocity
j_i^*	diffusion flux relative to the molar average velocity
$j_1^{(2)}$	diffusion flux of solute (1) relative to velocity of solvent (2)
j_i^a	diffusion flux of solute i relative to reference velocity a
J_s	entropy flux (Section 7.2)
J_T	total solute flux in different chemical forms (Section 6.2)
k	mass transfer coefficient based on a concentration driving force
k_p	mass transfer coefficient based on a partial pressure driving force (Table 8.2-2)
k_x, k_y	mass transfer coefficients based on mole fraction driving forces in liquid and gas, respectively (Table 8.2-2)
k_B	Boltzmann's constant
k_T	thermal conductivity (Chapters 20–21)
k^0	mass transfer coefficient at low transfer rate (Section 9.5)
k^0	mass transfer coefficient without chemical reaction (Chapter 17)
k'	capacity factor (Sections 4.4 and 15.1)
K	equilibrium constant for chemical reaction
K_G, K_L	overall mass transfer coefficients based on concentration driving force in gas or liquid, respectively
K_p	overall mass transfer coefficient based on partial pressure difference in gas
K_x, K_y	overall mass transfer coefficient based on mole fraction driving force in liquid or gas, respectively
Kn	Knudsen number (Section 6.4)
l	length, e.g., of a membrane
L	length, e.g., of a pipe
L	molar flux of liquid
L''	mass flux of liquid (Sections 10.2 and 13.1)
L'	molar flux of liquid in stripping section (Sections 12.3 and 13.3)
L_{ij}	Onsager phenomenological coefficient (Section 7.2)
L_p	solvent permeability (Section 18.3)
m	partition coefficient relating mole fractions in gas and liquid
M	mass
M	total solute (Sections 4.2 and 5.5)
$\tilde{M}_i$	molecular weight of species i

n	micelle aggregation number or hydration number (Section 6.2)
$\mathbf{n}_i$	flux of species i relative to fixed coordinates
N	number of ideal stages
$\tilde{N}$	Avogadro's number
N_i	flux of species i at an interface
N_i	number of moles of species i
NTU	number of transfer units
p	pressure
P	power
P	membrane permeability (Chapter 18)
P_{ij}	weighting factor (Section 7.3)
q	scattering vector (Section 5.6)
q	feed quality (Sections 12.3 and 13.3)
q	solute concentration per volume adsorbent (Chapter 15)
$\mathbf{q}$	energy flux (Chapters 7, 20, and 21)
r	radius
r, r_i	rate of chemical reaction
R	gas constant
R_D	reflux ratio (Chapters 12 and 13)
R_0	characteristic radius
s	distance from pipe wall (Section 9.4)
$\hat{S}$	specific entropy (Chapter 7)
$\bar{S}_i$	partial specific entropy of species i
t	time
$\mathbf{t}$	modal matrix (Section 7.3)
t_i	transference number of ion i (Section 6.1)
$t_{1/2}$	reaction half-life
T	temperature
u_i	ionic mobility (Section 6.1)
U	overall heat transfer coefficient
$\hat{U}$	specific internal energy
v_r, v_θ	velocities in the r and θ directions
v_x, v_y	velocities in the x and y directions
$\boldsymbol{v}$	mass average velocity
$\boldsymbol{v}^a$	velocity relative to reference frame a
$\boldsymbol{v}^o$	volume average velocity
$\boldsymbol{v}'$	velocity fluctuation (Sections 4.3 and 17.4)
$\boldsymbol{v}^*$	molar average velocity
$\boldsymbol{v}_i$	velocity of species i
V	volume
$\bar{V}_i$	partial molar or specific volume of species i
V_{ij}	fraction of molecular volume (Section 5.1)
W	width
W	work (Section 20.2)
W_s	shaft work (Section 20.2)
x	mole fraction in liquid of more volatile species (Chapters 12 and 13)

x_B, x_D, x_F	mole fractions of more volatile species in bottoms, distillate and feed, respectively (Chapters 12 and 13)
x_i	mole fraction of species i, especially in a liquid or solid phase
$\mathbf{X}_i$	generalized force causing diffusion (Section 7.2)
y	mole fraction in vapor of more volatile species (Chapters 12 and 13)
y_i	mole fraction of species i in a gas
z	position
$|z|$	magnitude of charge (Section 6.1)
z_i	charge on species i
α	thermal diffusivity (Chapters 20 and 21)
α	thermal diffusion factor (Section 21.5)
α	flake aspect ratio (Sections 6.4 and 9.5)
α_{ij}	conversion factor (Section 7.1)
β	diaphragm cell calibration constant (Sections 2.2 and 5.5)
β	pervaporation selectivity (Section 18.4)
γ	interfacial influence (Section 6.3)
γ	surface tension (Section 6.4)
γ_i	activity coefficient of species i
δ	thickness of thin layer, especially a boundary layer
$\delta(z)$	Dirac function of z
δ_{ij}	Kronecker delta
ε	void fraction
ε	enhancement factor (Section 17.1)
ε_{ij}	interaction energy between colliding molecules (Sections 5.1 and 20.4)
ζ	combined variable
η	Murphree efficiency (Section 13.4)
η	effectiveness factor (Section 17.1)
θ	dimensionless concentration
θ	fraction of unused adsorption bed (Section 15.3)
θ	fraction of surface elements (Section 9.2)
κ_i, κ_{-i}	forward and reverse reaction rate constants respectively of reaction i
λ	length ratio (Section 6.4)
λ	heat of vaporization (Sections 12.3 and 13.3)
λ_i	equivalent ionic conductance of species i (Section 6.1)
Λ	equivalent conductance
μ	viscosity
μ_i	chemical potential of species i
μ_i	partial specific Gibbs free energy of species i, i.e., the chemical potential divided by the molecular weight (Section 7.2)
ν	kinematic viscosity
ν	stoichiometric coefficient (Sections 16.5 and 17.2)
ξ	dimensionless position
ξ	correlation length (Section 6.3)
Π	osmotic pressure (Section 18.3)
ρ	total density, i.e., total mass concentration
ρ_i	mass concentration of species i
σ	rate of entropy production (Section 7.2)

σ	standard deviation (Sections 5.5 and 15.4)
σ, σ'	reflection coefficients (Section 18.3)
σ	Soret coefficient (Section 21)
$\boldsymbol{\sigma}$	diagonal matrix of eigenvalues (Chapter 7)
σ_i	eigenvalue (Section 7.3)
σ_{ij}	collision diameter
τ	characteristic time
τ	tortuosity (Section 6.4)
τ	residence time for surface element (Section 9.2)
τ	shear stress (Chapter 21)
τ_0	shear stress at wall (Section 9.4)
ϕ	Thiele modulus (Section 17.1)
ϕ_i	volume fraction of species i
ψ	electrostatic potential
$\boldsymbol{\psi}$	combined concentration (Section 7.3)
ω	jump frequency (Section 5.3)
ω	regular solution parameter (Section 6.3)
ω	coefficient of solute permeability (Section 18.3)
ω_i	mass fraction of species i
Ω	collision integral in Chapman–Enskog theory (Section 5.1)

Preface to the Third Edition

Like its earlier editions, this book has two purposes. First, it presents a clear description of diffusion, the mixing process caused by molecular motion. Second, it explains mass transfer, which controls the cost of processes like chemical purification and environmental control. The first of these purposes is scientific, explaining how nature works. The second purpose is more practical, basic to the engineering of chemical processes.

While diffusion was well explained in earlier editions, this edition extends and clarifies this material. For example, the Maxwell–Stefan alternative to Fick's equation is now treated in more depth. Brownian motion and its relation to diffusion are explicitly described. Diffusion in composites, an active area of research, is reviewed. These topics are an evolution of and an improvement over the material in earlier editions.

Mass transfer is much better explained here than it was earlier. I believe that mass transfer is often poorly presented because it is described only as an analogue of heat transfer. While this analogue is true mathematically, its overemphasis can obscure the simpler physical meaning of mass transfer. In particular, this edition continues to emphasize dilute mass transfer. It gives a more complete description of differential distillation than is available in other introductory sources. This description is important because differential distillation is now more common than staged distillation, normally the only form covered. This edition gives a much better description of adsorption than has been available. It provides an introduction to mass transfer applied in biology and medicine.

The result is an engineering book which is much more readable and understandable than other books covering these subjects. It provides much more physical insight than conventional books on unit operations. It explores the interactions between mass transfer and chemical reaction, which are omitted by many books on transport phenomena. The earlier editions are good, but this one is better.

The book works well as a text either for undergraduates or graduate students. For a one-semester undergraduate chemical engineering course of perhaps 45 lectures plus recitations, I cover Chapter 2, Sections 3.1 to 3.2 and 5.1 to 5.2, Chapters 8 to 10, 12 to 15, and 21. If there is time, I add Sections 16.1 to 16.3 and Sections 17.1 to 17.3. If this course aims at describing separation processes, I cover crystallization before discussing membrane separations. We have successfully taught such a course here at Minnesota for the last 10 years.

For a one semester graduate course for students from chemistry, chemical engineering, pharmacy, and food science, I plan for 45 lectures without recitations. This course covers Chapters 2 to 9 and Chapters 16 to 19. It has been a mainstay at many universities for almost 30 years.

This description of academic courses should not restrict the book's overall goal. Diffusion and mass transfer are often interesting because they are slow. Their rate controls many processes, from the separation of air to the spread of pollutants to the size of a human sperm. The study of diffusion is thus important, but it is also fun. I hope that this book catalyzes that fun for you.

Preface to Second Edition

The purpose of this second edition is again a clear description of diffusion useful to engineers, chemists, and life scientists. Diffusion is a fascinating subject, as central to our daily lives as it is to the chemical industry. Diffusion equations describe the transport in living cells, the efficiency of distillation, and the dispersal of pollutants. Diffusion is responsible for gas absorption, for the fog formed by rain on snow, and for the dyeing of wool. Problems like these are easy to identify and fun to study.

Diffusion has the reputation of being a difficult subject, much harder than, say, fluid mechanics or solution thermodynamics. In fact, it is relatively simple. To prove this to yourself, try to explain a diffusion flux, a shear stress, and chemical potential to some friends who have little scientific training. I can easily explain a diffusion flux: It is how much diffuses per area per time. I have more trouble with a shear stress. Whether I say it is a momentum flux or the force in one direction caused by motion in a second direction, my friends look blank. I have never clearly explained chemical potentials to anyone.

However, past books on diffusion have enhanced its reputation as a difficult subject. These books fall into two distinct groups that are hard to read for different reasons. The first group is the traditional engineering text. Such texts are characterized by elaborate algebra, very complex examples, and turgid writing. Students cheerfully hate these books; moreover, they remember what they have learned as scattered topics, not an organized subject.

The second group of books consists of texts on transport processes. These books present diffusion by analogy with fluid flow and heat transfer. They are much more readable than the traditional texts, especially for the mathematically adroit. They do have two significant disadvantages. First, topics important to diffusion but not to fluid flow tend to be omitted or deemphasized. Such cases include simultaneous diffusion and chemical reaction. Second, these books usually present diffusion last, so that fluid mechanics and heat transfer must be at least superficially understood before diffusion can be learned. This approach effectively excludes students outside of engineering who have little interest in these other phenomena. Students in engineering find difficult problems emphasized because the simple ones have already been covered for heat transfer. Whether they are engineers or not, all conclude that diffusion must be difficult.

In the first edition, I tried to describe diffusion clearly and simply. I emphasized physical insight, sometimes at the loss of mathematical rigor. I discussed basic concepts in detail, without assuming prior knowledge of other phenomena. I aimed at the scope of the traditional texts and at the clarity of books on transport processes. This second edition is evidence that I was partly successful. Had I been completely successful, no second edition would be needed. Had I been unsuccessful, no second edition would be wanted.

In this second edition, I've kept the emphasis on physical insight and basic concepts, but I've expanded the book's scope. Chapters 1–7 on diffusion are largely unchanged, though some description of diffusion coefficients is abridged. Chapter 8 on mass transfer is expanded to even more detail, for I found many readers need more help. Chapters 9–12, a description of traditional chemical processes are new. The remaining seven chapters, a spectrum of topics, are either new or significantly revised. The result is still useful broadly, but deeper on engineering topics.

I have successfully used the book as a text for both undergraduate and graduate courses, of which most are in chemical engineering. For an undergraduate course on unit operations, I first review the mass transfer coefficients in Chapter 8, for I find that students' memory of these ideas is motley. I then cover the material in Chapters 9–12 in detail, for this is the core of the subject. I conclude with simultaneous heat and mass transfer, as discussed in Chapters 19–20. The resulting course of 50 classes is typical of many offered on this subject. On their own, undergraduates have used Chapters 2–3 and 8–9 for courses on heat and mass transfer, but this book's scope seems too narrow to be a good text for that class.

For graduate students, I give two courses in alternate years. Neither requires the other as a prerequisite. In the first graduate course, on diffusion, I cover Chapters 1–7, plus Chapter 17 (on membranes). In the second graduate course, on mass transfer, I cover Chapters 8–9, Chapters 13–16, and Chapter 20. These courses, which typically have about 35 lectures, are an enormous success, year after year. For nonengineering graduate students and for various short courses, I've usually used Chapters 2, 8, 15–16, and any other chapters specific to a given discipline. For example, for those in the drug industry, I might cover Chapters 11 and 18.

I am indebted to many who have encouraged me in this effort. My overwhelming debt is to my colleagues at the University of Minnesota. When I become disheartened, I need simply to visit another institution to be reminded of the advantages of frank discussion without infighting. My students have helped, especially Sameer Desai and Diane Clifton, who each read large parts of the final manuscript. Mistakes that remain are my fault. Teresa Bredahl typed most of the book, and Clover Galt provided valuable editorial help. Finally, my wife Betsy gives me a wonderful rich life.

CHAPTER 1

Models for Diffusion

If a few crystals of a colored material like copper sulfate are placed at the bottom of a tall bottle filled with water, the color will slowly spread through the bottle. At first the color will be concentrated in the bottom of the bottle. After a day it will penetrate upward a few centimeters. After several years the solution will appear homogeneous.

The process responsible for the movement of the colored material is diffusion, the subject of this book. Diffusion is caused by random molecular motion that leads to complete mixing. It can be a slow process. In gases, diffusion progresses at a rate of about 5 cm/min; in liquids, its rate is about 0.05 cm/min; in solids, its rate may be only about 0.00001 cm/min. In general, it varies less with temperature than do many other phenomena.

This slow rate of diffusion is responsible for its importance. In many cases, diffusion occurs sequentially with other phenomena. When it is the slowest step in the sequence, it limits the overall rate of the process. For example, diffusion often limits the efficiency of commercial distillations and the rate of industrial reactions using porous catalysts. It limits the speed with which acid and base react and the speed with which the human intestine absorbs nutrients. It controls the growth of microorganisms producing penicillin, the rate of the corrosion of steel, and the release of flavor from food.

In gases and liquids, the rates of these diffusion processes can often be accelerated by agitation. For example, the copper sulfate in the tall bottle can be completely mixed in a few minutes if the solution is stirred. This accelerated mixing is not due to diffusion alone, but to the combination of diffusion and stirring. Diffusion still depends on random molecular motions that take place over smaller distances. The agitation or stirring is not a molecular process, but a macroscopic process that moves portions of the fluid over much larger distances. After this macroscopic motion, diffusion mixes newly adjacent portions of the fluid. In other cases, such as the dispersal of pollutants, the agitation of wind or water produces effects qualitatively similar to diffusion; these effects, called dispersion, will be treated separately.

The description of diffusion involves a mathematical model based on a fundamental hypothesis or "law." Interestingly, there are two common choices for such a law. The more fundamental, Fick's law of diffusion, uses a diffusion coefficient. This is the law that is commonly cited in descriptions of diffusion. The second, which has no formal name, involves a mass transfer coefficient, a type of reversible rate constant.

Choosing between these two models is the subject of this chapter. Choosing Fick's law leads to descriptions common to physics, physical chemistry, and biology. These descriptions are explored and extended in Chapters 2–7. Choosing mass transfer coefficients produces correlations developed explicitly in chemical engineering and used implicitly in chemical kinetics and in medicine. These correlations are described in Chapters 8–15. Both approaches are used in Chapters 16–21.

We discuss the differences between the two models in Section 1.1 of this chapter. In Section 1.2 we show how the choice of the most appropriate model is determined.

In Section 1.3 we conclude with additional examples to illustrate how the choice between the models is made.

1.1 The Two Basic Models

In this section we want to illustrate the two basic ways in which diffusion can be described. To do this, we first imagine two large bulbs connected by a long thin capillary (Fig. 1.1-1). The bulbs are at constant temperature and pressure and are of equal volumes. However, one bulb contains carbon dioxide, and the other is filled with nitrogen.

To find how fast these two gases will mix, we measure the concentration of carbon dioxide in the bulb that initially contains nitrogen. We make these measurements when only a trace of carbon dioxide has been transferred, and we find that the concentration of carbon dioxide varies linearly with time. From this, we know the amount transferred per unit time.

We want to analyze this amount transferred to determine physical properties that will be applicable not only to this experiment but also in other experiments. To do this, we first define the flux:

$$(\text{carbon dioxide flux}) = \left(\frac{\text{amount of gas removed}}{\text{time (area capillary)}}\right) \tag{1.1-1}$$

In other words, if we double the cross-sectional area, we expect the amount transported to double. Defining the flux in this way is a first step in removing the influences of our particular apparatus and making our results more general. We next assume that the flux is proportional to the gas concentration:

$$(\text{carbon dioxide flux}) = k \left(\begin{array}{c}\text{carbon dioxide}\\\text{concentration}\\\text{difference}\end{array}\right) \tag{1.1-2}$$

The proportionality constant k is called a mass transfer coefficient. Its introduction signals one of the two basic models of diffusion. Alternatively, we can recognize

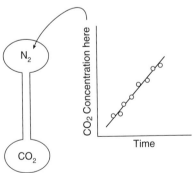

Fig. 1.1-1. A simple diffusion experiment. Two bulbs initially containing different gases are connected with a long thin capillary. The change of concentration in each bulb is a measure of diffusion and can be analyzed in two different ways.

that increasing the capillary's length will decrease the flux, and we can then assume that

$$(\text{carbon dioxide flux}) = D\left(\frac{\text{carbon dioxide concentration difference}}{\text{capillary length}}\right) \quad (1.1\text{-}3)$$

The new proportionality constant D is the diffusion coefficient. Its introduction implies the other model for diffusion, the model often called Fick's law.

These assumptions may seem arbitrary, but they are similar to those made in many other branches of science. For example, they are similar to those used in developing Ohm's law, which states that

$$\begin{pmatrix} \text{current, or} \\ \text{area times flux} \\ \text{of electrons} \end{pmatrix} = \left(\frac{1}{\text{resistance}}\right) \begin{pmatrix} \text{voltage, or} \\ \text{potential} \\ \text{difference} \end{pmatrix} \quad (1.1\text{-}4)$$

Thus, the mass transfer coefficient k is analogous to the reciprocal of the resistance. An alternative form of Ohm's law is

$$\begin{pmatrix} \text{current density} \\ \text{or flux of} \\ \text{electrons} \end{pmatrix} = \left(\frac{1}{\text{resistivity}}\right)\begin{pmatrix} \text{potential} \\ \text{difference} \\ \text{length} \end{pmatrix} \quad (1.1\text{-}5)$$

The diffusion coefficient D is analogous to the reciprocal of the resistivity.

Neither the equation using the mass transfer coefficient k nor that using the diffusion coefficient D is always successful. This is because of the assumptions made in their development. For example, the flux may not be proportional to the concentration difference if the capillary is very thin or if the two gases react. In the same way, Ohm's law is not always valid at very high voltages. But these cases are exceptions; both diffusion equations work well in most practical situations, just as Ohm's law does.

The parallels with Ohm's law also provide a clue about how the choice between diffusion models is made. The mass transfer coefficient in Eq. 1.1-2 and the resistance in Eq. 1.1-4 are simpler, best used for practical situations and rough measurements. The diffusion coefficient in Eq. 1.1-3 and the resistivity in Eq. 1.1-5 are more fundamental, involving physical properties like those found in handbooks. How these differences guide the choice between the two models is the subject of the next section.

1.2 Choosing Between the Two Models

The choice between the two models outlined in Section 1.1 represents a compromise between ambition and experimental resources. Obviously, we would like to express our results in the most general and fundamental ways possible. This suggests working with diffusion coefficients. However, in many cases, our experimental measurements will dictate a more approximate and phenomenological approach. Such approximations often imply mass transfer coefficients, but they usually still permit us to reach our research goals.

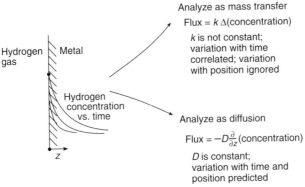

Fig. 1.2-1. Hydrogen diffusion into a metal. This process can be described with either a mass transfer coefficient k or a diffusion coefficient D. The description with a diffusion coefficient correctly predicts the variation of concentration with position and time, and so is superior.

This choice and the resulting approximations are best illustrated by two examples. In the first, we consider hydrogen diffusion in metals. This diffusion substantially reduces a metal's ductility, so much so that parts made from the embrittled metal frequently fracture. To study this embrittlement, we might expose the metal to hydrogen under a variety of conditions and measure the degree of embrittlement versus these conditions. Such empiricism would be a reasonable first approximation, but it would quickly flood us with uncorrelated information that would be difficult to use effectively.

As an improvement, we can undertake two sets of experiments. First, we can saturate metal samples with hydrogen and determine their degrees of embrittlement. Thus we know metal properties versus hydrogen concentration. Second, we can measure hydrogen uptake versus time, as suggested in Fig. 1.2-1, and correlate our measurements as mass transfer coefficients. Thus we know average hydrogen concentration versus time.

To our dismay, the mass transfer coefficients in this case will be difficult to interpret. They are anything but constant. At zero time, they approach infinity; at large time, they approach zero. At all times, they vary with the hydrogen concentration in the gas surrounding the metal. They are an inconvenient way to summarize our results. Moreover, the mass transfer coefficients give only the *average* hydrogen concentration in the metal. They ignore the fact that the hydrogen concentration very near the metal's surface will reach saturation but the concentration deep within the metal will remain zero. As a result, the metal near the surface may be very brittle but that within may be essentially unchanged.

We can include these details in the diffusion model described in the previous section. This model assumed that

$$\left(\begin{array}{c}\text{hydrogen} \\ \text{flux}\end{array}\right) = D \frac{\left(\begin{array}{c}\text{hydrogen} \\ \text{concentration at } z=0\end{array}\right) - \left(\begin{array}{c}\text{hydrogen} \\ \text{concentration at } z=l\end{array}\right)}{(\text{thickness at } z=l) - (\text{thickness at } z=0)}$$

$$(1.2\text{-}1)$$

or, symbolically,

$$j_1 = D \frac{c_1|_{z=0} - c_1|_{z=l}}{l - 0} \tag{1.2-2}$$

where the subscript 1 symbolizes the diffusing species. In these equations, the distance l is that over which diffusion occurs. In the previous section, the length of the capillary was appropriately this distance; but in this case, it seems uncertain what the distance should be. If we assume that it is very small,

$$j_1 = D \lim_{l \to 0} \frac{c_1|_{z=z} - c_1|_{z=z+l}}{z|_{z+l} - z|_z} = -D \frac{dc_1}{dz} \tag{1.2-3}$$

We can use this relation and the techniques developed later in this book to correlate our experiments with only one parameter, the diffusion coefficient D. We then can correctly predict the hydrogen uptake versus time and the hydrogen concentration in the gas. As a dividend, we get the hydrogen concentration at all positions and times within the metal.

Thus the model based on the diffusion coefficient gives results of more fundamental value than the model based on mass transfer coefficients. In mathematical terms, the diffusion model is said to have distributed parameters, for the dependent variable (the concentration) is allowed to vary with all independent variables (like position and time). In contrast, the mass transfer model is said to have lumped parameters (like the average hydrogen concentration in the metal).

These results would appear to imply that the diffusion model is superior to the mass transfer model and so should always be used. However, in many interesting cases the models are equivalent. To illustrate this, imagine that we are studying the dissolution of a solid drug suspended in water, as schematically suggested by Fig. 1.2-2. The dissolution of this drug is known to be controlled by the diffusion of the dissolved drug away from the solid surface of the undissolved material. We measure the drug concentration versus time as shown, and we want to correlate these results in terms of as few parameters as possible.

One way to correlate the dissolution results is to use a mass transfer coefficient. To do this, we write a mass balance on the solution:

$$\begin{pmatrix} \text{accumulation} \\ \text{of drug in} \\ \text{solution} \end{pmatrix} = \begin{pmatrix} \text{total rate of} \\ \text{dissolution} \end{pmatrix}$$

$$V \frac{dc_1}{dt} = A j_1$$

$$= Ak[c_1(\text{sat}) - c_1] \tag{1.2-4}$$

where V is the volume of solution, A is the total area of the drug particles, $c_1(\text{sat})$ is the drug concentration at saturation and at the solid's surface, and c_1 is the concentration in the bulk solution. Integrating this equation allows quantitatively fitting our results with one parameter, the mass transfer coefficient k. This quantity is independent of drug solubility, drug area, and solution volume, but it does vary with physical properties like stirring rate and solution viscosity. Correlating the effects of these properties turns out to be straightforward.

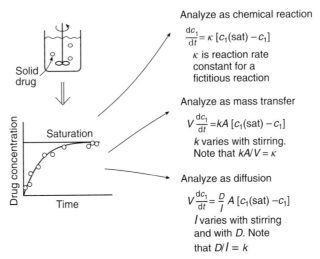

Fig. 1.2-2. Rates of drug dissolution. In this case, describing the system with a mass transfer coefficient k is best because it easily correlates the solution's concentration versus time. Describing the system with a diffusion coefficient D gives a similar correlation but introduces an unnecessary parameter, the film thickness l. Describing the system with a reaction rate constant k also works, but this rate constant is a function not of chemistry but of physics.

The alternative to mass transfer is diffusion theory, for which the mass balance is

$$V\frac{dc_1}{dt} = A\left(\frac{D}{l}\right)[c_1(\text{sat}) - c_1] \tag{1.2-5}$$

in which l is an unknown parameter, equal to the average distance across which diffusion occurs. This unknown, called a film or unstirred layer thickness, is a function not only of flow and viscosity but also of the diffusion coefficient itself.

Equations 1.2-4 and 1.2-5 are equivalent, and they share the same successes and short-comings. In the former, we must determine the mass transfer coefficient experimentally; in the latter, we determine instead the thickness l. Those who like a scientific veneer prefer to measure l, for it genuflects toward Fick's law of diffusion. Those who are more pragmatic prefer explicitly recognizing the empirical nature of the mass transfer coefficient.

The choice between the mass transfer and diffusion models is thus often a question of taste rather than precision. The diffusion model is more fundamental and is appropriate when concentrations are measured or needed versus both position and time. The mass transfer model is simpler and more approximate and is especially useful when only average concentrations are involved. The additional examples in section 1.3 should help us decide which model is appropriate for our purposes.

Before going on to the next section, we should mention a third way to correlate the results other than the two diffusion models. This third way is to assume that the disso-lution shown in Fig. 1.2-2 is a first-order, reversible chemical reaction. Such a reaction might be described by

$$\frac{dc_1}{dt} = \kappa c_1(\text{sat}) - \kappa c_1 \tag{1.2-6}$$

In this equation, the quantity $\kappa c_1(\text{sat})$ represents the rate of dissolution, κc_1 stands for the rate of precipitation, and κ is a rate constant for this process. This equation is mathematically identical with Eqs. 1.2-4 and 1.2-5 and so is equally successful. However, the idea of treating dissolution as a chemical reaction is flawed. Because the reaction is hypothetical, the rate constant is a composite of physical factors rather than chemical factors. We do better to consider the physical process in terms of a diffusion or mass transfer model.

1.3 Examples

In this section, we give examples that illustrate the choice between diffusion coefficients and mass transfer coefficients. This choice is often difficult, a juncture where many have trouble. I often do. I think my trouble comes from evolving research goals, from the fact that as I understand the problem better, the questions that I am trying to answer tend to change. I notice the same evolution in my peers, who routinely start work with one model and switch to the other model before the end of their research.

We shall not solve the following examples. Instead, we want only to discuss which diffusion model we would initially use for their solution. The examples given certainly do not cover all types of diffusion problems, but they are among those about which I have been asked in the last year.

Example 1.3-1: Ammonia scrubbing   Ammonia, the major material for fertilizer, is made by reacting nitrogen and hydrogen under pressure. The product gas can be washed with water to dissolve the ammonia and separate it from other unreacted gases. How can you correlate the dissolution rate of ammonia during washing?

Solution   The easiest way is to use mass transfer coefficients. If you use diffusion coefficients, you must somehow specify the distance across which diffusion occurs. This distance is unknown unless the detailed flows of gases and the water are known; they rarely are (see Chapters 8 and 9).

Example 1.3-2: Reactions in porous catalysts   Many industrial reactions use catalysts containing small amounts of noble metals dispersed in a porous inert material like silica. The reactions on such a catalyst are sometimes slower in large pellets than in small ones. This is because the reagents take longer to diffuse into the pellet than they do to react. How should you model this effect?

Solution   You should use diffusion coefficients to describe the simultaneous diffusion and reaction in the pores in the catalyst. You should not use mass transfer coefficients because you cannot easily include the effect of reaction (see Sections 16.1 and 17.1).

Example 1.3-3: Corrosion of marble   Industrial pollutants in urban areas like Venice cause significant corrosion of marble statues. You want to study how these pollutants penetrate marble. Which diffusion model should you use?

Solution   The model using diffusion coefficients is the only one that will allow you to predict pollutant concentration versus position in the marble. The model using mass transfer coefficients will only correlate how much pollutant enters the statue, not what happens to the pollutant (see Sections 2.3 and 8.1).

Example 1.3-4: Protein size in solution You are studying a variety of proteins that you hope to purify and use as food supplements. You want to characterize the size of the proteins in solution. How can you use diffusion to do this?

    Solution Your aim is determining the molecular size of the protein molecules. You are not interested in the protein mass transfer except as a route to these molecular properties. As a result, you should measure the protein's diffusion coefficient, not its mass transfer coefficient. The protein's diffusion coefficient will turn out to be proportional to its radius in solution (see Section 5.2).

Example 1.3-5: Antibiotic production Many drugs are made by fermentations in which microorganisms are grown in a huge stirred vat of a dilute nutrient solution or "beer." Many of these fermentations are aerobic, so the nutrient solution requires aeration. How should you model oxygen uptake in this type of solution?

    Solution Practical models use mass transfer coefficients. The complexities of the problem, including changes in air bubble size, flow effects of the non-Newtonian solution, and foam caused by biological surfactants, all inhibit more careful study (see Chapter 8).

Example 1.3-6: Facilitated transport across membranes Some membranes contain a mobile carrier, a reactive species that reacts with diffusing solutes, facilitating their transport across the membrane. Such membranes can be used to concentrate copper ions from industrial waste and to remove carbon dioxide from coal gas. Diffusion across these membranes does not vary linearly with the concentration difference across them. The diffusion can be highly selective, but it is often easily poisoned. Should this diffusion be described with mass transfer coefficients or with diffusion coefficients?

    Solution This system includes not only diffusion but also chemical reaction. Diffusion and reaction couple in a nonlinear way to give the unusual behavior observed. Understanding such behavior will certainly require the more fundamental model of diffusion coefficients (see Section 18.5).

Example 1.3-7: Flavor retention When food products are spray-dried, they lose a lot of flavor. However, they lose less than would be expected on the basis of the relative vapor pressures of water and the flavor compounds. The reason apparently is that the drying food often forms a tight gellike skin across which diffusion of the flavor compounds is inhibited. What diffusion model should you use to study this effect?

    Solution Because spray-drying is a complex, industrial-scale process, it is usually modeled using mass transfer coefficients. However, in this case you are interested in the inhibition of diffusion. Such inhibition will involve the sizes of pores in the food and of molecules of the flavor compounds. Thus you should use the more basic diffusion model, which includes these molecular factors (see Section 6.4).

Example 1.3-8: The smell of marijuana   Recently, a large shipment of marijuana was seized in the Minneapolis–St. Paul airport. The police said their dog smelled it. The owners claimed that it was too well wrapped in plastic to smell and that the police had conducted an illegal search without a search warrant. How could you tell who was right?

      Solution   In this case, you are concerned with the diffusion of odor across the thin plastic film. The diffusion rate is well described by either mass transfer or diffusion coefficients. However, the diffusion model explicitly isolates the effect of the solubility of the smell in the film, which dominates the transport. This solubility is the dominant variable (see Section 2.2). In this case, the search was illegal.

Example 1.3-9: Scale-up of wet scrubbers   You want to use a wet scrubber to remove sulfur oxides from the flue gas of a large power plant. A wet scrubber is essentially a large piece of pipe set on its end and filled with inert ceramic material. You pump the flue gas up from the bottom of the pipe and pour a lime slurry down from the top. In the scrubber, there are various reactions, such as

$$CaO + SO_2 \rightarrow CaSO_3 \tag{1.2-6}$$

The lime reacts with the sulfur oxides to make an insoluble precipitate, which is discarded. You have been studying a small unit and want to use these results to predict the behavior of a larger unit. Such an increase in size is called a scale-up. Should you make these predictions using a model based on diffusion or mass transfer coefficients?

      Solution   This situation is complex because of the chemical reactions and the irregular flows within the scrubber. Your first try at correlating your data should be a model based on mass transfer coefficients. Should these correlations prove unreliable, you may be forced to use the more difficult diffusion model (see Chapters 9, 16, and 17).

1.4 Conclusions

      This chapter discusses the two common models used to describe diffusion and suggests how you can choose between these models. For fundamental studies where you want to know concentration versus position and time, use diffusion coefficients. For practical problems where you want to use one experiment to tell how a similar one will behave, use mass transfer coefficients. The former approach is the distributed-parameter model used in chemistry, and the latter is the lumped-parameter model used in engineering. Both approaches are used in medicine and biology, but not always explicitly.

      The rest of this book is organized in terms of these two models. Chapters 2–4 present the basic model of diffusion coefficients, and Chapters 5–7 review the values of the diffusion coefficients themselves. Chapters 8–15 discuss the model of mass transfer coefficients, including their relation to diffusion coefficients. Chapters 16–19 explore the coupling of diffusion with heterogeneous and homogeneous chemical reactions, using both models. Chapters 20–21 explore the simpler coupling between diffusion and heat transfer.

In the following chapters, keep both models in mind. People involved in basic research tend to be overcommitted to diffusion coefficients, whereas those with broader objectives tend to emphasize mass transfer coefficients. Each group should recognize that the other has a complementary approach that may be more helpful for the case in hand.

Questions for Discussion

1. What are the dimensions in mass M, length L, and time t of a diffusion coefficient?
2. What are the dimensions of a mass transfer coefficient?
3. What volume is implied by Ficks's law?
4. What volume is implied when defining a mass transfer coefficient?
5. Can the diffusion coefficient ever be negative?
6. Give an example for a diffusion coefficient which is the same in all directions. Give an example when it isn't.
7. When a silicon chip is doped with boron, does the doping involve diffusion?
8. Does the wafting of smells of a pie baking in the oven involve diffusion?
9. How does breathing involve diffusion?
10. How is a mass transfer coefficient related to a reaction rate constant?
11. Will a heat transfer coefficient and a mass transfer coefficient be related?
12. Will stirring a suspension of sugar in water change the diffusion coefficient? Will it change the density? Will it change the mass transfer coefficient?

PART I

Fundamentals of Diffusion

CHAPTER 2

Diffusion in Dilute Solutions

In this chapter, we consider the basic law that underlies diffusion and its application to several simple examples. The examples that will be given are restricted to dilute solutions. Results for concentrated solutions are deferred until Chapter 3.

This focus on the special case of dilute solutions may seem strange. Surely, it would seem more sensible to treat the general case of all solutions and then see mathematically what the dilute-solution limit is like. Most books use this approach. Indeed, because concentrated solutions are complex, these books often describe heat transfer or fluid mechanics first and then teach diffusion by analogy. The complexity of concentrated diffusion then becomes a mathematical cancer grafted onto equations of energy and momentum.

I have rejected this approach for two reasons. First, the most common diffusion problems do take place in dilute solutions. For example, diffusion in living tissue almost always involves the transport of small amounts of solutes like salts, antibodies, enzymes, or steroids. Thus many who are interested in diffusion need not worry about the complexities of concentrated solutions; they can work effectively and contentedly with the simpler concepts in this chapter.

Second and more important, diffusion in dilute solutions is easier to understand in physical terms. A diffusion flux is the rate per unit area at which mass moves. A concentration profile is simply the variation of the concentration versus time and position. These ideas are much more easily grasped than concepts like momentum flux, which is the momentum per area per time. This seems particularly true for those whose backgrounds are not in engineering, those who need to know about diffusion but not about other transport phenomena.

This emphasis on dilute solutions is found in the historical development of the basic laws involved, as described in Section 2.1. Sections 2.2 and 2.3 of this chapter focus on two simple cases of diffusion: steady-state diffusion across a thin film and unsteady-state diffusion into an infinite slab. This focus is a logical choice because these two cases are so common. For example, diffusion across thin films is basic to membrane transport, and diffusion in slabs is important in the strength of welds and in the decay of teeth. These two cases are the two extremes in nature, and they bracket the behavior observed experimentally. In Section 2.4 and Section 2.5, these ideas are extended to other examples that demonstrate mathematical ideas useful for other situations.

2.1 Pioneers in Diffusion

2.1.1 Thomas Graham

Our modern ideas on diffusion are largely due to two men, Thomas Graham and Adolf Fick. Graham was the elder. Born on December 20, 1805, Graham was the son of a successful manufacturer. At 13 years of age he entered the University of Glasgow with the intention of becoming a minister, and there his interest in science was stimulated by Thomas Thomson.

Graham's research on the diffusion of gases, largely conducted during the years 1828 to 1833, depended strongly on the apparatus shown in Fig. 2.1-1 (Graham, 1829; Graham, 1833). This apparatus, a "diffusion tube," consists of a straight glass tube, one end of which is closed with a dense stucco plug. The tube is filled with hydrogen, and the end is sealed with water, as shown. Hydrogen diffuses through the plug and out of the tube, while air diffuses back through the plug and into the tube.

Because the diffusion of hydrogen is faster than the diffusion of air, the water level in this tube will rise during the process. Graham saw that this change in water level would lead to a pressure gradient that in turn would alter the diffusion. To avoid this pressure gradient, he continually lowered the tube so that the water level stayed constant. His experimental results then consisted of a volume-change characteristic of each gas originally held in the tube. Because this volume change was characteristic of diffusion, "the diffusion or spontaneous intermixture of two gases in contact is effected by an interchange of position of infinitely minute volumes, being, in the case of each gas, inversely proportional to the square root of the density of the gas" (Graham, 1833, p. 222). Graham's original experiment was unusual because the diffusion took place at constant pressure, not at constant volume (Mason, 1970).

Graham also performed important experiments on liquid diffusion using the equipment shown in Fig. 2.1-2 (Graham, 1850); in these experiments, he worked with dilute solutions. In one series of experiments, he connected two bottles that contained solutions at different concentrations; he waited several days and then separated the bottles and analyzed their contents. In another series of experiments, he placed a small bottle containing a solution of known concentration in a larger jar containing only water. After waiting several days, he removed the bottle and analyzed its contents.

Graham's results were simple and definitive. He showed that diffusion in liquids was at least several thousand times slower than diffusion in gases. He recognized that the diffusion process got still slower as the experiment progressed, that "diffusion must

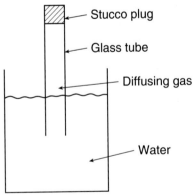

Fig. 2.1-1. Graham's diffusion tube for gases. This apparatus was used in the best early study of diffusion. As a gas like hydrogen diffuses out through the plug, the tube is lowered to ensure that there will be no pressure difference.

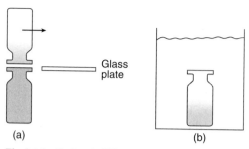

Fig. 2.1-2. Graham's diffusion apparatus for liquids. The equipment in (a) is the ancestor of free diffusion experiments; that in (b) is a forerunner of the capillary method.

Table 2.1-1 *Graham's results for liquid diffusion*

Weight percent of sodium chloride	Relative flux
1	1.00
2	1.99
3	3.01
4	4.00

Source: Data from Graham (1850).

necessarily follow a diminishing progression." Most important, he concluded from the results in Table 2.1-1 that "the quantities diffused appear to be closely in proportion ... to the quantity of salt in the diffusion solution" (Graham, 1850, p. 6). In other words, the flux caused by diffusion is proportional to the concentration difference of the salt.

2.1.2 *Adolf Fick*

The next major advance in the theory of diffusion came from the work of Adolf Eugen Fick. Fick was born on September 3, 1829, the youngest of five children. His father, a civil engineer, was a superintendent of buildings. During his secondary schooling, Fick was delighted by mathematics, especially the work of Poisson. He intended to make mathematics his career. However, an older brother, a professor of anatomy at the University of Marburg, persuaded him to switch to medicine.

In the spring of 1847, Fick went to Marburg, where he was occasionally tutored by Carl Ludwig. Ludwig strongly believed that medicine, and indeed life itself, must have a basis in mathematics, physics, and chemistry. This attitude must have been especially appealing to Fick, who saw the chance to combine his real love, mathematics, with his chosen profession, medicine.

In the fall of 1849, Fick's education continued in Berlin, where he did a considerable amount of clinical work. In 1851 he returned to Marburg, where he received his degree. His thesis dealt with the visual errors caused by astigmatism, again illustrating his determination to combine science and medicine (Fick, 1852). In the fall of 1851, Carl Ludwig became professor of anatomy in Zurich, and in the spring of 1852 he brought Fick along as a prosector. Ludwig moved to Vienna in 1855, but Fick remained in Zurich until 1868.

Paradoxically, the majority of Fick's scientific accomplishments do not depend on diffusion studies at all, but on his more general investigations of physiology (Fick, 1903). He did outstanding work in mechanics (particularly as applied to the functioning of muscles), in hydrodynamics and hemorheology, and in the visual and thermal functioning of the human body. He was an intriguing man. However, in this discussion we are interested only in his development of the fundamental laws of diffusion.

In his first diffusion paper, Fick (1855a) codified Graham's experiments through an impressive combination of qualitative theories, casual analogies, and quantitative experiments. His paper, which is refreshingly straightforward, deserves reading today. Fick's introduction of his basic idea is almost casual: "[T]he diffusion of the dissolved material ... is left completely to the influence of the molecular forces basic to the same law ... for the spreading of warmth in a conductor and which has already been applied with such great success to the spreading of electricity" (Fick, 1855a, p. 65). In other words, diffusion can be described on the same mathematical basis as Fourier's law for heat conduction or Ohm's law for electrical conduction. This analogy remains a useful pedagogical tool.

Fick seemed initially nervous about his hypothesis. He buttressed it with a variety of arguments based on kinetic theory. Although these arguments are now dated, they show physical insights that would be exceptional in medicine today. For example, Fick recognized that diffusion is a dynamic molecular process. He understood the difference between a true equilibrium and a steady state, possibly as a result of his studies with muscles (Fick, 1856). Later, Fick became more confident as he realized his hypothesis was consistent with Graham's results (Fick, 1855b).

Using this basic hypothesis, Fick quickly developed the laws of diffusion by means of analogies with Fourier's work (Fourier, 1822). He defined a total one-dimensional flux J_1 as

$$J_1 = Aj_1 = -AD\frac{\partial c_1}{\partial z} \tag{2.1-1}$$

where A is the area across which diffusion occurs, j_1 is the flux per unit area, c_1 is concentration, and z is distance. This is the first suggestion of what is now known as Fick's law. The quantity D, which Fick called "the constant depending of the nature of the substances," is, of course, the diffusion coefficient. Fick also paralleled Fourier's development to determine the more general conservation equation

$$\frac{\partial c_1}{\partial t} = D\left(\frac{\partial^2 c_1}{\partial z^2} + \frac{1}{A}\frac{\partial A}{\partial z}\frac{\partial c_1}{\partial z}\right) \tag{2.1-2}$$

When the area A is a constant, this becomes the basic equation for one-dimensional unsteady-state diffusion, sometimes called Fick's second law.

Fick next had to prove his hypothesis that diffusion and thermal conduction can be described by the same equations. He was by no means immediately successful. First, he tried to integrate Eq. 2.1-2 for constant area, but he became discouraged by the numerical effort required. Second, he tried to measure the second derivative experimentally. Like many others, he found that second derivatives are difficult to measure: "the second difference increases exceptionally the effect of [experimental] errors."

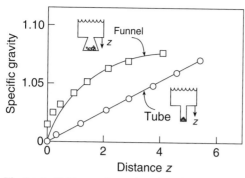

Fig. 2.1-3. Fick's experimental results. The crystals in the bottom of each apparatus saturate the adjacent solution, so that a fixed concentration gradient is established along the narrow, lower part of the apparatus. Fick's calculation of the curve for the funnel was his best proof of Fick's law.

His third effort was more successful. He used a glass cylinder containing crystalline sodium chloride in the bottom and a large volume of water in the top, shown as the lower apparatus in Fig. 2.1-3. By periodically changing the water in the top volume, he was able to establish a steady-state concentration gradient in the cylindrical cell. He found that this gradient was linear, as shown in Fig. 2.1-3. Because this result can be predicted either from Eq. 2.1-1 or from Eq. 2.1-2, this was a triumph.

But this success was by no means complete. After all, Graham's data for liquids anticipated Eq. 2.1-1. To try to strengthen the analogy with thermal conduction, Fick used the upper apparatus shown in Fig. 2.1-3. In this apparatus, he established the steady-state concentration profile in the same manner as before. He measured this profile and then tried to predict these results using Eq. 2.1-2, in which the funnel area A available for diffusion varied with the distance z. When Fick compared his calculations with his experimental results, he found good agreement. These results were the initial verification of Fick's law.

2.1.3 Forms of Fick's Law

Useful forms of Fick's law in dilute solutions are shown in Table 2.1-2. Each equation closely parallels that suggested by Fick, that is, Eq. 2.1-1. Each involves the same phenomenological diffusion coefficient. Each will be combined with mass balances to analyze the problems central to the rest of this chapter.

One must remember that these flux equations imply no convection in the same direction as the one-dimensional diffusion. They are thus special cases of the general equations given in Table 3.2-1. This lack of convection often indicates a dilute solution. In fact, the assumption of a dilute solution is more restrictive than necessary, for there are many concentrated solutions for which these simple equations can be used without inaccuracy. Nonetheless, for the novice, I suggest thinking of diffusion in a dilute solution.

2.2 Steady Diffusion Across a Thin Film

In the previous section we detailed the development of Fick's law, the basic relation for diffusion. Armed with this law, we can now attack the simplest example: steady

Table 2.1-2 *Fick's law for diffusion without convection*

For one-dimensional diffusion in Cartesian coordinates	$-j_1 = D\dfrac{dc_1}{dz}$
For radial diffusion in cylindrical coordinates	$-j_1 = D\dfrac{dc_1}{dr}$
For radial diffusion in spherical coordinates	$-j_1 = D\dfrac{dc_1}{dr}$

Note: More general equations are given in Table 3.2-1.

diffusion across a thin film. In this attack, we want to find both the diffusion flux and the concentration profile. In other words, we want to determine how much solute moves across the film and how the solute concentration changes within the film.

This problem is very important. It is one extreme of diffusion behavior, a counterpoint to diffusion in an infinite slab. Every reader, whether casual or diligent, should try to master this problem now. Many may be superficial because film diffusion is so simple mathematically. Please do not dismiss this important problem; it is mathematically straightforward but physically subtle. Think about it carefully.

2.2.1 The Physical Situation

Steady diffusion across a thin film is illustrated schematically in Fig. 2.2-1. On each side of the film is a well-mixed solution of one solute, species 1. Both these solutions are dilute. The solute diffuses from the fixed higher concentration, located at $z \leq 0$ on the left-hand side of the film, into the fixed, less concentrated solution, located at $z \geq l$ on the right-hand side.

We want to find the solute concentration profile and the flux across this film. To do this, we first write a mass balance on a thin layer Δz, located at some arbitrary position z within the thin film. The mass balance in this layer is

$$\begin{pmatrix} \text{solute} \\ \text{accumulation} \end{pmatrix} = \begin{pmatrix} \text{rate of diffusion} \\ \text{into the layer at } z \end{pmatrix} - \begin{pmatrix} \text{rate of diffusion} \\ \text{out of the layer} \\ \text{at } z + \Delta z \end{pmatrix}$$

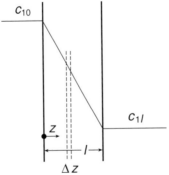

Fig. 2.2-1. Diffusion across a thin film. This is the simplest diffusion problem, basic to perhaps 80% of what follows. Note that the concentration profile is independent of the diffusion coefficient.

Because the process is in steady state, the accumulation is zero. The diffusion rate is the diffusion flux times the film's area A. Thus

$$0 = A\left(j_1|_z - j_1|_{z+\Delta z}\right) \tag{2.2-1}$$

Dividing this equation by the film's volume, $A\Delta z$, and rearranging,

$$0 = -\left(\frac{j_1|_{z+\Delta z} - j_1|_z}{(z+\Delta z) - z}\right) \tag{2.2-2}$$

When Δz becomes very small, this equation becomes the definition of the derivative

$$0 = -\frac{d}{dz}j_1 \tag{2.2-3}$$

Combining this equation with Fick's law,

$$-j_1 = D\frac{dc_1}{dz} \tag{2.2-4}$$

we find, for a constant diffusion coefficient D,

$$0 = D\frac{d^2c_1}{dz^2} \tag{2.2-5}$$

This differential equation is subject to two boundary conditions:

$$z = 0, \quad c_1 = c_{10} \tag{2.2-6}$$

$$z = l, \quad c_1 = c_{1l} \tag{2.2-7}$$

Again, because this system is in steady state, the concentrations c_{10} and c_{1l} are independent of time. Physically, this means that the volumes of the adjacent solutions must be much greater than the volume of the film.

2.2.2 Mathematical Results

The desired concentration profile and flux are now easily found. First, we integrate Eq. 2.2-5 twice to find

$$c_1 = a + bz \tag{2.2-8}$$

The constants a and b can be found from Eqs. 2.2-6 and 2.2-7, so the concentration profile is

$$c_1 = c_{10} + (c_{1l} - c_{10})\frac{z}{l} \tag{2.2-9}$$

This linear variation was, of course, anticipated by the sketch in Fig. 2.2-1. The flux is found by differentiating this profile:

$$j_1 = -D\frac{dc_1}{dz} = \frac{D}{l}(c_{10} - c_{1l}) \tag{2.2-10}$$

Because the system is in steady state, the flux is a constant.

As mentioned earlier, this case is easy mathematically. Although it is very important, it is often underemphasized because it seems trivial. Before you conclude this, try some of the examples that follow to make sure you understand what is happening.

Example 2.2-1: Membrane diffusion   Derive the concentration profile and the flux for a single solute diffusing across a thin membrane. As in the preceding case of a film, the membrane separates two well-stirred solutions. Unlike the film, the membrane is chemically different from these solutions.

   Solution   As before, we first write a mass balance on a thin layer Δz:

$$0 = A\left(j_1|_z - j_1|_{z+\Delta z}\right)$$

This leads to a differential equation identical with Eq. 2.2-5:

$$0 = D\frac{d^2 c_1}{dz^2}$$

However, this new mass balance is subject to somewhat different boundary conditions:

$$z = 0, \quad c_1 = HC_{10}$$

$$z = l, \quad c_1 = HC_{1l}$$

where H is a partition coefficient, the concentration in the membrane divided by that in the adjacent solution. This partition coefficient is an equilibrium property, so its use implies that equilibrium exists across the membrane surface. In many cases, it can be about equal to the relative solubility within the film compared with that outside. For a film containing pores, H may just be the void fraction of the film.

   The concentration profile that results from these relations is

$$c_1 = HC_{10} + H(C_{1l} - C_{10})\frac{z}{l}$$

which is analogous to Eq. 2.2-9. This result looks harmless enough. However, it suggests concentration profiles likes those in Fig. 2.2-2, which contain sudden discontinuities at the interface. If the solute is more soluble in the membrane than in the surrounding

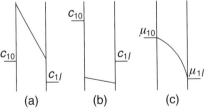

(a)   (b)   (c)

Fig. 2.2-2. Concentration profiles across thin membranes. In (a), the solute is more soluble in the membrane than in the adjacent solutions; in (b), it is less so. Both cases correspond to a chemical potential gradient like that in (c).

solutions, then the concentration increases. If the solute is less soluble in the membrane, then its concentration drops. Either case produces enigmas. For example, at the left-hand side of the membrane in Fig. 2.2-2(a), solute diffuses from the solution at C_{10} into the membrane at *higher* concentration.

This apparent quandary is resolved when we think carefully about the solute's diffusion. Diffusion often can occur from a region of low concentration into a region of high concentration; indeed, this is the basis of many liquid–liquid extractions. Thus the jumps in concentration in Fig. 2.2-2 are not as bizarre as they might appear; rather, they are graphical accidents that result from using the same scale to represent concentrations inside and outside membrane.

This type of diffusion can also be described in terms of the solute's energy or, more exactly, in terms of its chemical potential. The solute's chemical potential does not change across the membrane's interface, because equilibrium exists there. Moreover, this potential, which drops smoothly with concentration, as shown in Fig. 2.2-2(c), is the driving force responsible for the diffusion. The exact role of this driving force is discussed more completely in Sections 6.3 and 7.2.

The flux across a thin membrane can be found by combining the foregoing concentration profile with Fick's law:

$$j_1 = \frac{[DH]}{l}(C_{10} - C_{1l})$$

This is parallel to Eq. 2.2-10. The quantity in square brackets in this equation is called the permeability, and it is often reported experimentally. The quantity ($[DH]/l$) is called the permeance. The partition coefficient H is often found to vary more widely than the diffusion coefficient D, so differences in diffusion tend to be less important than the differences in solubility.

Example 2.2-2: Membrane diffusion with fast reaction   Imagine that while a solute is diffusing steadily across a thin membrane, it can rapidly and reversibly react with other immobile solutes fixed within the membrane. Find how this fast reaction affects the solute's flux.

Solution   The answer is surprising: The reaction has no effect. This is an excellent example because it requires careful thinking. Again, we begin by writing a mass balance on a layer Δz located within the membrane:

$$\begin{pmatrix} \text{solute} \\ \text{accumulation} \end{pmatrix} = \begin{pmatrix} \text{solute diffusion in} \\ \text{minus that out} \end{pmatrix} + \begin{pmatrix} \text{amount produced} \\ \text{by chemical reaction} \end{pmatrix}$$

Because the system is in steady state, this leads to

$$0 = A(j_1|_z - j_1|_{z+\Delta z}) - r_1 A\Delta z$$

or

$$0 = -\frac{d}{dz}j_1 - r_1$$

where r_1 is the rate of disappearance of the mobile species 1 in the membrane. A similar mass balance for the immobile product 2 gives

$$0 = -\frac{d}{dz}j_2 + r_1$$

But because the product is immobile, j_2 is zero, and hence r_1 is zero. As a result, the mass balance for species 1 is identical with Eq. 2.2-3, leaving the flux and concentration profile unchanged.

This result is easier to appreciate in physical terms. After the diffusion reaches a steady state, the local concentration is everywhere in equilibrium with the appropriate amount of the fast reaction's product. Because these local concentrations do not change with time, the amounts of the product do not change either. Diffusion continues unaltered.

This case in which a chemical reaction does not affect diffusion is unusual. For almost any other situation, the reaction can engender dramatically different mass transfer. If the reaction is irreversible, the flux can be increased many orders of magnitude, as shown in Section 17.1. If the diffusion is not steady, the apparent diffusion coefficient can be much greater than expected, as discussed in Example 2.3-2. However, in the case described in this example, the chemical reaction does not affect diffusion.

Example 2.2-3: Concentration-dependent diffusion The diffusion coefficient is remarkably constant. It varies much less with temperature than the viscosity or the rate of a chemical reaction. It varies surprisingly little with solute: for example, most diffusion coefficients of solutes dissolved in water fall within a factor of ten.

Diffusion coefficients also rarely vary with solute concentration, although there are some exceptions. For example, a small solute like water may show a concentration-dependent diffusion when diffusing into a polymer. To explore this, assume that

$$D = \frac{D_0 c_1}{c_{10}}$$

Then calculate the concentration profile and the flux across a thin film.

Solution Finding the concentration profile is a complete parallel to the simpler case of constant diffusion coefficient discussed at the start of this section. We again begin with a steady-state mass balance on a differential volume $A\Delta z$:

$$0 = A(j_1|_z - j_1|_{z+\Delta z})$$

Dividing by this volume, taking the limit as Δz goes to zero and combining with Fick's law gives

$$0 = -\frac{dj_1}{dz} = -\frac{d}{dz}\left(-\frac{D_0 c_1}{c_{10}}\frac{dc_1}{dz}\right)$$

This parallels Equation 2.2-5, but with a concentration-dependent D. This mass balance is subject to the boundary conditions

$$z = 0, \quad c_1 = c_{10}$$

$$z = l, \quad c_1 = 0$$

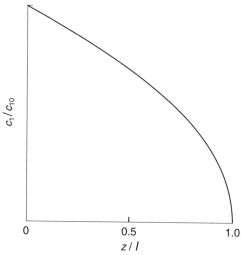

Fig. 2.2-3. Concentration-dependent diffusion across a thin film. While the steady-state flux is constant, the concentration gradient changes with position.

These are a special case of Eqs. 2.2-6 and 2.2-7. Integration gives the concentration profile

$$c_1 = c_{10}\left(1 - \frac{z}{l}\right)^{1/2}$$

This profile can be combined with Fick's law to give the flux

$$j_1 = -D\frac{dc_1}{dz} = -\frac{D_0 c_1}{c_{10}}\frac{dc_1}{dz} = \frac{D_0 c_{10}}{2l}$$

The flux is half that of the case of a constant diffusion coefficient D_0.

The meaning of this result is clearer if we consider the concentration profile shown in Fig. 2.2-3. The profile is nonlinear; indeed, its slope at $z = l$ is infinite. However, at that boundary, the diffusion coefficient is zero because the concentation is zero. The product of this infinite gradient and a zero coefficient is the constant flux, with an apparent diffusion coefficient equal to $(D_0/2)$. This unexceptional average value illustrates why Fick's law works so well.

Example 2.2-4: Diaphragm-cell diffusion One easy way to measure diffusion coefficients is the diaphragm cell shown in Fig. 2.2-4. These cells consist of two well-stirred volumes separated by a thin porous barrier or diaphragm. In the more accurate experiments, the diaphragm is often a sintered glass frit; in many successful experiments, it is just a piece of filter paper (see Section 5.5). To measure a diffusion coefficient with this cell, we fill the lower compartment with a solution of known concentration and the upper compartment with solvent. After a known time, we sample both upper and lower compartments and measure their concentrations.

Find an equation that uses the known time and the measured concentrations to calculate the diffusion coefficient.

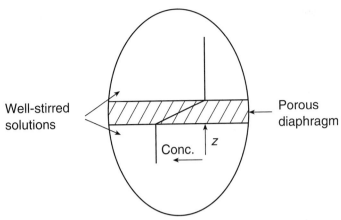

Fig. 2.2-4. A diaphragm cell for measuring diffusion coefficients. Because the diaphragm has a much smaller volume than the adjacent solutions, the concentration profile within the diaphragm has essentially the linear, steady-state value.

Solution An exact solution to this problem is elaborate and unnecessary. The useful approximate solution depends on the assumption that the flux across the diaphragm quickly reaches its steady-state value. This steady-state flux is approached even though the concentrations in the upper and lower compartments are changing with time. The approximations introduced by this assumption will be considered later.

In this pseudosteady state, the flux across the diaphragm is that given for membrane diffusion:

$$j_1 = \left[\frac{DH}{l}\right](C_{1,\text{lower}} - C_{1,\text{upper}})$$

Here, the quantity H includes the fraction of the diaphragm's area that is available for diffusion. We next write an overall mass balance on the adjacent compartments:

$$V_{\text{lower}} \frac{dC_{1,\text{lower}}}{dt} = -Aj_1$$

$$V_{\text{upper}} \frac{dC_{1,\text{upper}}}{dt} = +Aj_1$$

where A is the diaphragm's area. If these mass balances are divided by V_{lower} and V_{upper}, respectively, and the equations are subtracted, one can combine the result with the flux equation to obtain

$$\frac{d}{dt}(C_{1,\text{lower}} - C_{1,\text{upper}}) = D\beta(C_{1,\text{upper}} - C_{1,\text{lower}})$$

in which

$$\beta = \frac{AH}{l}\left(\frac{1}{V_{\text{lower}}} + \frac{1}{V_{\text{upper}}}\right)$$

is a geometrical constant characteristic of the particular diaphragm cell being used. This differential equation is subject to the obvious initial condition

$$t = 0, \quad C_{1,\text{lower}} - C_{1,\text{upper}} = C_{1,\text{lower}}^0 - C_{1,\text{upper}}^0$$

If the upper compartment is initially filled with solvent, then its initial solute concentration will be zero.

Integrating the differential equation subject to this condition gives the desired result:

$$\frac{C_{1,\text{lower}} - C_{1,\text{upper}}}{C_{1,\text{lower}}^0 - C_{1,\text{upper}}^0} = e^{-\beta D t}$$

or

$$D = \frac{1}{\beta t} \ln \left(\frac{C_{1,\text{lower}}^0 - C_{1,\text{upper}}^0}{C_{1,\text{lower}} - C_{1,\text{upper}}} \right)$$

We can measure the time t and the various concentrations directly. We can also determine the geometric factor β by calibration of the cell with a species whose diffusion coefficient is known. Then we can determine the diffusion coefficients of unknown solutes.

There are two major ways in which this analysis can be questioned. First, the diffusion coefficient used here is an effective value altered by the tortuosity in the diaphragm. Theoreticians occasionally assert that different solutes will have different tortuosities, so that the diffusion coefficients measured will apply only to that particular diaphragm cell and will not be generally usable. Experimentalists have cheerfully ignored these assertions by writing

$$D = \frac{1}{\beta' t} \ln \left(\frac{C_{1,\text{lower}}^0 - C_{1,\text{upper}}^0}{C_{1,\text{lower}} - C_{1,\text{upper}}} \right)$$

where β' is a new calibration constant that includes any tortuosity. So far, the experimentalists have gotten away with this: Diffusion coefficients measured with the diaphragm cell do agree with those measured by other methods.

The second major question about this analysis comes from the combination of the steady-state flux equation with an unsteady-state mass balance. You may find this combination to be one of those areas where superficial inspection is reassuring, but where careful reflection is disquieting. I have been tempted to skip over this point, but have decided that I had better not. Here goes:

The adjacent compartments are much larger than the diaphragm itself because they contain much more material. Their concentrations change slowly, ponderously, as a result of the transfer of a lot of solute. In contrast, the diaphragm itself contains relatively little material. Changes in its concentration profile occur quickly. Thus, even if this profile is initially very different from steady state, it will approach a steady state before the concentrations in the adjacent compartments can change much. As a result, the profile across the diaphragm will always be close to its steady value, even though the compartment concentrations are time dependent.

These ideas can be placed on a more quantitative basis by comparing the relaxation time of the diaphragm, l^2/D, with that of the compartments, $1/(D\beta)$. The analysis used here will be accurate when (Mills, Woolf, and Watts, 1968)

$$1 \gg \frac{l^2/D_{\text{eff}}}{1/(\beta D_{\text{eff}})} = V_{\substack{\text{diaphragm} \\ \text{voids}}} \left(\frac{1}{V_{\text{lower}}} + \frac{1}{V_{\text{upper}}} \right)$$

This type of "pseudosteady-state approximation" is common and underlies most mass transfer coefficients discussed later in this book.

The examples in this section show that diffusion across thin films can be difficult to understand. The difficulty does not derive from mathematical complexity; the calculation is easy and essentially unchanged. The simplicity of the mathematics is the reason why diffusion across thin films tends to be discussed superficially in mathematically oriented books. The difficulty in thin-film diffusion comes from adapting the same mathematics to widely varying situations with different chemical and physical effects. This is what is difficult to understand about thin-film diffusion. It is an understanding that you must gain before you can do creative work on harder mass transfer problems. Remember: this case is the base for perhaps 80 percent of the diffusion problems in this book.

2.3 Unsteady Diffusion in a Semi-infinite Slab

We now turn to a discussion of diffusion in a semi-infinite slab, which is basic to perhaps 10 percent of the problems in diffusion. We consider a volume of solution that starts at an interface and extends a long way. Such a solution can be a gas, liquid, or solid. We want to find how the concentration varies in this solution as a result of a concentration change at its interface. In mathematical terms, we want to find the concentration and flux as a function of position and time.

This type of mass transfer is sometimes called free diffusion simply because this is briefer than "unsteady diffusion in a semi-infinite slab." At first glance, this situation may seem rare because no solution can extend an infinite distance. The previous thin-film example made more sense because we can think of many more thin films than semi-infinite slabs. Thus we might conclude that this semi-infinite case is not common. That conclusion would be a serious error.

The important case of a semi-infinite slab is common because any diffusion problem will behave as if the slab is infinitely thick at short enough times. For example, imagine that one of the thin membranes discussed in the previous section separates two identical solutions, so that it initially contains a solute at constant concentration. Everything is quiescent, at equilibrium. Suddenly the concentration on the left-hand interface of the membrane is raised, as shown in Fig. 2.3-1. Just after this sudden increase, the concentration near this left interface rises rapidly on its way to a new steady state. In these first few seconds, the concentration at the right interface remains unaltered, ignorant of the turmoil on the left. The left might as well be infinitely far away; the membrane, for these first few seconds, might as well be infinitely thick. Of course, at larger times, the system will slither into the steady-state limit in Fig. 2.3-1(c). But in those first seconds, the membrane does behave like a semi-infinite slab.

This example points to an important corollary, which states that cases involving an infinite slab and a thin membrane will bracket the observed behavior. At short times,

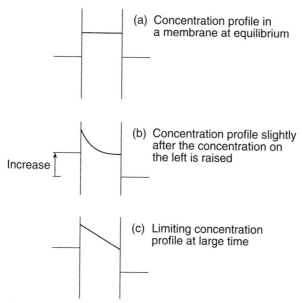

(a) Concentration profile in a membrane at equilibrium

Increase

(b) Concentration profile slightly after the concentration on the left is raised

(c) Limiting concentration profile at large time

Fig. 2.3-1. Unsteady- versus steady-state diffusion. At small times, diffusion will occur only near the left-hand side of the membrane. As a result, at these small times, the diffusion will be the same as if the membrane was infinitely thick. At large times, the results become those in the thin film.

diffusion will proceed as if the slab is infinite; at long times, it will occur as if the slab is thin. By focussing on these limits, we can bracket the possible physical responses to different diffusion problems.

2.3.1 The Physical Situation

The diffusion in a semi-infinite slab is schematically sketched in Fig. 2.3-2. The slab initially contains a uniform concentration of solute $c_{1\infty}$. At some time, chosen as time zero, the concentration at the interface is suddenly and abruptly increased, although the solute is always present at high dilution. The increase produces the time-dependent concentration profile that develops as solute penetrates into the slab.

We want to find the concentration profile and the flux in this situation, and so again we need a mass balance written on the thin layer of volume $A\Delta z$:

$$\begin{pmatrix} \text{solute accumulation} \\ \text{in volume } A\Delta z \end{pmatrix} = \begin{pmatrix} \text{rate of diffusion} \\ \text{into the layer at } z \end{pmatrix} - \begin{pmatrix} \text{rate of diffusion} \\ \text{out of the layer} \\ \text{at } z + \Delta z \end{pmatrix} \qquad (2.3\text{-}1)$$

In mathematical terms, this is

$$\frac{\partial}{\partial t}(A\Delta z c_1) = A\left(j_1|_z - j_1|_{z+\Delta z}\right) \qquad (2.3\text{-}2)$$

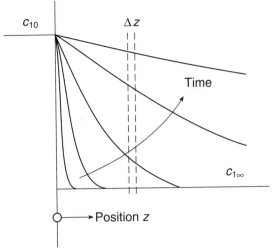

Fig. 2.3-2. Free diffusion. In this case, the concentration at the left is suddenly increased to a higher constant value. Diffusion occurs in the region to the right. This case and that in Fig. 2.2-1 are basic to most diffusion problems.

We divide by $A\Delta z$ to find

$$\frac{\partial c_1}{\partial t} = -\left(\frac{j_1|_{z+\Delta z} - j_1|_z}{(z+\Delta z) - z}\right) \tag{2.3-3}$$

We then let Δz go to zero and use the definition of the derivative

$$\frac{\partial c_1}{\partial t} = -\frac{\partial j_1}{\partial z} \tag{2.3-4}$$

Combining this equation with Fick's law and assuming that the diffusion coefficient is independent of concentration, we get

$$\frac{\partial c_1}{\partial t} = D\frac{\partial^2 c_1}{\partial z^2} \tag{2.3-5}$$

This equation is sometimes called Fick's second law, or "the diffusion equation." In this case, it is subject to the following conditions:

$$t = 0, \quad \text{all } z, \quad c_1 = c_{1\infty} \tag{2.3-6}$$

$$t > 0, \quad z = 0, \quad c_1 = c_{10} \tag{2.3-7}$$

$$z = \infty, \quad c_1 = c_{1\infty} \tag{2.3-8}$$

Note that both $c_{1\infty}$ and c_{10} are taken as constants. The concentration $c_{1\infty}$ is constant because it is so far from the interface as to be unaffected by events there; the concentration c_{10} is kept constant by adding material at the interface.

2.3.2 Mathematical Solution

The solution of this problem is easiest using the method of "combination of variables." This method is easy to follow, but it must have been difficult to invent. Fourier, Graham, and Fick failed in the attempt; it required Boltzmann's tortured imagination (Boltzmann, 1894).

The trick to solving this problem is to define a new variable

$$\zeta = \frac{z}{\sqrt{4Dt}} \tag{2.3-9}$$

The differential equation can then be written as

$$\frac{dc_1}{d\zeta}\left(\frac{\partial\zeta}{\partial t}\right) = D\frac{d^2c_1}{d\zeta^2}\left(\frac{\partial\zeta}{\partial z}\right)^2 \tag{2.3-10}$$

or

$$\frac{d^2c_1}{d\zeta^2} + 2\zeta\frac{dc_1}{d\zeta} = 0 \tag{2.3-11}$$

In other words, the partial differential equation has been almost magically transformed into an ordinary differential equation. The magic also works for the boundary conditions: from Eq. 2.3-7,

$$\zeta = 0, \quad c_1 = c_{10} \tag{2.3-12}$$

and from Eqs. 2.3-6 and 2.3-8,

$$\zeta = \infty, \quad c_1 = c_{1\infty} \tag{2.3-13}$$

With the method of combination of variables, the transformation of the initial and boundary conditions is often more critical than the transformation of the differential equation.

The solution is now straightforward. One integration of Eq. 2.3-11 gives

$$\frac{dc_1}{d\zeta} = ae^{-\zeta^2} \tag{2.3-14}$$

where a is an integration constant. A second integration and use of the boundary conditions give

$$\frac{c_1 - c_{10}}{c_{1\infty} - c_{10}} = \text{erf }\zeta \tag{2.3-15}$$

where

$$\text{erf }\zeta = \frac{2}{\sqrt{\pi}}\int_0^{\zeta} e^{-s^2}\,ds \tag{2.3-16}$$

is the error function of ζ. This is the desired concentration profile giving the variation of concentration with position and time.

In many practical problems, the flux in the slab is of greater interest than the concentration profile itself. This flux can again be found by combining Fick's law with Eq. 2.3-15:

$$j_1 = -D\frac{\partial c_1}{\partial z} = \sqrt{D/\pi t}\, e^{-z^2/4Dt}(c_{10} - c_{1\infty})$$
(2.3-17)

One particularly useful limit is the flux across the interface at $z = 0$:

$$j_1|_{z=0} = \sqrt{D/\pi t}\,(c_{10} - c_{1\infty})$$
(2.3-18)

This flux is the value at the particular time t and not that averaged over time. This distinction will be important in Section 9.2.

At this point, I have the same pedagogical problem I had in the previous section: I must convince you that the apparently simple results in Eqs. 2.3-15 and 2.3-18 are valuable. These results are exceeded in importance only by Eqs. 2.2-9 and 2.2-10. Fortunately, the mathematics may be difficult enough to spark thought and reflection; if not, the examples that follow should do so.

Example 2.3-1: Diffusion across an interface The picture of the process in Fig. 2.3-2 implies that the concentration at $z = 0$ is continuous. This would be true, for example, if when $z \geq 0$ there was a highly swollen gel, and when $z < 0$ there was a stirred solution.

A much more common case occurs when there is a gas–liquid interface at $z = 0$. Ordinarily, the gas at $z < 0$ will be well mixed, but the liquid will not. How will this interface affect the results given earlier?

    Solution Basically, it will have no effect. The only change will be a new boundary condition, replacing Eq. 2.3-7:

$$z = 0, \quad c_1 = cx_1 = c\frac{p_{10}}{H}$$

where c_1 is the concentration of solute in the liquid, x_1 is its mole fraction, p_{10} is its partial pressure in the gas phase, H is the Henry's law constant, and c is the total molar concentration in the liquid.

The difficulties caused by a gas–liquid interface are another result of the plethora of units in which concentration can be expressed. These difficulties require concern about units, but they do not demand new mathematical weapons. The changes required for a liquid–liquid interface can be similarly subtle.

Example 2.3-2: Free diffusion with fast chemical reaction In many problems, the diffusing solutes react rapidly and reversibly with surrounding material. The surrounding material is stationary and cannot diffuse. For example, in the dyeing of wool, some dyes can react quickly and reversibly with the wool as dye diffuses into the fiber. How does such a rapid chemical reaction change the concentration profile and the flux?

    Solution In this case, the chemical reaction can radically change the process by reducing the apparent diffusion coefficient and increasing the interfacial flux of solute. These radical changes stand in stark contrast to the steady-state result, where the chemical reaction produces no effect.

To solve this example, we first recognize that the solute is effectively present in two forms: (1) free solute that can diffuse and (2) reacted solute fixed at the point of reaction. If this reaction is reversible and faster than diffusion,

$$c_2 = Kc_1$$

where c_2 is the concentration of the solute that has already reacted, c_1 is the concentration of the unreacted solute that can diffuse, and K is the equilibrium constant of the reaction. If the reaction is minor, K will be small; as the reaction becomes irreversible, K will become very large.

With these definitions, we now write a mass balance for each solute form. These mass balances should have the form

$$\begin{pmatrix} \text{accumulation} \\ \text{in } A\Delta z \end{pmatrix} = \begin{pmatrix} \text{diffusion in} \\ \text{minus that out} \end{pmatrix} + \begin{pmatrix} \text{amount produced by} \\ \text{reaction in } A\Delta z \end{pmatrix}$$

For the diffusing solute, this is

$$\frac{\partial}{\partial t}[A\Delta z c_1] = A(j_1|_z - j_1|_{z+\Delta z}) - r_1 A\Delta z$$

where r_1 is the rate of disappearance per volume of species 1, the diffusing solute. By arguments analogous to Eqs. 2.2-2 to 2.2-5, this becomes

$$\frac{\partial c_1}{\partial t} = D\frac{\partial^2 c_1}{\partial z^2} - r_1$$

The term on the left-hand side is the accumulation; the first term on the right is the diffusion in minus the diffusion out; the term r_1 is the effect of chemical reaction.

When we write a similar mass balance on the second species, we find

$$\frac{\partial}{\partial t}[A\Delta z c_2] = -r_1 A\Delta z$$

or

$$\frac{\partial c_2}{\partial t} = r_1$$

We do not get a diffusion term because the reacted solute cannot diffuse. We get a reaction term that has a different sign but the same magnitude, because any solute that disappears as species 1 reappears as species 2.

To solve these questions, we first add them to eliminate the reaction term:

$$\frac{\partial}{\partial t}(c_1 + c_2) = D\frac{\partial^2 c_1}{\partial z^2}$$

We now use the fact that the chemical reaction is at equilibrium:

$$\frac{\partial}{\partial t}(c_1 + Kc_1) = D\frac{\partial^2 c_1}{\partial z^2}$$

$$\frac{\partial c_1}{\partial t} = \frac{D}{1 + K}\frac{\partial^2 c_1}{\partial z^2}$$

This result is subject to the same initial and boundary conditions as before in Eqs. 2.3-6, 2.3-7, and 2.3-8. As a result, the only difference between this example and the earlier problem is that $D/(1 + K)$ replaces D.

This is intriguing. The chemical reaction has left the mathematical form of the answer unchanged, but it has altered the apparent diffusion coefficient. The concentration profile now is

$$\frac{c_1 - c_{10}}{c_{1\infty} - c_{10}} = \text{erf}\frac{z}{\sqrt{4[D/(1 + K)]t}}$$

and the interfacial flux is

$$j_1|_{z=0} = \sqrt{D(1 + K)/\pi t}(c_{10} - c_{1\infty})$$

The flux has been increased by the chemical reaction.

These effects of chemical reaction can easily be several orders of magnitude. As will be detailed in Chapter 5, diffusion coefficients tend to fall in fairly narrow ranges. Those coefficients for gases are around 0.1 cm^2/sec; those in ordinary liquids cluster about 10^{-5} cm^2/sec. Deviations from these values of more than an order of magnitude are unusual. However, differences in the equilibrium constant K of a million or more occur frequently. Thus a fast chemical reaction can tremendously influence the unsteady diffusion process.

Example 2.3-3: Determining diffusion coefficients from free diffusion experiments   Diffusion in an infinite slab is the geometry used for the most accurate measurements of diffusion coefficients. These most accurate measurements determine the concentration profile by interferometry. One relatively simple method, the Rayleigh interferometer, uses a rectangular cell in which there is an initial step function in refractive index. The decay of this refractive index profile is followed by collimated light through the cell to give interference fringes. These fringes record the refractive index versus camera position and time.

Find equations that allow this information to be used to calculate diffusion coefficients.

Solution   The concentration profiles established in the diffusion cell closely approach the profiles calculated earlier for a semi-infinite slab. The cell now effectively contains two semi-infinite slabs joined together at $z = 0$. The concentration profile is unaltered from Eq. 2.3-15:

$$\frac{c_1 - c_{10}}{c_{1\infty} - c_{10}} = \text{erf}\frac{z}{\sqrt{4Dt}}$$

where $c_{10}[=(c_{1\infty}+c_{1-\infty})/2]$ is the average concentration between the two ends of the cell. How accurate this equation is depends on how exactly the initial change in concentration can be realized; in practice, this change can be within 10 seconds of a true step function.

We must convert the concentration and cell position into the experimental measured refractive index and camera position. The refractive index n is linearly proportional to the concentration:

$$n = n_{\text{solvent}} + bc_1$$

where n_{solvent} is the refractive index of the solvent and b is a constant determined from experiment. Each position in the camera Z is proportional to a position in the diffusion cell:

$$Z = az$$

where a is the magnification of the apparatus. It is experimentally convenient not to measure the position of one fringe but rather to measure the intensity minima between the many fringes. These minima occur when

$$\frac{n - n_0}{n_\infty - n_0} = \frac{j}{J/2}$$

where n_∞ and n_0 are the refractive indices at $z = \infty$ and $z = 0$, respectively; J is the total number of interference fringes, and j is an integer called the fringe number. This number is most conveniently defined as zero at $z = 0$, the center of the cell. Combining these equations,

$$\frac{j}{J/2} = \text{erf}\frac{Z_j}{a\sqrt{4Dt}}$$

where Z_j is the intensity minimum associated with the jth fringe. Because a and t are experimentally accessible, measurements of Z_j (j, J) can be used to find the diffusion coefficient D. While the accuracy of interferometric experiments like this remains unrivaled, the use of these methods has declined because they are tedious.

2.4 Three Other Examples

The two previous sections describe diffusion across thin films and in semi-infinite slabs. In this section, we turn to discussing mathematical variations of diffusion problems. This mathematical emphasis changes both the pace and the tone of this book. Up to now, we have consistently stressed the physical origins of the problems, constantly harping on natural effects like changing liquid to gas. Now we shift to the more common text book composition, a sequence of equations sometimes as jarring as a twelve-tone concerto.

In these examples, we have three principal goals:

(1) We want to show how the differential equations describing diffusion are derived.
(2) We want to examine the effects of spherical and cylindrical geometries.
(3) We want to supply a mathematical primer for solving these different diffusion equations.

In all three examples, we continue to assume dilute solutions. The three problems examined next are physically important and will be referred to again in this book. However, they are introduced largely to achieve mathematical goals.

2.4.1 Decay of a Pulse (Laplace Transforms)

As a first example, we consider the diffusion away from a sharp pulse of solute. This example is the third truly important problem for diffusion. It complements the cases of a thin film and the semi-infinite slab to form the basis of perhaps 95 percent of all the diffusion problems which are encountered. The initially sharp concentration gradient relaxes by diffusion in the z direction into the smooth curves shown in Fig. 2.4-1. We want to calculate the shape of these curves. This calculation illustrates the development of a differential equation and its solution using Laplace transforms.

As usual, our first step is to make a mass balance on the differential volume $A\Delta z$ as shown:

$$\begin{pmatrix} \text{solute} \\ \text{accumulation} \\ \text{in } A\Delta z \end{pmatrix} = \begin{pmatrix} \text{solute} \\ \text{diffusion into} \\ \text{this volume} \end{pmatrix} - \begin{pmatrix} \text{solute} \\ \text{diffusion out of} \\ \text{this volume} \end{pmatrix} \qquad (2.4\text{-}1)$$

In mathematical terms, this is

$$\frac{\partial}{\partial t}[A\Delta z c_1] = Aj_1|_z - Aj_1|_{z+\Delta z} \qquad (2.4\text{-}2)$$

Dividing by the volume and taking the limit as Δz goes to zero gives

$$\frac{\partial c_1}{\partial t} = -\frac{\partial j_1}{\partial z} \qquad (2.4\text{-}3)$$

Combining this relation with Fick's law of diffusion,

$$\frac{\partial c_1}{\partial t} = D\frac{\partial^2 c_1}{\partial z^2} \qquad (2.4\text{-}4)$$

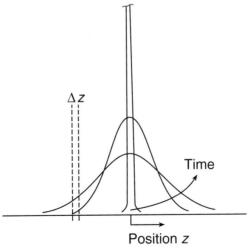

Fig. 2.4-1. Diffusion of a pulse. The concentrated solute originally located at $z = 0$ diffuses as the Gaussian profile shown. This is the third of the three most important cases, along with those in Figs. 2.2-1 and 2.3-2.

This is the same differential equation basic to diffusion in a semi-infinite slab and considered in the previous section. The boundary conditions on this equation are different as follows. First, far from the pulse, the solute concentration is zero:

$$t > 0, \quad z = \infty, \quad c_1 = 0 \tag{2.4-5}$$

Second, because diffusion occurs at the same speed in both directions, the pulse is symmetric:

$$t > 0, \quad z = 0, \quad \frac{\partial c_1}{\partial z} = 0 \tag{2.4-6}$$

This is equivalent to saying that at $z = 0$, the flux has the same magnitude in the positive and negative directions.

The initial condition for the pulse is more interesting in that all the solute is initially located at $z = 0$:

$$t = 0, \quad c_1 = \frac{M}{A} \delta(z) \tag{2.4-7}$$

where A is the cross-sectional area over which diffusion is occurring, M is the total amount of solute in the system, and $\delta(z)$ is the Dirac function. This can be shown to be a reasonable condition by a mass balance:

$$\int_{-\infty}^{\infty} c_1 A dz = \int_{-\infty}^{\infty} \frac{M}{A} \delta(z) A dz = M \tag{2.4-8}$$

In this integration, we should remember that $\delta(z)$ has dimensions of (length)$^{-1}$.

To solve this problem, we first take the Laplace transform of Eq. 2.4-4 with respect to time:

$$s\bar{c}_1 - c_1(t = 0) = D \frac{d^2\bar{c}_1}{dz^2} \tag{2.4-9}$$

where $\bar{c}_1$ is the transformed concentration. The boundary conditions are

$$z = 0, \quad \frac{d\bar{c}_1}{dz} = -\frac{M/A}{2D} \tag{2.4-10}$$

$$z = \infty, \quad \bar{c}_1 = 0 \tag{2.4-11}$$

The first of these reflects the properties of the Dirac function, but the second is routine. Equation 2.4-9 can then easily by integrated to give

$$\bar{c}_1 = ae^{\left(\sqrt{s/D}\, z\right)} + be^{\left(-\sqrt{s/D}\, z\right)} \tag{2.4-12}$$

where a and b are integration constants. Clearly, a is zero by Eq. 2.4-11. Using Eq. 2.4-10, we find b and hence $\bar{c}_1$:

$$\bar{c}_1 = \frac{M/A}{2D} \sqrt{D/s}\, \exp\left(-\sqrt{s/D}\, z\right) \tag{2.4-13}$$

The inverse Laplace transform of this function gives

$$\bar{c}_1 = \frac{M/A}{\sqrt{4\pi Dt}}e^{-z^2/4Dt} \tag{2.4-14}$$

which is a Gaussian curve. You may wish to integrate the concentration over the entire system to check that the total solute present is M.

This solution can be used to solve many unsteady diffusion problems that have unusual initial conditions. More important, it is often used to correlate the dispersion of pollutants, especially in the air, as discussed in Chapter 4.

2.4.2 Steady Dissolution of a Sphere (Spherical Coordinates)

Our second example, which is easier mathematically, is the steady dissolution of a spherical particle, as shown in Fig. 2.4-2. The sphere is of a sparingly soluble material, so that the sphere's size does not change much. However, this material quickly dissolves in the surrounding solvent, so that the solute's concentration at the sphere's surface is saturated. Because the sphere is immersed in a large fluid volume, the concentration far from the sphere is zero.

The goal is to find both the dissolution rate and the concentration profile around the sphere. Again, the first step is a mass balance. In contrast with the previous examples, this mass balance is most conveniently made in spherical coordinates originating from the center of the sphere. Then we can make a mass balance on a spherical shell of thickness Δr located at some arbitrary distance r from the sphere. This spherical shell is like the rubber of a balloon of surface area $4\pi r^2$ and thickness Δr.

A mass balance on this shell has the same general form as those used earlier:

$$\begin{pmatrix} \text{solute accumulation} \\ \text{within the shell} \end{pmatrix} = \begin{pmatrix} \text{diffusion} \\ \text{into the shell} \end{pmatrix} - \begin{pmatrix} \text{diffusion} \\ \text{out of the shell} \end{pmatrix} \tag{2.4-15}$$

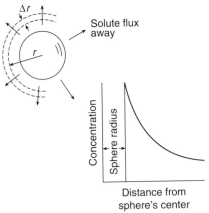

Fig. 2.4-2. Steady dissolution of a sphere. This problem represents an extension of diffusion theory to a spherically symmetric situation. In actual physical situations, this dissolution can be complicated by free convection caused by diffusion.

In mathematical terms, this is

$$\frac{\partial}{\partial t}\left(4\pi r^2 \Delta r c_1\right) = 0 = \left(4\pi r^2 j_1\right)_r - \left(4\pi r^2 j_1\right)_{r+\Delta r} \tag{2.4-16}$$

The accumulation on the left-hand side of this mass balance is zero because diffusion is steady, not varying with time. Novices frequently make a serious error at this point by canceling the r^2 out of both terms on the right-hand side. This is wrong. The term $r^2 j_1$ is evaluated at r in the first term; that is, it is $r^2(j_1|_r)$. The term is evaluated at $(r + \Delta r)$ in the second term; so it equals $(r + \Delta r)^2(j_1|_{r + \Delta r})$.

If we divide both sides of this equation by the spherical shell's volume and take the limit as $\Delta r \to 0$, we find

$$0 = -\frac{1}{r^2}\frac{d}{dr}\left(r^2 j_1\right) \tag{2.4-17}$$

Combining this with Fick's law and assuming that the diffusion coefficient is constant,

$$0 = \frac{D}{r^2}\frac{d}{dr}\left(r^2 \frac{dc_1}{dr}\right) \tag{2.4-18}$$

This basic differential equation is subject to two boundary conditions:

$$r = R_0, \quad c_1 = c_1(\text{sat}) \tag{2.4-19}$$

$$r = \infty, \quad c_1 = 0 \tag{2.4-20}$$

where R_0 is the sphere radius. If the sphere were dissolving in a partially saturated solution, this second condition would be changed, but the basic mathematical structure would remain unaltered. One integration of Eq. 2.4-18 yields

$$\frac{dc_1}{dr} = \frac{a}{r^2} \tag{2.4-21}$$

where a is an integration constant. A second integration gives

$$c_1 = b - \frac{a}{r} \tag{2.4-22}$$

Use of the two boundary conditions gives the concentration profile

$$c_1 = c_1(\text{sat})\frac{R_0}{r} \tag{2.4-23}$$

The dissolution flux can then be found from Fick's law:

$$j_1 = -D\frac{dc_1}{dr} = \frac{DR_0}{r^2}c_1(\text{sat}) \tag{2.4-24}$$

which, at the sphere's surface, is

$$j_1 = \frac{D}{R_0}c_1(\text{sat}) \tag{2.4-25}$$

This example is a mainstay of the analysis of diffusion. It is a good mathematical introduction of spherical coordinates, and it gives a result which is much like that for steady diffusion across a thin film. After all, Eq. 2.4-25 is the complete parallel of Eq. 2.2-10, but with the sphere radius R_0 replacing the film thickness l. Thus most teachers repeat this example as gospel.

Unfortunately, this result is only rarely supported by experiment. The reason is that the dissolution of the sphere almost always causes a density difference in the surrounding solution, which in turn causes flow by free convection. This flow accelerates the dissolution rate. For example, for dissolution in water, a density difference of 10^{-6} g/cm^3, almost too small to measure, causes a 400 percent increase in the dissolution expected from Eqs. 2.4-25. Students should beware: don't trust your teacher on this point.

2.4.3 Unsteady Diffusion into Cylinders (Cylindrical Coordinates and Separation of Variables)

The final example, probably the hardest of the three, concerns the diffusion of a solute into the cylinder shown in Fig. 2.4-3. The cylinder initially contains no solute. At

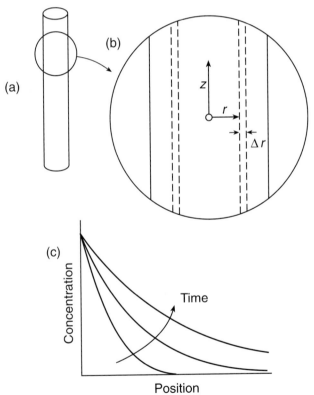

Fig. 2.4-3. Waterproofing a fence post. This problem is modeled as diffusion in an infinite cylinder, and so represents an extension to a cylindrically symmetric situation. In reality, the ends of the post must be considered, especially because diffusion with the grain is faster than across the grain.

time zero, it is suddenly immersed in a well-stirred solution that is of such enormous volume that its solute concentration is constant. The solute diffuses into the cylinder symmetrically. Problems like this are important in the chemical treatment of wood.

We want to find the solute's concentration in this cylinder as a function of time and location. As in the previous examples, the first step is a mass balance; in contrast, this mass balance is made on a cylindrical shell located at r, of area $2\pi Lr$, and of volume $2\pi Lr\Delta r$, where L is the cylinder's length. The basic balance

$$\begin{pmatrix} \text{solute accumulation} \\ \text{in this cylindrical shell} \end{pmatrix} = \begin{pmatrix} \text{solute diffusion} \\ \text{into the shell} \end{pmatrix} - \begin{pmatrix} \text{solute diffusion} \\ \text{out of the shell} \end{pmatrix} \quad (2.4\text{-}26)$$

becomes in mathematical terms

$$\frac{\partial}{\partial t}(2\pi rL\Delta rc_1) = (2\pi rLj_1)_r - (2\pi rLj_1)_{r+\Delta r} \quad (2.4\text{-}27)$$

We can now divide by the shell's volume and take the limit as Δr becomes small:

$$\frac{\partial}{\partial t}c_1 = -\frac{1}{r}\frac{\partial}{\partial r}(rj_1) \quad (2.4\text{-}28)$$

Combining this expression with Fick's law gives the desired mass balance

$$\frac{\partial c_1}{\partial t} = \frac{D}{r}\frac{\partial}{\partial r}\left(r\frac{\partial c_1}{\partial r}\right) \quad (2.4\text{-}29)$$

which is subject to the following conditions:

$$t \leqslant 0, \quad \text{all } r, \quad c_1 = 0 \quad (2.4\text{-}30)$$

$$t > 0, \quad r = R_0, \quad c_1 = c_1(\text{surface}) \quad (2.4\text{-}31)$$

$$r = 0, \quad \frac{\partial c_1}{\partial r} = 0 \quad (2.4\text{-}32)$$

In these equations, $c_1(\text{surface})$ is the concentration at the cylinder's surface and R_0 is the cylinder's radius. The first of the boundary conditions results from the large volume of surrounding solution, and the second reflects the symmetry of the concentration profiles.

Problems like this are often algebraically simplified if they are written in terms of dimensionless variables. This is standard practice in many advanced textbooks. I often find this procedure confusing, because for me it produces only a small gain in algebra at the expense of a large loss in physical insight. Nonetheless, we shall follow this procedure here to illustrate the simplification possible. We first define three new variables:

$$\text{dimensionless concentration}: \theta = 1 - \frac{c_1}{c_1(\text{surface})} \quad (2.4\text{-}33)$$

$$\text{dimensionless position}: \xi = \frac{r}{R_0} \quad (2.4\text{-}34)$$

$$\text{dimensionless time}: \tau = \frac{Dt}{R_0^2} \quad (2.4\text{-}35)$$

The differential equation and boundary conditions now become

$$\frac{\partial \theta}{\partial \tau} = \frac{1}{\xi}\frac{\partial}{\partial \xi}\left(\xi\frac{\partial \theta}{\partial \xi}\right) \tag{2.4-36}$$

subject to

$$\tau = 0, \quad \text{all } \xi, \quad \theta = 1 \tag{2.4-37}$$

$$\tau > 0, \quad \xi = 1, \quad \theta = 0 \tag{2.4-38}$$

$$\xi = 0, \quad \frac{\partial \theta}{\partial \xi} = 0 \tag{2.4-39}$$

For the novice, this manipulation can be more troublesome than it looks.

To solve these equations, we first assume that the solution is the product of two functions, one of time and one of radius:

$$\theta(\tau, \xi) = g(\tau)\, f(\xi) \tag{2.4-40}$$

When Eqs. 2.4-36 and 2.4-40 are combined, the resulting tangle of terms can be separated by division with $g(\tau)f(\xi)$:

$$f(\xi)\frac{dg(\tau)}{d\tau} = \frac{g(\tau)}{\xi}\frac{d}{d\xi}\xi\frac{df(\xi)}{d\xi}$$

$$\frac{1}{g(\tau)}\frac{dg(\tau)}{d\tau} = \frac{1}{\xi f(\xi)}\frac{d}{d\xi}\xi\frac{df(\xi)}{d\xi} \tag{2.4-41}$$

Now, if one fixes ξ and changes τ, $f(\xi)$ remains constant but $g(\tau)$ varies. As a result,

$$\frac{1}{g(\tau)}\frac{dg(\tau)}{d\tau} = -\alpha^2 \tag{2.4-42}$$

where α is a constant. Similarly, if we hold τ constant and let ξ change, we realize

$$\frac{1}{\xi f(\xi)}\frac{d}{d\xi}\xi\frac{df(\xi)}{d\xi} = -\alpha^2 \tag{2.4-43}$$

Thus the partial differential Eq. 2.4-36 has been converted into two ordinary differential Eqs. 2.4-42 and 2.4-43.

The solution of the time-dependent part of this result is easy:

$$g(\tau) = a'e^{-\alpha^2\tau} \tag{2.4-44}$$

where a' is an integration constant. The solution for $f(\xi)$ is more complicated, but straightforward:

$$f(\xi) = aJ_0(\alpha\xi) + bY_0(\alpha\xi) \tag{2.4-45}$$

where J_0 and Y_0 are Bessel functions and a and b are two more constants. From Eq. 2.4-39 we see that $b = 0$. From Eq. 2.4-38, we see that

$$0 = aJ_0(\alpha) \tag{2.4-46}$$

Because a cannot be zero, we recognize that there must be an entire family of solutions for which

$$J_0(\alpha_n) = 0 \tag{2.4-47}$$

The most general solution must be the sum of all solutions of this form found for different integral values of n:

$$\theta(\tau, \xi) = \sum_{n=1}^{\infty} (aa'_n) J_0(\alpha_n \xi) e^{-\alpha_n^2 \tau} \tag{2.4-48}$$

We now use the initial condition Eq. 2.4-37 to find the remaining integration constant $(aa')_n$:

$$1 = \sum_{n=1}^{\infty} (aa')_n J_0(\alpha_n \xi) \tag{2.4-49}$$

We multiply both sides of this equation by $\xi J_0(\alpha_n \xi)$ and integrate from $\xi = 0$ to $\xi = 1$ to find (aa'). The total result is then

$$\theta = \sum_{n=1}^{\infty} \left[\frac{2}{\alpha_n J_1(\alpha_n)} \right] J_0(\alpha_n \xi) e^{-\alpha_n^2 \tau} \tag{2.4-50}$$

or, in terms of our original variables,

$$\frac{c_1}{c_1(\text{surface})} = 1 - 2 \sum_{n=1}^{\infty} \frac{e^{-D\alpha_n^2 t / R_0^2} J_0(\alpha_n r / R_0)}{\alpha_n J_1(\alpha_n r / R_0)} \tag{2.4-51}$$

This is the desired result, though the α_n must still be found from Eq. 2.4-47.

This problem clearly involves a lot of work. The serious reader should certainly work one more problem of this type to get a feel for the idea of separation of variables and for the practice of evaluating integration constants. Even the serious reader probably will embrace the ways of avoiding this work described in the next chapter.

2.5 Convection and Dilute Diffusion

In many practical problems, both diffusion and convective flow occur. In some cases, especially in fast mass transfer in concentrated solutions, the diffusion itself causes the convection. This type of mass transfer, a subject of Chapter 3, requires more complicated physical and mathematical analyses.

There is another group of important problems in which diffusion and convection can be more easily handled. These problems arise when diffusion and convection occur normal to each other. In other words, diffusion occurs in one direction, and convective flow occurs in a perpendicular direction. Three of these problems are examined in this section. The first, steady diffusion across a thin flowing film, parallels Section 2.2; the second, diffusion into a liquid film, is a less obvious analogue to Section 2.3. These two examples tend to bracket the observed experimental behavior, and they are basic to theories relating diffusion and mass transfer coefficients (see Chapter 9).

2.5.1 Steady Diffusion Across a Falling Film

The first of the problems of concern here, sketched in Fig. 2.5-1, involves steady-state diffusion across a thin, moving liquid film. The concentrations on both sides of this film are fixed by electrochemical reactions, but the film itself is moving steadily. I have chosen this example not because it occurs often but because it is simple. I ask that readers oriented toward the practical will wait with later examples for results of greater applicability.

To solve this problem, we make three key assumptions:

(1) The liquid solution is dilute. This assumption is the axiom for this entire chapter.
(2) The liquid is the only resistance to mass transfer. This implies that the electrode reactions are fast.
(3) Mass transport is by diffusion in the z direction and by convection in the x direction. Transport by the other mechanisms is negligible.

It is the last of these assumptions that is most critical. It implies that convection is negligible in the z direction. In fact, diffusion in the z direction automatically generates convection in this direction, but this convection is small in a dilute solution. The last assumption also suggests that there is no diffusion in the x direction. There is such diffusion, but it is assumed much slower and hence much less important in the x direction than convection.

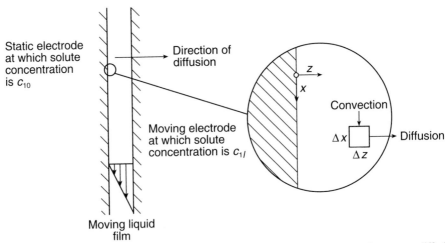

Fig. 2.5-1. Steady diffusion in a moving film. This case is mathematically the same as diffusion across a stagnant film, shown in Fig. 2.2-1. It is basic to the film theory of mass transfer described in Section 9.1.

This problem can be solved by writing a mass balance on the differential volume $W\Delta x\Delta z$, where W is the width of the liquid film, normal to the plane of the paper:

$$\begin{pmatrix} \text{solute accumulation} \\ \text{in } W\Delta x\Delta z \end{pmatrix} = \begin{pmatrix} \text{solute diffusing in at } z \text{ minus} \\ \text{solute diffusing out at } z + \Delta z \end{pmatrix}$$

$$+ \begin{pmatrix} \text{solute flowing in at } x \text{ minus} \\ \text{solute flowing out at } x + \Delta x \end{pmatrix} \tag{2.5-1}$$

or, in mathematical terms,

$$\frac{\partial}{\partial t}(c_1 W\Delta x\Delta z) = [(j_1 W\Delta x)_z - (j_1 W\Delta x)_{z+\Delta z}]$$
$$+ [(c_1 v_x W\Delta z)_x - (c_1 v_x W\Delta z)_{x+\Delta x}] \tag{2.5-2}$$

The term on the left-hand side is zero because of the steady state. The second term in square brackets on the right-hand side is also zero, because neither c_1 nor v_x changes with x. The concentration c_1 does not change with x because the film is long, and there is nothing that will cause the concentration to change in the x direction. The velocity v_x certainly varies with how far we are across the film (i.e., with z), but it does not vary with how far we are along the film (i.e., with x).

After dividing by $W\Delta x\Delta z$ and taking the limit as this volume goes to zero, the mass balance in Eq. 2.5-2 becomes

$$0 = -\frac{dj_1}{dz} \tag{2.5-3}$$

This can be combined with Fick's law to give

$$0 = D\frac{d^2 c_1}{dz^2} \tag{2.5-4}$$

This equation is subject to the boundary conditions

$$z = 0, \quad c_1 = c_{10} \tag{2.5-5}$$

$$z = l, \quad c_1 = c_{1l} \tag{2.5-6}$$

When these results are combined with Fick's law, we have exactly the same problem as that in Section 2.2. The answers are

$$c_1 = c_{10} + (c_{1l} - c_{10})\frac{z}{l} \tag{2.5-7}$$

$$j_1 = \frac{D}{l}(c_{10} - c_{1l}) \tag{2.5-8}$$

The flow has no effect. Indeed, the answer is the same as if the fluid was not flowing.

This answer is typical of many problems involving diffusion and flow. When the solutions are dilute, the diffusion and convection often are perpendicular to each other and the solution is straightforward. You may almost feel gypped; you girded yourself for a difficult problem and found an easy one. Rest assured that more difficult problems follow.

2.5.2 Unsteady Diffusion into a Falling Film

The second problem of interest is illustrated schematically in Fig. 2.5-2. A thin liquid film flows slowly and without ripples down a flat surface. One side of this film wets the surface; the other side is in contact with a gas, which is sparingly soluble in the liquid. We want to find out how much gas dissolves in the liquid.

To solve this problem, we again go through the increasingly familiar litany: we write a mass balance as a differential equation, combine this with Fick's law, and then integrate this to find the desired result. We do this subject to four key assumptions:

   (1)   The solution are always dilute.
   (2)   Mass transport is by z diffusion and x convection.
   (3)   The gas is pure.
   (4)   The contact between gas and liquid is short.

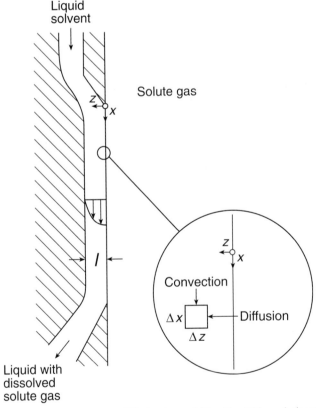

Fig. 2.5-2. Unsteady-state diffusion into a falling film. This analysis turns out to be mathematically equivalent to free diffusion (see Fig. 2.3-2). It is basic to the penetration theory of mass transfer described in Section 11.2.

The first two assumptions are identical with those given in the earlier example. The third means that there is no resistance to diffusion in the gas phase, only in the liquid. The final assumption simplifies the analysis.

We now make a mass balance on the differential volume ($W\Delta x\Delta z$), shown in the inset in Fig. 2.5-2:

$$
\begin{pmatrix} \text{mass accumulation} \\ \text{within } W\Delta x\Delta z \end{pmatrix} = \begin{pmatrix} \text{mass diffusing in at } z \text{ minus} \\ \text{mass diffusing out at } z + \Delta z \end{pmatrix}
$$

$$
+ \begin{pmatrix} \text{mass flowing in at } x \text{ minus} \\ \text{mass flowing out at } x + \Delta x \end{pmatrix} \tag{2.5-9}
$$

where W is the width, taken perpendicular to the paper. This result is parallel to those found in earlier sections:

$$
\left[\frac{\partial}{\partial t}(c_1 W\Delta x\Delta z) \right] = [(W\Delta x j_1)_z - (W\Delta j_1)_{z+\Delta z}]
$$

$$
+ [(W\Delta z c_1 v_x)_x - (W\Delta z c_1 v_x)_{x+\Delta x}] \tag{2.5-10}
$$

When the system is at steady state, the accumulation is zero. Therefore, the left-hand side of the equation is zero. No other terms are zero, because j_1 and c_1 vary with both z and x. If we divide by the volume $W\Delta x\Delta z$ and take the limit as this volume goes to zero, we find

$$
0 = -\frac{\partial j_1}{\partial z} - \frac{\partial}{\partial x} c_1 v_x \tag{2.5-11}
$$

We now make two further manipulations. First, we combine this with Fick's law. Second, we set v_x equal to its maximum value, a constant. This second change reflects the assumption of short contact times. At such times, the solute barely has a chance to cross the interface, and it diffuses only slightly into the fluid. In this interfacial region, the fluid velocity reaches the maximum suggested in Fig. 2.5-2, so the use of a constant value is probably not a serious assumption. Thus the mass balance is

$$
\frac{\partial c_1}{\partial (x/v_{max})} = D\frac{\partial^2 c_1}{\partial z^2} \tag{2.5-12}
$$

The left-hand side of this equation represents the solute flow out minus that in; the right-hand side is the diffusion in minus that out.

This mass balance is subject to the following conditions:

$$
x = 0, \quad \text{all } z, \quad c_1 = 0 \tag{2.5-13}
$$

$$
x > 0, \quad z = 0, \quad c_1 = c_1(\text{sat}) \tag{2.5-14}
$$

$$
z = l, \quad c_1 = 0 \tag{2.5-15}
$$

where $c_1(\text{sat})$ is the concentration of dissolved gas in equilibrium with the gas itself, and l is the thickness of the falling film in Fig. 2.5-2. The last of these three boundary conditions is replaced with

$$
x > 0, \quad z = \infty, \quad c_1 = 0 \tag{2.5-16}
$$

This again reflects the assumption that the film is exposed only a very short time. As a result, the solute can diffuse only a short way into the film. Its diffusion is then unaffected by the exact location of the other wall, which, from the standpoint of diffusion, might as well be infinitely far away.

This problem is described by the same differential equation and boundary conditions as diffusion in a semi-infinite slab. The sole difference is that the quantity x/v_{max} replaces the time t. Because the mathematics is the same, the solution is the same. The concentration profile is

$$\frac{c_1}{c_1(\text{sat})} = 1 - \text{erf}\frac{z}{\sqrt{4Dx/v_{max}}} \tag{2.5-17}$$

and the flux at the interface is

$$j_1|_{z=0} = \sqrt{Dv_{max}/\pi x}\, c_1(\text{sat}) \tag{2.5-18}$$

These are the answers to this problem.

These answers appear abruptly because we can adopt the mathematical results of Section 2.3. Those studying this material for the first time often find this abruptness jarring. Stop and think about this problem. It is an important problem, basic to the penetration theory of mass transfer discussed in Section 9.2. To supply a forum for further discussion, we shall now consider this problem from another viewpoint.

The alternative viewpoint involves changing the differential volume on which we make the mass balance. In the foregoing problem, we chose a volume fixed in space, a volume through which liquid was flowing. This volume accumulated no solute, so its use led to a steady-state differential equation. Alternatively, we can choose a differential volume floating along with the fluid at a speed v_{max}. The use of this volume leads to an unsteady-state differential equation like Eq. 2.3-5. Which viewpoint is correct?

The answer is that both are correct; both eventually lead to the same answer. The fixed-coordinate method used earlier is often dignified as "Eulerian," and the moving-coordinate picture is described as "Lagrangian." The difference between them can be illustrated by the situation of watching fish swimming upstream in a fast-flowing river. If we watch the fish from a bridge, we may see only slow movement, but if we watch the fish from a freely floating canoe, we realize that the fish are moving rapidly.

2.5.3 Free Convection Caused by Diffusion

A third, much more difficult, example of convection and diffusion occurs in the apparatus shown schematically in Fig. 2.5-3. The apparatus consists of two well-stirred reservoirs. The upper reservoir contains a dense solution, but the lower one is filled with less dense solvent. Because solution and solvent are miscible, solute diffuses from the upper reservoir into the lower one.

We want to know if the difference in densities between solution and solvent will cause flow. From our experience, we expect that flow will occur if the tube diameter is large. After all, gin tends to rise to the surface of a summer's gin-and-tonic without completely mixing, and vinegar falls below oil in salad dressing. Intuitively, we expect that such flows will cease if the tube diameter becomes small. More speculatively, we might guess

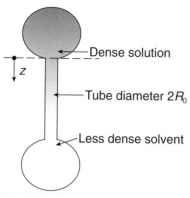

Fig. 2.5-3. Free convection in a vertical tube. A dense solution will not flow when the tube diameter is small. Diffusion damps the tendency to flow.

that whether or not flow occurs depends inversely on viscosity; high viscosity means less chance to flow.

To analyze this problem more completely, we write a mass balance on the solute, an overall mass balance on all species present, and a momentum balance to describe the flow. We then imagine small perturbations in the concentration or in the flow. If our balances indicate that these small perturbations get smaller with time, then the system is stable. If these perturbations grow with time, then the system is unstable, and free convection will occur.

We first write these balances for the unperturbed system in which no free convection exists. These are

$$0 = D\nabla^2 \bar{c}_1 - \bar{v} \cdot \nabla \bar{c}_1 \tag{2.5-19}$$

$$0 = -\nabla \cdot \bar{v} \tag{2.5-20}$$

$$0 = -\nabla p + \bar{\rho} \mathbf{g} \tag{2.5-21}$$

where p is the pressure, ρ is the density, and $\mathbf{g}$ is the acceleration due to gravity; the overbars refer to the unperturbed system. The solution of these equations for the situation shown in Fig. 2.5-3 is that expected:

$$\frac{\bar{c}_1 - c_{10}}{c_{1l} - c_{10}} = \frac{\bar{\rho} - \rho_0}{\rho_l - \rho_0} = \frac{z}{l} \tag{2.5-22}$$

$$\bar{v} = 0 \tag{2.5-23}$$

$$\bar{p} = p_0 + \int_0^z \bar{\rho} g \, dz \tag{2.5-24}$$

Although we do not need the details of these solutions in the following, I find them reassuring.

The corresponding equations for an incompressible but perturbed system are

$$\frac{\partial c_1}{\partial t} = D\nabla^2 c_1 - v \cdot \nabla c_1 \tag{2.5-25}$$

$$0 = -\nabla \cdot v \tag{2.5-26}$$

$$\rho\frac{\partial v}{\partial t} = \mu\nabla^2 v - \nabla p + \rho g \tag{2.5-27}$$

where μ is the viscosity. We now rewrite these relations in terms of the perturbations themselves. For example, for the mass balance, we define

$$c_1 = c_1 + c_1' \tag{2.5-28}$$

$$v = v' \tag{2.5-29}$$

where the primes signify perturbations from stable values of Eqs. 2.5-22 through 2.5-24. Remember that the stable value of the velocity is zero. We then insert these definitions in Eq. 2.5-25, subtract Eq. 2.5-19, and neglect terms involving the squares of perturbations:

$$\frac{\partial c_1'}{\partial t} = D\nabla^2 c_1' - v_z'\frac{d\bar{c}_1}{dz} \tag{2.5-30}$$

Equation 2.5-30 is subject to the boundary condition that the tube walls are solid:

$$r = R_0, \quad \frac{\partial c_1'}{\partial r} = 0 \tag{2.5-31}$$

Similar arguments lead to a modified momentum balance:

$$\left(\mu\nabla^2 - \rho\frac{\partial}{\partial t}\right)v' = -\nabla p' + \rho' g = -\nabla p' + g\beta c_1' \tag{2.5-32}$$

in which the primed quantities are again perturbations and $\beta(= \partial\rho/\partial c_1)$ describes the density increase caused by the solute. Equation 2.5-32 is subject to the condition

$$r = R_0, \quad v_z' = 0 \tag{2.5-33}$$

This says that there is no vertical flow at the wall.

Equations 2.5-30 and 2.5-32 must now be solved simultaneously. A simple solution requires two chief assumptions. The first is that the time derivatives in these equations can be neglected; this is equivalent to the assertion that marginal stability can exist. The second assumption is that the perturbations have their largest effects normal to the z direction; this implies that any convection cells that occur will be long. I find these assumptions reasonable, but hardly obvious. Because they are justified by experiment, they are tributes to the genius of G. I. Taylor (1954), who had the gall to present the answers to this problem without derivation.

Taylor found that for the perturbations to grow, the Rayleigh number Ra must be

$$Ra = \left(\frac{gR_0^4}{\mu D}\right)\frac{\partial\rho}{\partial z} > 67.94 \tag{2.5-34}$$

This value corresponds to the case in which solution is falling down one side of the tube and solvent is rising through the other side.

This critical value of the Rayleigh number provides the limit for the stability in the vertical tube. When the density difference is small enough so that the Rayleigh number is less than 67.94, free convection will not occur. When the density difference is so large that the Rayleigh number exceeds 67.94, then free convection does occur. This result also supports our intuitive speculations at the beginning of this section. The chances for free convection decrease sharply as the tube diameter decreases. They also decrease as the viscosity or the diffusion coefficient increases. In every case, the change of a given variable required to spark free convection can be predicted from this critical Rayleigh number.

2.6 A Final Perspective

This chapter is very important, a keystone of this book. It introduces Fick's law for dilute solutions and shows how this law can be combined with mass balances to calculate concentrations and fluxes. The mass balances are made on thin shells. When these shells are very thin, the mass balances become the differential equations necessary to solve the various problems. Thus the bricks from which this chapter is built are largely mathematical: shell balances, differential equations, and integrations in different coordinate systems.

However, we must also see a different and broader blueprint based on physics, not mathematics. This blueprint includes the two limiting cases of diffusion across a thin film and diffusion in a semi-infinite slab. Most diffusion problems fall between these two limits. The first, the thin film, is a steady-state problem, mathematically easy and some-times physically subtle. The second, the unsteady-state problem of the thick slab, is harder to calculate mathematically and is the limit at short times.

In many cases, we can use a simple criterion to decide which of the two central limits is more closely approached. This criterion hinges on the magnitude of the Fourier number

$$\frac{(\text{length})^2}{\left(\dfrac{\text{diffusion}}{\text{coefficient}}\right)(\text{time})}$$

This variable is the argument of the error function of the semi-infinite slab, it determines the standard deviation of the decaying pulse, and it is central to the time dependence of diffusion into the cylinder. In other words, it is a key to all the foregoing unsteady-state problems. Indeed, it can be easily isolated by dimensional analysis.

This variable can be used to estimate which limiting case is more relevant. If it is much larger than unity, we can assume a semi-infinite slab. If it is much less than unity, we should expect a steady state or an equilibrium. If it is approximately unity, we may be forced to make a fancier analysis. For example, imagine that we are testing a membrane for an industrial separation. The membrane is 0.01 centimeters thick, and the diffusion coefficient in it is 10^{-7} cm^2/sec. If our experiments take only 10 seconds, we have an unsteady-state problem like the semi-infinite slab; if they take three hours, we approach a steady-state situation.

In unsteady-state problems, this same variable may also be used to estimate how far or how long mass transfer has occurred. Basically, the process is significantly advanced when this variable equals unity. For example, imagine that we want to guess how far gasoline has evaporated into the stagnant air in a glass-fiber filter. The evaporation has been going on about 10 minutes, and the diffusion coefficient is about 0.1 cm²/sec. Thus

$$\frac{(\text{length})^2}{(0.1\,\text{cm}^2/\text{sec})(600\,\text{sec})} = 1; \text{ length} = 8\,\text{cm}$$

Alternatively, suppose we find that hydrogen has penetrated about 0.1 centimeter into nickel. Because the diffusion coefficient in this case is about $10^{-8}\,\text{cm}^2/\text{sec}$, we can estimate how long this process has been going on:

$$\frac{(10^{-1}\,\text{cm}^2)}{(10^{-8}\,\text{cm}^2/\text{sec})(\text{time})} = 1; \text{ time} = 10\,\text{d}$$

This sort of heuristic argument is often successful.

A second important perspective between these two limiting cases results from comparing their interfacial fluxes given in Eqs. 2.2-10 and 2.3-18:

$$j_1 = \frac{D}{l}\Delta c_1 \quad \text{(thin film)}$$

$$j_1 = \sqrt{D/\pi t}\Delta c_1 \quad \text{(thick slab)}$$

Although the quantities D/l and $(D/\pi t)^{1/2}$ vary differently with diffusion coefficients, they both have dimensions of velocity; in fact, in the life sciences, they sometimes are called "the velocity of diffusion." In later chapters, we shall discover that these quantities are equivalent to the mass transfer coefficients used at the beginning of this book.

Questions for Discussion

1. If the concentration difference for diffusion across a thin film is doubled, what happens to the flux?
2. If it is doubled for diffusion into a semi-infinite slab, what happens to the flux?
3. If the diffusion coefficient across a thin film is doubled, what happens to the flux?
4. If it is doubled for diffusion into a semi-infinite slab, what happens to the flux?
5. What is the average flux into a semi-infinite slab over a time t_1?
6. What are some different ways in which an effective diffusion coefficient in a porous medium could be defined?
7. Explain Fig. 2.2-2 to someone without scientific training.
8. Explain why the funnel data in Fig. 2.1-3 curve downwards.
9. Imagine that you have a thin film separating two identical well-stirred solutions. At time zero, the solute concentration in one solution is doubled. Sketch the concentration profiles in the film vs. position and time.
10. Estimate the flux in the previous question at a very short time.
11. Estimate it at a very long time.
12. How would the width of a spreading pulse change if the diffusion coefficient doubled?

13. Would the steady flux across a thin film increase if there was fast reaction producing a mobile product?

14. Imagine you have two films clamped together. The diffusion coefficient in one film is constant, but that in the other depends on concentration. If you reverse the concentration difference, the flux will reverse. Will its magnitude change?

Problems

1. Water evaporating from a pond does so as if it were diffusing across an air film 0.15 cm thick. The diffusion coefficient of water in 20 °C air is about $0.25 \text{ cm}^2/\text{sec}$. If the air out of the film is fifty percent saturated, how fast will the water level drop in a day? *Answer:* 1.24 cm/d.

2. In 1765, Benjamin Franklin made a variety of experiments on the spreading of oils on the pond in Clapham Common, London. Franklin estimated the thickness of the oil layers to be about 25 Å. Many more recent scientists have tried to use similar layers of fatty acids and alcohols to retard evaporation from ponds and reservoirs in arid regions. The monolayers used today usually are characterized by a resistance around 2 sec/cm. Assuming that they are the thickness of Franklin's layer and that they can dissolve up to 1.8% water, estimate the diffusion coefficient across the monolayers. *Answer:* $7 \cdot 10^{-6} \text{ cm}^2/\text{sec}$.

3. The diffusion coefficient of NO_2 into stagnant water can be measured with the apparatus shown below. Although the water is initially pure, the mercury drop moves to show that 0.82 cm^3 of NO_2 is absorbed in 3 minutes. The gas–liquid interface has an area of 36.3 cm^2, the pressure is 0.93 atm, the temperature is 16 °C, and the Henry's law constant is $37{,}000 \text{ cm}^3 \text{ atm/mol}$. What is the desired diffusion coefficient? (J. Kopinsky) *Answer:* $5 \cdot 10^{-6} \text{ cm}^2/\text{sec}$.

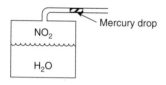

4. About 85.6 cm^2 of a flexible polymer film 0.051 cm thick is made into a bag, filled with distilled water, and hung in an oven at 35 °C and 75% relative humidity. The bag is weighed, giving the following data:

Time (d)	Bag weight (g)
0	14.0153
1	13.9855
4	13.9104
7	13.8156
8	13.7710
12	13.6492
14	13.5830
16	13.5256

What is the permeability (DH) of the polymer film? (R. Contravas) *Answer:* 2.2 · 10⁵ cm²/sec.

5. Diaphragm cells are frequently calibrated by allowing 1-M potassium chloride to diffuse into pure water. The average diffusion coefficient in this case is 1.859 · 10⁻⁵ cm²/sec. Your cell has compartment volumes of 42.3 cm³ and 40.8 cm³; the diaphragm is a glass frit 2.51 cm in diameter, 0.16 cm thick, and of porosity 0.34. In one calibration experiment, the concentration difference at 36 hr 6 min is 49.2% of that originally present. (a) What is the cell's calibration constant? *Answer:* 0.294 cm⁻². (b) What is the effective length of the diaphragm's pores? *Answer:* 0.28 cm. (c) The current pores are about 2 · 10⁻⁴ cm in diameter. What is the effect of increasing the pore diameter ten times at constant porosity?

6. Diffusion coefficients in gases can be measured by injecting a solute gas into a solvent gas in laminar plug flow and measuring the concentration with a thermistor placed downstream. The concentration downstream is given by

$$c_1 = \frac{Q}{4\pi Dz} e^{-r^2 v / 2 Dz}$$

where Q is the solute injection rate, z is the distance downstream, r is the distance away from the z axis, and v is the gas flow. One series of measurements involves the diffusion of helium in nitrogen at 25 °C and 1 atm. In one particular measurement, the maximum concentration of helium is 0.48 wt% when z is 1.031 cm and Q is 0.045 cm³/sec. What is the diffusion coefficient? (H. Beesley) *Answer:* 0.11 cm²/sec.

7. Low-carbon steel can be hardened for improved wear resistance by carburizing. Steel is carburized by exposing it to a gas, liquid, or solid that provides a high carbon concentration at the surface. The figure below [D. S. Clark and W. R. Varney, *Physical Metallurgy for Engineers*. Princeton, N.J.: Van Nostrand (1962)] shows carbon content versus depth in steel carburized at 930 °C. Estimate D from this graph, assuming diffusion without reaction between carbon and iron. (H. Beesley) *Answer:* 5.3 · 10⁻⁷ cm²/sec.

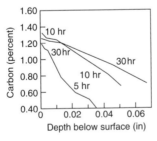

8. The twin-bulb method of measuring diffusion is shown below. The bulbs, which are stirred and of equal volume, initially contain binary gas mixtures of different compositions. At time zero, the valve is opened; at time t, the valve is closed, and the bulk contents are analyzed. Explain how this information can be used to calculate the diffusion coefficient in this binary gas mixture.

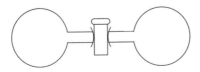

9. Find the steady-state flux out of a pipe with a porous wall. The pipe has an inner radius R_i and an outer radius R_o. The solute has a fixed, finite concentration c_{1i} inside of the pipe, but is essentially at zero concentration outside. As a result, solute diffuses through the wall with a diffusion coefficient D. When you have found the result, compare it with the results for steady-state diffusion across a thin slab and away from a dissolving sphere. *Answer:* $Dc_{1i}[R_o \ln(R_o/R_i)]^{-1}$.

10. Controlled release is important in agriculture, especially for insect control. One common example involves the pheromones, sex attractants released by insects. If you mix this attractant with an insecticide, you can wipe out all of one sex of a particular insect pest. A device for releasing one pheromone is shown schematically below. This pheromone does not subline instantaneously, but at a rate of

$$ r_0 = 6 \cdot 10^{-17}[1 - (1.10 \cdot 10^7 \text{cm}^3/\text{mol})c_1] \text{ mol/sec} $$

where c_1 is the concentration in the vapor. The permeability of this material through the polymer (DH) is $1.92 \cdot 10^{-12}$ cm^2/sec. The concentration of pheromone outside of the device is essentially zero. (a) What is the concentration (moles per cubic centimeter) of pheromone in the vapor? (b) How fast is the pheromone released by this device?

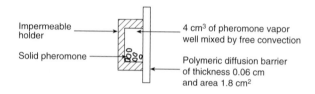

Impermeable holder

Solid pheromone

4 cm^3 of pheromone vapor well mixed by free convection

Polymeric diffusion barrier of thickness 0.06 cm and area 1.8 cm^2

11. Antique glass objects can be dated by measuring the amount of hydration near the object's surface. This amount can be measured using ^{15}N nuclear magnetic resonance [W. A. Lanford, *Science*, **196**, 975 (1977)]. Derive equations for the total amount of hydration, assuming that water reacts rapidly and reversibly with the glass to produce an immobile hydrate. Discuss how this amount can provide a measure of the age of the object.

12. One type of packaging film with thickness "l" has an immobile sacrificial reagent at initial concentration c_{20} within the wall of the package. A solute at concentration c_{10} outside of the film, like water, oxygen, or radioactive cesium, diffuses into the film, reacting irreversibly with the sacrificial reagent as it goes. The product may be mobile or immobile; since the reaction is irreversible, it does not matter. This reaction shows the solute's penetration across the film. (a) Write mass balances for the solute and the immobile reagent. (b) Write possible initial and boundary conditions for these equations. (c) If the reaction were infinitely fast, how would your equations change?

13. Researchers in microelectronics have found that a slight scratch on the surface of gallium arsenide causes a zinc dopant to diffuse into the arsenide. Apparently, this occurs because the scratch increases crystal defects and hence the local diffusion coefficient. When these devices are later baked at 850 °C, the small pulse may spread, for its diffusion coefficient at this high temperature is about 10^{-11} cm^2/sec. If it spreads enough to increase the zinc concentration to 10 percent of the maximum at

$4 \cdot 10^{-4}$ cm away from the scratch, the device is ruined. How long can we bake the device? (S. Balloge) *Answer:* 30 min.

14. Adolf Fick made the experiments required to determine the diffusion coefficient using the equipment shown in Fig. 2.1-3. In these devices, he assumed that the salt concentration reached saturation in the bottom and that it was always essentially zero in the large solvent bath. As a result, the concentration profiles eventually reached steady state. Calculate these profiles.

15. Consider a layer of bacteria contained between two semipermeable membranes that allow the passage of a chemical solute S, but do not allow the passage of bacteria. The movement of the bacteria B is described with a flux equation roughly parallel to a diffusion equation:

$$j_B = -D_0 \frac{d}{dz}[B] + \chi[B]\frac{d}{dz}[S]$$

where D_0 and χ are constant transport coefficients. In other words, the bacterial flux is affected by [S], although the bacteria neither produce or consume S. If the concentrations of S are maintained at $[S]_0$ and 0 at the upper and lower surfaces of the bacterial suspension, (a) determine $[S](z)$, and (b) determine $[B](z)$.

16. Extraction of sucrose from food materials is often correlated in terms of diffusion coefficients. The diffusion coefficients can be calculated assuming short times and an infinite slab:

$$D = \left(\frac{\pi}{4t}\right)\left(\frac{M}{c_{10}}\right)^2$$

where M is the total extracted per area and c_{10} is the sucrose concentration at saturation. However, the diffusion coefficients found are not constant, as shown below (H. G. Schwartzberg and R. Y. Chao, *Food Technology,* Feb. 1982, p. 73). The reason the diffusion coefficient is not constant is not because of the failure of the approximation of an infinite slab; it reflects the fact that beets and cane are not homogeneous. Instead, they have a network of cells connected by vascular channels. Diffusion across the cell wall is slow, and it dominates behavior in thin slices; diffusion through vascular channels is much faster and supplements the flux for thick slices. Develop equations

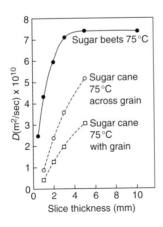

that justify these qualitative arguments. These equations will contain the diffusion coefficient across cell walls D_w, the diffusion coefficient in channels D_c, and the fraction of channels ε.

Further Reading

Bird, R. B., Stewart, W. E., and Lightfoot, E. N. (2002). *Transport Phenomena*. New York: Wiley.

Boltmann, L. (1894). *Annalen der Physik und Chemie*, **53**, 959.

Crank, J. (1975). *The Mathematics of Diffusion*, 2nd ed. Oxford: Clarendon Press.

Fick, A. E. (1852). *Zeitschrift für Rationelle Medicin*, **2**, 83.

Fick, A. E. (1855a). *Poggendorff's Annalen der Physik*, **94**, 59.

Fick, A. E. (1855b). *Philisophical Magazine*, **10**, 30.

Fick, A. E. (1856). *Medizinische Physik*. Brunswick.

Fick, A. E. (1903). *Gesammelte Abhandlungen*. Würzburg.

Fourier, J. B. (1822). *Théorie analytique de la chaleur*. Paris.

Graham, T. (1829). *Quarterly Journal of Science, Literature and Art*, **27**, 74.

Graham, T. (1833). *Philosophical Magazine*, 2, **175**, 222, 351.

Graham, T. (1850). *Philosophical Transactions of the Royal Society of London*, **140**, 1.

Mason, E. A. (1970). *Philosophical Journal*, **7**, 99.

CHAPTER 3

Diffusion in Concentrated Solutions

Diffusion causes convection. To be sure, convective flow can have many causes. For example, it can occur because of pressure gradients or temperature differences. However, even in isothermal and isobaric systems, diffusion will always produce convection. This was clearly stated by Maxwell in 1860: "Mass transfer is due partly to the motion of translation and partly to that of agitation." In more modern terms, we would say that any mass flux may include both convection and diffusion.

This combination of convection and diffusion can complicate our analysis. The easier analyses occur in dilute solutions, in which the convection caused by diffusion is vanishingly small. The dilute limit provides the framework within which most people analyze diffusion. This is the framework presented in Chapter 2.

In some cases, however, our dilute-solution analyses do not successfully correlate our experimental observations. Consequently, we must use more elaborate equations. This elaboration is best initiated with the physically based examples given in Section 3.1. This is followed by a catalogue of flux equations in Section 3.2. These flux equations form the basis for the simple analyses of diffusion and convection in Section 3.3 that parallel those in the previous chapter.

After simple analyses, we move in Section 3.4 to general mass balances, sometimes called the general continuity equations. These equations involve the various coordinate systems introduced in Chapter 2. They allow solutions for the more difficult problems that arise from the more complicated physical situation. Fortunately, the complexities inherent in these examples can often be dodged by effectively exploiting selected readings. A guide to these readings is given in Section 3.5.

The material in this chapter is more complicated than that in Chapter 2 and is unnecessary for many who are not trying to pass exams in advanced courses. Nonetheless, this material has fascinating aspects, as well as some tedious ones. Those studying these aspects often tend to substitute mathematical manipulation for thought. Make sure that the intellectual framework in Chapter 2 is secure before starting this more advanced material.

3.1 Diffusion With Convection

The statement by Maxwell quoted earlier suggests that diffusion and convection always occur together, that one cannot occur without the other. This fact sets diffusion apart from many other phenomena. For example, thermal conduction can certainly occur without convection. In contrast, diffusion generates its own convection, so that understanding the process can be much more complicated, especially in concentrated solutions.

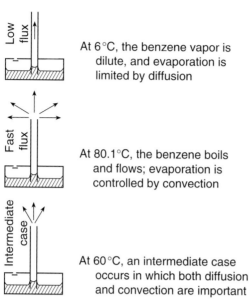

At 6°C, the benzene vapor is dilute, and evaporation is limited by diffusion

At 80.1°C, the benzene boils and flows; evaporation is controlled by convection

At 60°C, an intermediate case occurs in which both diffusion and convection are important

Fig. 3.1-1. Evaporation of benzene. This process is dominated by diffusion in dilute solutions, but it includes both diffusion and convection in concentrated solutions.

3.1.1 A Qualitative Example

To illustrate how diffusion and convection are interrelated, we consider the example shown in Fig. 3.1-1. The physical system consists of a large reservoir of benzene connected to a large volume of air by means of a capillary tube. Benzene evaporates and moves through the capillary into the surrounding air.

At room temperature, not much benzene evaporates because its vapor pressure is low. Benzene vapor moves slowly up the tube because of Brownian motion, that is, because of thermally induced agitation of the molecules. This is the process basic to diffusion studied in the previous chapter.

At the boiling point, the situation is completely different. The liquid benzene boils into vapor, and the vapor rushes up the capillary. This rush is clearly a pressure-driven flow, a convection caused by the sharply increased volume of the vapor as compared with the liquid. It has little to do with diffusion.

At intermediate temperatures, both diffusion and convection will be important, because the processes take place simultaneously. To understand such intermediate cases, we must look at how mass transport works.

3.1.2 Separating Convection From Diffusion

The complete description of mass transfer requires separating the contributions of diffusion and convection. The usual way of effecting this separation is to assume that these two effects are additive:

$$\begin{pmatrix} \text{total mass} \\ \text{transported} \end{pmatrix} = \begin{pmatrix} \text{mass transported} \\ \text{by diffusion} \end{pmatrix} + \begin{pmatrix} \text{mass transported} \\ \text{by convection} \end{pmatrix} \qquad (3.1\text{-}1)$$

In more exact terms, we define the total mass flux n_1 as the mass transported per area per time relative to fixed coordinates. This flux, in turn, is used to define an average solute velocity v_1:

$$n_1 = c_1 v_1 \tag{3.1-2}$$

where c_1 is the local concentration. We then divide v_1 into two parts:

$$n_1 = c_1(v_1 - v^a) + c_1 v^a = j_1^a + c_1 v^a \tag{3.1-3}$$

where v^a is some convective "reference" velocity. The first term j_1^a on the right-hand side of this equation represents the diffusion flux, and the second term $c_1 v^a$ describes the convection.

Interestingly, there is no clear choice for what this convective reference velocity should be. It might be the mass average velocity that is basic to the equations of motion, which in turn are a generalization of Newton's second law. It might be the velocity of the solvent, because that species is usually present in excess. We cannot automatically tell. We only know that we should choose v^a so that v^a is zero as frequently as possible. By doing so, we eliminate convection essentially by definition, and we are left with a substantially easier problem.

To see which reference velocity is easiest to use, we consider the diffusion apparatus shown in Fig. 3.1-2. This apparatus consists of two bulbs, each of which contains a gas or liquid solution of different composition. The two bulbs are connected by a long, thin capillary containing a stopcock. At time zero, the stopcock is opened; after an experimentally desired time, the stopcock is closed. The solutions in the two bulbs are then analyzed, and the concentrations are used to calculate the diffusion coefficient. The equations used in these calculations are identical with those used for the diaphragm cell.

Here, we examine this apparatus to elucidate the interaction of diffusion and convection, not to measure the diffusion coefficient. The examination is easiest for the special cases of gases and liquids. For gases, we imagine that one bulb is filled with nitrogen and the other with hydrogen. During the experiment, the number of moles in the left bulb always equals the number of moles in the identical right bulb because isothermal and isobaric ideal gases have a constant number of moles per volume. The volume of the left bulb equals the volume of the right bulb because the bulbs are rigid. Thus the average velocity of the moles v^* and the average velocity of the volume v^0 are both zero.

In contrast, the average velocity of the mass v in this system is not zero. To see why this is so, imagine balancing the apparatus on a knife edge. This edge will initially be located left of center, as in Fig. 3.1-2(b), because the nitrogen on the left is heavier than the hydrogen on the right. As the experiment proceeds, the knife edge must be shifted toward the center because the densities in the two bulbs will become more nearly equal.

Thus, in gases, the molar and volume average velocities are zero but the mass average velocity is not. Therefore, the molar and volume average velocities allow a simpler description in gases than the mass average velocity.

We now turn to the special case of liquids, shown in Fig. 3.1-2(c). The volume of the solution is very nearly constant during diffusion, so that the volume average velocity is very nearly zero. This approximation holds whenever there is no significant volume change after mixing. In my experience, this is true except for some alcohol–water systems, and even in those systems it is not a bad approximation.

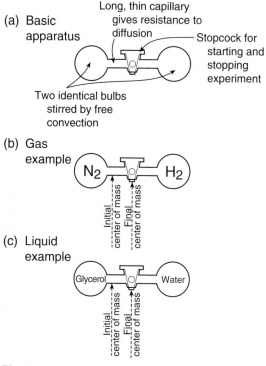

(a) Basic apparatus

Long, thin capillary gives resistance to diffusion

Stopcock for starting and stopping experiment

Two identical bulbs stirred by free convection

(b) Gas example

N₂   H₂

Initial center of mass — Final center of mass

(c) Liquid example

Glycerol   Water

Initial center of mass — Final center of mass

Fig. 3.1-2. An example of reference velocities. Descriptions of diffusion imply reference to a velocity relative to the system's mass or volume. Whereas the mass usually has a nonzero velocity, the volume often shows no velocity. Hence diffusion is best referred to the volume's average velocity.

The other two velocities are more difficult to estimate. To estimate these velocities for one case, imagine allowing 50-weight percent glycerol to diffuse into water. The volume changes less than 0.1 percent during this mixing, so that the volume average velocity is very nearly zero. The glycerol solution has a density of about 1.1 g/cm³, as compared with water at 1 g/cm³, so that the mass density changes about 10 percent. In contrast, the glycerol solution has a molar density of about 33 mol/l, as compared with water at 55 mol/l; so the molar concentration changes about fifty percent. Thus the mass average velocity will be nearer to zero than the molar average velocity.

Thus in this set of experiments, the molar and volume average velocities are zero for ideal gases and the volume and mass average velocities are close to zero for liquids. The mass average velocity is often inappropriate for gases, and the molar average velocity is rarely used for liquids. The volume average velocity is appropriate most frequently, and so it will be emphasized in this book.

3.2 Different Forms of the Diffusion Equation

The five most common forms of diffusion equations are given in Table 3.2-1. Each of these forms uses a different way to separate diffusion and convection. Of course,

Table 3.2-1 *Different forms of the diffusion equation*

Choice	Total flux (diffusion + convection)	Diffusion equation	Reference velocity	Where best used
Mass	$n_1 = j_1^m + \rho_1 v$	$j_1^m = \rho_1(v_1 - v)$ $= -D\rho\nabla\omega_1$	$v = \omega_1 v_1 + \omega_2 v_2$ $\rho v = n_1 + n_2$	Constant-density liquids; coupled mass and momentum transport
Molar	$n_1 = j_1^* + c_1 v^*$	$j_1^* = c_1(v_1 - v^*)$ $= -Dc\nabla y_1$	$v^* = y_1 v_1 + y_2 v_2$ $cv^* = n_1 + n_2$	Ideal gases where the total molar concentration c is constant
Volume	$n_1 = j_1 + c_1 v^0$	$j_1 = c_1(v_1 - v^0)$ $= -D\nabla c_1$	$v^0 = c_1 \bar{V}_1 v_1 + c_2 \bar{V}_2 v_2$ $= \bar{V}_1 n_1 + \bar{V}_2 n_2$	Best overall; good for constant-density liquids and for ideal gases; may use either mass or mole concentration
Solvent	$n_1 = j_1^{(2)} + c_1 v_2$	$j_1^{(2)} = c_1(v_1 - v_2)$ $= -D_1\nabla c_1$	v_2	Rare except for membranes; note that $D_1 \neq D_2 \neq D$
Maxwell–Stefan		$\nabla y_1 = \dfrac{y_1 y_2}{D'}(v_2 - v_1)$	None	Written for ideal gases; difficult to use in practice

in many cases, such a separation is obvious. If we have some salt in the bottom of a jar covered with stagnant water, then the movement of salt upwards is due to diffusion. If we pump salt water through a pipe, the dissolved salt moves by convection. These cases are straightforward.

However, in a few cases, the separation of diffusion and convection is more subtle. One of these cases, the evaporation of benzene vapor, was detailed in the previous section. To deal with these cases, we can use one of two strategies:

1. *We can describe diffusion in ways which parallel Fick's law.* This strategy retains the split between diffusion and convection and benefits from the physical insight which results. However, it requires defining convection carefully.
2. *We can describe diffusion in general ways which avoid reference to convection.* This strategy postpones the need for careful definition of convection, but destroys some of the physical insight possible. It is often preferred by those who seek a mathematically elegant description.

We give details of these strategies in the following paragraphs.

3.2.1 Fick's Law Parallels

The common forms of diffusion equations given in Table 3.2-1 are those listed in most books. The first, which gives diffusion relative to the mass average velocity, is preferred in texts on fluid flow, where the mechanics of the situation is key. The second, which gives diffusion relative to the molar average velocity, appears in descriptions for the kinetic theory of gases, where the fact that ideal gases all have the same molar concentration at the same pressure and temperature is the basic precept.

The most valuable of these forms is that defining diffusion relative to the volume average velocity. This is because for systems of constant density, the volume average velocity equals the mass average velocity. For systems of constant molar concentration, the volume average velocity equals the molar average velocity. Thus the volume average velocity includes the two commonly given analyses as special cases in a more general form.

To prove these assertions, we begin with the volume and mass average velocities. We find it convenient to describe the concentration in these systems as ρ_i, the mass of species "i" per volume; and as $\bar{V}_i$, the partial specific volume. Then

$$\rho_i \bar{V}_i = \rho_i \left(\frac{\partial V}{\partial m_i} \right)_{m_{j \neq i}} \tag{3.2-1}$$

where m_i is the mass of species "i." This derivative is the change in volume with a change in mass of species "i". If the system has constant density, this change is merely the reciprocal of the density ρ, so

$$\rho_i \bar{V}_i = \rho_i / \rho = \omega_i \tag{3.2-2}$$

Thus, for constant ρ,

$$v^0 = \sum_{i=1}^{2} \rho_i \bar{V}_i v_i = \sum_{i=1}^{2} \omega_i v_i = v \tag{3.2-3}$$

The volume and mass average velocities are the same for a system of constant density.

The volume average velocity is equal to the molar average velocity for ideal gases. Here, we find it convenient to describe concentration c_i as the moles of species "i" per volume; and as $\bar{V}_i$, the partial molar volume. Then

$$c_1 \bar{V}_1 = c_1 \left(\frac{\partial V}{\partial N_1} \right)_{N_2} = c_1 \left[\frac{\partial}{\partial N_1} \left(\frac{RT(N_1 + N_2)}{p} \right) \right]_{N_2}$$

$$= c_1 \left(\frac{RT}{p} \right) = \frac{c_1}{c} = y_1 \tag{3.2-4}$$

For constant total molar concentration c,

$$v^0 = \sum_{i=1}^{2} c_i \bar{V}_i v_i = \sum_{i=1}^{2} y_i v_i = v^* \tag{3.2-5}$$

The volume and molar average velocities are the same for systems that, like the ideal gas, have constant molar concentration.

Finally, we consider the diffusion relative to the solvent average velocity, for which from Table 3.2-1

$$j_1^{(2)} = -D_1 \nabla c_1 + c_1 v_2$$

This expression is used almost exclusively for transport across membranes. To see why, imagine that we are interested in transport of oxygen across a polyamide film. We choose oxygen as species "1"; we choose the polymer as the solvent species "2." Because the membrane is normally stationary, its velocity v_2 is zero, simplifying our analysis. If the membrane is nonporous, D will truly represent diffusion. If it has large pores, then D will really be a measure of flow in the porous membrane. If the membrane has small pores, smaller than the mean free path in the gaseous oxygen, then D will represent Knudsen diffusion. Some details for membranes are given in Chapter 18. We now turn to examples.

Example 3.2-1: One binary diffusion coefficient   Prove that if the partial molar volumes are constant, there is only one binary diffusion coefficient defined relative to the volume average velocity. In other words, because we define

$$n_1 = -D_1 \nabla c_1 + c_1 v^0$$
$$n_2 = -D_2 \nabla c_2 + c_2 v^0$$

prove D_1 equals D_2.

   Solution   We begin by multiplying the first equation by $\bar{V}_1$ and the second by $\bar{V}_2$. We then add these equations to find

$$[\bar{V}_1 n_1 + \bar{V}_2 n_2] = -D_1 \nabla c_1 \bar{V}_1 - D_2 \nabla c_2 \bar{V}_2 + [(c_1 \bar{V}_1 + c_2 \bar{V}_2)v^0]$$

The quantity in square brackets on the left equals that in square brackets on the right. Moreover, since $(c_1 \bar{V}_1 + c_2 \bar{V}_2) = 1$,

$$\nabla c_1 \bar{V}_1 = -\nabla c_2 \bar{V}_2$$

Thus

$$D_1 = D_2 = D$$

There is one binary diffusion coefficient relative to the volume average velocity. This result can be shown from the Gibbs–Duhem equation to be valid even when the partial molar volumes are not constant.

Example 3.2-2: Two flux equations with the same diffusion coefficient   If the partial molar volumes are constant, rearrange the flux equation written in molar concentrations

$$n_1 = -D \nabla c_1 + c_1 v^0$$

into the form

$$n_1 = -Dc \nabla x_1 + c_1 v^*$$

Do *not* assume that the total concentration, c, is constant in this rearrangement.

Solution We begin by writing the flux equation for the second species

$$n_2 = -D\nabla c_2 + c_2 v^0$$

Adding the flux equations, we find

$$(n_1 + n_2) = -D\nabla(c_1 + c_2) + (c_1 + c_2)v^0$$

$$cv^* = -D\nabla c + cv^0$$

Now we rewrite the flux equation for the first species as

$$n_1 = -D\nabla(x_1 c) + c_1 v^0$$

$$= -Dc\nabla x_1 + x_1(-D\nabla c + cv^0)$$

By combining with our earlier result, we find

$$n_1 = -Dc\nabla x_1 + c_1 v^*$$

which is what we seek. A similar analysis for the diffusion equation relative to the mass average velocity is possible for constant partial specific volumes.

Example 3.2-3: Different binary diffusion coefficients Some authors use flux equations of the form

$$n_1 = -D_1\nabla\rho_1 + \rho_1 v$$

where ρ_1 is the mass of species 1 per volume. Show that the coefficient D_1 is not equal to the D used in Table 3.2-1 and that this binary system involves two different diffusion coefficients.

Solution To solve this example, we must rewrite the concentration ρ_1 in terms of the mass fraction ω_1. By definition

$$\rho_1 = \omega_1 \rho$$

$$= \omega_1(\rho_1 + \rho_2)$$

But

$$\omega_1 \bar{V}_1 + \omega_2 \bar{V}_2 = 1$$

where the $\bar{V}_i$ are the partial specific volumes, taken as constants. We now eliminate ρ_2 to find, after some rearrangement,

$$\rho_1 \bar{V}_2 = \frac{\omega_1}{1 - \omega_1(1 - \bar{V}_1/\bar{V}_2)}$$

Thus

$$\nabla\rho_1 = \frac{1}{\bar{V}_2[1 - \omega_1(1 - \bar{V}_1/\bar{V}_2)]^2}\nabla\omega_1$$

By combining this with the flux equation above, and comparing the result with the first diffusion equation in Table 3.2-1, we see that

$$D_1 = D\rho \bar{V}_2[1 - \rho_1(1 - \bar{V}_1/\bar{V}_2)]^2$$

The coefficients D_1 and D are equal only if both partial specific volumes are equal to the reciprocal of the density. By rotating subscripts, we see that D_1 doesn't equal D_2. By starting with molar concentrations and the molar average velocity, we can derive similar expressions in terms of the molar average velocity. All these expressions are more complicated than those referred to volume average velocity.

Example 3.2-4: Diffusion-engendered flow   In the diffusion apparatus shown in Fig. 3.1-2(b), one bulb contains nitrogen and the other hydrogen. The temperature and pressure are such that the diffusion coefficient is 0.1 cm^2/sec. The length l is 10 cm. Find v^0, v^*, and v at the average concentration in the system.

    Solution   The volume in this system does not move, so v^0 is zero. If the gases are ideal, then the molar concentration is constant everywhere and $v^* = 0$. Because of this, we can use the thin-film results from Section 2.2:

$$j_1 = c_1 v_1 = \frac{D}{l}(c_{10} - c_{1l})$$

If species 1 is nitrogen at an average concentration of 0.5c,

$$v_1 = \left[\frac{D}{l}\right]\left(\frac{c_{10} - c_{1l}}{c_1}\right) = \left[\frac{0.1\,\text{cm}^2/\text{sec}}{10\,\text{cm}}\right]\left(\frac{1 - 0}{0.5}\right) = 0.02\,\text{cm/sec}$$

By similar arguments, for hydrogen,

$$v_2 = -0.02\,\text{cm/sec}$$

Note that these velocities vary as the average concentration c_1 varies.
    We next find the mass fractions of each species:

$$\omega_1 = \frac{c_1 \tilde{M}_1}{c_1 \tilde{M}_1 + c_2 \tilde{M}_2} = \frac{0.5(28)}{0.5(28) + 0.5(2)} = 0.933$$

where $\tilde{M}_i$ is the molecular weight of species i. Similarly,

$$\omega_2 = 0.067$$

Then the mass average velocity is

$$v = \omega_1 v_1 + \omega_2 v_2 = 0.933 \times (0.020) + 0.067(-0.020) = 0.017\,\text{cm/sec}$$

The result is dominated by the nitrogen because of its higher molecular weight.

3.2.2 Maxwell–Stefan Equations

We next want to describe an analysis of diffusion which avoids carefully defining convection. After all, the reference velocities explored above are difficult, requiring careful thought. We may appreciate that the volume average velocity equals molar average velocity for ideal gases, and that the volume average velocity equals the mass average velocity for systems of constant density. We may recognize that these complications normally vanish for dilute solutions. Still, we may yearn for a description of diffusion which avoids these complexities, which sends reference velocities to the same intellectual pergatory as standard-state chemical potentials and pressure-dependent fugacity coefficients.

Such an apparently simpler description is provided by the Maxwell–Stefan equations, the last result in Table 3.2-1. For a binary system, these may be written as

$$\nabla y_1 = \frac{y_1 y_2}{D'} (v_2 - v_2) \tag{3.2-6}$$

where y_i and v_i are the mole fraction and velocity of species "i," normally in the gas phase, and D' is a new Maxwell–Stefan diffusion coefficient. The corresponding result, for nonideal liquid solutions is

$$\nabla \mu_1 = \frac{RT x_2}{D''} (v_2 - v_1) \tag{3.2-7}$$

where μ_i and x_i are the chemical potential and liquid mole fraction of species "i," and D'' is another diffusion coefficient. A variety of other, similar forms have also been suggested and have achieved some popularity, especially in Europe.

These Maxwell–Stefan equations have three significant advantages over the Fick's law parallels described earlier in this section. First, for dilute solutions, they quickly reduce to the normal form of Fick's law, so that all earlier dilute solution results can be used without worry. Second, they avoid the issue of reference velocities by using the velocity difference $(v_2 - v_1)$. After our intellectual struggles with these references we may find this a blessed relief. Third, for the special case of ideal gases, these equations are easily generalized to multicomponent systems, as detailed in Section 7.1. These are three significant advantages.

At the same time, this alternative formulation obscures any convection in the system. As a result, it reduces the physical insight possible for many, including me. This loss of insight can make solving simple problems harder. Some who can think clearly in abstract mathematical terms will find the Maxwell–Stefan form innately superior. I am not one of this group. I will use the Fick's law parallels because I need all the physical insight that I can get.

Example 3.2-5: Comparing diffusion coefficients Show how D' in Equation 3.2-6 is related to D defined relative to the volume average velocity.

Solution Equation 3.2-6 may be rewritten as follows

$$\nabla y_1 = \frac{y_1 y_2}{D'} (v_2 - v_1)$$

$$= \frac{1}{cD'} (y_1 n_2 - y_2 n_1)$$

where c is the total concentration. By definition

$$v^0 = c_1 \bar{V}_1 v_1 + c_2 \bar{V}_2 v_2$$
$$= \bar{V}_1 n_1 + \bar{V}_2 n_2$$

Thus

$$n_2 = \frac{v^0 - \bar{V}_1 n_1}{\bar{V}_2}$$

Combining this with the flux equation, rearranging, and remembering that $(y_1 \bar{V}_1 + y_2 \bar{V}_2)$ is $(1/c)$, we find

$$n_1 = -D' c^2 \bar{V}_2 \nabla y_1 + c_1 v^0$$

By comparing this with the results of Example 3.2-2, we see that

$$D = D' c \bar{V}_2$$

For an ideal gas, the partial molar volume of every gas is equal to the reciprocal of the total molar concentration c. Thus $c \bar{V}_2$ is one and

$$D = D'$$

The two diffusion coefficients are the same.

Example 3.2-6: The effect of non-ideal solutions   Show how D' in Equation 3.2-6 is related to D'' in Equation 3.2-7.
     Solution   The chemical potential μ_1 is given by

$$\mu_1 = \mu_1^0 + RT \ln \gamma_1 x_1$$

where μ_1^0 is a reference value and γ_1 is an activity coefficient. Thus at constant temperature

$$\nabla \mu_1 = RT \left(1 + \frac{\partial \ln \gamma_1}{\partial \ln x_1} \right) \nabla \ln x_1$$

$$= \frac{RT}{x_1} \left(1 + \frac{\partial \ln \gamma_1}{\partial \ln x_1} \right) \nabla x_1$$

Combining with Equation 3.2-6, we find

$$\nabla x_1 = \frac{x_1 x_2}{D'' \left(1 + \dfrac{\partial \ln \gamma_1}{\partial \ln x_1} \right)} (v_2 - v_1)$$

Thus

$$D' = D'' \left(1 + \frac{\partial \ln \gamma_1}{\partial \ln x_1} \right)$$

The diffusion coefficients are related by an activity correction.

The idea that diffusion is better described by a chemical potential gradient than by a concentration gradient appears frequently. It is basic to estimates of diffusion in liquids, as proposed by Einstein and discussed in Section 5.2. It is correct for dilute solutions of electrolytes, covered in Section 6.1. However, it is certainly wrong near spinodals or consolute points (Section 6.3), and it seems untested for other, highly nonideal solutions. I use it confidently for dilute solutions but cautiously elsewhere.

3.3 Parallel Diffusion and Convection

We now want to combine the equations developed above with mass balances to calculate fluxes and concentration profiles. This is, of course, the same objective as in Chapter 2. The difference here is that both diffusion and convection are significant. The analysis of the more complicated problems of diffusion and convection is aided by the parallels in the case of a thin film and an infinite slab around which Chapter 2 is organized. Such parallels produce powerful pedagogy.

3.3.1 Fast Diffusion Through a Stagnant Film

The first problem that we consider involves the same rapid evaporation that was used as the key example in Section 3.1. We recall that at intermediate temperatures, the evaporation rate depends on both diffusion and convection up the tube.

We want to calculate the flux and the concentration profile where both diffusion and convection are important. To make this calculation, we must parallel our earlier scheme, but with a more exact physical understanding and a more complicated mathematical analysis. Just as before, the scheme starts with a mass balance, combines this balance with Fick's law, and then runs through the math to the desired result.

This mass balance is written on the differential volume $A\Delta z$ shown in Fig. 3.3-1:

$$\begin{pmatrix} \text{solute accumulated} \\ \text{in volume } A\Delta z \end{pmatrix} = \begin{pmatrix} \text{solute transported} \\ \text{in at } z \end{pmatrix} - \begin{pmatrix} \text{solute transported} \\ \text{out at } z + \Delta z \end{pmatrix} \quad (3.3\text{-}1)$$

In mathematical terms, this is

$$\frac{\partial}{\partial t}(A\Delta z c_1) = An_1|_z - An_1|_{z+\Delta z} \quad (3.3\text{-}2)$$

If we divide by the volume $A\Delta z$ and take the limit as this volume goes to zero, we find

$$\frac{\partial c_1}{\partial t} = -\frac{\partial n_1}{\partial z} \quad (3.3\text{-}3)$$

At steady state, there is no accumulation, so

$$0 = \frac{\partial n_1}{\partial z} \quad (3.3\text{-}4)$$

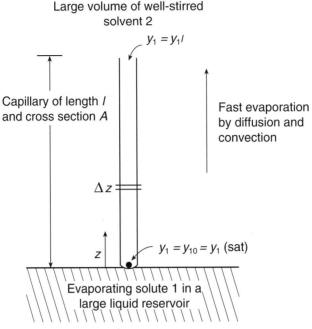

Fig. 3.3-1. Fast evaporation in a thin capillary. This problem is analogous to that shown in Fig. 2.2-1, but for a concentrated solution.

This is easily integrated to show that n_1 is constant. This sensibly says that at steady state, the total flux up the tube is constant. Note that we have not shown that the diffusion flux is constant.

We now want to combine this result with Fick's law. However, because we are dealing with fast evaporation and a potentially concentrated solution, we must consider both diffusion and convection. For simplicity, we choose the volume average velocity v^0 from Table 3.2-1.

$$n_1 = j_1 + c_1 v^0 = -D\frac{dc_1}{dz} + c_1(c_1 V_1 v_1 + c_2 V_2 v_2) \tag{3.3-5}$$

By definition, $c_1 v_1$ equals n_1, and $c_2 v_2$ equals n_2. If the solvent vapor is stagnant, its flux n_2 and its velocity v_2 must be zero. Thus

$$n_1 = -D\frac{dc_1}{dz} + c_1 \bar{V}_1 n_1 \tag{3.3-6}$$

Moreover, if the vapors in the capillary are ideal, then the total molar concentration is a constant and $\bar{V}_1$ equals $1/c$ (see Eq. 3.2-4). Thus the differential equation we seek is

$$n_1(1 - y_1) = -Dc\frac{dy_1}{dz} \tag{3.3-7}$$

This is subject to the two boundary conditions

$$z = 0, \quad y_1 = y_{10} \tag{3.3-8}$$

$$z = l, \quad y_1 = y_{1l} \tag{3.3-9}$$

There are two boundary conditions for the first-order differential equation because n_1 is an unknown integration constant.

The flux and concentration profiles are now routinely found. The concentration profile is exponential:

$$\frac{1 - y_1}{1 - y_{10}} = \left(\frac{1 - y_{1l}}{1 - y_{10}}\right)^{z/l} \tag{3.3-10}$$

The total flux is constant and logarithmic:

$$n_1 = \frac{Dc}{l} \ln\left(\frac{1 - y_{1l}}{1 - y_{10}}\right) \tag{3.3-11}$$

Note that doubling the concentration difference no longer automatically doubles the total flux. Like the total flux, the diffusion flux is logarithmic, but it is not constant:

$$j_1 = -Dc\frac{dy_1}{dz} = Dc\left(\frac{1 - y_{10}}{l}\right)\left(\frac{1 - y_{1l}}{1 - y_{10}}\right)^{z/l} \ln\left(\frac{1 - y_{1l}}{1 - y_{10}}\right) \tag{3.3-12}$$

The diffusion flux is smallest at the bottom of the capillary. It steadily rises to its largest value at the top of the capillary.

If the solution is dilute, we can simplify these results. To do this, we first remember that for small y_1,

$$(1 - y_1)^a \doteq 1 - ay_1 + \cdots \tag{3.3-13}$$

$$\frac{1}{1 - y_1} \doteq 1 + y_1 + \cdots \tag{3.3-14}$$

and

$$\ln(1 - y_1) \doteq -y_1 + \cdots \tag{3.3-15}$$

The concentration profile in Eq. 3.3-10 thus becomes

$$1 - y_1 = (1 - y_{10})(1 - y_{1l} + y_{10} - \cdots)^{z/l}$$

$$= 1 - y_{10} + \frac{z}{l}(y_{10} - y_{1l}) + \cdots \tag{3.3-16}$$

This can be rewritten in more familiar terms by multiplying both sides of the equation by the total concentration c and rearranging:

$$c_1 = c_{10} + (c_{1l} - c_{10})\frac{z}{l} \tag{3.3-17}$$

In other words, the concentration profile becomes linear, not exponential, as the solution becomes dilute.

The total flux in dilute solution can be simplified in a similar fashion:

$$n_1 = \frac{Dc}{l}[\ln(1 - y_{1l}) - \ln(1 - y_{10})]$$

$$\doteq \frac{Dc}{l}(y_{10} - y_{1l})$$

$$\doteq \frac{D}{l}(c_{10} - c_{1l}) \qquad\qquad (3.3\text{-}18)$$

which is, of course, the simple relation derived earlier in Eq. 2.2-10. The diffusion flux j_1 equals n_1 in this dilute limit. Thus Eqs. 3.3-10 and 3.3-11 are equivalent to Eqs. 2.2-9 and Eqns. 2.2-10 in dilute solution.

The analysis above is not hard to understand one line at a time, but it may be hard to understand in total. To supply this total understanding, we consider a special case of benzene liquid at 60 °C evaporating through a capillary into pure air. At this temperature, the partial pressure of benzene is 400 mm Hg, so the mole fraction of benzene at the liquid vapor interface y_{10} is $400/760 = 0.53$. The mole fraction at the other end of the capillary y_{1l} is zero. Thus we can find the concentration profile from Eq. 3.3-10, the total flux from Eq. 3.3-11, and the diffusion flux from Eq. 3.3-12.

The meaning of these results is much clearer from Fig. 3.3-2. The total flux is a constant from $z = 0$ to $z = l$. The diffusion flux is smallest at $z = 0$, the liquid–vapor interface, but rises to equal the total flux at $z = l$, where diffusion is the only mass transfer mechanism operating within the capillary. The concentration profile is nonlinear,

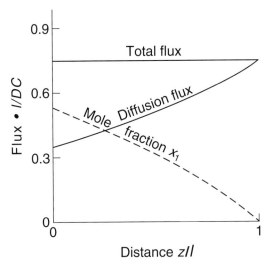

Fig. 3.3-2. Concentration and flux in concentrated diffusion. The concentration profile is no longer linear, as in Fig. 2.2-1. The constant total flux is the sum of diffusion and convection, each of which varies.

but its slope is still proportional to the diffusion flux. This slope is always negative, smallest at $z = 0$ where mass transfer by convection is greatest. Finally, note that the vertical distance between the total flux and the diffusion flux is the convective flux $c_1 v^0$; it is largest at the liquid–vapor interface, where $z = 0$, and equals zero at the end of the capillary, where $z = l$. Please think about this figure carefully because it can help you understand diffusion-induced convection.

3.3.2 Fast diffusion Into a Semi-infinite Slab

The second problem considered in this section is illustrated schematically in Fig. 3.3-3. In this problem, a volatile liquid solute evaporates into a long gas-filled capillary. The solvent gas in the capillary initially contains no solute. As solute evaporates, the interface between the vapor and the liquid solute drops. However, the gas is essentially insoluble in the liquid. We want to calculate the solute's evaporation rate, including the effect of diffusion-induced convection and the effect of the moving interface (Arnold, 1944).

In this problem, we first choose the origin of our coordinate system ($z = 0$) as the liquid–vapor interface. We then write a mass balance for the solute 1 on the differential volume $A\Delta z$, shown in Fig. 3.3-3:

$$\begin{pmatrix} \text{solute} \\ \text{accumulation} \\ \text{in } A\Delta z \end{pmatrix} = \begin{pmatrix} \text{solute} \\ \text{transport} \\ \text{in} \end{pmatrix} - \begin{pmatrix} \text{solute} \\ \text{transport} \\ \text{out} \end{pmatrix} \tag{3.3-19}$$

or, in symbolic terms,

$$\frac{\partial}{\partial t}(c_1 A\Delta z) = (An_1)_z - (An_1)_{z+\Delta z} \tag{3.3-20}$$

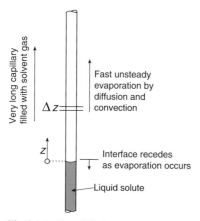

Fig. 3.3-3. Fast diffusion in a semi-infinite slab. This problem is analogous to that shown in Fig. 2.3-2, but for a concentrated solution. Because of this higher concentration, the liquid–vapor interface moves significantly, complicating the situation.

Dividing by the differential volume and taking the limit as this volume goes to zero,

$$\frac{\partial c_1}{\partial t} = -\frac{\partial n_1}{\partial z} \tag{3.3-21}$$

We then split the diffusion and convection:

$$\frac{\partial c_1}{\partial t} = D\frac{\partial^2 c_1}{\partial z^2} - \frac{\partial}{\partial z}c_1 v^0 \tag{3.3-22}$$

By definition, the volume average velocity is

$$v^0 = c_1\bar{V}_1v_1 + c_2\bar{V}_2v_2 = \bar{V}_1n_1 + \bar{V}_2n_2 \tag{3.3-23}$$

In the steady-state case treated earlier, we argued that the solvent was stagnant, so that n_2 was zero and the problem was simple. Here, in an unsteady case, the solvent flux varies with position and time; therefore, no easy simplification is possible.

We must write a continuity equation for the solvent gas 2:

$$\frac{\partial c_2}{\partial t} = -\frac{\partial n_2}{\partial z} \tag{3.3-24}$$

If we multiply Eqs. 3.3-21 and 3.3-24 by the appropriate partial molar volumes and add them, we find

$$\frac{\partial}{\partial t}(\bar{V}_1c_1 + \bar{V}_2c_2) = -\frac{\partial}{\partial z}(\bar{V}_1n_1 + \bar{V}_2n_2) \tag{3.3-25}$$

But the quantity $\bar{V}_1c_1 + \bar{V}_2c_2$ always equals unity, making the left-hand side of this equation zero; thus $\bar{V}_1n_1 + \bar{V}_2n_2$ must be independent of z. However, at the interface, n_2 is zero because the solvent gas 2 is insoluble in the liquid. Thus

$$\bar{V}_1n_1 + \bar{V}_2n_2 = \bar{V}_1n_1|_{z=0} = \bar{V}_1\left(-D\frac{\partial c_1}{\partial z}\bigg|_{z=0} + c_1\bar{V}_1n_1|_{z=0}\right) \tag{3.3-26}$$

When we combine this with Eq. 3.3-22 we find

$$\frac{\partial c_1}{\partial t} = D\frac{\partial^2 c_1}{\partial z^2} + \left(\frac{D\bar{V}_1(\partial c_1/\partial z)}{1 - c_1\bar{V}_1}\right)_{z=0}\frac{\partial c_1}{\partial z} \tag{3.3-27}$$

subject to the conditions

$$t = 0, \quad \text{all } z > 0, \quad c_1 = 0 \tag{3.3-28}$$

$$t > 0, \quad\quad z = 0, \quad c_1 = c_1(\text{sat}) \tag{3.3-29}$$

$$z = \infty, \quad c_1 = 0 \tag{3.3-30}$$

The solute concentration $c_1(\text{sat})$ is that in the vapor in equilibrium with the liquid.

Like the problem of dilute diffusion in a semi-infinite slab, this problem is solved by defining the combined variable

$$\zeta = z/\sqrt{4Dt} \tag{3.3-31}$$

The differential equation now becomes

$$\frac{d^2c_1}{d\zeta^2} + 2(\zeta - \Phi)\frac{dc_1}{d\zeta} = 0 \tag{3.3-32}$$

subject to the conditions

$$\zeta = 0, \quad c_1 = c_1(\text{sat}) \tag{3.3-33}$$

$$\zeta = \infty, \quad c_1 = 0 \tag{3.3-34}$$

and in which

$$\Phi = -\frac{1}{2}\left(\frac{\bar{V}_1(dc_1/d\zeta)}{1 - c_1\bar{V}_1}\right)_{z=0} \tag{3.3-35}$$

The constant Φ, a dimensionless velocity, characterizes both the convection engendered by diffusion and the movement of the interface. If Φ is zero, convection effects are zero.

Equation 3.3-32 can be integrated once to give

$$\frac{dc_1}{d\zeta} = (\text{constant})e^{-(\zeta - \Phi)^2} \tag{3.3-36}$$

A second integration and evaluation of the boundary conditions give

$$\frac{c_1}{c_1(\text{sat})} = \frac{1 - \text{erf}(\zeta - \Phi)}{1 + \text{erf}\,\Phi} \tag{3.3-37}$$

We can calculate Φ from this result and Eq. 3.3-35:

$$\bar{V}_1 c_1(\text{sat}) = \left(1 + \frac{1}{\sqrt{\pi}(1 + \text{erf}\,\Phi)\Phi e^{\Phi^2}}\right)^{-1} \tag{3.3-38}$$

A plot of Φ versus concentration is shown in Fig. 3.3-4. Note that when $c_1(\text{sat})$ is small, Φ goes to zero. In other words, when the solution is dilute, convection is unimportant.

We also want to calculate the interfacial flux N_1. To find this, we must again split diffusion and convection, using Fick's law:

$$N_1 = n_1|_{z=0} = -\left(\frac{D(\partial c_1/\partial z)}{1 - c_1\bar{V}_1}\right)_{z=0}$$

$$= \sqrt{D/\pi t}\left(\frac{1}{1 - \bar{V}_1 c_1(\text{sat})}\right)\frac{e^{-\Phi^2}}{1 + \text{erf}\,\Phi}c_1(\text{sat}) \tag{3.3-39}$$

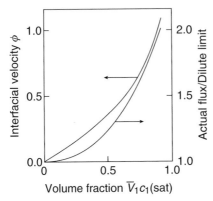

Fig. 3.3-4. Flux and interfacial movement. As the solution becomes dilute, the interfacial concentration $c_1(\text{sat})$ becomes small, the actual flux approaches the dilute-solution limit (see Eq. 2.3-18), and the velocity Φ becomes zero.

where Φ is still found from Eq. 3.3-38 or Fig. 3.3-4. The increase of this flux beyond that in a dilute solution is also given in this figure.

Example 3.3-1: Errors caused by neglecting convection   Consider the experiments shown in Fig. 3.3-5. How much error is caused by calculating the rate of benzene evaporation if only diffusion is considered?

    Solution   The sizes of the errors depend on the concentrations and thus on the temperature. At 6 °C, the vapor pressure of benzene is about 37 mm Hg. If the total pressure is one atmosphere,

$$y_1 = \frac{c_1}{c} = \frac{p_1(\text{sat})}{p} = \frac{37}{760} = 0.049$$

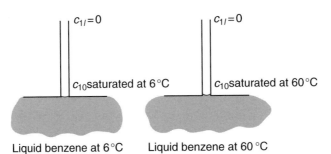

Fig. 3.3-5. Examples of benzene diffusion and convection. In the dilute solution at the left, the exact results are close to the approximate ones in Eq. 2.2-10. In the concentrated case at the right, they are not.

The total flux is, from Eq. 3.3-11,

$$n_1 = \frac{Dc}{l} \ln\left(\frac{1-0}{1-0.049}\right)$$

$$= \frac{0.050Dc}{l}$$

The flux, assuming a dilute solution, calculated from Eq. 2.2-10, is

$$n_1 = j_1 = \frac{Dc}{l}(0.049 - 0)$$

This two percent difference is well within the needs of most practical calculations. Thus the dilute solution equations are more than adequate here.

At 60 °C, the choice is less obvious because the vapor pressure is about 395 mm Hg. When we calculate the mole fraction in the same way, we find

$$n_1 = \frac{Dc}{l} \ln\left(\frac{1-0}{1-(395/760)}\right) = 0.73\frac{Dc}{l}$$

The dilute-solution estimate is

$$n_1 = j_1 = \frac{395}{760}\frac{Dc}{l} = 0.52\frac{Dc}{l}$$

The dilute-solution equations underestimate the flux by a significant error of about forty percent.

3.4 Generalized Mass Balances

As the problems that we discuss in this chapter become more and more complex, the development of the differential equations becomes more and more tedious. Such tedium can be avoided by using the generalized mass balances developed in this section. These mass balances automatically include both steady- and unsteady-state situations. They imply the usual variety of coordinate systems, and they reflect the vectorial nature of mass fluxes. They are excellent weapons.

However, like most weapons, the generalized mass balances can injure those trying to use them. Effective use requires uncommon skill in connecting the mathematical ideal and the physical reality. Some seem born with this skill; more seem to develop it over time. If you have trouble applying these equations, return to the shell balance method. It may take longer, but it is safer. You can check your equations by later comparing them with those found from the generalized results.

To find the generalized mass balances, we consider the small differential volume located at (x, y, z) shown in Fig. 3.4-1. We want to write a mass balance on this volume:

$$\left(\begin{array}{c}\text{mass of species 1} \\ \text{accumulating in} \\ \Delta x \Delta y \Delta z\end{array}\right) = \left(\begin{array}{c}\text{mass flux of} \\ \text{species 1} \\ \text{in minus that out}\end{array}\right) + \left(\begin{array}{c}\text{mass produced by} \\ \text{homogeneous} \\ \text{chemical reaction}\end{array}\right)$$

$$(3.4-1)$$

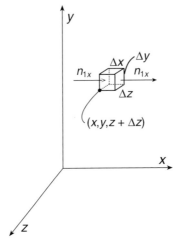

Fig. 3.4-1. The arbitrary volume for deriving the generalized mass balances. The fluxes in the x direction are shown in this figure; fluxes in other directions are also included in the derivation. The results are shown in Tables 3.4-1 and 3.4-2.

The mass fluxes relative to fixed coordinates include transport in all three directions. For example, the mass flux out of the volume in the x direction, shown in Fig. 3.4-1, is $n_{1x}\Delta y \Delta z$, where $\Delta y \Delta z$ is the area across which this flux occurs. In mathematical terms, the mass balance is then

$$\frac{\partial}{\partial t}(c_1 \Delta x \Delta y \Delta z) = (n_{1_x}\Delta y \Delta z)_x - (n_{1_x}\Delta y \Delta z)_{x+\Delta x}$$
$$+ (n_{1_y}\Delta x \Delta z)_y - (n_{1_y}\Delta x \Delta z)_{y+\Delta y}$$
$$+ (n_{1_z}\Delta x \Delta y)_z - (n_{1_z}\Delta x \Delta y)_{z+\Delta z}$$
$$+ r_1 \Delta x \Delta y \Delta z \tag{3.4-2}$$

where r_1 is the rate per unit volume of a homogeneous chemical reaction producing solute 1. Dividing by the differential volume $\Delta x \Delta y \Delta z$ and taking the limit as this volume goes to zero gives

$$\frac{\partial}{\partial t}c_1 = -\frac{\partial}{\partial x}n_{1_x} - \frac{\partial}{\partial y}n_{1_y} - \frac{\partial}{\partial z}n_{1_z} + r_1 \tag{3.4-3}$$

or, in vectorial notation,

$$\frac{\partial}{\partial t}c_1 = -\nabla \cdot \mathbf{n}_1 + r_1 \tag{3.4-4}$$

We can also write the flux in terms of diffusion and convection:

$$\mathbf{n}_1 = -D\nabla c_1 + c_1 \mathbf{v}^0 \tag{3.4-5}$$

where v^0 is the volume average velocity. Combining,

$$\frac{\partial c_1}{\partial t} = D\nabla^2 c_1 - \nabla \cdot c_1 v^0 + r_1 \tag{3.4-6}$$

This equation is the general form of all the shell balances derived to date.

The species mass balance represented by Eq. 3.4-6 is often effectively complemented by the overall mass balance:

$$\begin{pmatrix} \text{total mass} \\ \text{accumulation} \\ \text{in } \Delta x \Delta y \Delta z \end{pmatrix} = \begin{pmatrix} \text{total mass} \\ \text{flux in minus} \\ \text{that out} \end{pmatrix} \tag{3.4-7}$$

This can be written in terms similar to those used earlier:

$$\begin{aligned}
\frac{\partial}{\partial t}(\rho \Delta x \Delta y \Delta z) = {}& (\rho v_x \Delta y \Delta z)_x - (\rho v_x \Delta y \Delta z)_{x+\Delta x} \\
&+ (\rho v_y \Delta x \Delta z)_y - (\rho v_y \Delta x \Delta z)_{y+\Delta y} \\
&+ (\rho v_z \Delta x \Delta y)_z - (\rho v_z \Delta x \Delta y)_{z+\Delta z}
\end{aligned} \tag{3.4-8}$$

in which v_x, v_y, and v_z are components of the mass average velocity. Dividing by the volume $\Delta x \Delta y \Delta z$ and taking the limit as each difference becomes small, we find

$$\frac{\partial \rho}{\partial t} = -\frac{\partial}{\partial x}\rho v_x - \frac{\partial}{\partial y}\rho v_y - \frac{\partial}{\partial z}\rho v_z \tag{3.4-9}$$

In vectorial notation, this is

$$\frac{\partial \rho}{\partial t} = -\nabla \cdot \rho v \tag{3.4-10}$$

This result, called the continuity equation, has no reaction term because no total mass is generated or destroyed by nonnuclear chemical reactions.

We would like to use the continuity equation to simplify the species mass balance. We cannot do so directly because the continuity equation contains the mass average velocity, and the species mass balance involves the volume average velocity. Although some investigators fuss about this difference, we should recognize that we can solve many problems where these velocities are the same. They are the same at constant density, as shown by Eq. 3.2-3.

If we assume constant density, the overall continuity equation becomes

$$0 = -\nabla \cdot v = -\nabla \cdot v^0 \tag{3.4-11}$$

We then multiply this equation by c_1 and subtract the result from Eq. 3.4-6:

$$\frac{\partial c_1}{\partial t} + v^0 \cdot \nabla c_1 = D\nabla^2 c_1 + r_1 \tag{3.4-12}$$

This result is frequently useful for problems of diffusion and convection.

This generalized equation is shown in different coordinate systems in Tables 3.4-1 and Tables 3.4-2. The overall mass balance is given in Table 3.4-3. These equations include

Table 3.4-1 *Mass balance for species 1 in various coordinate systems*

Rectangular coordinates

$$\frac{\partial c_1}{\partial t} = -\frac{\partial n_{1x}}{\partial x} - \frac{\partial n_{1y}}{\partial y} - \frac{\partial n_{1z}}{\partial z} + r_1 \tag{A}$$

Cylindrical coordinates

$$\frac{\partial c_1}{\partial t} = -\frac{1}{r}\frac{\partial}{\partial r}(rn_{1r}) - \frac{1}{r}\frac{\partial n_{1\theta}}{\partial \theta} - \frac{\partial n_{1z}}{\partial z} + r_1 \tag{B}$$

Spherical coordinates

$$\frac{\partial c_1}{\partial t} = -\frac{1}{r^2}\frac{\partial}{\partial r}(r^2 n_{1r}) - \frac{1}{r\sin\theta}\frac{\partial}{\partial\theta}(n_{1\theta}\sin\theta) - \frac{1}{r\sin\theta}\frac{\partial n_{1\phi}}{\partial\phi} + r_1 \tag{C}$$

Note: The rate r_1 is for the production of species 1 per volume.

Table 3.4-2 *Mass balance for species 1 combined with Fick's law*

Rectangular coordinates

$$\frac{\partial c_1}{\partial t} + v_x^0\frac{\partial c_1}{\partial x} + v_y^0\frac{\partial c_1}{\partial y} + v_z^0\frac{\partial c_1}{\partial z} = D\left(\frac{\partial^2 c_1}{\partial x^2} + \frac{\partial^2 c_1}{\partial y^2} + \frac{\partial^2 c_1}{\partial z^2}\right) + r_1 \tag{A}$$

Cylindrical coordinates

$$\frac{\partial c_1}{\partial t} + v_r^0\frac{\partial c_1}{\partial r} + \frac{v_\theta^0}{r}\frac{\partial c_1}{\partial\theta} + v_z^0\frac{\partial c_1}{\partial z} = D\left[\frac{1}{r}\frac{\partial}{\partial r}\left(r\frac{\partial c_1}{\partial r}\right) + \frac{1}{r^2}\frac{\partial^2 c_1}{\partial\theta^2} + \frac{\partial^2 c_1}{\partial z^2}\right] + r_1 \tag{B}$$

Spherical coordinates

$$\frac{\partial c_1}{\partial t} + v_r^0\frac{\partial c_1}{\partial r} + v_\theta^0\frac{\partial c_1}{\partial\theta} + \frac{v_\theta^0}{r\sin\theta}\frac{\partial c_1}{\partial\phi} = D\left[\frac{1}{r^2}\frac{\partial}{\partial r}\left(r^2\frac{\partial c_1}{\partial r}\right) + \frac{1}{r^2\sin\theta}\frac{\partial}{\partial\theta}\left(\sin\theta\frac{\partial c_1}{\partial\theta}\right)\right.$$

$$\left. + \frac{1}{r^2\sin^2\theta}\frac{\partial^2 c_1}{\partial\phi^2}\right] + r_1 \tag{C}$$

Note: The diffusion coefficient D and the density ρ are assumed constant. In this case, the mass average and volume average velocities are equal. Again, r_1 is the rate of production of species 1 per volume.

the effects of chemical reaction, convection, and concentration-driven diffusion. However, they are not quite as general as their title suggests. For example, they do not include the effects of electric or magnetic forces. Nonetheless, they often provide a useful route to the differential equations for diffusion, as shown by the following examples.

Example 3.4-1: Fast diffusion through a stagnant film and into a semi-infinite slab   Find differential equations describing these two situations from the general equations in Tables 3.4-1 to 3.4-3. Compare your results with the shell-balance results in the previous section.

   Solution   The first of these cases, sketched in Fig. 3.1-1 or Fig. 3.3-1, concerns the fast evaporation of a liquid solute through a stagnant vapor. This evaporation is in

Table 3.4-3 *Total mass balance in several coordinate systems*

Rectangular coordinates

$$\frac{\partial \rho}{\partial t} = -\frac{\partial}{\partial x}(\rho v_x) - \frac{\partial}{\partial y}(\rho v_y) - \frac{\partial}{\partial z}(\rho v_z) \tag{A}$$

Cylindrical coordinates

$$\frac{\partial \rho}{\partial t} = -\frac{1}{r}\frac{\partial}{\partial r}(\rho r v_r) - \frac{1}{r}\frac{\partial}{\partial \theta}(\rho v_\theta) - \frac{\partial}{\partial z}(\rho v_z) \tag{B}$$

Spherical coordinates

$$\frac{\partial \rho}{\partial t} = -\frac{1}{r^2}\frac{\partial}{\partial r}(\rho r^2 v_r) - \frac{1}{r \sin\theta}\frac{\partial}{\partial \theta}(\rho v_\theta \sin\theta) - \frac{1}{r \sin\theta}\frac{\partial}{\partial \phi}(\rho v_\phi) \tag{C}$$

Note: The velocity here is the mass average and not the volume average commonly used with Fick's law.

steady state, has no chemical reaction, and occurs only in the z direction. Thus Eq. A in Table 3.4-1 becomes

$$0 = \frac{\partial n_{1z}}{\partial z}$$

Alternatively, for constant density, Eq. A in Table 3.4-2 becomes

$$v_z^0 \frac{\partial}{\partial z}c_1 = D\frac{\partial^2 c_1}{\partial z^2}$$

Either of these equations leads to a solution of the problem like that in Section 3.3.

The second example, shown schematically in Fig. 2.3-2, depends on the unsteady evaporation of a liquid solute into a solvent gas. Again, the process is one-dimensional, without chemical reaction. From Eq. A in Table 3.4-1, we find

$$\frac{\partial c_1}{\partial t} = -\frac{\partial n_{1z}}{\partial z}$$

Alternatively, for constant density, Eq. A in Table 3.4-2 becomes

$$\frac{\partial c_1}{\partial t} + v_z^0 \frac{\partial}{\partial z}c_1 = D\frac{\partial^2 c_1}{\partial z^2}$$

The first term on the left-hand side of this result represents accumulation and the second is convection. The right-hand side represents diffusion. Again, the solution to these equations parallels that in the previous section.

The reader whose primary interest is in diffusion may question why these generalized equations are necessary and why the shell balances used before are not sufficient. I share this skepticism, and I prefer the physical insight supplied by the shell-balance technique.

At the same time, students often plead to be taught the material in this section, even though they may later question its utility. The students' plea originates not from considerations of mass transfer but from their studies of fluid mechanics. In fluid mechanics, the generalized equations are extremely helpful, especially in cases of curved streamlines.

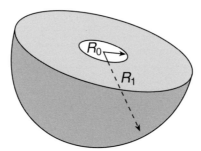

Fig. 3.4-2. Dissolution out of a coated hemisphere. The impermeable coating stops the diffusion except out of an uncoated hole of radius R_0.

Analogues of curved streamlines do not occur frequently in diffusion. Thus the mathematics of diffusion is easier than that of fluid mechanics, but the physical chemistry is more difficult.

Example 3.4-2: Dissolution out of a coated hemisphere   Orally taken drugs often result in drug concentrations which oscillate dramatically with time. Soon after a pill is swallowed, the drug concentration may be high, even toxic; four hours later, the concentration may be below that needed to be effective. Thus many have sought pills which would give a more even drug release vs. time, a topic detailed in Chapter 19.

One such pill, shown schematically in Figure 3.4-2, consists of a coated hemisphere of radius R_1 with a central hole of radius R_0. The hemisphere contains a solid drug at concentration c_{20}, which can dissolve to form a saturated solution at $c_1(\text{sat})$. Solid drug is immobile, but dissolved drug moves with a diffusion coefficient D. The entire hemisphere is coated with an impermeable layer, except for the hole. At small times, diffusion coming out the hole is reasonable because the drug doesn't have far to go. At larger times, diffusion is still reasonable: while the distance to diffuse is bigger, the area supplying drug is bigger, too.

Develop differential equations describing this drug release.

Solution   Diffusion in this case has spherical symmetry, with concentration gradients only in the r-directions. If the drug is dilute, there is no convective flow. Thus from Table 3.4-2 Eq. C we obtain for dissolved drug

$$\frac{\partial c_1}{\partial t} = \frac{D}{r^2} \frac{\partial}{\partial r}\left(r^2 \frac{\partial c_1}{\partial r}\right) + r_1$$

For undissolved drug,

$$\frac{\partial c_2}{\partial t} = -r_1$$

This is subject to the constraints

$$t = 0, \quad \text{all } r, \quad c_1 = c_1\,(\text{sat})$$
$$t > 0, \quad r = R_0, \quad c_1 = 0$$
$$r = R_1, \quad \frac{\partial c_1}{\partial r} = 0$$

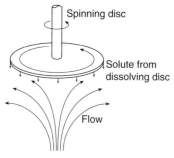

Fig. 3.4-3. Diffusion near a spinning disc. The amount dissolving per unit area is found to be the same everywhere on the disc's surface. Such simplicity makes this disc a powerful experimental tool.

We will also need a condition on the total mass of undissolved drug present

$$t = 0, \quad \text{all } r, \quad c_2 = c_{20}$$

We must also specify the kinetics of the dissolution. These equations can be solved numerically, but provide little insight without the solution.

An alternative strategy is to assume that, like many drugs the one used here dissolves rapidly relative to diffusion. In this case, the dissolved drug concentration c_1 will equal $c_1(\text{sat})$ everywhere that solid drug is present, i.e., where $c_2 > 0$. In this case, the problem is now simpler mathematically: the mass balance becomes

$$0 = \frac{D}{r^2} \frac{\partial}{\partial r} \left(r^2 \frac{\partial c_1}{\partial r} \right)$$

subject to

$$r = R_o, \quad c_1 = 0$$
$$r = R', \quad c_1 = c_1(\text{sat})$$

This is easily solved analytically. We must then find the variation of R' with time from

$$\int_0^t j_1|_{r=R_0} \left(r\pi R_0^2 \right) dt = \int_{R_1}^{R_0} c_{20} \left(4\pi r^2 \right) dr$$

This approximate solution should be used until we are forced by our experimental data to solve the more difficult and more complete problem. In my experience, many working on diffusion use mathematics which is more elaborate than their data justify. I urge you to use the simplest description that you can until you have good reasons to need elaboration.

Example 3.4-3: The flux near a spinning disc   The final example in this section is the spinning disc shown in Fig. 3.4-3. The disc is made of a sparingly soluble solute that slowly dissolves in the flowing solvent. This dissolution rate is diffusion-controlled. Calculate the rate at which the disc dissolves.

Solution This problem requires both mathematical skill and physical intuition. The dissolution will reach a steady state only when the disc is rotating; if the disc is not rotating, the problem will be equivalent to the semi-infinite slab discussed in Example 3.4-1. To solve the rotating disc problem, we choose cylindrical coordinates centered on the disc. The steady-state mass balance is found from Eq. B in Table 3.4-2:

$$v_r^0 \frac{\partial c_1}{\partial r} + \frac{v_\theta^0}{r} \frac{\partial c_1}{\partial \theta} + v_z^0 \frac{\partial c_1}{\partial z} = D \left[\frac{1}{r} \frac{\partial}{\partial r} \left(r \frac{\partial c_1}{\partial r} \right) + \frac{1}{r^2} \frac{\partial^2 c_1}{\partial \theta^2} + \frac{\partial^2 c_1}{\partial z^2} \right]$$

We recognize that the problem is angularly symmetric, so c_1 does not vary with θ. We also assume that the disc is infinitely wide, so that the concentration is a function only of z. I find this assumption mind-boggling, but it is justified by the success of the following calculations.

With these simplifications, the mass balance becomes

$$v_z \frac{dc_1}{dz} = D \frac{d^2 c_1}{dz^2}$$

subject to the conditions

$$z = 0, \quad c_1 = c_1(\text{sat})$$
$$z = \infty, \quad c_1 = 0$$

The first of these conditions implies equilibrium across the solid–fluid interface. Integration of the preceding equation gives

$$c_1 = a \int_0^z e^{-(1/D) \left[\int_0^r v_z(s) ds \right]} dr + b$$

where a and b are integration constants. From the foregoing conditions, b equals $c_1(\text{sat})$, so that

$$\frac{c_1}{c_1(\text{sat})} = 1 - \frac{\int_0^z e^{-(1/D) \left[\int_0^r v_z(s) ds \right]} dr}{\int_0^\infty e^{-(1/D) \left[\int_0^r v_z(s) ds \right]} dr}$$

If we know the velocity $v_z(z)$, we can find the concentration profile. We then use Fick's law to find the reaction rate.

The calculation of $v_z(z)$ is a problem in fluid mechanics beyond the scope of this book, but given in detail in the literature. When the values found for $v_z(z)$ are inserted into the previous equations, the result is

$$\frac{c_1}{c_1(\text{sat})} = \frac{\int_0^\Omega e^{-u^3} du}{\int_0^\infty e^{-u^3} du}$$

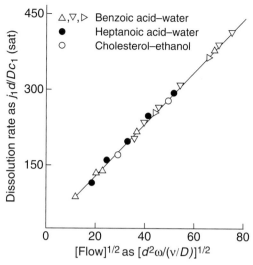

Fig. 3.4-4. Dissolution rate versus flow for a spinning disc. The dissolution rate and flow are described as Sherwood and Reynolds numbers, respectively. The data fit the form predicted, which should be valid over the unshaded region.

in which

$$\Omega = z \left(\frac{1.82 D^{1/3} \nu^{1/6}}{\omega^{1/2}} \right)^{-1}$$

and ν is the kinematic viscosity of the fluid and ω is the angular velocity of the disc. The diffusion flux is then

$$j_1|_{z=0} = -D \frac{\partial c_1}{\partial z}\Big|_{z=0} = 0.62 \left(\frac{D^{2/3} \omega^{1/2}}{\nu^{1/6}} \right) c_1(\text{sat})$$

This result is often written in terms of dimensionless groups:

$$-j_1 = \left[0.62 \frac{D}{d} \left(\frac{d^2 \omega \rho}{\mu} \right)^{1/2} \left(\frac{\mu}{\rho D} \right)^{1/3} \right] c_1(\text{sat})$$

where d is the disc diameter. The first term in parentheses is the Reynolds number, and the second is the Schmidt number.

To my delight, this analysis is verified by experiment. The dissolution varies with the square root of the Reynolds number, as shown in Fig. 3.4-4. As a result, the assumption that the flux is a function only of z is justified. Because the flux is independent of disc diameter, it has the same value near the disc's center and near its edge. Such a constant flux is uncommon, and it makes the interpretation of experimental results unusually straightforward. It is this feature that makes the rotating disc a popular experimental tool.

3.5 A Guide to Previous Work

In many cases, detailed solutions to diffusion problems can be adapted from calculations that have already been published, thus avoiding the mathematical detail presented in the earlier sections in this chapter. These calculations involve the same differential equations, with relatively minor changes of boundary conditions. They include elaborate but straightforward manipulations, like the integration of concentration profiles to find average concentrations.

Unfortunately, the published results are limited because they are often based on mathematical analogies with thermal conduction. Such analogies have merit; indeed, they provided the original stimulus for Fick's law of diffusion. However, in thermal conduction, there is no analogy for diffusion-induced convection and rarely an analogy for an effect like chemical reaction. On the other hand, in diffusion, there is no effect parallel to thermal radiation. These differences are commonly ignored by teachers because they want the pedagogical benefits of analogy. In fact, convection, chemical reaction, and radiation are frequently central in the problems studied.

Even with these limitations, the published solutions can be used to save considerable effort. Besides individual papers, there are two important books that have collected and compared this literature. The first, Crank's *The Mathematics of Diffusion* (1975), discusses aspects of chemical reactions. The second, Carslaw and Jaeger's *The Conduction of Heat in Solids* (1986), must be used by analogy, but it includes a more complete selection of boundary conditions. The notation used in these books is compared with that used here in Table 3.5-1.

In the remainder of this section we give examples illustrating how this literature can be used effectively.

Example 3.5-1: Diffusion through a polymer film   Imagine that we are studying a polymer film that is permeable to olefins like ethylene but much less permeable to aliphatic hydrocarbons. Such a film could be used for selectively separating the ethylene produced by dehydrogenation reactions. As part of our study, we use the diaphragm cell shown in

Table 3.5-1 *Comparisons of notation between this book and two major references*

Variable	Our notation	Crank's notation	Carslaw and Jaeger analogue
Time	t	t	t
Position	x, y, z, r	x, y, z, r	x, y, z, r
Concentration	c_1	C	Temperature υ
Concentration at boundary	$c_{10}, c_{1i}, C_{10}, \ldots$	C_1, C_0	Temperature at boundary φ
Binary diffusion coefficient	D	D	"Thermometric conductivity" κ
Flux relative to reference velocity	j_1	F	Heat flux f
Flux relative to fixed coordinates	n_1	F	Heat flux f
Flux at boundary	N_1 or $n_1\vert_{z=0}$	–	Heat flux at boundary F_0
Total amount diffusing from time 0 to t	M_t	M_t	–

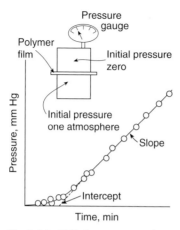

Fig. 3.5-1. Diffusion across a polymer film. When the pressure in the top compartment is determined as a function of time, the slope and intercept are measures of diffusion and solubility of gas in the polymer.

Fig. 3.5-1. This diaphragm cell consists of two compartments separated by the polymer film of interest. The top compartment is initially evacuated, but the lower one is filled with ethylene. We measure the ethylene concentration in the upper compartment as a function of time.

The data obtained for ethylene transport are exemplified by those shown in the figure. Initially, the pressure in the upper compartment varies in a complex way, but it will eventually approach that in the lower compartment. At the moderate times of most of our experiment, the pressure in the upper compartment is proportional to time, with a known slope and a definite intercept. How are this slope and intercept related to diffusion in the polymer film?

Solution The basic differential equation for this problem is that for a slab:

$$\frac{\partial c_1}{\partial t} = D\frac{\partial^2 c_1}{\partial z^2}$$

subject to the conditions

$$t = 0, \quad \text{all } z, \quad c_1 = 0$$
$$t > 0, \quad z = 0, \quad c_1 = Hp_0$$
$$z = l, \quad c_1 = Hp_l \doteq 0$$

in which l is the film's thickness and H is a Henry's law coefficient relating ethylene pressure in the gas to ethylene concentration in the film. The solution to this equation and the boundary conditions are given by Crank (1975, p. 50, Eq. 4-22):

$$\frac{c_1}{Hp_0} = 1 - \frac{z}{l} - \frac{2}{\pi}\sum_{n=1}^{\infty}\left(\frac{\sin(n\pi z/l)}{n}\right)e^{-Dn^2\pi^2 t/l^2}$$

Thus, almost before we have started, we have the concentration profile that we need.

We now must cast the problem in terms of the actual experiment variables we are using. First, from a mole balance on the top compartment,

$$\frac{dN_1}{dt} = \frac{V}{RT}\frac{dp}{dt} = -AD\frac{\partial c_1}{\partial z}\bigg|_{z=l}$$

in which V and p are the volume and pressure of the upper compartment and A is the film's area. Combining this with the concentration profile, we integrate subject to the condition that the upper compartment's pressure is initially zero:

$$p = \frac{ARTp_0}{Vl}\left[HDt + \frac{2Hl^2}{\pi^2}\sum_{n=1}^{\infty}\frac{\cos n\pi}{n^2}(1 - e^{-Dn^2\pi^2 t/l^2})\right]$$

At large time, the exponential terms become small, and this result becomes

$$p = \left\{\frac{ARTp_0}{Vl}\right\}(HD)\left[t - \frac{l^2}{6D}\right]$$

The quantity in braces is known experimentally. Thus the intercept of the data in Fig. 3.5-1 is related to the diffusion coefficient D. The slope of these data is related to the permeability HD. I am always delighted that an experiment like this gives both an equilibrium and a transport property.

This example has value well beyond the specific case studied. It shows how the mathematical complexities inherent in the problem can be circumvented by carefully using the literature. This circumventure focuses attention on the real difficulty of the problem, which is connecting the specific physical situation with the more general mathematical abstraction. This is the connection where most of you will have trouble. You can learn how to use the mathematics involved; you must think harder about connecting them with the actual situation.

Example 3.5-2: Diffusion through an orifice   As a second example, we consider an orifice of radius R in a thin film. Diffusion is occurring through the orifice from a large volume of high concentration c_{10} to a second large volume at zero concentration. Like the case of a thin film, the diffusion is in steady state, so there are no unsteady complications as in the previous example. Unlike the case of a thin film, however, the diffusion is not one-dimensional, but necks down to pass through the orifice, as shown in Figure 3.5-2. Because the film is extremely thin, there is no concentration gradient in the orifice itself.

Calculate the steady state flux through this orifice.

   Solution   We first note that in the z-direction, the concentration changes going into the orifice mirror those going out of the orifice. We thus calculate the flux from one volume at concentration c_{10} to the orifice itself, taken to be a sink at concentration $c_1 = 0$. We then will recognize that the flux we want will just be half that which we have calculated.

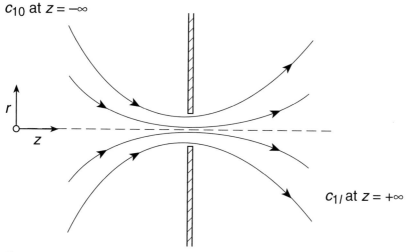

c_{10} at $z = -\infty$

r

z

c_{1l} at $z = +\infty$

Fig. 3.5-2. Diffusion through an orifice. The diffusing solute necks down to pass through a circular hole.

For this problem, diffusion is best described in terms of cylindrical coordinates, given as Eq. B in Table 3.4-2. There is no flow, no change with time, no reaction, and no angular variation, so this solute mass balance becomes

$$0 = \frac{D}{r}\frac{\partial}{\partial r}\left(r\frac{\partial c_1}{\partial r}\right) + \frac{\partial^2 c_1}{\partial z^2}$$

This is subject to the boundary conditions

$$z = 0, \quad r \leqslant R, \qquad c_1 = 0$$
$$r > R, \quad \frac{dc_1}{dz} = 0$$
$$z \geqslant 0, \quad r = \infty, \qquad c_1 = c_{10}$$
$$z = \infty, \quad \text{all } r, \qquad c_1 = c_{10}$$

The concentration profile for this problem is given by (Crank, p. 43)

$$c_1 = c_{10}\left(1 - \frac{2}{\pi}\tan^{-1}\frac{R}{\sqrt{\frac{1}{2}\left[\left(r^2 + z^2 - R^2\right)\right] + \left\{\left(r^2 + z^2 - R^2\right)^2 + 4z^2 R^2\right\}^{1/2}}}\right)$$

The concentration gradient at the orifice itself is also given

$$\frac{\partial c_1}{\partial z}\bigg|_{z=0} = \frac{2c_{10}}{\pi\sqrt{R^2 - r^2}}$$

where $r < R$ is within the orifice itself. To find the total flux J_1, we must integrate over the entire surface of the orifice

$$J_1 = \int_0^{L_1 T} \int_0^R \left(-D \frac{\partial c_1}{\partial z}\Big|_{z=0} \right) r\, dr\, d\theta$$

$$= -4DRc_{10}$$

The flux is negative because it is in the $(-z)$ direction. The flux per orifice area j_1 is

$$j_1 = \frac{J_1}{\pi R^2} = -\frac{4Dc_{10}}{\pi R}$$

As the pore becomes smaller, the flux per area becomes larger, though the total amount becomes smaller. Finally, the flux in and out of an orifice is

$$j_1 = -\frac{2Dc_{10}}{\pi R}$$

The flux into and out of the orifice is just half that into the orifice.

   This example has two characteristics which are worth noting. First, while the problem is carefully solved in the literature, the result given (the concentration profile) is not the result we seek (the flux). This is often the case. The literature will reduce our mathematical burden, but it will not supply exactly what we want.

   The second characteristic of this result is its strong parallel with the diffusion across a thin film and diffusion away from a dissolving sphere. For diffusion across a thin film of thickness l from a solution at c_{10} to a pure solvent with $c_1 = 0$, we found in Equation 2.2-10 that

$$j_1 = \frac{Dc_{10}}{l}$$

For dissolution of a solute sphere of radius R which has a concentration at saturation of $c_1(\text{sat})$, in a solution of pure solvent, we found in Equation 2.4-25 that

$$j_1 = \frac{Dc_{10}}{R}$$

These equations differ only from that derived for the orifice because of the factor $(2/\pi)$, a different characteristic length l or R, and a different direction for diffusion. Thus we infer correctly that all steady-state diffusion problems will give very similar results. We will use this inference to develop theories of mass transfer in more complicated geometries, as detailed in Chapter 9.

Example 3.5-3: Effective diffusion coefficients in a porous catalyst pellet   Imagine that we have a porous catalyst pellet containing a dilute gaseous solution. We want to measure the effective diffusion of solute by dropping this pellet into a small, well-stirred bath of a solvent gas and measuring how fast the solute appears in this bath. How can we plot these measurements to find the effective diffusion coefficient?

   Solution   Again, we begin with a mass balance, combine this with Fick's law and the appropriate boundary conditions, and then adapt the available mathematical hoopla to find the result. The only feature different from before is that we must do so for both the pellet and the bath.

With the pellet, a mass balance on a spherical shell or one taken from Table 3.4-2 yields

$$\frac{\partial c_1}{\partial t} = \frac{D_{\text{eff}}}{r^2} \frac{\partial}{\partial r} r^2 \frac{\partial c_1}{\partial r}$$

This implicitly lumps any tortuous multidimensional diffusion into an "effective" one-dimensional diffusion coefficient D_{eff}. This equation is subject to

$$t = 0, \quad \text{all } r, \qquad c_1 = c_{10}$$
$$t > 0, \quad r = 0, \qquad \frac{\partial c_1}{\partial r} = 0$$
$$r = R_0, \qquad c_1 = C_1(t)$$

where R is the pellet radius and $C_1(t)$ is the bath concentration, a function of time. It is this coupling of the sphere and bath concentrations that makes this problem interesting.

We now make a mass balance on the solute in the bath of volume V_B:

$$V_B \frac{dC_1}{dt} = 4\pi R^2 n_1|_{r=R} = -4\pi R^2 D_{\text{eff}} \frac{\partial c_1}{\partial r}\Big|_{r=R}$$

subject to

$$t = 0, \quad C_1 = 0$$

This mass balance contains no diffusion term because the bath is well mixed.

Problems that are mathematically analogous to this one are discussed by Carslaw and Jaeger (1986) and Crank (1975). The most useful result given is that for the concentration in the bath:

$$C_1 = \frac{c_{10}}{1 + B} - 6Bc_{10} \sum_{n=1}^{\infty} \frac{e^{-D_{\text{eff}} \alpha_n^2 t}}{B^2 R^2 \alpha_n^2 + 9(B + 1)}$$

in which

$$\tan(R_0 \alpha_n) = \frac{3R\alpha_n}{3 + BR^2 \alpha_n^2}$$

and

$$B = \frac{V_B}{(4/3)\pi R^3 \varepsilon}$$

where ε is the void fraction in the sphere.

The results are plotted in Fig. 3.5-3. To find the diffusion coefficient, we first calculate B and $C_1(1 + B)/c_{10}$. We then read $D_{\text{eff}}t/R^2$ from the figure and calculate D_{eff}.

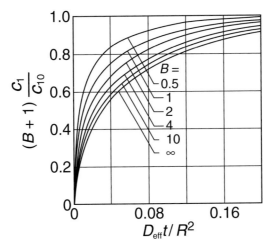

Fig. 3.5-3. Bath concentration versus time. A porous catalyst pellet containing a solute gas is dropped into a stirred bath of solvent gas. The solute concentration in the bath measured versus time provides a value for diffusion in the pellet. A similar graph for heat conduction is given by Carslaw and Jaeger (1986, p. 241).

3.6 Conclusions

Diffusion in concentrated solutions is complicated by the convection caused by the diffusion process. This convection must be handled with a more complete form of Fick's law, often including a reference velocity. The best reference velocity is the volume average, for it is most frequently zero. The results in this chapter are valid for both concentrated and dilute solutions; so they are more complete than the limits of dilute solutions given in Chapter 2.

Nonetheless, those who study diffusion routinely think and work in terms of the dilute-solution limit. You should also. The dilute limit is easier to understand and easier to use for quick, qualitative calculations. It is the basis for finding how diffusion is related to chemical reaction, dispersion, or mass transfer coefficients. You should be aware of the problems that arise in nondilute cases; you should be able to work through them if necessary; but you need not recall their details. Think dilute.

Questions for Discussion

1. How does the total flux n_1 differ from the diffusion flux j_1?
2. When does the volume average velocity v^0 equal the mass average velocity v?
3. When does v^0 equal the molar average velocity v^*?
4. Is there convection in distillation?
5. Suggest a problem where the Fick's law form of diffusion equation is easiest to use.
6. Suggest one where the Maxwell–Stefan form of the diffusion equation is easiest.
7. What is the physical significance of each term in the first equation in Table 3.4-2?

8. Write the equations for steady-state diffusion across a flat membrane, across the wall of a tube, and out of a spherical shell.
9. Show that these equations reduce to the same limit when the membrane, wall, and shell are thin.
10. Heat and mass transfer are often said to be equivalent processes. What heat transfer property corresponds to the diffusion coefficient?
11. What heat transfer property corresponds to chemical reaction?
12. What mass transfer variable corresponds to thermal radiation?
13. We usually expect that doubling the concentration difference in a single phase will double the diffusion flux. When will this not be true?
14. Cooking times in minutes for a single brand of pasta are as follows

Capellini	2
Linguini	11
Fettucini	7
Spaghetti	12
Lasagna	9

Since all are made from the same flour, why are they different?

Problems

1. Dry ice is placed in the bottom of a capillary tube 6.2 cm long. Air is blown across the top of the tube. Calculate the ratio of the total flux to the diffusion flux halfway up the capillary for the following conditions: (a) A temperature of $-124\,°C$, where the vapor pressure is 5 mm Hg. *Answer:* 1.00. (b) A temperature of $-86\,°C$, where the vapor pressure is 400 mm Hg. *Answer:* 1.45.

2. A gas-oil feedstock is irreversibly and very rapidly cracked on a heated metal plate in an experimental reactor. The cracking reduces the molecular weight by an average factor of three. Calculate the rate of this process, assuming that the gas oil diffuses through a thin unstirred film of thickness l near the plate. Note that the reagent must be constantly diffusing against product moving away from the plate. Compare this rate with that for diffusion through a thin film and with evaporation through a stagnant solvent.

3. Imagine a long tube partially filled with liquid benzene at $60\,°C$. Beginning at time zero, the benzene evaporates into the initially pure air with a diffusion coefficient of about $0.104\ cm^2/sec$. How fast does the liquid–vapor interface move with time? *Answer:* $4 \cdot 10^{-4}\ cm/sec$ at 1 second.

4. One interesting membrane reactor uses a homogeneous catalyst that cannot pass through an ultrafiltration membrane. Reagents flow continuously toward the membrane, but the catalyst is injected only at the start of the experiment. It forms the

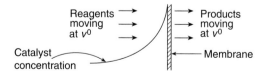

concentration profile shown above. If the catalyst injected per membrane area is M/A, find the concentration profile of the catalyst.

5. The diffusion coefficient relative to the volume average velocity is defined by

$$-j_1 = D\nabla c_1$$

That relative to the molar average velocity is defined by

$$-j_1^* = D*c\nabla x_1$$

Show that the diffusion coefficients in these two definitions are equal, even if the molar concentration c is not constant.

6. Imagine a thick spherical shell of a relatively impermeable polymer. Inside the shell, at radial positions less than R_{in}, there is a drug solution at c_1 (sat), kept constant by the presence of crystals of solid drug. This inside solution is well mixed by the relatively rapid diffusion. Outside the shell, at radial positions greater than R_{out}, the drug concentration is always essentially zero.

    (a) Calculate the drug's flux out of the spherical shell using the Fick's law description of diffusion.

    (b) Calculate this flux using the Maxwell–Stefan description of diffusion.

    (c) Implicitly, we have assumed a binary form of diffusion equation, i.e., we have assumed drug (species "1") diffusing through the polymer shell (species "2"). In fact, water (species "3") will also diffuse through this shell, from the outside into the shell. Thus we are dealing with a ternary solution, to be discussed in detail in Chapter 7. Anticipating this discussion, describe how drug diffusion should be described, both with Fick's law and with the Maxwell–Stefan equations.

7. You want to measure the permeability of an artificial membrane to oxygen. Such membranes are often suggested as a possible means of separating air. To make this measurement, you clamp a section of the membrane in the apparatus shown below. The membrane section is 3 cm in diameter and only 35 mm thick. It is attached to a backing layer that gives it mechanical stability, but it means that only 17.3% of the membrane surface is available for diffusion. To begin an experiment, the gas volume of 68 cm^3 is evacuated to less than 10^{-5} torr (1 torr $\equiv$ 1 mm Hg.) The pressure is then measured and found to be

$$p(\text{mm Hg}) = 88(t - 2.3)$$

where t is the total elapsed time in seconds. Find the Henry's law coefficient and the diffusion coefficient for oxygen in this membrane. *Answer:* $D = 9 \cdot 10^{-7}$ cm^2/sec.

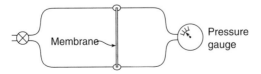

8. Kerkhof and Geboers in their paper "Toward a Unified Theory of Isotropic Molecular Transport Phenomena" (*AIChE Journal* (2005), **51**, 81) give the following equation:

$$\frac{\partial c_1}{\partial t} = -v_z \frac{\partial c_1}{\partial z} + \frac{D}{r}\frac{\partial}{\partial r}\left(r\frac{\partial c_1}{\partial r}\right) + D\frac{\partial^2 c_1}{\partial z^2}$$

Please answer the following: (a) What physical system is implied? (b) What is the differential volume on which this mass balance is written? (c) What is the meaning of each of the four terms?

9. You want to measure the effectiveness of a porous solid desiccant. To do so, you attach a slab of the desiccant $0.5 \times 20 \times 20$ cm to a thin wire. You then attach the wire to an analytical balance and suspend the slab in a chamber at $45\,^{\circ}\mathrm{C}$ and twenty percent

Time(min)	Slab weight(g)
0	166.25
10	167.03
20	167.59
30	168.07
40	168.48
50	168.88
60	169.25

relative humidity. You find that the slab weight varies with time as follows: Find the permeability of water vapor in this desiccant. You will find that the data fit neither a finite slab nor an infinite slab. One good alternative model is to postulate pores in the slab. *Answer:* $3.5\ \mathrm{cm}^2/\mathrm{sec}$.

10. Copper dispersed in porous low-grade ore pellets 0.2 cm in diameter is leached with 4-M H_2SO_4. The copper dissolves quickly, but diffuses slowly out of the pellets. Because the ore is low grade, the porosity can be assumed constant, and the copper concentration will be low in the acid outside of the pellets. Estimate how long it will take to remove eighty percent of the copper if the effective diffusion coefficient of the copper is $2.5 \cdot 10^{-6}\ \mathrm{cm}^2/\mathrm{sec}$. *Answer:* 10 min.

11. A large polymer slab initially containing traces of solvent is exposed to excess fresh air to allow solvent to escape. Find the concentration of solvent in the slab as a function of position and time. Assume that the diffusion coefficient is a constant, but discuss how you might expect it to vary. Try to solve this problem yourself, but compare your answers with those in the literature.

12. One method of studying diffusion in liquids used by Thomas Graham is that shown in Fig. 2.1-2(b). It consists of a small bottle of solution immersed in a large bath of solvent. Calculate the solute concentration in the bath as a function of time, and show how this variation can be used to determine the diffusion coefficient.

13. Wool is dyed by dropping it into a dyebath that contains dye at a concentration C_{10} and that has a volume V. The dye diffuses into the wool, so that its concentration in the dyebath drops with time. You can measure this concentration change. You can also measure the equilibrium uptake of the dye. How can you use measurements of this change at small increments of time to find the diffusion coefficient of the dye in the wool?

14. Sows love to dig truffles, the mushroom that shows up as a condiment in French
    cooking. Apparently, sows do this because they smell in the truffles the sex attractant
    or pheromone 5a-androst-16-en-3-ol, which is secreted by boars and by human males
    [R. Claus, H. O. Hoppen, and H. Karg, *Experimentia*, **37**, 1178 (1981); M. Kirk-Smith,
    D. A. Booth, D. Carroll, and P. Davies, *Res. Comm. Psychol. Psychiat. Behav.*, **3**, 379
    (1978)]. Imagine that the truffle is a point source located a distance *d* below the surface
    of the ground. Calculate the flux of pheromone leaving the ground above.

15. Imagine that two immiscible substances containing a common dilute solute are
    brought into contact. Solute then diffuses from one of these substances into the other.
    Calculate the concentration profiles of the solute in each of the substances, assuming
    that each substance behaves as a semi-infinite slab. (S. Gehrke)

16. Find the steady-state flux away from a rapidly dissolving drop that produces a concen-
    trated solution. Compare your result with that found for a sparingly soluble sphere and
    with the various results for diffusion across a stagnant film.

Further Reading

Arnold, J. H. (1944). *Transactions of the American Institute of Chemical Engineers*, **40**, 361.

Carslaw, H. S. and Jaeger, J. C. (1986). *The Conduction of Heat in Solids*, 2nd ed. Oxford:
    Clarendon Press.

Crank, J. (1975). *The Mathematics of Diffusion*, 2nd ed. Oxford: Clarendon Press.

Levich, V. (1962). *Physicochemical Hydrodynamics*. New York: Prentice-Hall.

Maxwell, J. C. (1860). *Philosophical Magazine*, **19**, 19; **20**, 21.

Maxwell, J. C. (1952). *Scientific Papers Vol. 2*, ed. W. D. Niven, p. 629. New York: Dover.

CHAPTER 4

Dispersion

All thoughtful persons are justifiably concerned with the presence of chemicals in the environment. In some cases, chemicals like pesticides and perfumes are deliberately released; in other cases, chemicals like hydrogen sulfide and carbon dioxide are discharged as the result of manufacturing; in still others, chemicals like styrene and dioxin can be accidentally spilled. In all cases, everyone worries about the long-term effects of such chemical challenges.

Public concern has led to legislation at federal, state, and local levels. This legislation often is phrased in terms of regulation of chemical concentrations. These regulations take different forms. The maximum allowable concentration may be averaged over a day or over a year. The acid concentration (as pH) can be held within a particular range, or the number and size of particles going up a stack can be restricted. Those working with chemicals must be able to anticipate whether or not these chemicals can be adequately dispersed. They must consider the problems involved in locating a chemical plant on the shore of a lake or at the mouth of a river.

The theory for dispersion of these chemicals is introduced in this short chapter. As might be expected, dispersion is related to diffusion. The relation exists on two very different levels. First, dispersion is a form of mixing, and so on a molecular level it involves diffusion of molecules. This molecular dispersion is not understood in detail, but it takes place so rapidly that it is rarely the most important feature of the process. Second, dispersion and diffusion are described with very similar mathematics. This means that analyses developed for diffusion can often correlate results for dispersion.

In Section 4.1, we give a simple example of dispersion to illustrate the similarities to and differences from diffusion. We discuss dispersion coefficients for environmental and industrial situations in Section 4.2. In Section 4.3, we discuss how diffusion and flow interact to produce dispersion in turbulent flow. In Section 4.4, we make similar calculations for laminant flow. Overall, the material is presented at an elementary level, partly because it is unevenly understood at any other level and partly because more detail seems outside the scope of this book.

4.1 Dispersion From a Stack

Everyone has seen smoke pouring from a smokestack. On a cold, clear day, the plume will climb high into the sky, spreading and fading. In a high wind, the plume will be quickly dispersed, almost as if it never existed.

We want to explain these differences in dispersion so that we can anticipate the effects of wind, weather, and different amounts of smoke. To do so, we need to model the dispersion. Such a model should recognize the characteristics of the smoke as it moves downwind. For example, we might find characteristics like those in Fig. 4.1-1 for a plume

95

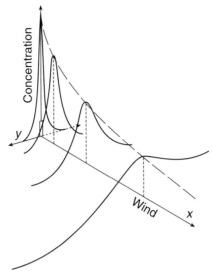

Fig. 4.1-1. Dispersion of smoke. Smoke discharged continuously from a stack has an average concentration that is approximately Gaussian. This shape can be predicted from a diffusion theory. However, the smoke is dispersed much more rapidly than would be expected from diffusion coefficients.

in a 15-km/hr wind. The smoke concentration has roughly a Gaussian shape and has a width of about 1 km when it is 10 km downwind.

Before we begin to model this plume, we should consider what we mean by "smoke concentration." Such a concentration is clearly some arbitrary average over all components, be they present as molecules or as small particles. Such a concentration may affect people in different ways. For example, if the smoke has an odor, doubling the smoke concentration will make the odor less than twice as strong. If the smoke contains poisons, doubling its concentration may more than double its toxicity. We should remember to consider the effects of smoke concentration carefully.

The obvious model for a plume like that in Fig. 4.1-1 is that developed in Section 2.4 for the one-dimensional decay from a pulse. In this model, we assume that x is the wind direction and z is the horizontal direction normal to both the wind and ground. As a first approximation, we assume that the smoke is well mixed in the vertical y direction. On this basis, we can extend the solution given in Section 2.4 to the case of a steady release of smoke S:

$$c_1 = \frac{S}{4\pi D_{app}x}e^{-z^2/4D_{app}t} \tag{4.1-1}$$

where c_1 is an average smoke concentration, with dimensions M/L^3; S is the smoke release rate, M/t; and D_{app} is an apparent diffusion coefficient for the smoke.

This model does a good job of predicting the general shape of the smoke plume. It predicts that the maximum smoke concentration $(S/4\pi D_{app}x)$ does drop as x increases. It does predict that the smoke spreads out in a roughly Gaussian profile, just as is observed. Thus diffusion theory apparently can be applied successfully to the release of pollutants.

However, this model is a disaster at predicting how much the plume spreads. From observation, we know that it actually has spread about 1 kilometer. From the arguments in Section 2.4, we know that the width of this peak l should be about

$$l = \sqrt{4Dt} \tag{4.1-2}$$

In gases, diffusion coefficients are about $0.1 \text{ cm}^2/\text{sec}$, and the time is about 10 km/(15 km/hr), or 40 minutes. On this basis, l should be about 30 centimeters, 3,000 times less than the observed width of 1 kilometer. A factor of 3,000 is a big error, even for engineers.

The explanation for this major discrepancy is the wind. In previous chapters, mixing occurred by diffusion caused by molecular motion. Here, mixing occurs as the wind blows the plume over woods, around hills, and across lakes. This mixing is more rapid than diffusion because of the flow.

We now are in something of a quandary. We have a good diffusion model in Eq. 4.1-1 that explains most of the qualitative features of the plume, but this model grossly underpredicts the effects. To resolve this, we assume that mass transport in the plume is described by the flux equation

$$-j_1 = D\frac{\partial c_1}{\partial z} + E\frac{\partial c_1}{\partial z} \tag{4.1-3}$$

where D is the actual diffusion coefficient; and E is a dispersion coefficient caused by the wind. In the smoke-stack case, the diffusion term in this equation must be small relative to the dispersion term. However, the mass balance will have the same mathematical form as before, subject to the same boundary conditions as before. Thus it will have the same mathematical solution as Equation 4.1-1, but with the new dispersion coefficient E replacing the diffusion coefficient D.

The new dispersion coefficient must usually be measured experimentally. Like the diffusion coefficient, the dispersion coefficient has dimensions of (L^2/t). Unlike the diffusion coefficient, the dispersion coefficient is largely independent of chemistry. It will not be a strong function of molecular weight or chemical structure, but will have close to the same values for carbon monoxide, styrene, and smoke. Unlike the diffusion coefficient, the dispersion coefficient will be a strong function of position. It will have different values in different directions. Thus dispersion may look like diffusion, and it may be described by the same kinds of equations, but it is a different effect.

The foregoing arguments may strike you as silly, a casual invention with a veneer of equations. After all, diffusion is based on a "law." To try to describe dispersion with a diffusion equation seems like cheating.

Nonetheless, this is how dispersion is described. In the rest of this chapter, we explore the details of this description more carefully. These details often lead to less accurate predictions than those possible for diffusion. However, dispersion can be very important, so that even an approximate solution can have considerable practical value.

4.2 Dispersion Coefficients

Dispersion coefficients are very different for turbulent and laminar flow. For turbulent flow, we expect that the dispersion coefficient should be a function of the fluid's velocity v and some characteristic length l. On dimensional grounds, we then expect that the Péclet number for dispersion is

$$\frac{lv}{E} = \text{constant} \tag{4.2-1}$$

This turns out to be approximately true for turbulent dispersion in pipelines of diameter d, for which

$$\frac{dv}{E} = 2 \tag{4.2-2}$$

This result implies that dispersion is not a function of the diffusion coefficient D, which is verified by experiment. However, it also says that E is proportional to v. This is only approximately true; in fact, E increases with v to a power slightly greater than one.

In contrast, the result for laminar flow on pipelines found both from theory and experiment is

$$\frac{E}{dv} = \frac{D}{dv} + \frac{dv}{192D} \tag{4.2-3}$$

This sensibly says that at very low flow, the dispersion coefficient equals the diffusion coefficient. However, at most nonzero flows, the second term on the right-hand side of Equation 4.2-3 is dominant, and E becomes proportional to the flow v. Under these circumstances, E is inversely proportional to D. For laminar flow, a small diffusion coefficient results in large dispersion, and a large diffusion coefficient produces small dispersion. Why this counterintuitive result is true is explained in Section 4.4.

Other geometries combine the results of laminar and turbulent flow. In general, they suggest that E varies linearly with velocity v and becomes independent of D at high flow. For example, dispersion coefficients in packed beds are most often presented as the sum of the contributions of diffusion and flow:

$$E = \beta_1 D + \beta_2 dv \tag{4.2-4}$$

where β_1 and β_2 are constants. While β_1 is sometimes described as the reciprocal of a tortuosity, its common value of around 0.7 is inconsistent with more direct experimental measures of this quantity. The common values of β_2 cluster around 0.5, especially for the dispersion of gases in beds of larger particles. Values of β_2 rise for particles smaller than 0.2 cm, possibly because of polydisperse diameters. Some data for packed beds, presented in dimensionless form, are shown in Figure 4.2-1.

The quantity (dv/D) is the common Péclet number for diffusion, which in analytical chemistry is usually called the "reduced velocity." For fast flows, dispersion in gases and liquids is similar, but at lower flows, dispersion in liquids is larger.

We can use these concepts of dispersion to describe a variety of problems. In this description, we will normally know that our data have the form of diffusion from a pulse, or of diffusion into a semi-infinite slab. We will try to write equations involving

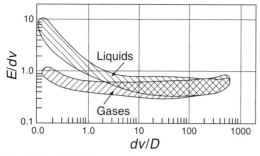

Fig. 4.2-1. Axial dispersion in packed beds. Again, at high flow, the Péeclet number is about constant.

a dispersion coefficient that parallel those which use a diffusion coefficient. When I first tried to solve problems like this, I felt that I was somehow stealing from the analysis of diffusion. I was stealing, but in the same sense Fick stole Fourier's description of heat conduction in order to originally describe diffusion. Such "stealing," which is an attempt at understanding, often supplies new physical insight. Try it yourself by working the following examples, and you can get a better understanding of both diffusion and dispersion.

Example 4.2-1: Cyanide dispersion A metal stamping company has inadvertently spilled cyanide-containing waste into a small creek. Behaving responsibly, they notified the local environmental authorities who arrived promptly to analyze the creek water. These authorities find that the concentration 2 km downstream has a maximum of 860 ppm and a concentration 50 m from the maximum of 410 ppm. The stream is flowing at 0.6 km/hr. (a) What dispersion coefficient is implied by these results? How does it compare with the diffusion coefficient? (b) What will the maximum concentration be 15 km downstream?

Solution We begin our analysis with a mass balance on the differential slice of creek. However, we choose this slice as located near the maximum concentration but moving at the average flow v. Thus

$$
\begin{bmatrix} \text{mass} \\ \text{accumulation} \end{bmatrix} = \begin{bmatrix} \text{mass in} - \text{that out} \\ \text{by diffusion} \\ \text{and dispersion} \end{bmatrix}
$$

$$
\frac{\partial c_1}{\partial t} = -\frac{\partial}{\partial z} j_1
$$

where c_1 is the cyanide concentration and j_1 is the flux relative to the flow. Note that the position z is the actual location minus (vt), i.e., it is the position relative to the moving fluid. Combining with Equation 4.1-2,

$$
\frac{\partial c_1}{\partial t} = E \frac{\partial^2 c_1}{\partial z^2}
$$

We have assumed the dispersion coefficient E is much larger than the diffusion coefficient D. This mass balance is subject to the conditions

$$t = 0, \quad \text{all } z, \qquad c_1 = (M/A)\delta(z)$$

$$z = \infty, \qquad c_1 = 0$$

$$z = 0, \qquad \frac{\partial c_1}{\partial z} = 0$$

The first condition describes the pulse caused by the spill. The second implies that the creek contained a negligible concentration of cyanide before the spill. The third condition says that the concentration is largest when the moving coordinate z is zero. We are now ready to solve the problem.

(a) *Dispersion coefficient.* The mass balance and its constraints are the same as those for diffusion near a pulse, which was discussed in Section 2.4. Because the mathematical description is the same, the solution is also the same but with D replaced by E:

$$c_1 = \left[\frac{M/A}{\sqrt{4\pi Et}} \right] e^{-\frac{z^2}{4Et}}$$

Note that the quantity in square brackets is the maximum concentration; remember that z is the distance along the river away from that maximum. Thus inserting the values given

$$410 \text{ ppm} = 860 \text{ ppm e}^{-\frac{(80\,\text{m})^2}{4E\left(\frac{-2\,\text{km}}{0.6\,\text{km/hr}}\right)\frac{3600\text{sec}}{\text{hr}}}}$$

$$E = 700 \text{ cm}^2/\text{sec}$$

This is much greater than the diffusion coefficient, which is about 10^{-5} cm^2/sec. This large difference underscores the difference between the physical origins of diffusion and dispersion. Diffusion depends on molecular motion, but dispersion depends on velocity fluctuations.

b) *Maximum concentration.* This concentration is easily found from a ratio of concentrations

$$\frac{c_1(\text{max at } t_2)}{c_1(\text{max at } t_1)} = \sqrt{\frac{t_1}{t_2}}$$

$$\frac{c_1(\text{max at 15 km})}{860 \text{ ppm}} = \sqrt{\frac{2 \text{ km}}{15 \text{ km}}}$$

$$c_1(\text{max at 15 km}) = 314 \text{ ppm}$$

The solution to pollution is dilution.

Example 4.2-2: Dispersion in a pipeline   We have a 10-cm pipeline 3 km long for moving reagent gases at 5 m/sec from our wharf to our plant. We want to use this pipeline for different gases, one after the other. How much will the gases mix?

Solution   Imagine that we initially have the pipe filled with one gas and then we suddenly start to pump in a second gas. Because the pipe has a much greater length than diameter, we can expect its contents to be well mixed radially. However, we do expect that there will be significant concentration changes in the axial direction. To describe these, we choose a coordinate system originally located at the initial interface between the gases but moving with the average gas velocity. We then write a mass balance around this moving point

$$\frac{\partial c_1}{\partial t} = E \frac{\partial^2 c_1}{\partial z^2}$$

This mass balance is subject to the conditions

$$t = 0, \quad z > 0, \quad c_1 = c_{1\infty}$$
$$t > 0, \quad z = 0, \quad c_1 = c_{10}$$
$$z = \infty, \quad c_1 = c_{1\infty}$$

in which c_{10} is the average concentration between the gases. The derivation of these relations is a complete parallel to that in Section 2.3. Indeed, the entire problem is mathematically identical with this earlier one, although the diffusion coefficient D used before is now replaced with the dispersion coefficient E. The results are, by analogy,

$$\frac{c_1 - c_{10}}{c_{1\infty} - c_{10}} = \text{erf} \frac{z}{\sqrt{4Et}}$$

The value for E is estimated from Equation 4.2-2

$$E = \frac{1}{2} dv$$

$$E = 0.5(10\,\text{cm})(500\,\text{cm/sec}) = 2,500\,\text{cm}^2/\text{sec}$$

The concentration change is significant when

$$z = \sqrt{4Et}$$

$$= \sqrt{4\left(2500\,\text{cm}^2/\text{sec}\right)[(3\,\text{km})/(500\,\text{cm/sec})](1,000\,\text{m/km})(1\,\text{m}/100\,\text{cm})}$$

$$= 24\,\text{m}$$

About one percent of the pipeline will contain mixed gases.

4.3 Dispersion in Turbulent Flow

We now recognize that dispersion can be described by the mathematics of diffusion but that it requires flow. When such flow exists, dispersion is much faster than diffusion. It has a different physical origin than the small-scale, Brownian motion of molecules. Interestingly, its physical origin is completely different for dispersion in turbulent flow than in laminar flow.

In this section we discuss the origins of dispersion in turbulent flow. This discussion is especially relevant to problems common in environmental engineering, problems like pollutant dilution in rivers or the spreading of plumes. Not surprisingly, the origin of the effect turns out to be a consequence of turbulent fluctuations in velocity and concentration. The coupling between these fluctuations is the cause of dispersion. In more informal terms, gusts and eddies cause dispersion.

To show how turbulence affects dispersion, we return to the mass balances developed in general terms in Section 3.4. For example, for flow described in Cartesian coordinates, we have from Table 3.4-2:

$$\frac{\partial c_1}{\partial t} = D\left(\frac{\partial^2 c_1}{\partial x^2} + \frac{\partial^2 c_1}{\partial y^2} + \frac{\partial^2 c_1}{\partial z^2}\right) - \frac{\partial}{\partial x}c_1 v_x - \frac{\partial}{\partial y}c_1 v_y - \frac{\partial}{\partial z}c_1 v_z - \kappa c_1 c_2 \qquad (4.3\text{-}1)$$

The left-hand side of this equation is the accumulation within a differential volume. The first three terms on the right-hand side describe the amount that enters by diffusion minus the amount that leaves by diffusion. The next three describe the same thing for convection. The last term on the right-hand side is the amount of solute consumed by a second-order chemical reaction, included for reasons that will become evident later. The quantity κ is the chemical rate constant of this reaction.

In turbulent flow, we expect both velocity and concentration to fluctuate. For the smoke plume, the velocity fluctuations are the wind gusts, and the concentration fluctuations can be reflected as sudden changes in odor. To rewrite this equation to include these fluctuations, we define

$$c_1 = \bar{c}_1 + c_1' \qquad (4.3\text{-}2)$$

where c_1' is the fluctuation and $\bar{c}_1$ is the average value:

$$\bar{c}_1 = \frac{1}{\tau}\int_0^\tau c_1 dt \qquad (4.3\text{-}3)$$

Note that the time average of c_1' is zero. By similar definitions,

$$v_x = \bar{v}_x + v_x' \qquad (4.3\text{-}4)$$

where v_x' is the fluctuation, and

$$\bar{v}_x = \frac{1}{\tau}\int_0^\tau v_x dt \qquad (4.3\text{-}5)$$

Again, the average of the fluctuations is zero. Definitions for v_y' and v_z' are similar.

We now insert these definitions into Eq. 4.3-1 and average this equation over the short time interval τ. In some cases, such a substitution is dull:

$$\frac{1}{\tau}\int_0^\tau \left(D\frac{\partial^2 c_1}{\partial x^2}\right) dt = \frac{D}{\tau}\frac{\partial^2}{\partial x^2}\int_0^\tau c_1 dt = D\frac{\partial^2 \bar{c}_1}{\partial x^2}$$

In other cases, it is intriguing:

$$\frac{1}{\tau}\int_0^\tau \kappa c_1 c_2 dt = \frac{\kappa}{\tau}\int_0^\tau (\bar{c}_1 + c_1')(\bar{c}_2 + c_2')dt$$

$$= \frac{\kappa}{\tau}\int_0^\tau (\bar{c}_1\bar{c}_2 + \bar{c}_1 c_2' + c_1'\bar{c}_2 + c_1' c_2')dt$$

$$= \frac{\kappa}{\tau}\left[\bar{c}_1\bar{c}_2\tau + 0 + 0 + \int_0^\tau (c_1' c_2')dt\right]$$

$$= \kappa\left(\bar{c}_1\bar{c}_2 + \overline{c_1' c_2'}\right) \tag{4.3-6}$$

where the new term $\overline{c_1' c_2'}$ represents the time average of the product of the fluctuations. In practice, this new term may be almost as large as the term $\bar{c}_1\bar{c}_2$, but of opposite sign. In a similar fashion,

$$\frac{1}{\tau}\int_0^\tau \frac{\partial}{\partial x}\bar{v}_x c_1 dt = \frac{\partial}{\partial x}\bar{v}_x\bar{c}_1 + \frac{1}{\tau}\frac{\partial}{\partial x}\int_0^\tau v_x' c_1' dt$$

$$= \frac{\partial}{\partial x}\bar{v}_x\bar{c}_1 + \frac{\partial}{\partial x}\overline{v_x' c_1'} \tag{4.3-7}$$

Again, we have the prospect of coupled fluctuations, analogous to the Reynolds stresses that are basic to theories of turbulent flow.

When we combine these averaged terms, we get the following mass balance:

$$\frac{\partial\bar{c}_1}{\partial t} = D\left(\frac{\partial^2\bar{c}_1}{\partial x^2} + \frac{\partial^2\bar{c}_1}{\partial y^2} + \frac{\partial^2\bar{c}_1}{\partial z^2}\right) - \left(\frac{\partial}{\partial x}\bar{v}_x\bar{c}_1 + \frac{\partial}{\partial y}\bar{v}_y\bar{c}_1 + \frac{\partial}{\partial z}\bar{v}_z\bar{c}_1\right)$$

$$- \left(\frac{\partial}{\partial x}\overline{v_x' c_1'} + \frac{\partial}{\partial y}\overline{v_y' c_1'} + \frac{\partial}{\partial z}\overline{v_z' c_1'}\right) - \kappa\bar{c}_1\bar{c}_2 - \kappa\overline{c_1' c_2'} \tag{4.3-8}$$

Most of the terms are like those in Eq. 4.3-1, and they have the same physical significance. The underlined terms are new. The last one deals with changes in reaction rate effected by the fluctuations. The other three describe the mixing caused by turbulent flow, that is, by the dispersion. They are the focus of this section.

We next remember the origin of the diffusion terms, that

$$D\frac{\partial^2\bar{c}_1}{\partial x^2} = -\frac{\partial}{\partial x}\bar{j}_{1x} \tag{4.3-9}$$

or, more basically,

$$\bar{j}_{1x} = -D\frac{\partial\bar{c}_1}{\partial x} \tag{4.3-10}$$

By analogy, because the flux $\overline{v'_x c'_1}$ has a physical meaning similar to the diffusion flux $\bar{j}_1$, we may define

$$\overline{v'_x c'_1} = -E_x \frac{\partial \bar{c}_1}{\partial x} \tag{4.3-11}$$

Definitions for other directions are made in similar ways. This always seems intellectually arrogant to me because I know that we define E_x, E_y, and E_z so that we will get results that are mathematically parallel to diffusion. It seems a rationalization, jerry-built on top of diffusion theory. It is all these things. It also is the best first approximation of turbulent dispersion, the basis from which other theories proceed.

4.4 Dispersion in Laminar Flow: Taylor Dispersion

In earlier sections of this chapter we saw how components in smoke plumes and pipelines sometimes spread much more rapidly than expected. The concentrations of these component pulses could be described by diffusion equations but by using new dispersion coefficients. In turbulent flow, these dispersion coefficients were the result of coupled fluctuations of concentration and velocity.

Dispersion can also occur in laminar flow but for completely different reasons. This is not surprising because laminar flow has no sudden concentration or velocity fluctuations. In this section we discuss one example of dispersion in laminar flow. This leads to an accurate prediction of the dispersion coefficient. This particular example is so instructive that it is worth including in detail.

The specific example concerns the fate of a sharp pulse of solute injected into a long, thin tube filled with solvent flowing in laminar flow (Fig. 4.4-1). As the solute pulse moves through the tube, it is dispersed. We want to calculate the concentration profile resulting from this dispersion.

Because the complete analysis of this problem is complicated, we first give the results and then the derivation. The concentration of the pulse averaged across the tube's cross-section will be shown to be

$$\bar{c}_1 = \frac{M/\pi R^2}{\sqrt{4\pi Et}} e^{-(z-vt)^2/4Et} \tag{4.4-1}$$

in which M is the total solute in the pulse, R is the tube's radius, z is the distance along the tube, v is the fluid's velocity, and t is the time. This equation is a close parallel to Eq. 2.4-14, except that the diffusion coefficient D is replaced by the dispersion coefficient E. This can be shown explicitly to be

$$E = \frac{(Rv)^2}{48D} \tag{4.4-2}$$

Note that E depends *inversely* on the diffusion coefficient.

This fascinating result indicates that rapid diffusion leads to small dispersion and that slow diffusion produces large dispersion (Fig. 4.4-1). The reasons why this occurs are sketched in Fig. 4.4-2. The initial pulse is sharp, like that shown in (a). The laminar flow

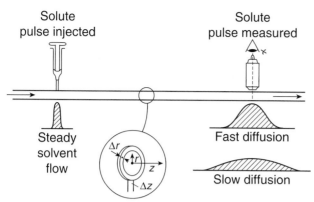

Fig. 4.4-1. Taylor dispersion. In this case, solvent is passing in steady laminar flow through a long, thin tube. A pulse of solute is injected near the tube's entrance. This pulse is dispersed by the solvent flow, as shown.

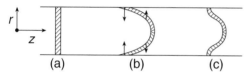

Fig. 4.4-2. Causes of Taylor dispersion. In Taylor dispersion, fast diffusion unexpectedly produces little dispersion, and vice versa. The reasons for this are shown here. The initial solute pulse (a) is deformed by flow (b). In fast-flowing regions, diffusion occurs outward, and in the slow flow near the wall, diffusion occurs inward. Thus diffusion in the radial direction inhibits dispersion caused by axial flow (c).

quickly distorts the pulse, as in (b). If there is no diffusion, the distortion continues unabated, and the pulse is widely dispersed. If, instead, there is rapid diffusion, material in the center of the tube tends to diffuse outward, into a region of solvent that is moving more slowly. Simultaneously, material that is left behind near the tube walls tends to diffuse toward the center, into a region of faster flow. This radial diffusion thus inhibits the dispersion induced by axial convection.

4.4.1 Analyzing Taylor Dispersion

To apply these ideas quantitatively, we again write a mass balance, add Fick's law, and manipulate the result mathematically. In this instance, I am reminded of a cartoon by Thomas Nast, showing a virtuous soul laden with debt and responsibility, staggering along a tortuous path. To the left of the path, the ground drops away into ignorance; to the right, the ground disappears into chaos. In going through this next analysis, you may feel like that poor soul, treading a very narrow path.

We begin this analysis with three assumptions:

(1) The solutions are dilute. This is assumed true even for the initial pulse.
(2) The laminar flow is unchanged by the pulse. This means that the velocity varies only with radius.

(3)   Mass transport is by radial diffusion and axial convection. Other transport
mechanisms are negligible.

The most important assumption is the last one, for it separates diffusion and convection.
It is accurate if

$$7.2 \left(\frac{LD}{R^2 v} \right) \gg 1 \tag{4.4-3}$$

where L is the tube length. This condition is valid for long, thin tubes.

We now make a mass balance on the washer-shaped element shown in the inset in Fig.
4.4-1 to find

$$\frac{\partial c_1}{\partial t} = -\frac{1}{r}\frac{\partial}{\partial r}(rj_1) - \frac{\partial}{\partial z}(c_1 v_z) \tag{4.4-4}$$

The velocity v_z is the laminar result and so is independent of z:

$$v_z = 2v \left[1 - \left(\frac{r}{R} \right)^2 \right] \tag{4.4-5}$$

When Eqs. 4.4-4 and 4.4-5 are combined with Fick's law,

$$\frac{\partial c_1}{\partial t} = \frac{D}{r}\frac{\partial}{\partial r} r \frac{\partial c_1}{\partial r} - 2v \left[1 - \left(\frac{r}{R} \right)^2 \right] \frac{\partial c_1}{\partial z} \tag{4.4-6}$$

This is subject to the conditions

$$t = 0, \quad \text{all } z, \quad c_1 = \left(\frac{M}{\pi R^2} \right) \delta(z) \tag{4.4-7}$$

$$t > 0, \quad r = R, \quad \partial c_1 / \partial r = 0 \tag{4.4-8}$$

$$r = 0, \quad \partial c_1 / \partial r = 0 \tag{4.4-9}$$

The initial condition is like that for the decay of a pulse.

We next define the new coordinates

$$\eta = \frac{r}{R} \tag{4.4-10}$$

$$\zeta = (z - vt)/R_0 \tag{4.4-11}$$

In terms of these quantities, Eq. 4.4-6 becomes

$$\frac{D}{\eta}\frac{\partial}{\partial \eta} \left(\eta \frac{\partial c_1}{\partial \eta} \right) = 2vR \left(\frac{1}{2} - \eta^2 \right) \frac{\partial c_1}{\partial \zeta} \tag{4.4-12}$$

One solution to Eq. 4.4-12 that satisfies Eq. 4.4-8 is

$$c_1 = c_{1|\eta=0} + \frac{1}{4} \left[\frac{vR}{D} \left(\frac{\partial c_1}{\partial \zeta} \right) \Big|_{\eta=0} \right] \left(\eta^2 - \frac{1}{2}\eta^4 \right) \tag{4.4-13}$$

However, we want not the local concentration but the average across the tube:

$$\bar{c}_1(z) = \frac{1}{\pi R^2} \int_0^R 2\pi r c_1(r, z) dr$$

$$= 2 \int_0^1 \eta c_1 d\eta \tag{4.4-14}$$

Because of the pulse, the radial variations of concentration are small relative to the axial ones, so

$$\frac{\partial c_1}{\partial \zeta} \doteq \frac{\partial \bar{c}_1}{\partial \zeta} \tag{4.4-15}$$

We now can write a new overall mass balance in terms of this average concentration:

$$\frac{\partial \bar{c}_1}{\partial t} = -\frac{\partial J_1}{\partial (\zeta R)} \tag{4.4-16}$$

in which J_1 is the averaged flux in the direction of flow

$$J_1 = \frac{1}{\pi R^2} \int_0^R 2\pi r (v_z - v) \left(c_1 - c_1|_{\eta=0} \right) dr \tag{4.4-17}$$

Equation 4.4-16 can be written as

$$\frac{\partial \bar{c}_1}{\partial (tv/R)} = \frac{\partial \bar{c}_1}{\partial \tau} = -\frac{\partial (J_1/v)}{\partial \zeta}$$

$$= -\frac{\partial}{\partial \zeta} \left[4 \int_0^1 \eta \left(\frac{1}{2} - \eta^2 \right) c_1 d\eta \right] \tag{4.4-18}$$

Combining this result with Eqs. 4.4-13 and 4.4-14, we find, after some work, that

$$\frac{\partial \bar{c}_1}{\partial \tau} = \left(\frac{vR}{48D} \right) \frac{\partial^2 \bar{c}_1}{\partial \zeta^2} \tag{4.4-19}$$

The quantity in parentheses is a Péclet number, giving the relative importance of axial convection and radial diffusion. The conditions are now

$$\tau = 0, \quad \text{all } \zeta, \quad \bar{c}_1 = \frac{M}{\pi R^3} \delta(\zeta) \tag{4.4-20}$$

$$\tau > 0, \quad \zeta = \infty, \quad \bar{c}_1 = 0 \tag{4.4-21}$$

$$\zeta = 0, \quad \frac{\partial \bar{c}_1}{\partial \zeta} = 0 \tag{4.4-22}$$

Equations 4.4-19 to 4.4-22 for Taylor dispersion have exactly the same mathematical form as those for the decay of a pulse in Section 2.4. As a result, they must have the same solution. This solution is that given in Eq. 4.4-1.

4.4.2 Chromatography

Taylor dispersion has an important extension in chromatography. Chromatography is a separation method often used for chemical analysis of complex mixtures. In this analysis, a pulse of mixed solutes is injected into one end of a packed bed of absorbent (the stationary phase) and washed through the bed with solvent (the mobile phase). Because the solutes are absorbed to different degrees, they are washed out of the bed (eluted) at different times.

The analysis of chromatography is usually empirical, a consequence of the normally complex geometry of the absorbent. One special case where analysis is more exact involves a solute pulse injected into fluid in laminar flow in a cylindrical tube, just like the solute pulse shown in Fig. 4.4-1. Now, however, the walls of the tube are coated with a thin film of absorbent. The injected solute is retarded by absorption in that thin layer.

Our goal is to determine the shape of the pulse eluted from this absorbent-coated tube. To do so, we first recognize that the tube's contents are subject to the mass balance

$$\frac{\partial c_1}{\partial t} = D \left[\frac{1}{r} \frac{\partial}{\partial r} r \frac{\partial c_1}{\partial r} + \frac{\partial^2 c_1}{\partial z^2} \right] - 2v \left[1 - \left(\frac{r}{R} \right)^2 \right] \frac{\partial c_1}{\partial z} \tag{4.4-23}$$

This mass balance is like that in Eq. 4.4-6 except that we have not neglected axial diffusion. It is subject to the conditions:

$$t = 0, \quad \text{all } z, \qquad c_1 = \frac{M}{\pi R^2} \delta(z) \tag{4.4-24}$$

$$t > 0, \quad r = 0, \quad \partial c_1/\partial r = 0 \tag{4.4-25}$$

$$r = R_0, \qquad c_1' = H c_1 \tag{4.4-26}$$

$$D' \frac{\partial c_1'}{\partial r} = D \frac{\partial c_1}{\partial r} \tag{4.4-27}$$

where c_1' and D' are the concentration and the diffusion coefficient of the solute in the absorbent layer and H is an equilibrium constant between the tube's contents and the absorbent. Eqs. 4.4-24 and 4.4-25 are the same as Eqs. 4.4-7 and 4.4-9, respectively, but Eqs. 4.4-26 and 4.4-27 are new, a reflection of the interaction between the tube's contents and the absorbent. Because of this interaction, we need a mass balance on the absorbent as well:

$$\frac{\partial c_1'}{\partial t} = \frac{D'}{r} \frac{\partial}{\partial r} r \frac{\partial c_1'}{\partial r} \tag{4.4-28}$$

Note that we are neglecting both convection and axial diffusion in the stationary absorbent. This mass balance in the absorbent is subject to the following conditions:

$$t = 0, \quad \text{all } r, \qquad c_1' = 0 \qquad (4.4\text{-}29)$$

$$t > 0, \quad r = R+\delta, \quad \frac{\partial c_1'}{\partial z} = 0 \qquad (4.4\text{-}30)$$

where δ is the thickness of the absorbent layer.

The most useful solution to this extended form of Taylor dispersion is the limit where the absorbed layer is thin. This limit, called the Golay equation, is

$$\bar{c}_1 = \frac{M/\pi R^2}{\sqrt{4\pi E t_0}} e^{-\frac{(z - v t_0)^2}{4 E t_0}} \qquad (4.4\text{-}31)$$

where

$$t_0 = \frac{L}{v}(1 + k') \qquad (4.4\text{-}32)$$

$$k' = H\delta/R \qquad (4.4\text{-}33)$$

and where the dispersion coefficient E is now more complex

$$E = D(1+k') + \frac{R^2(v)^2}{48D}\left(\frac{1 + 6k' + 11(k')^2}{1 + k'}\right) + \left(\frac{\delta^2(v)^2}{3D'}\right)\left(\frac{k'}{1 + k'}\right) \qquad (4.4\text{-}34)$$

The physical significance of the terms in Eqs. 4.4-32 and 4.4-33 is straightforward: The retention time t_0 is the average residence time of the solute, and the capacity factor k' is the equilibrium ratio of solute held in the absorbent to that inside the tube itself.

However, the physical significance of the dispersion coefficient E given by Eq. 4.4-34 is by far the most interesting. The first term on the right-hand side of this equation represents the dispersion caused by axial diffusion. Note that a small diffusion coefficient contributes little to axial dispersion, and a large diffusion coefficient contributes more. While this effect is neglected in the analysis leading to Eq. 4.4-2, it can be significant in chromatography and so is included here.

The second term on the right-hand side of Eq. 4.4-34 is due to Taylor dispersion, i.e., to coupled radial diffusion and axial convection. The dispersion from this source, which is usually much larger than that caused by axial diffusion, is inversely proportional to the diffusion coefficient. Thus a small diffusion coefficient contributes a lot to dispersion and a large diffusion coefficient contributes less. This source of dispersion also depends on the square of the tube's radius, so making the tube 10 times smaller can reduce dispersion 100 times. This is why chromatography often uses absorbents with small channels.

The third term on the right-hand side of Eq. 4.4-34 represents dispersion caused by retardation in the absorbent layer. If diffusion in the absorbent is very fast, the absorbent won't affect the dispersion much; if the layer is very thin ($\delta/R \ll 1$), the absorbent won't have much effect on dispersion either. Remember that the absorbent may not directly affect dispersion but will still indirectly dominate the separation if k' is much greater than one.

We now can see from Eqs. 4.4-31 and 4.4-34 how chromatography can promise a good separation and how this good separation can be compromised by dispersion. Remember that in a chromatographic separation, a pulse of mixed solutes is injected at one end of a packed column and then eluted out the other end by the mobile solvent phase. These injected solutes will be eluted at different retention times t_0 when their absorption is different. The amount by which the retention times differ is largely controlled by the difference in the capacity factors k'.

At the same time, the separation of these solutes can be compromised by dispersion. If the dispersion coefficient E were near zero, then each solute would be eluted as a sharp pulse. Because the dispersion coefficient is not zero, the solutes are eluted as broader pulses. When these pulses overlap, our separation is compromised.

We can see how to reduce dispersion and aid our separation by considering the various terms in Eq. 4.4-34. As a general rule, we can't change the diffusion coefficients much; we're stuck with the physical properties of our solute and our absorbents. We can use low velocities, which reduce Taylor dispersion and absorbent-caused dispersion. We can use small channels – small values of R – though this often means large pressure drops. We must recognize that even as v and R become very small, we will always have dispersion from axial diffusion.

The cases of laminar flow in a straight tube are exceptions because we can calculate the dispersion coefficient exactly. In some ways, they are like the friction factor for laminar flow in a pipe, which also can be calculated explicitly. In general, we should not expect such exact results, just as we do not expect to calculate a priori the friction factors for laminar flow in packed beds or for turbulent flow in a pipe. We usually will be forced to treat dispersion empirically.

4.5 Conclusions

This chapter discusses dispersion, an important effect caused by the coupling of concentration differences and fluid flow. Dispersion frequently can be described by the same mathematics used so effectively for diffusion; in this sense, this chapter represents special cases of diffusion theory.

If you use the materials in this chapter, you should always remember that diffusion and dispersion have very different physical origins and proceed at very different speeds. Remembering this difference is especially important because some refer to both processes as "diffusion." Physicians speak of diffusion of drugs in the bloodstream, and environmental engineers discuss diffusion of pollutants. Some of these processes may include the narrower definition of molecular diffusion used in this book, but the process dynamics cannot be predicted from diffusion theory alone. Be careful.

Questions for Discussion

1. What are the dimensions of the dispersion coefficient?
2. What is the difference between a diffusion coefficient and a dispersion coefficient?
3. How will the maximum concentration in a stream vary with the distance traveled?

4. Why is the dispersion coefficient in turbulent flow independent of the diffusion coefficient?

5. Why does the dispersion coefficient in laminar flow depend inversely on the diffusion coefficient?

6. How does the dispersion coefficient vary with viscosity?

7. In a packed bed, radial dispersion is usually much larger than axial dispersion. Why?

8. At what velocity will dispersion in turbulent flow be smallest?

9. At what velocity will dispersion in laminar flow be smallest?

10. When in chromatography a solute is absorbed, how will its elution time change?

11. When in chromatography a solute is absorbed, how will its dispersion change?

12. What is the physical meaning of the capacity factor k', defined by Equation 4.4-33?

13. What are the limits of the Golay equation (Equation 4.4-31) when the absorption is strong?

Problems

1. A dyeing plant is continuously discharging an aqueous waste saturated with xylene into a river flowing at 0.16 m/sec. About 200 m downstream the maximum xylene concentration is 130 ppm. Estimate the maximum value 2 km downstream.

2. You are pumping 1.7 kg/sec of a cold stream of monomer through 72 m of 2.5-cm-diameter pipe to a reactor. At the entrance of the pipe, you inject 30 pulses per second of catalyst with a small piston pump. When this stream reaches the reactor, the total stream is quickly heated, and polymerization begins. The cold stream has a specific gravity of 0.83 and a viscosity of 3.7 centipoises. How well will the catalyst be mixed by flow through the pipe?

3. You are studying dispersion in a small air-lift fermentor. This fermentor is 1.6 m tall, with a 10-cm diameter. Air and pure water are fed into the bottom at superficial velocities of 11 and 0.78 cm/sec; under these conditions the gas bubbles occupy 45% of the column volume. You continuously add 15 cm^3/min of 1-M NaCl solution near the top of the column. You find by conductance that the salt concentration halfway down the column is $2.32 \cdot 10^{-3}$ M. What is the dispersion coefficient? *Answer:* 54 cm^2/sec.

4. The best marathon in Minnesota is run by Grandma's, a reformed brothel in Duluth. In the 1981 race, 3,202 persons finished. One-quarter of the runners finished within 3 hr 6 min and half within 3 hr 26 min. If I ran the race in 2 hr 54 min 42 sec, what place did I come in? *Answer:* 460 by experiment.

5. A handful of pheromone-impregnated pellets are being used to give an overall release of 1.3 mol/hr into a 15-km/hr wind blowing in the z direction. In this case, the pheromone concentration is given by

$$c_1 = \frac{S}{2\pi Ez} e^{-vr^2/4Ez}$$

where r is the width of release. Gypsy moths respond to this release over an area 25 km long, with a maximum width of 8 km. What is the dispersion coefficient E of the pheromone?

6. Harvest ants inform each other of danger by releasing a pulse of pheromone. The dispersion of this pheromone can be modeled using results like those in this chapter. For harvest ants, the maximum distance over which this chemical alarm is effective is 6 cm; this occurs at a time of 32 sec. The alarm is no longer effective after 35 sec [E. O. Wilson, *Psyche*, **65**, 41 (1958)]. Assume that the pheromone is dispersed in a hemispherical volume, so its concentration is

$$c_1 = \frac{2M}{(4\pi Et)^{3/2}} e^{-r^2/4Et}$$

Also assume that neighboring ants respond only when c_1 exceeds c_{10}. Then show that

$$R = \left[6Dt \ln \left(\frac{t_{\text{final}}}{t} \right) \right]^{1/2}$$

where R is the radius of communication at time t, and t_{final} is the time when the signal is ignored. Discuss how R varies with t. Estimate the dispersion coefficient E from the values given, and compare it with your guess of a diffusion coefficient. *Answer: $E = 2$ cm^2/sec.*

7. In 1905, five muskrats escaped in Bohemia. These animals quickly spread over Europe as shown below [J. G. Skellam, *Biometrika*, **38**, 196 (1951)]:

Year	Area inhabited
1905	0
1909	50
1911	120
1915	300
1920	670
1927	1,720

Show that these results are consistent with a two-dimensional dispersion model

$$c_1 = \frac{M_0}{4\pi ET} e^{\alpha t - r^2/4Et}$$

where

$$\left(\begin{array}{c} \text{growth rate} \\ \text{of } M \end{array} \right) = \alpha M$$

E is the dispersion coefficient, and M_0 is the original number of animals.

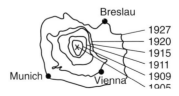

8. To measure backmixing in a tubular packed-bed chemical reactor, you inject a pulse of carbon dioxide into nitrogen flowing through the reactor and measure the carbon dioxide concentration in the effluent. The effluent concentration fits the equation

$$\frac{c}{c_0} = e^{-\frac{(t - 16 \text{ min})^2}{32 \text{ min}}}$$

The reactor is 3.3 meters long. What is the dispersion coefficient?

9. Glacial moraines can be dated by their shape, described by the continuity equation:

$$\frac{\partial z}{\partial t} = -\frac{\partial q}{\partial x}$$

where z is the vertical height of the moraine, x is the horizontal axis, and q is the soil flux [B. Hallet and J. Putkonen, *Science*, **265**, 937 (1994)]. The soil flux is in turn given by

$$q = -E \frac{\partial z}{\partial x}$$

where E is the dispersion coefficient. For slopes a few meters tall, E is typically between 10^{-4} and 10^{-2} m²/yr. For taller slopes, however, E varies with distance:

$$E = A + Bx$$

where A and B are 10^{-2} m²/yr and 10^{-4} m/yr, respectively. Use these values to estimate how the shape of a moraine, which is originally a step 100 m high, changes over time.

Further Reading

Aris, R. (1956). *Proceedings of the Royal Society of London, Series A*, **235**, 67.

Crank, J. (1975). *The Mathematics of Diffusion*. Oxford: Clarendon Press.

Golay, M. J. E. (1958). *Theory of Gas Chromatography*. In *Gas Chromatography*, ed. C.H. Desty. London: Butterworth.

Guichon, G., F. Attila, D.G. Shirazi, and A. M. Katti (2006). *Fundamentals of Preparative and Non Linear Chromatography*. Boston: Academic Press.

Liu, B., and G. Guiochon (2003). *Modeling for Preparative Chromatography*. Amsterdam: Elsevier.

Taylor, G. I. (1953). *Proceedings of the Royal Society of London, Series A*, **219**, 186.

Taylor, G. I. (1954). *Proceedings of the Royal Society of London, Series A*, **223**, 446.

PART II

Diffusion Coefficients

CHAPTER 5

Values of Diffusion Coefficients

Until now, we have treated the diffusion coefficient as a proportionality constant, the unknown parameter appearing in Fick's law. We have found mass fluxes and concentration profiles in a broad spectrum of situations using this law. Our answers have always contained the diffusion coefficient as an adjustable parameter.

Now we want to calculate values of the flux and the concentration profile. For this, we need to know the diffusion coefficients in these particular situations. We must depend largely on experimental measurements of these coefficients, because no universal theory permits their accurate a-priori calculation. Unfortunately, the experimental measurements are unusually difficult to make, and the quality of the results is variable. Accordingly, we must be able to evaluate how good these measurements are.

Before we begin, we should list the guidelines that tend to stick in everyone's mind. Diffusion coefficients in gases, which can be estimated theoretically, are about 0.1 cm^2/sec. Diffusion coefficients in liquids, which cannot be as reliably estimated, cluster around 10^{-5} cm^2/sec. Diffusion coefficients in solids are slower still, 10^{-30} cm^2/sec, and they vary strongly with temperature. Diffusion coefficients in polymers and glasses lie between liquid and solid values, say about 10^{-8} cm^2/sec, and these values can be strong functions of solute concentration.

The accuracy and origins of these guidelines are explored in this chapter. Gases, liquids, solids, and polymers are discussed in Sections 5.1 through 5.4, respectively. In these sections we give a selection of typical values, as well as one common method of estimating these values. After we sketch the sources of these estimations, we explore other concerns, like the pressure dependence of diffusion in gases or the concentration variations of diffusion in liquids. Section 5.5 summarizes Brownian motion, showing how random walks are related to diffusion. Section 5.6 discusses the common experimental methods of measuring diffusion coefficients.

5.1 Diffusion Coefficients in Gases

Diffusion coefficients in gases are illustrated by the values in Table 5.1-1. At one atmosphere and near room temperature, these values lie between 0.1 and 1 cm^2/sec. Indeed, given the variation of the chemistry, the values vary remarkably little. To a first approximation, the coefficients are inversely proportional to pressure, so doubling the pressure cuts the diffusion coefficient in half. They vary with the 1.5 to 1.8 power of the temperature, so an increase of 300 K triples the coefficients. They vary in a more complicated fashion with factors like molecular weight.

The physical significance of diffusion coefficients of this size is best illustrated by remembering unsteady-state diffusion problems like the semi-infinite slab discussed in Chapter 2. In these problems, the key experimental variable is $z^2/4Dt$. When this variable

117

Table 5.1-1 *Experimental values of diffusion coefficients in gases at one atmosphere*

Gas pair	Temperature (K)	Diffusion coefficient (cm^2/sec)
Air–benzene	298.2	0.096
Air–CH_4	282.0	0.196
Air–C_2H_5OH	273.0	0.102
Air–CO_2	282.0	0.148
Air–H_2	282.0	0.710
Air–H_2O	289.1	0.282
	298.2	0.260
	312.6	0.277
	333.2	0.305
Air–He	282.0	0.658
Air–*n*-hexane	294.0	0.080
Air–toluene	299.1	0.086
Air–aniline	299.1	0.074
Air–2-propanol	299.1	0.099
CH_4–He	298.0	0.675
CH_4–H_2	298.0	0.726
CH_4–H_2O	307.7	0.292
CO–N_2	295.8	0.212
^{12}CO–^{14}CO	373.0	0.323
CO–H_2	295.6	0.743
CO–He	295.6	0.702
CO_2–H_2	298.2	0.646
CO_2–N_2	298.2	0.165
CO_2–O_2	296.0	0.156
CO_2–He	298.4	0.597
CO_2–CO	315.4	0.185
CO_2–H_2O	307.4	0.202
CO_2–SO_2	263.0	0.064
$^{12}CO_2$–$^{14}CO_2$	312.8	0.125
CO_2–propane	298.1	0.087
H_2–N_2	297.2	0.779
H_2–O_2	316.0	0.891
H_2–He	317.0	1.706
H_2–Ar	317.0	0.902
H_2–Xe	341.2	0.751
H_2–SO_2	285.5	0.525
H_2–H_2O	307.1	0.915
H_2–NH_3	298.0	0.783
H_2–ethane	298.0	0.537
H_2–*n*-hexane	288.7	0.290
H_2–cyclohexane	288.6	0.319
H_2–benzene	311.3	0.404
N_2–O_2	316.0	0.230
	293.2	0.220
N_2–He	317.0	0.794
N_2–Ar	316.0	0.216
N_2–NH_3	298.0	0.230
N_2–H_2O	298.2	0.293
N_2–SO_2	263.0	0.104
N_2–ethane	298.0	0.148

(*Continued*)

Table 5.1-1 (*Continued*)

Gas pair	Temperature (K)	Diffusion coefficient (cm²/sec)
N_2–*n*-butane	298.0	0.096
N_2–isobutane	298.0	0.090
N_2–*n*-hexane	288.6	0.076
N_2–*n*-octane	303.1	0.073
N_2–2,2,4-trimethylpentane	303.3	0.071
N_2–benzene	311.3	0.102
O_2–He (He trace)	298.2	0.737
(O_2–trace)	298.2	0.718
O_2–He	317.0	0.822
O_2–H_2O	308.1	0.282
O_2–CCl_4	296.0	0.075
O_2–benzene	311.3	0.101
O_2–*n*-hexane	288.6	0.075
O_2–*n*-octane	303.1	0.071
O_2–2,2,4-trimethylpentane	303.0	0.071
He–Ar	298.0	0.742
He–H_2O	298.2	0.908
He–NH_3	297.1	0.842
Ar–Ne	303.0	0.327
Ar–Kr	303.0	0.140
Ar–Xe	329.9	0.137
Ne–Kr	273.0	0.223
Ethylene–H_2O	307.8	0.204

Source: Data from Hirschfelder *et al.* (1954), Marrero and Mason (1972), and Poling *et al.* (2001).

equals unity, the diffusion process has proceeded significantly. In other words, where z^2 equals $4Dt$, the diffusion has penetrated a distance z in the time t.

In gases, this penetration distance is much larger than in other phases. For example, the diffusion coefficient of water vapor diffusing in air is about 0.3 cm²/sec. In 1 second, the diffusion will penetrate 0.5 cm; in 1 minute, 4 cm; and in 1 hour, 30 cm.

5.1.1 Gaseous Diffusion Coefficients From the Chapman–Enskog Theory

The most common method for theoretical estimation of gaseous diffusion is that developed independently by Chapman and by Enskog (Chapman and Cowling, 1970). This theory, accurate to an average of about eight percent, leads to the equation

$$D = \frac{1.86 \cdot 10^{-3} T^{3/2} (1/\tilde{M}_1 + 1/\tilde{M}_2)^{1/2}}{p\sigma_{12}^2 \Omega} \tag{5.1-1}$$

in which D is the diffusion coefficient measured in cm²/sec, T is the absolute temperature in Kelvin, p is the pressure in atmospheres, and the $\tilde{M}_i$ are the molecular weights.

The quantities σ_{12} and Ω are molecular properties characteristic of the detailed theory. The collision diameter σ_{12}, given in angstroms, is the arithmetic average of the two species present:

$$\sigma_{12} = \frac{1}{2}(\sigma_1 + \sigma_2) \tag{5.1-2}$$

Values of σ_1 and σ_2 are listed in Table 5.1-2. The dimensionless quantity Ω is more complex, but usually of order one. Its detailed calculation depends on an integration of the interaction between the two species. This interaction is most frequently described by the Lennard–Jones 12-6 potential. The resulting integral varies with the temperature and the energy of interaction. This energy ε_{12} is a geometric average of contributions from the two species:

$$\varepsilon_{12} = \sqrt{\varepsilon_1 \varepsilon_2} \tag{5.1-3}$$

Values of the ε_{12}/k_B are also given in Table 5.1-2. Once ε_{12} is known, Ω can be found as a function of $k_B T/\varepsilon_{12}$ using the values in Table 5.1-3. The calculation of the diffusion coefficients now becomes straightforward if the σ_i and the ε_i are known.

5.1.2 The Nature of Kinetic Theories

The results of the Chapman–Enskog theory are based on detailed analyses of molecular motion in dilute gases. These analyses depend on the assumption that molecular interactions involve collisions between only two molecules at a time (Fig. 5.1-1). Such interactions are much simpler than the lattice interactions in solids or the less regular and still more complex interactions in liquids.

The nature of theories of this type is best illustrated for a gas of rigid spheres of very small molecular dimensions (Cunningham and Williams, 1980). For such a theory, the diffusion flux has the following form:

$$n_1 = -\frac{1}{3}\bar{v}l\frac{dc_1}{dz} + c_1 v^0 \tag{5.1-4}$$

The second term on the right represents convection and the first indicates diffusion. The diffusion term has three parts: $\bar{v}$, the average molecular velocity; l, the mean free path of the molecules; and dc_1/dz, the concentration gradient. This term makes physical sense: the flux will certainly increase if either the velocity of the molecules or the average distance they travel increases.

If we compare Eq. 5.1-4 with Fick's law, we find

$$D = \frac{1}{3}\bar{v}l \tag{5.1-5}$$

Both the average velocity $\bar{v}$ and the mean free path l of the rigid spheres can be calculated. The average velocity is

$$\bar{v} = \sqrt{2k_B T/m} \tag{5.1-6}$$

Table 5.1-2 *Lennard–Jones potential parameters found from viscosities*

Substance		$\sigma(\text{Å})$	$\varepsilon_{12}/k_B(\text{K})$
Ar	Argon	3.542	93.3
He	Helium	2.551	10.2
Kr	Krypton	3.655	178.9
Ne	Neon	2.820	32.8
Xe	Xenon	4.047	231.0
Air	Air	3.711	78.6
Br_2	Bromine	4.296	507.9
CCl_4	Carbon tetrachloride	5.947	322.7
$CHCl_3$	Chloroform	5.389	340.2
CH_2Cl_2	Methylene chloride	4.898	356.3
CH_3Cl	Methyl chloride	4.182	350.0
CH_3OH	Methanol	3.626	481.8
CH_4	Methane	3.758	148.6
CO	Carbon monoxide	3.690	91.7
CO_2	Carbon dioxide	3.941	195.2
CS_2	Carbon disulfide	4.483	467.0
C_2H_2	Acetylene	4.033	231.8
C_2H_4	Ethylene	4.163	224.7
C_2H_6	Ethane	4.443	215.7
C_2H_5Cl	Ethyl chloride	4.898	300.0
C_2H_5OH	Ethanol	4.530	362.6
CH_3OCH_3	Methyl ether	4.307	395.0
CH_2CHCH_3	Propylene	4.678	298.9
C_3H_8	Propane	5.118	237.1
$n\text{-}C_3H_7OH$	n-Propyl alcohol	4.549	576.7
CH_3COCH_3	Acetone	4.600	560.2
$n\text{-}C_4H_{10}$	n-Butane	4.687	531.4
iso-C_4H_{10}	Isobutane	5.278	330.1
$n\text{-}C_5H_{12}$	n-Pentane	5.784	341.1
C_6H_6	Benzene	5.349	412.3
C_6H_{12}	Cyclohexane	6.182	297.1
$n\text{-}C_6H_{14}$	n-Hexane	5.949	399.3
Cl_2	Chlorine	4.217	316.0
HBr	Hydrogen bromide	3.353	449.0
HCN	Hydrogen cyanide	3.630	569.1
HCl	Hydrogen chloride	3.339	344.7
HF	Hydrogen fluoride	3.148	330.0
HI	Hydrogen iodide	4.211	288.7
H_2	Hydrogen	2.827	59.7
H_2O	Water	2.641	809.1
H_2S	Hydrogen sulfide	3.623	301.1
Hg	Mercury	2.969	750.0
NH_3	Ammonia	2.900	558.3
NO	Nitric oxide	3.492	116.7
N_2	Nitrogen	3.798	71.4
N_2O	Nitrous oxide	3.828	232.4
O_2	Oxygen	3.467	106.7
SO_2	Sulfur dioxide	4.112	335.4

Note: Data from Hirschfelder *et al.* (1954).

Table 5.1-3 *The collision integral Ω*

k_BT/ε_{12}	Ω	k_BT/ε_{12}	Ω	k_BT/ε_{12}	Ω
0.30	2.662	1.65	1.153	4.0	0.8836
0.40	2.318	1.75	1.128	4.2	0.8740
0.50	2.066	1.85	1.105	4.4	0.8652
0.60	1.877	1.95	1.084	4.6	0.8568
0.70	1.729	2.1	1.057	4.8	0.8492
0.80	1.612	2.3	1.026	5.0	0.8422
0.90	1.517	2.5	0.9996	7	0.7896
1.00	1.439	2.7	0.9770	9	0.7556
1.10	1.375	2.9	0.9576	20	0.6640
1.30	1.273	3.3	0.9256	60	0.5596
1.50	1.198	3.7	0.8998	100	0.5130
1.60	1.167	3.9	0.8888	300	0.4360

Source: Data from Hirschfelder *et al.* (1954).

in which m is the molecular mass. The mean free path l is

$$l = \frac{k_BT/p}{\left(\frac{\pi}{4}\sigma^2\right)} \tag{5.1-7}$$

in which σ is the diameter of the spheres, and p/k_BT is the concentration of molecules per volume. Combining, we find

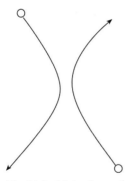

Fig. 5.1-1. Molecular motion in a dilute gas. In a gas, molecular collisions occur at low density, and so may be treated as bimolecular. This simplicity facilitates development of good kinetic theories for diffusion.

$$D = \left(\frac{4\sqrt{2}}{3\pi}\right)\frac{(k_BT)^{3/2}}{m^{1/2}p\sigma^2} \tag{5.1-8}$$

When we compare this result with Eq. 5.1-1, we see that the rigid-sphere theory predicts essentially the same dependence on temperature, pressure, molecular weight, and molecular size. The Chapman–Enskog theory is an improvement over the simple theory because the details of the collisions are explicitly included.

5.1.3 Gaseous Diffusion Coefficients From Empirical Correlations

Predictions from the Chapman–Enskog kinetic theory tend to be limited in two ways. First, the theory requires estimates of σ_{12} and ε_{12}; such estimates are not available for all gases. Second, the theory assumes nonpolar gases, and this excludes compounds like water and ammonia. These interactions depend on replacing the Lennard–Jones potential used to characterize the collision with more exact potentials. Such replacement is often complex.

Instead, many authors have developed empirical relations. One effective example (Fuller, Schettler, and Giddings, 1966) is

$$D = 10^{-3} \frac{T^{1.75}(1/\tilde{M}_1 + 1/\tilde{M}_2)^{1/2}}{p\left[\left(\sum_i V_{i1}\right)^{1/3} + \left(\sum_i V_{i2}\right)^{1/3}\right]^2} \tag{5.1-9}$$

in which T is in Kelvin, p is in atmospheres, and the V_{ij} are the volumes of parts of the molecule j, tabulated in Table 5.1-4. This correlation is about as successful as Eq. 5.1-1. To me, the impressive feature is the similarity between the two equations: the pressure and molecular-weight dependence are unchanged. The temperature dependence is not much different when we remember that Ω is a function of temperature. The term for diffusion volumes here parallels the term in σ^2. It is not surprising that the two equations have similar success.

5.1.4 Gas Diffusion at High Pressure

The equations given earlier in this chapter allow prediction of diffusion coefficients in dilute gases to within an average of eight percent. These predictions, which are about twice as accurate as those for liquids, are often hailed as a final answer. However,

Table 5.1-4 *Atomic diffusion volumes for use in Eq. 5.1-9*

Atomic and structural diffusion-volume increments V_{ij}		Diffusion volumes for simple molecules ΣV_{ij}	
C	16.5	H_2	7.07
H	1.98	He	2.88
O	5.48	N_2	17.9
(N)	5.69	O_2	16.6
(Cl)	19.5	Air	20.1
(S)	17.0	Ar	16.1
Aromatic ring	−20.2	Kr	22.8
Heterocyclic ring	−20.2	CO	18.9
		CO_2	26.9
		N_2O	35.9
		NH_3	14.9
		H_2O	12.7
		(Cl_2)	37.7
		(SO_2)	41.1

Note: Parentheses indicate that the value is uncertain.
Source: Adapted from Fuller, Schettler, and Giddings (1966).

I have the nagging suspicion that their success is promulgated by those who have worked hard on these methods or who have become intimidated by the intellectual edifice erected by Maxwell, Enskog, and others. In fact, although these equations agree with experiment at low pressures, they are much less successful at high pressures. At higher pressures, few binary data are available; for self-diffusion, a sensible empirical suggestion is

$$pD = p_0 D_0 \qquad\qquad (5.1\text{-}10)$$

in which the subscript 0 indicates values at low pressure at the same temperature. The inverse relation between diffusion and pressure, consistent with Eq. 5.1-1, is a good guideline.

Some more elaborate theories have attempted to correlate the product (pD) with the reduced pressure and temperature, that is, with the pressure and temperature relative to values at the critical point. Such a correlation, implicitly based on the theory of corresponding states, can be applied to transport phenomena by assuming that thermodynamic variables can be defined in nonequilibrium situations. We will make such an assumption in the irreversible thermodynamics arguments in Section 7.2. In the current case, however, this effort at correlation suggests significant corrections only when the reduced temperature is less than 1.4. Under these circumstances, Fick's law breaks down because diffusion occurs not as single solute molecules but as a cluster of solute molecules, as described in Section 6.3. In the face of this complexity, I would use Equation 5.1-10 with confidence when the temperature divided by the critical temperature is above 1.4 and make experiments at lower temperatures.

Some other aspects of gaseous diffusion remain unexplored. For example, diffusion of molecules of very different sizes, like hydrogen and high molecular weight *n*-alkanes, has not been sufficiently studied. Concentration-dependent diffusion in gases, although a common phenomenon, has been largely ignored. These aspects deserve careful inspection.

Example 5.1-1: Estimating diffusion with the Chapman–Enskog theory   Calculate the diffusion coefficient of argon in hydrogen at 1 atmosphere and 175 °C. The experimental value is 1.76 cm^2/sec.

   Solution   We first need to find σ_{12} and ε_{12}. From the values in Table 5.1-2,

$$\sigma_{12} = \frac{1}{2}(\sigma_1 + \sigma_2)$$

$$= \frac{1}{2}(3.54 + 2.83) = 3.18\text{Å}$$

and

$$\frac{\varepsilon_{12}}{k_B T} = \sqrt{(\varepsilon_1/k_B)(\varepsilon_2/k_B)}/T$$

$$= \frac{\sqrt{124(38.0)}}{448} = 0.166$$

From Table 5.1-3, we find that Ω is 0.81. Thus, from Eq. 5.1-1,

$$D = \frac{1.86 \cdot 10^{-3} T^{3/2}(1/\tilde{M}_1 + 1/\tilde{M}_2)^{1/2}}{p\sigma_{12}^2\Omega}$$

$$= \frac{1.86 \cdot 10^{-3}(448)^{3/2}(1/39.9 + 1/2.02)^{1/2}}{(1)(3.18)^2(0.81)} = 1.55\,\text{cm}^2/\text{sec}$$

The theoretical prediction is about ten percent below the experimental observation.

Example 5.1-2: Comparing two estimates of gas diffusion   Use the Chapman–Enskog theory and the empirical correlation in Equation 5.1-9 to estimate the diffusion of hydrogen in nitrogen at 21 °C and 2 atmospheres. The experimental value is $0.38\,\text{cm}^2/\text{sec}$.
   Solution   For the Chapman–Enskog theory, the key parameters are

$$\sigma_{12} = \frac{1}{2}(\sigma_{H_2} + \sigma_{N_2}) = \frac{1}{2}(2.92 + 3.68) = 3.30\,\text{Å}$$

and

$$\frac{\varepsilon_{12}}{k_B T} = \frac{\sqrt{(\varepsilon_{H_2}/k_B)(\varepsilon_{N_2}/k_B)}}{T} = \frac{\sqrt{(38.0)(91.5)}}{294} = 0.201$$

This second value allows interpolation from Table 5.1-3:

$$\Omega = 0.842$$

Combining these results with Eq. 5.1-1 gives

$$D = \frac{1.86 \cdot 10^{-3} T^{3/2}(1/\tilde{M}_{H_2} + 1/\tilde{M}_{N_2})^{1/2}}{p\sigma^2\Omega}$$

$$= \frac{1.86 \cdot 10^{-3}(294)^{3/2}(1/2.02 + 1/28.0)^{1/2}}{2(3.30)^2(0.842)} = 0.37\,\text{cm}^2/\text{sec}$$

The value is about three percent low, a very solid estimate.
   For the Fuller correlation, the appropriate volumes are found from Table 5.1-4. The results can then be combined with Eq. 5.1-9:

$$D = \frac{10^{-3} T^{1.75}(1/\tilde{M}_{H_2} + 1/\tilde{M}_{N_2})^{1/2}}{p\left[(V_{H_2})^{1/3} + (V_{N_2})^{1/3}\right]^2}$$

$$= \frac{10^{-3}(294)^{1.75}(1/2.02 + 1/28.0)^{1/2}}{2\left[(7.07)^{1/3} + (17.9)^{1/3}\right]^2} = 0.37\,\text{cm}^2/\text{sec}$$

Again, the error is about three percent.

Example 5.1-3: Diffusion in supercritical carbon dioxide   Carbon dioxide, above its critical point, may become an important industrial solvent because it is cheap, nontoxic, and nonexplosive. Estimate the diffusion of iodine in carbon dioxide at 0 °C and 33 atmospheres. The diffusion coefficient measured under these conditions is $7 \cdot 10^{-4}$ cm²/sec.

   Solution   The binary diffusion coefficient at 0 °C and 1 atmosphere can be found from Eq. 5.1-1:

$$D_0 = 0.043 \text{ cm}^2/\text{sec}$$

From Eq. 5.1-10,

$$D = 0.43 \text{ cm}^2/\text{sec} \left(\frac{1 \text{ atm}}{33 \text{ atm}} \right)$$
$$= 13 \cdot 10^{-4} \text{ cm}^2/\text{sec}$$

This is as accurate as we have any right to expect, especially because the critical point for carbon dioxide is close, at 30 °C and 72 atmospheres.

5.2   Diffusion Coefficients in Liquids

   Diffusion coefficients in liquids are exemplified by the values given in Tables 5.2-1 and 5.2-2. Most of these values fall close to 10^{-5} cm²/sec. This is true for common organic solvents, mercury, and even molten iron. Exceptions occur for high molecular-weight solutes like albumin and polystyrene, where diffusion can be 100 times slower. Actually, the range of these values is remarkably small. At 25 °C, almost none are faster than $10 \cdot 10^{-5}$ cm²/sec, and those significantly below 10^{-5} cm²/sec are macromolecules, like hemoglobin. The reasons for this narrow range is that the viscosity of simple liquids like water and hexane varies little, and that diffusion coefficients are only a weak function of solute size.

   Diffusion coefficients in liquids are about ten thousand times slower than those in dilute gases. To see what this means, we again calculate the penetration distance $\sqrt{4Dt}$, which was the distance we found central to unsteady diffusion. As an example, consider benzene diffusing into cyclohexane with a diffusion coefficient of about $2 \cdot 10^{-5}$ cm²/sec. At time zero, we bring the benzene and cyclohexane into contact. After 1 second, the diffusion has penetrated 0.004 cm, compared with 0.3 cm for gases; after 1 minute, the penetration is 0.03 cm, compared with 4 cm; after 1 hour, it is 0.3 cm, compared with 30 cm.

   The sloth characteristic liquid diffusion means that diffusion often limits the overall rate of processes occurring in liquids. In chemistry, diffusion limits the rate of acid–base reactions; in physiology, diffusion limits the rate of digestion; in metallurgy, diffusion can control the rate of surface corrosion; in the chemical industry, diffusion is responsible for the rates of liquid–liquid extractions. Diffusion in liquids is important because it is slow.

5.2.1   *Liquid Diffusion Coefficients From the Stokes–Einstein Equation*

   The most common basis for estimating diffusion coefficients in liquids is the Stokes–Einstein equation. Coefficients calculated from this equation are accurate to

Table 5.2-1 *Diffusion coefficients at infinite dilution in water at 25 °C*

Solute	$D(\cdot 10^{-5} \ cm^2/sec)$
Acetic acid	1.21
Acetone	1.16
Ammonia	1.64
Argon	2.00
Benzene	1.02
Benzoic acid	1.00
Bromine	1.18
Carbon dioxide	1.92
Carbon monoxide	2.03
Chlorine	1.25
Ethane	1.20
Ethanol	0.84
Ethylene	1.87
Glycine	1.06
Helium	6.28
Hemoglobin	0.069
Hydrogen	4.50
Hydrogen sulfide	1.41
Methane	1.49
Methanol	0.84
n-Butanol	0.77
Nitrogen	1.88
Oxygen	2.10
Ovalbumin	0.078
Propane	0.97
Sucrose	$(0.5228 - 0.265c_1)^a$
Urea	$(1.380 - 0.0782c_1 + 0.00464c_1^2)^a$
Urease	0.035
Valine	0.83

Note: [a]Known to very high accuracy, and so often used for calibration; c_1 is in moles per liter.
Source: Data from Cussler (1976) and Poling et al. (2001).

only about twenty percent (Poling *et al.*, 2001). Nonetheless, this equation remains the standard against which alternative correlations are judged.

The Stokes–Einstein equation is

$$D = \frac{k_B T}{f} = \frac{k_B T}{6\pi\mu R_0} \tag{5.2-1}$$

where f is the friction coefficient of the solute, k_B is Boltzmann's constant, μ is the solvent viscosity, and R_0 is the solute radius. The temperature variation suggested by this equation is apparently correct, but it is much smaller than effects of solvent viscosity and solute radius. A discussion of these larger effects follows.

The diffusion coefficient varies inversely with viscosity when the ratio of solute to solvent radius exceeds five. This behavior is reassuring because the Stokes–Einstein equation is derived by assuming a rigid solute sphere diffusing in a continuum of solvent. Thus, for a large solute in a small solvent, Eq. 5.2-1 seems correct.

Table 5.2-2 *Diffusion coefficients at infinite dilution nonaqueous liquids*

Solvent	Solute[a]	$D(\cdot 10^{-5} \text{ cm}^2/\text{sec})$
Chloroform	Acetone	2.35
	Benzene	2.89
	Ethyl alcohol (15 °C)	2.20
	Ethyl ether	2.14
	Ethyl acetate	2.02
Benzene	Acetic acid	2.09
	Benzoic acid	1.38
	Cyclohexane	2.09
	Ethyl alcohol (15 °C)	2.25
	n-Heptane	2.10
	Oxygen (29.6 °C)	2.89
	Toluene	1.85
Acetone	Acetic acid	3.31
	Benzoic acid	2.62
	Nitrobenzene (20 °C)	2.94
	Water	4.56
n-Heptane	Carbon tetrachloride	3.70
	Dodecane	2.73
	n-Hexane	4.21
	Propane	4.87
	Toluene	4.21
Ethanol	Benzene	1.81
	Iodine	1.32
	Oxygen (29.6 °C)	2.64
	Water	1.24
	Carbon tetrachloride	1.50
n-Butanol	Benzene	0.99
	p-Dichlorobenzene	0.82
	Propane	1.57
	Water	0.56
n-Heptane	Benzene	3.40

Note: [a]Temperature 25 °C except as indicated.
Source: Data from Poling *et al.* (2001).

When the solute radius is less than five times that of the solvent, Eq. 5.2-1 breaks down (Chen *et al.*, 1981). This failure becomes worse as the solute size becomes smaller and smaller. Errors are especially large in high-viscosity solvents; the diffusion seems to vary with a smaller power of viscosity often around (–0.7). In extremely high-viscosity materials, diffusion becomes independent of viscosity: the diffusion of sugar in jello is very nearly equal to the diffusion of sugar in water.

The reason for this altered viscosity dependence is that viscosity often depends on much longer range interactions than diffusion. For example, in jello, the polymeric collagen forms hydrogen bonds that form a three-dimensional elastic network, which of course has very high viscosity. However, sugar and salts diffusing through this network are much smaller than the distances between these hydrogen bonds, so these solutes behave just as if they are diffusing through water. As evidence of this, the concentration dependence of the diffusion coefficient of potassium chloride diffusing in water–polyethylene glycol mixtures is exactly the same as that in water. Diffusion reflects short-range interactions.

As the standard, the Stokes–Einstein equation has often been extended and adapted. Two adaptations deserve special mention. The first is for small solutes. For this case, the factor 6π in Eq. 5.2-1 is often replaced by a factor of 4π or of two. The substitution of 4π can be rationalized on mechanical grounds as signifying solvent slipping past the surface of the solute molecule (Sutherland, 1905). The factor of two can be supported with the theory of absolute reaction rates (Glasstone *et al.*, 1941). Neither substitution always works.

The second adaptation of the Stokes–Einstein equation is its use to estimate the size and shape of proteins in dilute aqueous solution. Unfortunately, these estimates are compromised in two ways. First, if the solute is hydrated, then the radius found will refer to the solute–water complex, not to the solute itself. Second, if the solute is not spherical, then the radius R_0 will represent some average shape. Specifically, if the solute is a prolate (football-shaped) ellipsoid, then (Perrin, 1936)

$$
D\left(\begin{array}{c}\text{prolate}\\\text{ellipsoid}\end{array}\right) = \frac{k_B T}{6\pi\mu\left[\dfrac{(a^2 - b^2)^{1/2}}{\ln\left(\dfrac{a+(a^2-b^2)^{1/2}}{b}\right)}\right]}
\tag{5.2-2}
$$

in which a and b are the major and minor axes of the ellipsoid. For an oblate (disc-shaped) ellipsoid,

$$
D\left(\begin{array}{c}\text{oblate}\\\text{ellipsoid}\end{array}\right) = \frac{k_B T}{6\pi\mu\left[\dfrac{(a^2 - b^2)^{1/2}}{\tan^{-1}\left[\left(\dfrac{a^2-b^2}{b^2}\right)^{1/2}\right]}\right]}
\tag{5.2-3}
$$

These relations reduce to Eq. 5.2-1 for spheres when a equals b.

These diffusion coefficients are for normal translational diffusion, the subject of this book. Ellipsoids can also rotate, a process described by a rotational diffusion coefficient. This rotation implies a conservation equation like

$$
\frac{\partial c_1}{\partial t} = D_{rot}\frac{\partial^2 c_1}{\partial\theta^2}
\tag{5.2-4}
$$

where c_1 is the concentration of solutes with a particular angular orientation θ, an orientation due, for example, to shear or due to an electrostatic potential. Note that the new rotational diffusion coefficient D_{rot} appearing in this equation has the dimensions of reciprocal time, not of (length)^2 per time, the normal units for translational diffusion. For a prolate ellipsoid with $a \gg b$, the rotational diffusion coefficient is

$$
D_{rot} = \frac{\left(3\ln\frac{2a}{b}\right)k_B T}{8\pi\eta a^3}
\tag{5.2-5}
$$

For an oblate ellipsoid with $a \ll b$, it is

$$D_{\text{rot}} = \frac{3 k_B T}{32 \eta b^3} \tag{5.2-6}$$

These results are sometimes used to infer the shape of proteins in solution.

5.2.2 Deriving the Stokes–Einstein Equation

To predict diffusion in liquids, we do not account for molecular motion as in the theories used for gases. Instead, we idealize our system as a single rigid solute sphere moving slowly through a continuum of solvent (Fig. 5.2-1). We expect that the net velocity of this sphere will be proportional to the force acting on it:

$$\text{force} = f \boldsymbol{v}_1 \tag{5.2-4}$$

where f is defined as the friction coefficient. Because the sphere moves slowly, this friction coefficient can be found from Stokes' law (first published in 1850) to be $6\pi\mu R_0$. The force was taken by Einstein to be the negative of the chemical potential gradient (Einstein, 1905). Thus Eq. 5.2-4 can be rewritten:

$$- \nabla \mu_1 = (6\pi\mu R_0) \boldsymbol{v}_1 \tag{5.2-5}$$

The chemical potential gradient, defined per molecule (not per mole), is often described as a "virtual force," a thermodynamic parallel to mechanical or electrostatic forces.

When the solution is dilute, we can assume that it is ideal:

$$\mu_1 = \mu_1^0 + k_B T \ln x_1 = \mu_1^0 + k_B T \ln \frac{c_1}{c_1 + c_2} \doteq \mu_1^0 + k_B T \ln c_1 - k_B T \ln c_2 \tag{5.2-6}$$

In this result, we recognize that solvent concentration c_2 far exceeds solute concentration c_1, so c_2 is approximately constant. The gradient is then

$$\nabla \mu_1 = \frac{k_B T}{c_1} \nabla c_1 \tag{5.2-7}$$

Combining this with Eq. 5.2-5, we find

$$j_1 \doteq n_1 = c_1 \boldsymbol{v}_1 = -\frac{k_B T}{6\pi\mu R_0} \nabla c_1 \tag{5.2-8}$$

Comparison with Fick's law produces the Stokes–Einstein equation, Eq. 5.2-1.

The interesting assumption in this analysis is the way in which the velocity or flux is assumed to vary with the chemical potential gradient. This type of assumption is made frequently in studies of diffusion. It is central to the development of irreversible thermodynamics, and so it is at the core of the theories of multicomponent diffusion described in Chapter 7. Interestingly, it is known experimentally to be wrong in the highly nonideal solutions near critical points (see Section 6.3).

(a) Actual situation

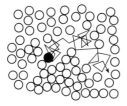

(b) Stokes–Einstein model

Fig. 5.2-1. Molecular motion in a liquid. In contrast with a gas, molecular motion in a liquid takes place at high density (a). Diffusion is complex, involving many interactions and vacancies. The available kinetic theories are good, but complex. To avoid this, many use the simple model of a solute sphere in a solvent continuum (b).

Because the Stokes–Einstein equation is limited to cases in which the solute is larger than the solvent, many investigators have developed correlations for cases in which solute and solvent are similar in size. The impressive aspect of these efforts is their similarly to the Stokes–Einstein equation. Almost all show the same temperature and viscosity dependence. All authors claim marginally better accuracy, but for such increased complexity that their results are rarely used. The exception is the Wilke–Chang correlation (1955), which predicts

$$D = \frac{7.4 \cdot 10^{-8} \left(\phi \tilde{M}_2 \right)^{1/2} T}{\mu \bar{V}_1^{0.6}} \qquad (5.2\text{-}9)$$

where D is the diffusion coefficient of solute "1," in cm^2/sec; $\tilde{M}_2$ is the molecular weight of solvent "2," in daltons; T is the temperature, in K; μ is the viscosity, in centipoises; and $\bar{V}_1$ is the molar volume of the solute, in cm^3/mol. The empirical parameter ϕ is 1 for most organic solvents, 1.5 for alcohols, and 2.6 for water. This result is widely used for fast estimates.

At this point, the common conclusion is to bemoan the accuracy of the predictions in liquids and to praise the accuracy of those in gases. In fact, the predictions in liquids are only twice as inaccurate as those in gases, even though the complexity of solute–solvent interactions in liquids is much greater. As a result, I do not share the frequent despair about these estimates, but feel that care and good judgment can lead to success.

5.2.3 Diffusion in Concentrated Solutions

The Stokes–Einstein equation and its empirical extensions are limited to infinitely dilute solutions. In fact, the diffusion coefficient in liquids varies with solute concentration, frequently by several hundred percent and sometimes with a maximum and minimum. We need a means of estimating these variations. Such estimations usually involve two steps. First, we assume that Eq. 5.2-4 can be written

$$- v_1 = \frac{1}{f} \nabla \mu_1 = \frac{D_0}{RT} \nabla \mu_1 \tag{5.2-10}$$

where D_0 is a new transport coefficient. For a nonideal solution,

$$\mu_1 = \mu_1^0 + k_B T \ln c_1 \gamma_1 \tag{5.2-11}$$

where γ_1 is an activity coefficient. Combining these two equations, we find

$$n_1 \doteq j_1 \doteq c_1 v_1 = - \left[D_0 \left(1 + \frac{\partial \ln \gamma_1}{\partial \ln c_1} \right) \right] \nabla c_1 \tag{5.2-12}$$

The quantity in brackets is the diffusion coefficient. This first step is a restatement of the idea that the velocity of diffusion varies with the gradient of chemical potential.

The second step consists of empirical estimates of the quantity D_0. These estimates are based on diffusion coefficients in dilute solutions. One of the most frequently cited estimates, used by Darken (1948), Hartley and Crank (1949), and others, is the arithmetic average:

$$D_0 = x_1 D_0(x_1 = 1) + x_2 D_0(x_2 = 1) \tag{5.2-13}$$

Another estimate, suggested by Vignes (1966) is the geometric average:

$$D_0 = [D_0(x_1 = 1)]^{x_1} [D_0(x_2 = 1)]^{x_2} \tag{5.2-14}$$

The geometric average seems more successful than the arithmetic one.

I am not convinced that these efforts to correct diffusion coefficients with activity coefficients are correct. I agree that some form of correction is indicated, and I admit that much of the correction must be empirical. However, I have found that the corrections suggested by Eq. 5.2-12 are usually too big. For example, if D drops with increasing concentration c_1, then the D_0 inferred from this equation tends to rise with increasing c_1. In the same sense, if D rises with increasing c_1, then D_0 drops over the same concentration range. Moreover, these corrections are wrong near the spinodal phase boundary, as detailed in Section 6.3. Thus I always treat these corrections with caution.

Example 5.2-1: Oxygen diffusion in water   Estimate the diffusion at 25 °C for oxygen dissolved in water using the Stokes–Einstein equation and the Wilke–Chang correlation. Compare your results with the experimental value of $1.8 \cdot 10^{-5}$ cm^2/sec.

Solution   For the Stokes–Einstein equation, the chief problem is to estimate the radius of the oxygen molecule. If we assume that this is half the collision diameter in the gas, then from Table 5.1-2,

$$R_0 = \frac{1}{2} \sigma_1 = 1.73 \cdot 10^{-8} \text{ cm}$$

When we insert this into the Stokes–Einstein equation,

$$D = \frac{k_B T}{6\pi \mu R_0} = \frac{(1.38 \cdot 10^{-16} \text{g cm}^2/\text{sec}^2 \text{ K})298 \text{ K}}{6\pi(0.01 \text{ g/cm sec})1.73 \cdot 10^{-8} \text{ cm}} = 1.3 \cdot 10^{-5} \text{cm}^2/\text{sec}$$

This value is thirty percent low. Replacing (6π) with (4π) gives a more accurate result; replacing (6π) with (2) gives too high a value. The Wilke–Chang correlation is somewhat better:

$$D = \frac{7.4 \cdot 10^{-8}\left(\phi \tilde{M}_{H_2O}\right)^{1/2} T}{\mu_{H_2O} \bar{V}_{O_2}^{0.6}} = \frac{7.4 \cdot 10^{-8}\left[2.6\left(18\,cm^3/mol\right)\right]^{1/2} 298\,K}{1\,cp(25\,cm^3/mol)^{0.6}}$$

$$= 2.2 \cdot 10^{-5}\,cm^2/sec$$

This is twenty percent high.

Example 5.2-2: Estimating molecular size from diffusion   Fibrinogen has a diffusion coefficient of about $2.0 \cdot 10^{-7}\,cm^2/sec$ at $37\,°C$. It is believed to be rod-shaped, about thirty times longer than it is wide. How large is the molecule?

   Solution   Because the molecule is rod-shaped, it can be approximated as a prolate ellipsoid. Thus, from Eq. 5.2-2,

$$D = \frac{k_B T}{6\pi\mu a\left[\dfrac{[1-(b/a)^2]^{1/2}}{\ln\left\{\dfrac{a}{b} + \left(\dfrac{a^2}{b^2}-1\right)^{1/2}\right\}}\right]}$$

$$2.0 \cdot 10^{-7}\,cm^2/sec = \frac{(1.38 \cdot 10^{-16}\,g\,cm^2/sec^2\,K)(310\,K)}{6\pi(0.00695\,g/cm\,sec)a\left[\dfrac{[1-(1/30)^2]^{1/2}}{\ln[30+(30^2-1)^{1/2}]}\right]}$$

Solving, we find that a equals 67 nm and b equals 2.2 nm. If fibrinogen were a sphere, its radius would be about 16 nm.

Example 5.2-3: Diffusion in an acetone–water mixture   Estimate the diffusion coefficient in a 50-mole% mixture of acetone (1) and water (2). This solution is highly nonideal, so that $[\partial \ln \gamma_1/\partial \ln c_1]$ equals -0.69. In pure acetone, the diffusion coefficient is $1.26 \cdot 10^{-5}\,cm^2/sec$; in pure water, it is $4.68 \cdot 10^{-5}\,cm^2/sec$. The experimental value in the mixture is $0.79 \cdot 10^{-5}\,cm^2$ sec, less than both limits.

   Solution   We first must estimate D_0. Because Eq. 5.2-14 is most often successful, we use it here:

$$D_0 = [D_0(x_1 = 1)]^{x_1}\,[D_0(x_2 = 1)]^{x_2}$$
$$= (1.26 \cdot 10^{-5}\,cm^2/sec)^{0.5}(4.86 \cdot 10^{-5}\,cm^2/sec)^{0.5}$$
$$= 2.43 \cdot 10^{-5}\,cm^2/sec$$

From Eq. 5.2-13,

$$D = D_0 \left(1 + \frac{\partial \ln \gamma_1}{\partial \ln c_1} \right)$$
$$= 2.43 \cdot 10^{-5} \, \text{cm}^2/\text{sec}(1 - 0.69)$$
$$= 0.75 \cdot 10^{-5} \, \text{cm}^2/\text{sec}$$

The agreement with the experimental value is unusually good.

5.3 Diffusion in Solids

Diffusion in solids is beyond the scope of this book. However, I want to give the briefest synopsis to provide a comparison with gases and liquids. Diffusion in solids is described by the same form of Fick's law as gases or liquids. The diffusion coefficients, however, are much, much smaller, as shown by the values in Table 5.3-1. These values do increase quickly with temperature. The exception is hydrogen. In metals, diatomic hydrogen first dissociates to form atomic hydrogen, which then loses its electron to the electron cloud within the metals. Thus in this case, "hydrogen diffusion" refers to the motion of naked protons, whose small size gives them an unusually large mobility.

The small value for diffusion coefficients in solids has two important consequences. First, the values are so small that almost all significant transport occurs through flaws and gaps in the solid, especially along grain boundaries. This is especially true for metals and crystals. Second, transport in solids almost always approaches the limit of a semi-infinite solid, rather than diffusion across a thin film. Again, hydrogen is the exception because it is so fast. For example, diffusion of hydrogen across thin membranes of palladium is sometimes suggested as a route to purify hydrogen.

The estimation of diffusion coefficients in solids is not accurate. In almost every case, one must use experimental results. Methods for rough estimates based on the theory for face-centered-cubic (FCC) metals are the standard by which other theories are judged,

Table 5.3-1 *Diffusion coefficients at 25 °C in some characteristic solids*

Solid	Solute	D (cm^2/sec)
Iron (α Fe; BCC)	Fe	$3 \cdot 10^{-48}$
	C	$6 \cdot 10^{-21}$
	H$_2$	$2 \cdot 10^{-9}$
Iron (α Fe; FCC)	Fe	$8 \cdot 10^{-55}$
	C	$3 \cdot 10^{-31}$
Copper	Cu	$8 \cdot 10^{-42}$
	Zn	$2 \cdot 10^{-38}$
SiO$_2$	H$_2$	$6 \cdot 10^{-13}$
	He	$4 \cdot 10^{-10}$

Note: In most cases, these values are extrapolated from values at higher temperatures.

just as the Stokes–Einstein equation is the standard for liquids. The diffusion coefficient in this case is

$$D = R_0^2 N \omega \qquad (5.3\text{-}1)$$

in which R_0 is the spacing between atoms; N is the fraction of sites vacant in the crystal; and ω is the jump frequency, the number of jumps per time from one position to the next. Values for R_0 are guessed from crystallographic data, and the fraction N is commonly estimated from the Gibbs free energy of mixing. The frequency ω is estimated by reaction-rate theories for the concentration of activated complexes, atoms midway between adjacent sites. The results of these estimations are commonly expressed as

$$D = D_0 e^{-\Delta H / RT} \qquad (5.3\text{-}2)$$

where D_0 and ΔH are estimated empirically. Values of ΔH are large, often above 100 kJ/mol, so that diffusion increases much more with temperature than for gases or for liquids.

Example 5.3-1: Diffusion of carbon in iron   Experiments show that the diffusion of carbon in body-centered cubic (BCC) iron is $2.4 \cdot 10^{-8}$ cm²/sec at 500 °C, but $1.7 \cdot 10^{-6}$ cm²/sec at 900 °C. Find an equation which allows estimating carbon diffusion at other temperatures.

Solution   The form of this relation is that of Equation 5.3-2

$$D = D_0 e^{-\Delta H / RT}$$

Inserting the values for D and T, we find

$$D = \left[6.2 \cdot 10^{-3} \text{ cm}^2/\text{sec} \right] \exp^{-(80 \,\text{kJ/mol})/RT}$$

The values for D_0 and for ΔH are slightly smaller than those commonly observed.

5.4 Diffusion in Polymers

Diffusion coefficients in high polymers are closer to those for liquids than to those for solids. This is true even for crystalline polymers, where the coefficients reflect transport around, not through, the small crystals. Typical values for synthetic high polymers are shown in Fig. 5.4-1. The values of these coefficients vary strongly with concentration. Naturally occurring polymers like proteins are not included in Fig. 5.4-1 because these species are best handled with the dilute-solution arguments in Section 5.2.

The results in Fig. 5.4-1 show that very different limits exist. The first of these limits occurs in dilute solution, where a polymer molecule is imagined as a solute sphere moving through a continuum of solvent. The second limit is in highly concentrated solution, where small solvent molecules squeeze through gaps in the polymer matrix.

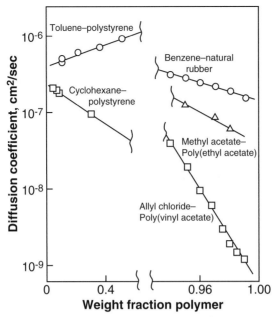

Fig. 5.4-1. Diffusion of high polymers. Diffusion in these systems has two interesting limits: at very low and very high polymer concentrations. Interestingly, the diffusion coefficients in these two limits may not be very different, even though the viscosity change is tremendous.

The third, not illustrated in Fig. 5.4-1, involves mixtures of two polymers. Each limit is discussed briefly below.

5.4.1 Polymer Solutes in Dilute Solution

A polymer molecule dissolved in a low-molecular-weight solvent is imagined as a necklace of spherical beads connected by a string that has no resistance to flow. The necklace is floating in a neutrally buoyant solvent continuum. If the solution is very dilute, the polymer molecules are greatly separated, so that they do not interact with each other, but only with the solvent. In some cases, the solvent will expand the polymer necklace in the solution; such a solvent is referred to as "good." In other cases the solvent and polymer will not strongly interact, and the polymer necklace will shrink into a small, introspective blob; such a solvent is called "poor."

Between these two extremes, the polymer and solvent can interact just enough so that the segments of the polymer necklace will be randomly distributed. This limit of a "random coil" of polymer is conventionally chosen as the "ideal" polymer solution, and a solvent showing these characteristics is called a θ solvent. Under these conditions, the diffusion of the polymer can be calculated as a correction to the Stokes–Einstein equation:

$$D = \frac{k_B T}{6\pi\mu R_e} \tag{5.4-1}$$

where R_e is the equivalent radius of the polymer. This radius is calculated to be

$$R_e = 0.676 \langle R^2 \rangle^{1/2} \tag{5.4-2}$$

in which $\langle R^2 \rangle^{1/2}$ is the root-mean-square radius of gyration, the common measure of the size of the polymer molecule in solution. This root-mean-square radius can be measured in a variety of ways; one common method is by light scattering. Equations 5.4-1 and 5.4-2 are confirmed by experiment. The measured ratio of equivalent radius to root-mean-square radius is 0.68, which is very close to the 0.676 suggested theoretically.

In good solvents and poor solvents, the diffusion coefficient still is estimated from the Stokes–Einstein equation, but the relation between the equivalent radius R_e and the root-mean-square radius $\langle R^2 \rangle^{1/2}$ seems less well known. Moreover, in good solvents, the diffusion coefficient can increase sharply with polymer concentration. This increase, which occurs in the face of rapidly increasing viscosity, is apparently the result of a highly nonideal solution. The increase is often estimated using parallels to Eq. 5.2-12. The accuracy of these estimates is uncertain.

5.4.2 Low Molecular Weight Solutes in a Polymer Solvent

The second limiting case of polymer diffusion occurs when a small dilute solute diffuses in a concentrated polymer solvent. Some examples are given on the right-hand side of Fig. 5.4-1. In addition to its scientific interest, this case has considerable practical value. It is important in devolatilization, that is, the removal of solvent and unreacted monomer from commercial polymers. This is especially important for polymers with consumer applications like food wrapping, because the volatile species may not be benign. Diffusion in this second case is also central to drying many solvent-based coatings. There, rapid solvent evaporation from the surface of the coating can produce a concentrated polymer skin. Slower diffusion through this skin then limits the coating's drying.

This case of diffusion in polymers is described by ideas drawn from both diffusion in liquids and diffusion in solids. The theoretical development takes place in two steps. First, the binary diffusion coefficient D is corrected for the nonideal solution

$$D = D_0 \left(1 + \frac{\partial \ln \gamma_1}{\partial \ln \phi_1} \right) \tag{5.4-3}$$

where D_0 is a new, "improved" coefficient; γ_1 is the activity coefficient of the small solute; and ϕ_1 is its volume fraction, the appropriate concentration variable to describe concentrations in a polymer solution. We should remember that the activity correction in parentheses has not often been critically examined. As stated above, it is often an over-correction when it is used to describe diffusion in conventional liquids.

We now turn to predicting the corrected coefficient D_0. We expect that this coefficient must include consideration of the solute's activation energy, which must be sufficient to overcome any attractive forces that constrain it near neighbouring polymer segments. We expect that this coefficient must vary with any space or "free volume" between the polymer chains. Only a fraction of this free volume will be accessible to the solute as a result of thermal fluctuations; it is this fraction which permits the diffusion.

While the details of these free-volume arguments are beyond the scope of this book, we can appreciate the arguments involved by looking at the form of the final prediction

$$D_0 = D_0' e^{-E/RT} \exp^{-[(\omega_1 V_{10} + \omega_2 V_{20})/(\omega_1 K_1 + \omega_2 K_2)]} \tag{5.4-4}$$

where D_0' is a constant preexponential factor, E is the solute–polymer attractive energy, and the second exponential is the effect of free volume. More specifically, the ω_i are mass fractions, the V_{i0} are specific critical free volumes, and the K_i are additional free volume parameters. These last parameters are strong functions of temperature. Equation 5.4-4 is successful in correlating experimental data, especially above the polymer's glass transition temperature.

One curious effect, called "non-Fickian diffusion" or "type II transport," sometimes occurs in the dissolution of high polymers by a good solvent. In these cases, diffusion may not follow Fick's law. For example, the speed with which the solvent penetrates into a thick polymer slab may not be proportional to the square root of time, which is the behavior expected from Fick's law (see Section 2.3).

This effect is believed to result from configurational changes in the polymer. As the solvent penetrates, the polymer molecules relax from their greatly hindered configuration as a partially crystalline solid into the more randomly coiled shape characteristic of a polymer dissolved in dilute solution. When this relaxation process is slower than the diffusion process, the dissolution is controlled by the relaxation kinetics, not by Fick's law. Although the process does not involve any phase boundaries, it is similar to a slow interfacial chemical reaction followed by fast diffusion. Again, it is common only in the case of fast dissolution in good solvent.

5.4.3 A Polymer Solute in a Polymer Solvent

In the third limiting case of polymer diffusion, both the solute and the solvent are polymers. This case has practical importance in adhesion, in material failure, and in polymer fabrication. In the simplest terms, this case includes why glue sticks.

Efforts to explain this case of polymer diffusion begin with a model, developed by Rouse, which represents the polymer chain as a linear series of beads connected by springs. The diffusion coefficient derived from this model is

$$D \text{ (Rouse)} = \frac{k_B T}{N\zeta} \tag{5.4-5}$$

where N is the degree of polymerization and ζ is a friction coefficient characteristic of the interaction of a bead with its surroundings. Because N is proportional to the molecular weight, this Rouse diffusion coefficient is proportional to the inverse of the polymer's molecular weight. In contrast, if the polymer were an untangled random coil, D would depend on the inverse square root of the molecular weight; if the polymer really condensed into one small sphere, D would vary with $\tilde{M}^{-1/3}$. The Rouse prediction is not verified experimentally except for polymers of low molecular weight. We need a better model.

The better model, called reptation, imagines the polymer chain confined within a curved tube (deGennes, 1979). Within this tube, the Rouse model governs the chain dynamics, but the polymer diffusion is governed by the time required to escape from the tube. Because motion in the tube is one-dimensional, this escape time τ is given by

$$L^2 = 2D(\text{Rouse})\tau \qquad (5.4\text{-}6)$$

where L is the tube length, proportional to the polymer's molecular weight. The macromolecular diffusion coefficient D, in the three dimensions, can be found from

$$\langle R^2 \rangle = 6D\tau \qquad (5.4\text{-}7)$$

where $\langle R^2 \rangle$ is again the root-mean-square radius of gyration, proportional to the square root of the molecular weight. Combining Eqs. 5.4-5 to 5.4-7, we find

$$D = \left(\frac{k_B T}{3\zeta}\right) \frac{\langle R^2 \rangle}{N L^2} \propto \tilde{M}^{-2} \qquad (5.4\text{-}8)$$

This result frequently comes close to predicting the molecular weight dependence of this case of polymer–polymer diffusion.

5.5 Brownian Motion

The diffusion coefficients listed above are easy to accept as experimentally valuable parameters, but they are harder to understand as a consequence of molecular motion. These coefficients are most often experimental values. In some cases, they are estimated from theories which imply models for the system involved. For gases, this is the model of gas molecules colliding in space. For liquids, they most often imply a solute sphere in a solvent soup. For solids, these estimates are based on a crystal lattice. In every case, the diffusion coefficients are not very directly related to random molecular motions.

In this short section, we want to reexamine these coefficients in terms of molecular motions. Such random "Brownian" motions were first observed in pollen grains by Robert Brown in June of 1827. He concluded that these motions "arose neither from currents in the fluid nor from gradual evaporation but from the particles [themselves]." In our terms, diffusion comes from random molecular motions. Such random motions are now widely studied, not only in physical science but in areas like fluctuations of exchange rates of currencies.

In this section, we describe these random motions in terms of probability theory, and so connect diffusion to this broader topic. Because we want a simple, easily understood connection, we consider only the simplest case of one-dimensional motion. This simplest case depends on three rules:

1.   Each particle moves either to the right or the left every τ seconds with a velocity v.
2.   The probability of moving right and that of moving left is 0.5. Moreover, the particles do not remember their earlier steps.

3.   Each particle moves independently of the others. This is again our old friend, the assumption of dilute solution.

The rules may be relaxed in many ways, but we are interested here only in this simplest limit.

These three rules have two consequences. First, the average position of a particle does not change. To demonstrate this, we consider a system of N independent particles. We then consider $z_i(n)$, the position of the ith particle after n steps. This particle must have arrived from a position either δ larger or δ smaller, i.e.,

$$z_i(n) = z_i(n-1) \pm \delta \tag{5.5-1}$$

Because these steps are random, the mean displacement of these particles after n steps is thus

$$
\begin{aligned}
\langle z(n) \rangle &= \frac{1}{N} \sum_{i=1}^{N} z(n) \\
&= \frac{1}{N} \sum_{i=1}^{N} [z(n-1) \pm \delta] \\
&= \frac{1}{N} \sum_{i=1}^{N} [z(n-1)] \\
&= \langle z(n-1) \rangle
\end{aligned}
\tag{5.5-2}
$$

The average position of the particles doesn't move. For example, if all the particles start at zero, their average position stays at zero.

The second consequence of the three rules given above is the estimation of how much the particles spread out. This can be described as the root mean square of the particle position $\langle z^2(n) \rangle^{1/2}$. To find this quantity, we note from Equation 5.5-1 that

$$z_i^2(n) = z_i^2(n-1) \pm 2\delta z_i(n-1) + \delta^2 \tag{5.5-3}$$

As before, we average this over all the N particles to find

$$
\begin{aligned}
\langle z^2(n) \rangle &= \frac{1}{N} \sum_{i=1}^{N} z_i^2(n) \\
&= \langle z^2(n-1) \rangle + \delta^2
\end{aligned}
\tag{5.5-4}
$$

Now imagine we have zero steps, so

$$\langle z^2(0) \rangle = 0 \tag{5.5-5}$$

For one step

$$\langle z^2(1) \rangle = \delta^2 \tag{5.5-6}$$

For two steps,

$$
\begin{aligned}
\langle z^2(2) \rangle &= \langle z^2(1) \rangle + \delta^2 \\
&= 2\delta^2
\end{aligned}
\tag{5.5-7}
$$

For n steps

$$\langle z^2(n)\rangle = n\delta^2 \tag{5.5-8}$$

As we allow more and more steps, the particle spread becomes more and more.

We want to connect this result with the diffusion coefficients used elsewhere in this book. To do so, we return to the example of the one-dimensional decay of a pulse, given in Equation 2.4-14 as

$$c_1 = \frac{M/A}{\sqrt{4\pi Dt}}\, e^{-z^2/4Dt} \tag{5.5-9}$$

For such a pulse, the standard deviation σ is defined as

$$\sigma^2 = 2Dt \tag{5.5-10}$$

But this standard deviation is exactly the same as the mean square of the particle position $\langle z^2(n)\rangle$. Moreover, the time t for the peak to spread is just $(n\tau)$. Thus

$$\langle z^2(n)\rangle = \left(\frac{t}{\tau}\right)\delta^2 = 2Dt \tag{5.5-11}$$

and

$$D = \frac{\langle z^2(n)\rangle}{2\tau} \tag{5.5-12}$$

The diffusion coefficient is the mean square particle displacement divided by twice the time for movement τ. Another way to write this result recognizes that the mean square distance per time is just the size of a step δ times the time-averaged velocity v:

$$D = \frac{\langle z^2(n)\rangle}{2\tau} = \frac{\delta v}{2} \tag{5.5-13}$$

This form is sometimes easier to apply than the previous equation. Like that previous equation, it is written for one-dimensional diffusion.

These results can be extended in many ways. If the result for one dimension is extended to diffusion in two dimensions

$$D = \frac{\langle z^2(n)\rangle}{4\tau} \tag{5.5-14}$$

For three dimensions, the result is

$$D = \frac{\langle z^2(n)\rangle}{6\tau} \tag{5.5-15}$$

More importantly, the small steps need not be by molecular diffusion but may also be from turbulent velocity fluctuations. In that case, the diffusion coefficient will be replaced by the dispersion coefficient as defined in Chapter 4. Alternatively, we can consider random motions under some sort of external force so that the probability of moving in one direction is different than that for movement in the opposite direction.

While in these cases, this type of calculation will give only rough estimates, the calculation is so easy that it may still be very useful.

Example 5.5-1: Self-diffusion in water   Estimate the diffusion at 25 °C of a trace of tritium-labeled water in regular water. Water molecules are about 0.26 nm in diameter, separated by 0.30 nm.

   Solution   The distance of a step will be $0.30 - 0.26 = 0.04$ nm.
   The velocity is given by

$$\frac{1}{2}mv^2 = k_\mathrm{B}T$$

$$\frac{1}{2}\left(\frac{20g}{6 \cdot 10^{23}}\right)v^2 = \left(1.38 \cdot 10^{-16}\,\frac{\mathrm{g\ cm}^2}{\mathrm{sec\ } K}\right)298K$$

$$v = 5 \cdot 10^4 \mathrm{\ cm/sec}$$

Thus from Eqs. 5.5-13 and 5.5-15

$$D = \frac{0.04 \cdot 10^{-7}\mathrm{\ cm}\ \left(5 \cdot 10^4 \mathrm{cm/sec}\right)}{6}$$

$$= 3 \cdot 10^{-5}\mathrm{cm}^2/\mathrm{sec}$$

This is close to the experimentally observed value.

Example 5.5-2: Random walks in a flake-filled film   We are studying random motions in a composite of aligned impermeable flakes like those shown in Figure 5.5-1(a). When random motions like these are averaged over many trajectories, we get the mean square displacement as a function of the total distance traveled, as shown in Figure 5.5-1(b). If the distance occurs in steps of a unit distance per second, what diffusion coefficient is inferred from these data?

   Solution   The key to this calculation is σ^2, the slope of the data in Figure 5.5-1(b), which is 0.014. The mean square displacement σ^2 varies linearly with the distance traveled, which in this case is numerically equal to the time in seconds. For example, from Equation 5.5-10, we get,

$$D = \frac{\sigma^2}{2t} = \frac{0.014}{2(150)} = 4.7 \cdot 10^{-5}$$

Note this is for diffusion vertically, detouring around the plates. Diffusion in the horizontal direction would give a different distance, and a different diffusion coefficient.

5.6 Measurement of Diffusion Coefficients

   In this section, we want to discuss the most convenient ways in which diffusion coefficients can be measured. This section is the counterpoint to the previous ones.

(a)

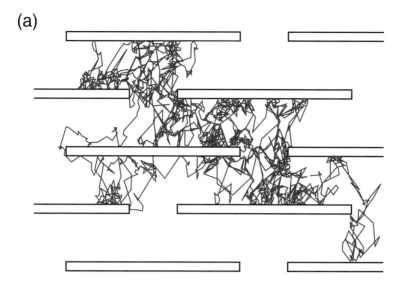

(b)

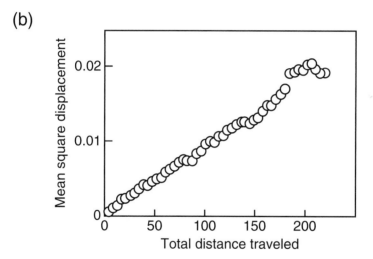

Fig. 5.5-1. Random walk in a flake-filled film. A typical path through the aligned flakes is shown in (a). The distance traveled perpendicular to the flakes is shown vs. the total distance travelled, which is proportional to time t.

Whereas the focus has been on using past experience to guide predictions, this section replaces the hope of prediction with the necessity of accurate measurements.

Measuring diffusion coefficients is reputed to be difficult. For example, Tyrell (1961) stated that "this is not an easy field of study in any sense. It took eighty years from the time when Thomas Graham worked on diffusion before precise data on diffusion coefficients began to be collected." This suggests that measurements of diffusion are a Holy Grail requiring noble knights who dedicate their lives to the quest.

In fact, although measurements are rarely routine, diffusion coefficients usually can be determined to within about five- or ten-percent accuracy without excessive effort. Because such accuracy is sufficient for most situations, we should always consider measuring the coefficients we need. The reputed difficulty of diffusion measurements stems from inherent masochists, like me, who make many of the experiments. We are never satisfied. When we attain coefficients accurate to 10%, we want 2%; when we achieve 2%, we want 0.5%.

If we have decided that measurements are essential, we must decide how to make them. There are many methods available, all described in glowing terms by their proponents. An exhaustive description of these methods could fill this book.

Instead of such an oppresive list, we shall consider only those methods of measuring diffusion that are reasonably accurate, that are easy to use, or that have some special advantage. I have tried below to state concisely the advantages and disadvantages of each method. I want to give the flavor of the laboratories themselves, and not just the polished publications that result.

The most useful methods of studying diffusion are shown in Table 5.6-1. The first three on this list are used most frequently. These three methods give accuracies sufficient for most practical purposes. They and the other methods will be described in greater detail in the following paragraphs.

5.6.1 Diaphragm Cell

The Stokes diaphragm cell is probably the best tool to start research on diffusion in gases or liquids or across membranes. It is inexpensive to build, rugged enough to use in an undergraduate lab, and yet capable of accuracies as high as 0.2%.

Diaphragm cells consist of two compartments separated either by a glass frit [Fig. 5.6-1(a)] or by a porous membrane [Fig. 5.6-1(b)] (Stokes *et al.*, 1950). The two compartments are most commonly stirred at about 60 rpm with a magnet rotating around the cell. Initially, the two compartments are filled with solutions of different concentrations. When the experiment is complete, the two compartments are emptied and the two solution concentrations are measured. The diffusion coefficient D is then calculated from the equation

$$D = \frac{1}{\beta t} \ln \left[\frac{(c_1, \text{bottom} - c_1, \text{top})_{\text{initial}}}{(c_1, \text{bottom} - c_1, \text{top})_{\text{at time } t}} \right] \tag{5.6-1}$$

in which β (in cm^{-2}) is a diaphragm-cell constant, t is the time, and c_1 is the solute concentration under the various conditions given. The detailed derivation of this equation is given in Example 2.2-4.

Four points about the diaphragm cell deserve emphasis. First, calculation of the diffusion coefficients requires accurate knowledge of the concentration differences, not the concentrations themselves. This means that very accurate chemical analyses may be required. For example, imagine we are measuring the diffusion of anthracene in hot decalin. Using gas chromatography, we measure the anthracene concentration as $5.1 \pm 0.1\%$ in the top solution and $6.1 \pm 0.1\%$ in the bottom solution. The concentration difference is then $1.0 \pm 0.2\%$, an error of twenty percent, even though our chemical

Table 5.6-1 *Characteristics of the best methods of measuring diffusion coefficients*

	Nature of diffusion	Apparatus expense	Apparatus construction	Concentrations difference required	Method of obtaining data	Overall value
The three best methods						
Diaphragm cell	Pseudosteady state	Small	Easy	Large	Concentration at known time; requires chemical analysis	Excellent; simple equipment outweights occasionally erratic results
Infinite couple	Unsteady in an infinite slab	Small	Easy	Large	Concentration vs. position at known time; requires chemical analysis	Excellent, but restricted to solids
Particle uptake	Unsteady into particles	Small	Easy	Large	Requires accurate chemical analysis	Good for fast, less accurate measurements
Three more expensive but important methods						
Taylor dispersion	Decay of a pulse	Moderate	Moderate	Average	Refractive index vs. time at known position	Excellent for dilute solutions
Nuclear magnetic resonance	Decay of a pulse	Large	Difficult	None	Change in nuclear spin	Good; works when other methods don't
Dynamic light scattering	Decay of a pulse	Large	Difficult	None	Doppler shift in scattered light	Very good for polymers
Other interesting methods						
Gouy interferometer	Unsteady in an infinite cell	Large	Moderate	Small	Refractive-index gradient vs. position and time is photographed	Very good; excellent data at great effort

(*Continued*)

Table 5.6-1 (*Continued*)

	Nature of diffusion	Apparatus expense	Apparatus construction	Concentrations difference required	Method of obtaining data	Overall value
Rayleigh or Mach–Zehnder interferometer	Unsteady in an infinite cell	Large	Difficult	Small	Refractive index vs. position and time is photographed	Very good; best for concentration-dependent diffusion
Capillary method	Unsteady out of finite cell	Small	Easy	Average	Concentration vs. time; usually requires radioactive counter	Good, but commonly used only with radioactive tracers
Spinning disc	Dissolution of solid or liquid	Small	Easy	Large	Concentration vs. time; requires chemical analysis	Good; requires diffusion-controlled dissolution, a stringent restraint
Wedge interferometer	Unsteady in an infinite cell	Moderate	Easy	Large	Refractive index vs. time is photographed	Fair; much harder to use than many authors suggest
Steady-state methods	Steady diffusion across known length	Moderate	Moderate	Large	Small concentration changes require exception analysis	Fair; easy analysis does not compensate for very difficult experiments

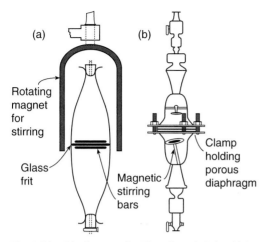

Fig. 5.6-1. Diaphragm cells. The cell on the left, which uses a porous glass frit as a diaphragm, is more accurate than that on the right, which used filter paper as a diaphragm. However, the cell with the glass frit requires a much longer experiment.

analyses are accurate to two percent. As a result, we might do better to use a differential refractometer to try to determine the concentration difference directly.

The second point about the diaphragm cell is the calibration constant β. This quantity is

$$\beta = \frac{A}{l}\left(\frac{1}{V_{\text{top}}} + \frac{1}{V_{\text{bottom}}}\right) \tag{5.6-2}$$

in which A is the area available for diffusion, l is the effective thickness of the diaphragm, and V_{top} and V_{bottom} are the volumes of the two cell compartments. We should note that A is the total area open for diffusion and so is not a strong function of the pore size in the diaphragm. As a rule, small pores are preferred. Large pores may give a slightly larger area, but they often allow accidental mixing caused by flow through the diaphragm. Because A and l are, as a rule, not exactly known, β must be found by experiment. In liquids, this calibration is commonly made with KCl–water or urea–water. Sucrose–water is less reliable because the solution often becomes contaminated by microorganisms. In gases, calibration depends on the method chosen to measure concentration.

The time required for diaphragm-cell measurements is determined by the value of β and hence by the nature of the diaphragm. For accurate work, the diaphragm should be a glass frit, and the experiments may take several days; for routine laboratory work, the diaphragm can be a piece of filter paper, and the experiments may take as little as a few hours. For studies of membrane transport, a piece of membrane can be used in place of the filter paper. For studies in gases, the entire diaphragm can be replaced by a long, thin capillary tube, like the apparatus in Fig. 3.1-2.

The third point is that diffusion should always take place vertically. In other words, the diaphragm should lie in the horizontal plane. If the diaphragm is vertical, free convection can be generated, leading to spurious results. Interestingly, if the diaphragm is horizontal, then placing the more dense solution in the upper compartment may be done without fear of free convection. Many investigators routinely do this, feeling that they get superior results. At the same time, most investigators have done away with the elaborate initial diffusion period suggested in early experiments. This period is significant only when the diaphragm volume is about one-sixth of the compartment volumes (Mills *et al.*, 1968).

The final point about this method is its occasional unreliability. Every good experimentalist subjectively judges the quality of his experiments as he goes along. Most can correctly estimate an experiment's success even without detailed analysis. With the diaphragm cell, however, I have never been able to guess. Experiments I expect to be erratic often are, but experiments that I think are correct sometimes give answers that are in error by an order of magnitude. One of my students minimized such unpleasant surprises by carefully wrapping his cells in a particular brand of plastic bag purchased from a particular store in Cleveland, Ohio. For him, this worked. I have never found a similar trick.

5.6.2 Infinite Couple

This experimental geometry, which is limited to solids, consists of two solid bars of differing compositions, as shown in Fig. 5.6-2. To start an experiment, the two bars are joined together and quickly raised to the temperature at which the experiment is to be made. After a known time, the bars are quenched, and the composition is measured as a function of position. In the past, this analysis was made by grinding off small amounts of bar and determining the composition by a series of wet chemical tests; now, the analysis is made more easily and quickly by an electron microprobe.

Because diffusion in solids is a slow process, the compositions at the ends of the solid bars away from the interface do not change with time. As a result, the concentration profile is that derived in Section 2.3:

$$\frac{c_1 - \bar{c}_1}{c_{1\infty} - \bar{c}_1} = \operatorname{erf}\left(\frac{z}{\sqrt{4Dt}}\right) \tag{5.6-3}$$

in which $c_{1\infty}$ is the concentration at that end of the bar where $z = \infty$ and $\bar{c}_1[= (c_{1\infty} + c_{1-\infty})/2]$ is the average concentration in the bars. The measured concentration profile is fit numerically to find the diffusion coefficient.

It must be remembered that diffusion in solids can be more complex than these paragraphs suggest. Some of this complexity stems from the different mechanisms by which diffusion in solids can occur. More subtle complexities arise from factors like residual stress in metal or the reference velocity on which diffusion is based. Such complexities dictate caution.

The infinite couple is a good method to measure diffusion in solids, but it is tedious. A faster though less accurate method is simply to drop some solid particles into a liquid solution and to measure the solution concentration c_1 as a function of time. At small times, the solute flux out of the solution and into the particles is given by

$$n_1 = j_1 = -\sqrt{\frac{D}{\pi t}} \, Hc_{10} \tag{5.6-4}$$

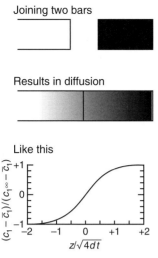

Joining two bars

Results in diffusion

Like this

Fig. 5.6-2. The infinite couple. In this method, two solid bars of different compositions are joined together at zero time. The concentration profiles shown develop with time and are measured chemically.

where c_{10} is the initial concentration in the solution (cf. Equation 2.3-18). From a mass balance on the solution,

$$V\frac{dc_1}{dt} = -An_1 = -A\sqrt{\frac{D}{\pi t}}Hc_{10} \tag{5.6-5}$$

where V is the total volume of solution, A is the total area of particles, and H is the partition coefficient between the solution and the solid particles. This mass balance is subject to the initial condition that

$$t = 0, \quad c_1 = c_{10} \tag{5.6-6}$$

Integrating, we find that

$$\frac{c_1}{c_{10}} = 1 - \left\{\left(\frac{A}{V}\right)\sqrt{\frac{4D}{\pi}}Hc_{10}\right\}\sqrt{t} \tag{5.6-7}$$

Thus a plot of (c_1/c_{10}) vs. the square root of time has a slope which is proportional to the diffusion coefficient D.

I have given this example to illustrate the mathematical approximations which are usually successful in making experimental measurements. In this case, three of these approximations are especially obvious

1. The particles are taken as semi-infinite slabs, so that the flux is accurately described by Equation 5.6-3. This is true only if the time for the experiments is much less than (particle size)$^2/D$.
2. The concentration c_{10} doesn't change during the experiment so that the flux remains that given by Equation 5.6-3. This assumption seems especially foolish because our experiment depends on measuring changes in c_1.

3.   The solution is well mixed so its concentration c_1 has the same value throughout the liquid, even right up to the solid particles. This is often true even if the liquid is not mixed because the diffusion coefficient in the liquid is so much greater than that in the solid.

Each of these three assumptions is serious and initially not obvious. If any one of these is not accurate, our calculations of diffusion using this method may be seriously in error.

However, in my experience the use of Eq. 5.6-7 does give accurate values of the diffusion coefficient. Thus the three assumptions above must be reasonably accurate, and the chief limitation of the experiment is accurately measuring the concentration c_1. This accuracy is essential because we are basing our calculation on a concentration difference $(c_{10} - c_1)$, a small difference between large numbers. This fact is the key for this experiment, as it was for the diaphragm cell.

In my experience, most novices measuring diffusion do not concentrate on this experimental measurement but rather on improving the mathematics behind Equation 5.6-7. These novices assume a finite slab and solve the diffusion equations for that case, getting results like those in Section 3.5. They include the variation of solution concentration with time, performing an analysis like that in Example 3.5-3. These novices are then dismayed that their results are poorly reproduceable, and they conclude that their mathematics is incorrect. It often isn't; it is unnecessary. The novices need instead to focus on their measurement of concentration.

The reason that so many novices make mistakes like this is that in their training, they practice harder and harder mathematics. They rarely practice better and better experimental accuracy. Thus this example has a moral: Please, when you start making measurements, use the simplest analysis possible until you are sure from experiment that it is inadequate.

5.6.3 *Taylor Dispersion*

We now turn to more complex and more expensive methods, which can also be easier to run or which give more accurate results. The first of these is Taylor dispersion, illustrated schematically in Fig. 5.6-3 (Ouano, 1972). This method, which is valuable for both gases and liquids, employs a long tube filled with solvent that slowly moves in laminar flow. A sharp pulse of solute is injected near one end of the tube. When this pulse comes out the other end, its shape is measured with a differential refractometer. Except for the refractometer, which can be purchased off the shelf, the apparatus is inexpensive and moderately easy to build. This apparatus can be used routinely by those with little training. It can be operated relatively easily at high temperature and pressure. It has the potential to give results accurate to better than one percent.

The concentration profile found in this apparatus is that for the decay of a pulse (see Section 4.2):

$$c_1 = \frac{M}{\pi R^2} \frac{\exp^{-(z - vt)^2/4Et}}{\sqrt{4\pi Et}} \tag{5.6-8}$$

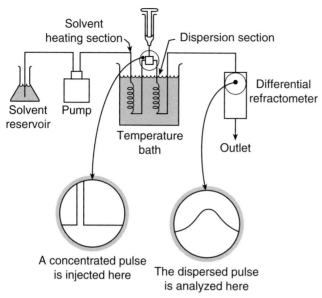

Fig. 5.6-3. The Taylor dispersion method. A sharp pulse is injected into a tube filled with flowing solvent. The dispersed pulse is measured at the tube's outlet. Interestingly, the pulse is dispersed more if the diffusion is slow.

where M is the total solute injected, R is the tube radius, v^0 is the average velocity of the flowing solvent, and E is a dispersion coefficient given by

$$E = \frac{(v^0 R)^2}{48\,D} \tag{5.6-9}$$

Because the refractive index varies linearly with the concentration, knowledge of the refractive-index profile can be used to find the concentration profile and the diffusion coefficient.

The fascinating aspect of this apparatus is the way in which the diffusion coefficient appears. Equation 5.6-8 has the same mathematical form as Eq. 2.4-14, but the dispersion coefficient E replaces the diffusion coefficient. So far, as good. However, E varies inversely with D, as explained in Section 4.4. Consequently, a widely spread pulse means a large E and a small D. A very sharp pulse indicates small dispersion and hence fast diffusion.

5.6.4 Spin Echo Nuclear Magnetic Resonance

The next two methods, spin echo nuclear magnetic resonance and dynamic light scattering, represent the adoption of expensive, complex equipment built to obtain molecular information to the new task of measuring diffusion. Because neither method tries only to measure diffusion coefficients, the accuracy is modest. Neither method requires an initial concentration difference, a major convenience in highly viscous systems. The real attraction of each system is the promise that existing equipment can be reapplied to the new objective of measuring diffusion.

Diffusion coefficients can be measured with nuclear magnetic resonance to an accuracy of around five percent. To do so, we first place a homogeneous sample in a large magnetic field. This external field aligns the magnetic moments of the atomic nuclei in the solute of interest. When the magnetic field is slightly perturbed, the atomic moments process, which can induce in an adjacent coil a small voltage of amplitude A oscillating with time:

$$A = A_0 \sin(t/\tau) \tag{5.6-10}$$

The period of this oscillation τ is normally the focus of interest, for it gives information about the local chemical environment.

Our interest is not in the period τ but in the amplitude A_0. To study this amplitude, we apply a second perturbation in the magnetic field. This second "pulsed gradient" is applied not in time, but in space. It is applied first in one direction and then – after a short time τ' – in the opposite direction. If the solute molecules were fixed in space, the two perturbations in space would produce no change in the amplitude A_0. However, these molecules aren't fixed but are moving by Brownian motion, so the amplitude A_0 is reduced.

We can measure this amplitude reduction as a function of the time τ' between the gradient pulses. The slope of this variation is a direct measure of the Brownian motion and hence of the diffusion coefficient. Thus if we make measurements on a solute of known diffusion coefficient and a solute of unknown diffusion coefficient, we can find the unknown as

$$\frac{D^*(\text{unknown})}{D^*(\text{known})} = \frac{(\partial A_0/\partial \tau')(\text{unknown})}{(\partial A_0/\partial \tau')(\text{known})} \tag{5.6-11}$$

Strictly speaking, such a measurement is not of the binary diffusion coefficient D but of the tracer diffusion coefficient D^* (cf. Section 7.5). In dilute solution, these have the same value.

5.6.5 Dynamic Light Scattering

Like nuclear magnetic resonance, dynamic light scattering uses expensive equipment for a relatively easy measurement of the diffusion coefficient. Like nuclear magnetic resonance, the measurement requires no initial concentration difference, and so is especially suited to viscous solutions. Unlike nuclear magnetic resonance, the measurement is of the binary coefficient, not the tracer diffusion coefficient.

Dynamic light scattering depends on measuring the autocorrelation function of scattered light as a function of scattering angle and time. To understand the method, we must first consider what happens to a wave of light traveling through the solution which we are studying. The wave will move in a constant direction until it strikes an inhomogeneity. Then part of the wave may be scattered by a changed impedence, that is, by an altered resistance to its motion that is proportional to the refractive index of the solution. How the light is scattered depends on how the inhomogeneities in the solution are organized. If the solution contains a completely random array of inhomogeneities, then the scattering will be the same in all directions. However, if the solution contains a perfectly ordered array of inhomogeneities, then the scattering will exist only at particular angles, called Bragg diffraction angles. At these angles, scattering results from constructive interference when scatterers are exactly an integral number of wavelengths apart. At all other angles, scattering produces destructive interference.

For the important case of concentrated polymer solutions, the scattering results from a solution that is between a random array and an ordered array. Each monomer unit can be considered a point scatterer; while the polymer molecules are randomly distributed in the solution, monomer units are not because they are part of polymer chains. However, the polymer molecules do move relative to each other because of Brownian motion. Hence any apparent order in the solution will decay with time.

This decay of order is measured as an autocorrelation function by the dynamic light scattering apparatus. Such a function gives the correlation between the solution's order at some arbitrary time zero and at some second time t. When t is near zero, the autocorrelation function is near one: The order hasn't changed much. When t becomes large the autocorrelation function is near zero: Any apparent order has vanished, replaced by a new apparent structure. In many cases, this decay can be described as a first-order exponential:

$$\langle A(0)A(t)\rangle \propto e^{-q^2 D t} \tag{5.6-12}$$

where $\langle A(0)A(t)\rangle$ is the autocorrelation function, D is the binary diffusion coefficient, t is the time, and q is the "scattering vector":

$$q = \frac{4\pi}{\lambda}\sin\left(\frac{\theta}{2}\right) \tag{5.6-13}$$

where λ is the wavelength of the scattered light and θ is the scattering angle.

Thus measurements of the autocorrelation function versus time allow calculation of the diffusion coefficient D. In practice, the range of diffusion coefficients that we can measure is determined by the scattering vector q, which has dimensions of reciprocal length. Roughly speaking, q^{-1} is a measure of the distance over which the measurement is being made. For visible light with a wavelength of 500 nanometers, we sample a distance of around 100 nm; for neutrons with a wavelength of 1 nanometer, we sample distances around 3 nm. Still, the important point is that the dynamic light scattering method provides a measurement of binary diffusion especially suitable for polymer solutions.

5.6.6 Some Very Accurate Methods

So far, we have discussed three easy methods and three more highly instrumented methods for measuring diffusion coefficients. Each of these six methods can give results accurate to a few percent, a suitable goal for most research. If higher accuracy is needed, we should turn to the interferometers shown in Fig. 5.6-4. These instruments depend on measuring an unsteady-state refractive index profile in a transparent system,

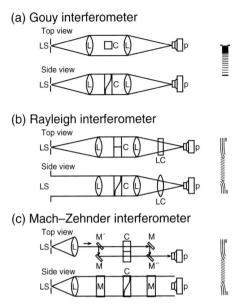

Fig. 5.6-4. Interferometers for accurate diffusion measurements. These three instruments can be expensive to build and hard to operate, but they give very accurate results. Each produces interference fringes like those shown at the right of each schematic. LS, light source; L, collimating lens; C, diffusion cell; LC, cylindrical lens; M, mirror; M', M'', half-silvered mirrors.

and so they are most useful for liquids. Their high accuracy is purchased at a great cost of both equipment and effort.

The interferometers differ optically. The Gouy interferometer, shown schematically in Fig. 5.6-4(a), is the more highly developed, accurate to better than 0.1%. It is relatively simple to build and easy to align. If one already has a method for measuring the interference fringes, this instrument is not particularly expensive. The Gouy method has been so highly developed that the extremely specialized jargon used in its operation may discourage newcomers. In fact, the experiments are simple to do; the hardest step is to understand the theory well enough to write the appropriate computer program. Average results with this instrument are at least equivalent to the best results obtained with any other device.

The Gouy interferometer measures the refractive-index gradient between two solutions that are diffusing into each other. The basic apparatus for measuring the gradient uses the lenses L to send parallel light rays from a light source LS through a diffusion cell C. If this cell contains a refractive-index profile, then light passing through the center of the cell will be deflected to produce an interference pattern of black horizontal lines, as shown at the right in Fig. 5.6-4(a). The amount of this deflection is proportional to the refractive-index gradient, a function of cell position and time.

The Mach–Zehnder and Rayleigh interferometers are solid alternatives to the Gouy interferometer. Although they are difficult to construct and adjust, they give information that is simpler to interpret. In the Mach–Zehnder apparatus, shown in Fig. 5.6-4(c), collimated light is split by the first half-silvered mirror M'. Half the light passes through each of the twin cells C and is recombined by the second half-silvered mirror M". In the Rayleigh apparatus, these mirrors are replaced by a cylindrical lens, shown in Fig. 5.6-4(b).

Both instruments measure refractive index versus cell position. If both cells contain homogeneous solutions, the interference fringes are sets of parallel vertical lines; if one cell contains a refractive-index gradient caused by diffusion, the interference fringes look like those shown at the right of Figs. 5.6-4(b) and (c). For both interferometers, these fringes can be used to calculate the diffusion coefficient.

5.6.7 Other Methods

The remaining common methods for measuring diffusion are listed in Table 5.6-1 roughly in order of their value. None of these methods is commonly superior to those described above, although each may be useful in specific cases.

The capillary method is most suitable for measurements with radioactive tracers. It uses a small diffusion cell made of precision-bore capillary tubing, perhaps 3 cm long and 0.05 cm in diameter. One end of this cell is sealed shut. After the cell is filled with a solution of known concentration, it is dropped into a large, stirred, thermostated solvent bath. At the end of the experiment, the cell is removed and the solute concentration within the cell is measured. The diffusion coefficient D can then be found from the equation

$$\frac{\bar{c}_1}{c_{10}} = \frac{8}{\pi^2} \sum_{n=1}^{\infty} \frac{1}{(2n-1)^2} \exp{-\pi^2(2n-1)^2(Dt/4l^2)} \tag{5.6-14}$$

in which c_{10} and $\bar{c}_1$ are the average concentrations in the cell at times zero and t, respectively, and l is the length of the cell.

Four characteristics of this method deserve mention. First, with careful technique it is accurate to better than 0.3%. The caveat is "careful technique"; it is unusually easy to fool yourself with this equipment, obtaining reproducible inaccurate results. Second, the small size of the diffusion cell dictates careful chemical analysis of very small volumes of solution. In practice, this suggests using either radioactive tracers or some other microanalytical method. Third, the power series in Eq. 5.6-14 converges rapidly. If you use reasonably long experiments, you can base your analysis on the first term in the series. Finally, for radioactive tracers this method may give an intradiffusion coefficient, not a binary coefficient (cf. Section 7.5).

The spinning-disc method depends on a solid or liquid disc of solute slowly rotating in a solvent volume (see Fig. 3.4-3). The solute concentrations in the solvent are analyzed versus time. If the disc's dissolution is diffusion-controlled, these concentrations allow calculation of the diffusion coefficient from Example 3.4-3 (Levich, 1962). If the disc's dissolution is not diffusion-controlled, we must choose another method.

The wedge interferometer is cheap and cute, a simple alternative to the expensive interferometers described earlier. It consists of two microscope slides separated at one edge with a coverslip. To start an experiment, one places drops of two different solutions next to each other on one slide. One then places the other slide and coverslip so that the drops are in contact in a wedge-shaped channel. When this wedge is put in a microscope, interference fringes indicate the concentration profile. Measuring the change of fringe position versus time allows calculation of the diffusion coefficient simply, cheaply, and approximately. Moreover, because only drops of solution are needed, one needs only very small amounts of solute.

The last entry in Table 5.6-1 refers to steady-state methods. These methods are like the diaphragm cell, but they replace the two well-stirred compartments with two flowing solutions. In principle such a replacement gives a true steady state, simplifying the analysis. In practice, the methods are a nightmare. The two solutions must flow at exactly the same rate, so expensive pumps and valves are needed. The experiments can consume huge amounts of solution. My advice is to choose a complex analysis and a simple unsteady experiment.

5.7 A Final Perspective

The characteristics of diffusion coefficients described in this chapter are summarized in Table 5.7-1. In general, diffusion coefficients in gases and in liquids can often be accurately estimated, but coefficients in solids and in polymers cannot. In gases, estimates based on the Chapman–Enskog kinetic theory are accurate to around ten percent. In liquids, estimates are based on the Stokes–Einstein equation or its empirical parallels. These estimates, accurate to around twenty percent, can be supplemented by a good supply of experimental data. In solids and polymers, theories allow coefficients to be correlated but rarely predicted.

These common generalizations help to solve only the routine problems with which we are faced. Many problems remain. For example, we may want to know the rate at which hydrochloric acid diffuses into oil-bearing sandstone. We may need to estimate the drying speed of lacquer. We may seek the rate of flavor release from lemon pie filling. All these examples depend on diffusion; none can be accurately estimated with the common generalizations.

Table 5.7-1 *A comparison of diffusion coefficients and their variations*

Phase	Typical value cm^2/sec	Variations with				Remarks
		Temperature	Pressure	Solute size	Viscosity	
Gases	10^{-1}	$T^{3/2}$	p^{-1}	(Diameter)$^{-2}$	μ^{+1}	Successful theoretical predictions
Liquids	10^{-5}	T	Small	(Radius)$^{-1}$	μ^{-1}	Can be concentration dependent
Solids	10^{-30}	Large	Small	(Lattice spacing)$^{+2}$	Not applicable	Wide range of values
Polymers	10^{-8}	Large	Small	(Molecular Weight)$^{(-0.5 \text{ to } -2)}$	Often small	Involves different special cases

Note: These heuristics summarize the more detailed discussions in this chapter.

In some cases, diffusion coefficients can be adequately estimated by more carefully considering the chemistry. Specific cases, discussed in the next chapter, include electrolytes and critical points. However, in most nonroutine problems the detailed chemistry is not known and experiments are essential. The primer on experiments given in this chapter should be your initiation.

Questions for Discussion

1. What are typical values of diffusion coefficients in gases, liquids, and solids?
2. If the diffusion of hydrogen in nitrogen gas is 0.78 cm^2/sec at 1 bar, what will it be at 50 bars?
3. Describe an experiment to measure the diffusion of oxygen in nitrogen. List any equipment needed.
4. Diffusion in liquids commonly assumes a rigid sphere in a continuum. When would this model be most accurate? When could it fail?
5. How would the diffusion coefficient of a protein vary with its molecular weight?
6. Describe an experiment to meaure the diffusion of glucose in water. List any equipment needed.
7. What are the limits of the diffusion of an ellipsoid as the ratio of axes (a/b) becomes very large?
8. Diffusion varies with viscosity to the ($+1$) power in gases but to the (-1) power in liquids. Why?
9. Why does hydrogen diffuse so much faster in metals than other solutes do?
10. Diffusion in metals often varies strongly with temperature in metals with an activation energy ΔH around 100 kJ/mol. What are the corresponding activation energies in gases and in liquids?

Problems

1. Estimate the diffusion coefficient of carbon dioxide in air at 740 mm Hg and 37 °C. How does this compare with the experimental value of 0.177 cm^2/sec? *Answer:* about 4% low.

2. As part of a course on diffusion, you are to measure the diffusion coefficient of ammonia in 25 °C air, using the two-bulb capillary apparatus shown in Fig. 3.1-2. In your apparatus, the bulbs have volumes of about 17 cm^3, and the capillary is 2.6 cm long and 0.083 cm in diameter. You are told that you should make your measurements when the concentration difference is about half the initial value. (a) Use the Chapman–Enskog theory to estimate how long you should run your experiment. *Answer:* 3.6 hrs (b) Why are you told to make your measurement near this particular concentration difference?

3. Estimate the diffusion coefficient at 25 °C of traces of ethanol in water and of traces of water in ethanol. Compare your estimates with the experimental values of $0.84 \cdot 10^{-5}$ cm^2/sec and $1.24 \cdot 10^{-5}$ cm^2/sec, respectively.

4. Tobacco mosaic virus has been shown by electron microscopy to be shaped like a cylinder 150 Å in diameter and 3,000 Å long. Its molecular weight is about 40 million, and its partial specific volume is 0.73 cm^3/g. Estimate the diffusion coefficient of this

material and compare with the experimental value at 25 °C of $3 \cdot 10^{-8}$ cm²/sec. *Answer:* $2.7 \cdot 10^{-8}$ cm²/sec.

5. Estimate the diffusion coefficient of lactic acid under each of the following conditions: (a) in air at room temperature and pressure; (b) in milk in the refrigerator; (c) through the wall of a plastic milk bottle.

6. In an experiment to determine the diffusion coefficient of urea in water at 25 °C with the diaphragm cell, you find that a density difference of 0.01503 g/cm³ decays to 0.01090 g/cm³ after a time of 16 hrs and 23 min. The cell's calibration constant is 0.397 cm⁻². If the density of these solutions varies linearly with concentration, what is the diffusion coefficient? Compare your answer with the value of $1.373 \cdot 10^{-5}$ cm²/sec obtained with the Gouy interferometer. *Answer:* $1.37 \cdot 10^{-5}$ cm²/sec.

7. The concentration profile of Ni_2SiO_4 diffusing into Mg_2SiO_4 is given below [M. Morioka, *Geochim Cosmochim Acta*, **45**, 1573 (1981)]. These data were found after 20 hrs using an infinite couple at 1,350 °C. Calculate the diffusion coefficient in this system. *Answer:* $1.2 \cdot 10^{-11}$ cm²/sec.

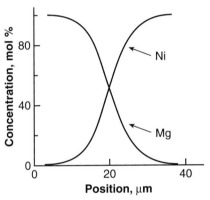

8. The ionic diffusion coefficient D or, more exactly, the ionic conductivity λ can frequently be described by the equation

$$\lambda = \frac{a_0}{T} exp^{-E/RT}$$

For β-alumina, the following values are obtained:

[G. C. Farrington and J. L. Briant, *Science*, **204**, 1371 (1979)]. (a) Calculate the ionic

	$R_0(\text{Å})$	$a_0(\text{K/ohm-cm})$	$E(\text{kcal/mol})$
Li^+	0.68	54	2.9
Na^+	0.98	2,500	2.4
H_3O^+	1.32	81,000	11.9
K^+	1.33	1,500	4.6

conductivity at 25 °C for each of these ions. (b) Show that these conductivities can be as large as that in 1-M KCl, in which the diffusion coefficient is $2.0 \cdot 10^{-5}$ cm²/sec. (c) Because we usually expect transport in solids to be much slower than transport in liquids, we recognize that β-alumina is an exceptional material. Discuss the factors that might cause this effect.

9. Jeng-Ping Yao and D. N. Bennion [*J. Phys. Chem.*, **75**, 3586 (1971)] measured the electrolytic conductance of aqueous solutions of tetra-*n*-amylammonium thiocyanate at 55 °C. The data are most easily presented graphically (see below). Note that this salt is a liquid at this temperature and is completely miscible with water; so the measurements go all the way from mass transfer at infinite dilution through to mass transfer in the molten salt. As detailed in Section 6.1, specific conductance is approximately equivalent to the diffusion coefficient times the ionic concentration. Use your knowledge of diffusion to suggest how the data at high salt concentration might be conveniently correlated.

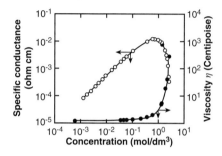

10. Diffusion in molten silicate deep within the earth is central to many of the chemical processes that take place there. However, the diffusion coefficients in such magma seem to vary widely. For example, for cesium ion dissolved in obsidian at 2 kilobars pressure,

$$D = 8 \cdot 10^{-2} \exp^{-49.9 \, \text{kcal}/RT} [=] \text{cm}^2/\text{sec}$$

For cesium ion dissolved in obsidian containing 6 wt% water,

$$D = 7 \cdot 10^{-5} \exp^{-19.52 \, \text{kcal}/RT} [=] \text{cm}^2/\text{sec}$$

[E. B. Watson, *Science*, **205**, 1259 (1979)]. (a) How much does the diffusion coefficient at 800 °C differ in the dry and the water-saturated samples? (b) The reason for this difference is not known. Assume that the water causes thin pores to form, and diffusion in the pores is that in bulk water. What is the pore area per obsidian area?

Further Reading

Barrer, R.M. (1941). *Diffusion in and through Solids.* New York: Macmillan.

Chapman, S. and Cowling, T.G. (1970). *The Mathematical Theory of Non-Uniform Gases*, 3rd ed. Cambridge: Cambridge University Press.

Chen, S.H., Davis, H.T., and Evans, D.F. (1981). *Journal of Physical Chemistry*, **75**, 1422.

Cunningham, R.E. and Williams, R.J.J. (1980). *Diffusion in Gases and Porous Media.* New York: Plenum.

Cussler, E.L. (1976). *Multicomponent Diffusion.* Amsterdam: Elsevier.

Darken, L.S. (1948). *Transactions of the American Institute of Mining, Metallurgical and Petroleum Engineers,* **175**, 184.

deGennes, P.G. (1979). *Scaling Concepts in Polymer Physics.* Ithaca: Cornell University Press.

Einstein, A. (1905). *Annalen der Physik,* **17**, 549.

Fuller, E.N., Schettler, P.D., and Giddings, J.C. (1966). *Industrial and Engineering Chemistry,* **58**, 19.

Glasstone, S., Laidles, K.J., and Eyring, H. (1941). *Theory of Rate Processes.* New York: McGraw-Hill.

Hartley, G.S. and Crank, J. (1949). *Transactions of the Faraday Society,* **45**, 801.

Hirschfelder, J., Curtiss, C.F., and Bird, R.B. (1954). *Molecular Theory of Gases and Liquids.* New York: Wiley.

Levich, V. (1962). *Physiochemical Hydrodynamics.* Englewood Cliffs, NJ: Prentice-Hall.

Marrero, T.R. and Mason, E.A. (1972). *Journal of Physical Chemistry Reference Data,* **1**, 1.

Ouano, A.C. (1972). *Industrial and Engineering Chemistry Fundamentals,* **11**, 268.

Perrin, F. (1936). *Journal de Physique et al Radium,* **7**, 1.

Poling, B.E., Prausnitz, J.M., and O'Connell, J.P. (2001). *The Properties of Gases and Liquids.* New York: McGraw-Hill.

Stokes, R.H. (1950). *Journal of the American Chemical Society,* **72**, 763, 2243; (1951) 73, 3528.

Sutherland, W. (1905). *Philosophical Magazine,* **9**, 781.

Tyrell, H.J.V. (1961). *Diffusion and Heat Flow in Liquids.* London: Butterworth.

Vignes, A. (1966). *Industrial and Engineering Chemistry Fundamentals,* **5**, 189.

Wilke, C.R. and Chang, P.C. (1955). *American Institute of Chemical Engineers Journal,* **1**, 264.

CHAPTER 6

Diffusion of Interacting Species

In this chapter, we turn to systems in which there are significant interactions between diffusing molecules. These interactions can strongly affect the apparent diffusion coefficients. In some cases, these effects produce unusual averages of the diffusion coefficients of different solutes; in others, they suggest a strong dependence of diffusion on concentration; in still others, they result in diffusion that is thousands of times slower than expected.

The discussion of these interactions involves a somewhat different strategy than that used earlier in this book. In Chapters 1–3, we treated the diffusion coefficient as an empirical parameter, an unknown constant that kept popping up in a variety of mathematical models. In more recent chapters, we have focused on the values of these coefficients measured experimentally. In the simplest cases, these values can be estimated from kinetic theory or from solute size; in more complicated cases, these values require experiments. In all these cases, the goal is to use our past experience to estimate the diffusion coefficients from which diffusion fluxes and the like can be calculated.

In this chapter, we consider the chemical interactions affecting diffusion much more explicitly, rather than hiding them as part of the empirically measured diffusion coefficient. The interactions affecting diffusion are conveniently organized into three groups. As a first group, we consider in Section 6.1 solute–solute interactions, particularly in strong electrolytes. We want to discover how sodium chloride diffusion is an average of the diffusion of sodium ions and of chloride ions. In Section 6.2, we turn to the transport of associating solutes like weak electrolytes and dyes. We want to know how the total diffusion of acetic acid varies from dilute solutions, where it is almost completely ionized, to concentrated solutions, where it is almost completely unionized.

The second group of interactions affecting diffusion involves solute–solvent interactions. In Section 6.3, we explore the extremely large solute–solvent interactions which occur near the spinodal limit, where phase separation is incipient. Diffusion in these regions leads to the phenomenon of spinodal decomposition, which is also discussed in Section 6.3.

In the last section of this chapter, we summarize diffusion affected by solute–boundary interactions, which is the third important group of interactions. Solute–boundary interactions occur in porous solids with fluid-filled pores. They include such diverse phenomena as Knudsen diffusion, capillary condensation, and molecular sieving. Because these phenomena promise high selectivity for separations, they are an active area for research. They and the other interactions illustrate the chemical factors that can be hidden in the diffusion coefficients which are determined by experiment.

6.1 Strong Electrolytes

Every high school chemistry student knows that when sodium chloride is dissolved in water, it is ionized. Sodium chloride in water does not diffuse as a single

161

molecule; instead, the sodium ions and chloride ions move separately through the solution. The movement of the ions means that a 0.1-M sodium chloride solution passes an electric current one million times more easily than water does. The large ion size relative to electrons means that such a solution passes current ten thousand times less easily than a metal does.

The diffusion of sodium chloride can be accurately described by a single diffusion coefficient. Somehow this does not seem surprising, because we always refer to sodium chloride as if it were a single solute and ignore the knowledge that it ionizes. We get away with this selective ignorance because the sodium and chloride ions diffuse at the same rate. If they did not do so, we could easily separate anions from cations.

Values of ionic diffusion coefficients are given in Table 6.1-1. These data, which are hidden in the literature of electrochemistry, are obtained by a variety of experimental methods, including tracer diffusion determinations. The table shows that different ions have different diffusion coefficients. The proton and the hydroxyl ion are unusually fast; big fat organic ions like tetrabutylammonium and tetraphenylborate are slow. Somewhat surprisingly, a potassium ion diffuses faster than a lithium ion does. This suggests that in aqueous solution, a potassium ion is smaller than a lithium ion. These sizes are unexpected from crystallographic measurements on the solid state that show the potassium ion is larger. The sizes in solution occur because the potassium ion is less strongly hydrated than the lithium, as discussed in Section 6.2-4.

The anomalously high value for protons merits discussion. This high value is inconsistent with the ion's size, which would suggest a more normal value. The reason for this behavior is that proton transport occurs by a different "Grotthus" mechanism. In this mechanism, shown schematically in Fig. 6.1-1, a proton does not move through water as an intact entity. Instead, it reacts with a water molecule, forcing a proton off the other side. This newly generated molecule reacts again to produce a third proton; this third proton continues the chain reaction. This transport may also involve proton tunnelling.

Another interesting result in Table 6.1-1 is that the sodium ion diffuses more slowly than the chloride ion. In other words, the sodium ion does not have the same diffusion

Table 6.1-1 *Diffusion coefficients of ions in water at 25 °C*

Cation	D	Anion	D
H^+	9.31	OH^-	5.28
Li^+	1.03	F^-	1.47
Na^+	1.33	Cl^-	2.03
K^+	1.96	Br^-	2.08
Rb^+	2.07	I^-	2.05
Cs^+	2.06	NO_3^-	1.90
Ag^+	1.65	CH_3COO^-	1.09
NH_4^+	1.96	$CH_3CH_2COO^-$	0.95
$N(C_4H_9)_4^+$	0.52	$B(C_6H_5)_4^-$	0.53
Ca^{2+}	0.79	SO_4^{2-}	1.06
Mg^{2+}	0.71	CO_3^{2-}	0.92
La^{3+}	0.62	$Fe(CN)_6^{3-}$	0.98

Note: Values at infinite dilution in 10^{-5} cm^2/sec. Calculated from data of Robinson and Stokes (1960).

Fig. 6.1-1. Proton diffusion in water. Proton diffusion occurs by the chain reaction shown between water molecules. Such a jump mechanism also exists in alcohol, but not in alcohol–water mixtures.

coefficient as the chloride ion. However, because sodium chloride diffuses with only one coefficient, the ionic diffusion coefficients must somehow be combined to give an average value. We shall now calculate this average, first for a simple 1-1 electrolyte like sodium chloride and then for more complicated electrolytes. With these results as a basis, we shall then briefly discuss electrical conductance.

6.1.1 Basic Arguments

Imagine a large, fat grandfather taking a small rambunctious girl for a walk. The rate at which the two travel will be largely determined by the grandfather. He will move slowly, even ponderously, toward their goal. The girl may run back and forth, taking many more steps and so covering more distance, but her progress will be dominated by her elder.

In the same way, the diffusion of a large, fat cation and a small, quick anion will be dominated by the slower ion. The diffusion will proceed as does the walk, and the smaller ion may move around more. However, the two ions are tied together electrostatically, and so their overall progress will be the same and will tend to be dominated by the slower ion (Fig. 6.1-2).

To examine this analogy more exactly, we must first write a flux equation for ion diffusion. In this effort, we consider only dilute solutions, like those in Chapter 2, and so ignore problems like the complicated reference velocities of Chapter 3. The obvious choice of a flux equation is the simplest form of Fick's law, which for a sodium ion will be

$$ -j_{Na} = D_{Na} \nabla c_{Na} \tag{6.1-1} $$

However, we quickly realize that this choice is inadequate, for it suggests that an electric field will not affect diffusion.

To include this electric field, we return to the argument used to derive the Stokes–Einstein equation in Section 5.2: that the ion velocity is proportional to the sum of all the forces acting on the ion. In symbolic terms, this is

$$ \left(\begin{matrix} \text{ion} \\ \text{velocity} \end{matrix} \right) = \left(\begin{matrix} \text{ion} \\ \text{mobility} \end{matrix} \right) \left(\begin{matrix} \text{chemical} \\ \text{forces} \end{matrix} + \begin{matrix} \text{electrical} \\ \text{forces} \end{matrix} \right) $$

$$ v_i = -u_i (\nabla \mu_i + z_i \mathcal{F} \nabla \psi) \tag{6.1-2} $$

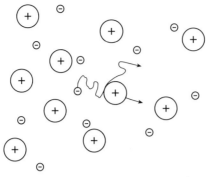

Fig. 6.1-2. Electrolyte diffusion. The two ions have the same charge and are present at the same local concentration. The larger cations (the positive ions) inherently move more slowly than the smaller anions (the negative ions). However, because of electroneutrality, both ions have the same net motion and hence the same flux.

where u_i is the ion mobility, z_i is the ionic charge (equal to $+1$ for Na^+), $\mathcal{F}$ is Faraday's constant, and ψ is the electrostatic potential.

Each of these terms deserves discussion. First, the mobility u_i is a physical property of the ion, a phenomenological coefficient that must be measured by experiment. This mobility is often taken to be $1/6\pi\mu R_0$, which, we recall, is a feature of the Stokes–Einstein equation. In fact, the use of this value simply restates our ignorance of mobility in terms of an effective ion radius, R_0.

Because the mobility is almost equivalent to the diffusion coefficient, it is something of a cultural artifact. It is included here because many papers dealing with electrolyte transport report their results in terms of mobilities, not in terms of diffusion coefficients. Faraday's constant is even more of a cultural artifact: it is a unit conversion factor explicitly included whenever this equation is written. The apparent supposition is that no one can properly use electrostatic units without a warning.

The charge and potential in Eq. 6.1-2 make explicit the electrical effects connecting the ions. Including the charge seems sensible; note that if the ion has a negative charge, the direction of the electrical effect is reversed. The potential also looks sensible. It has two distinct parts. One part includes the effect of any potential applied to the system, for example, by electrodes attached to a battery. A second part is the potential generated by the different diffusion rates of diffusion ions. For example, for sodium chloride, the potential includes the electrostatic interaction of the quicker chloride ions and the more sluggish sodium ions. It is thus the route by which we average ion diffusion coefficients.

To rewrite Eq. 6.1-2 as a flux relation, we take advantage of the fact that we are working in dilute solution and so assume that the solution is ideal:

$$\nabla\mu_i = \frac{RT}{c_i}\nabla c_i \qquad\qquad (6.1\text{-}3)$$

When this result is combined with Eq. 6.1-2, we get

$$-\,v_i = \frac{[u_i RT]}{c_i}\left(\nabla c_i + c_i z_i \frac{\mathcal{F}\nabla\psi}{RT}\right) \qquad\qquad (6.1\text{-}4)$$

which is equivalent to the flux equation

$$-j_i = -c_i v_i$$
$$= [RTu_i]\left(\nabla c_i + c_i z_i \frac{\mathcal{F} \nabla \psi}{RT}\right)$$
$$= [D_i]\left(\nabla c_i + c_i z_i \frac{\mathcal{F} \nabla \psi}{RT}\right) \tag{6.1-5}$$

These relations, sometimes called the Nernst–Planck equations (Bard and Faulkner, 2000), could be written down directly as a definition for D_i. If this were done, then the restriction to dilute solutions in Eq. 6.1-3 and the implicit neglect of a reference velocity in the first line of Eq. 6.1-5 would be hidden in the final flux equation, lumped into the experimental coefficient D_i. I find the derivation a sensible, reassuring rationalization, even though I know that it is arbitrary.

6.1.2 1-1 Electrolytes

We now want to describe the ion fluxes of a single strong 1-1 electrolyte. Such an electrolyte ionizes completely, producing equal numbers of cations and anions. Although the concentrations of anions and cations may vary through the solutions, the concentrations and the concentration gradients of these species are equal everywhere because of electroneutrality:

$$c_1 = c_2$$
$$\nabla c_1 = \nabla c_2 \tag{6.1-6}$$

where 1 and 2 refer to cation and anion, respectively. Like the ion concentrations, the ion fluxes are also related.

$$j_1 - j_2 = i/|z| \tag{6.1-7}$$

where $|z|$ is the magnitude of the ionic charge and i is the current density in appropriate units. This current density is defined as positive when it goes from positive to negative. To find the electrolyte flux, we first return to the basic flux equation for each ion:

$$-j_1 = D_1(\nabla c_1 + |z|c_1 \mathcal{F} \nabla \psi / RT) \tag{6.1-8}$$

$$-j_2 = D_2(\nabla c_2 - |z|c_2 \mathcal{F} \nabla \psi / RT) \tag{6.1-9}$$

These equations can be combined with Eq. 6.1-7 to find the current:

$$|z|i = D_2 \nabla c_2 - D_1 \nabla c_1 - (D_1 c_1 + D_2 c_2)|z|\mathcal{F} \nabla \psi / RT \tag{6.1-10}$$

But this equation now allows $\nabla \psi$ to be removed from the flux equations:

$$-j_1 = \frac{2D_1 D_2}{D_1 + D_2} \nabla c_1 - \frac{D_1}{D_1 + D_2}(i/|z|) \tag{6.1-11}$$

where we have used the fact that $c_1 = c_2$ to simplify the final expression. A similar equation for the anion flux j_2 can be derived.

Two important limits of the flux j_1 exist. First, when there is no current,

$$j_1 = j_2 = -D\nabla c_1 = -\left[\frac{2}{1/D_1 + 1/D_2}\right]\nabla c_1 \tag{6.1-12}$$

The quantity in brackets is the average diffusion coefficient of the electrolyte. Because it is a harmonic average of the diffusion coefficients of the individual ions, it is dominated by the slower ion. However, there is only one diffusion coefficient for the two diffusing ions because the ions are electrostatically coupled.

The second interesting limit of Eq. 6.1-11 occurs when the solution is well mixed, so that no gradients of anion and cation exist. In this case,

$$j_1 = [t_1](|z|i) = \left[\frac{D_1}{D_1 + D_2}\right](i/|z|) \tag{6.1-13}$$

$$j_2 = [t_2](-|z|i) = -\left[\frac{D_2}{D_1 + D_2}\right](i/|z|) \tag{6.1-14}$$

where the t_i, equal to the quantities in brackets, are the transference numbers, that is, the fractions of current transported by specific ions. Unlike the diffusion coefficient, these transference numbers are arithmetic averages of the ion diffusion coefficients. As a result, the transference numbers and the current in solution are both dominated by the faster ion.

Example 6.1-1: Diffusion of hydrogen chloride What is the diffusion coefficient at 25 °C for a very dilute solution of HCl in water? What is the transference number for the proton under these conditions?

Solution From the data in Table 6.1-1, the ionic diffusion coefficients are $9.31 \cdot 10^{-5}$ cm^2/sec for H^+ and $2.03 \cdot 10^{-5}$cm^2/sec for Cl^-. The electrolyte diffusion coefficient is given by Eq. 6.1-12:

$$D_{HCl} = \left[\frac{2}{1/D_{H^+} + 1/D_{Cl^-}}\right] = 3.3 \cdot 10^{-5} \text{cm}^2/\text{sec}$$

The slow ion dominates. The result is only 1.5 times greater than the chloride's diffusion coefficient, but it is 3.5 times less than the proton's diffusion coefficient.

The transference number, t_{H^+}, can be found in a straightforward manner from Eq. 6.1-13:

$$t_{H^+} = \frac{D_{H^+}}{D_{H^+} + D_{Cl^-}} = 0.82$$

The faster protons carry eighty-two percent of the current.

6.1.3 Non-1-1 Electrolytes

We now turn from the simple 1-1 electrolytes to more complicated electrolytes. Mathematical description of non-1-1 electrolytes is parallel to that developed earlier but more complex algebraically. The basic flux equation is the same as Eq. 6.1-5:

$$-j_i = D_i(\nabla c_i + c_i z_i \mathcal{F}\nabla\psi/RT) \tag{6.1-15}$$

The constraints on concentration and flux at zero current are

$$z_1c_1 + z_2c_2 = 0 \tag{6.1-16}$$

and

$$z_1j_1 + z_2j_2 = 0 \tag{6.1-17}$$

When the electrostatic potential is eliminated, the diffusion equation for ion 1 becomes

$$-j_1 = D\nabla c_1 = \left[\frac{D_1D_2(z_1^2c_1 + z_2^2c_2)}{D_1z_1^2c_1 + D_2z_2^2c_2}\right]\nabla c_1 \tag{6.1-18}$$

where the quantity in brackets is D, the diffusion coefficient of the electrolyte.

This equation can be somewhat misleading because of the unequal charge. For example, imagine that we are interested in the diffusion of very dilute solutions of calcium chloride. If the calcium is ion 1, then its flux will be half the flux of chloride. When only one electrolyte is present, we may wish to rewrite this equation in terms of the total electrolyte flux j_T and the total electrolyte concentration c_T, defined as

$$j_T = j_1/|z_2| = j_2/|z_1| \tag{6.1-19}$$

$$c_T = c_1/|z_2| = c_2/|z_1| \tag{6.1-20}$$

The diffusion equation for a single non-1-1 electrolyte now becomes

$$-j_T = D\nabla c_T = \left[\frac{|z_1| + |z_2|}{|z_2|/D_1 + |z_1|/D_2}\right]\nabla c_T \tag{6.1-21}$$

where the quantity in brackets is again the diffusion coefficient of the non-1-1 electrolyte.

This diffusion forms a curious contrast with the special case of a 1-1 electrolyte described by Eq. 6.1-12. Both equations involve a type of harmonic average of the ionic diffusion coefficients. Thus we might expect that both cases are more strongly influenced by the slower ion. However, if this slower ion has a much larger charge than the faster ion, the faster ion may come to dominate the diffusion, because the harmonic average is weighted by the ion charge. The effect of this weighting can be more clearly shown by examples.

Example 6.1-2: Diffusion of lanthanum chloride What is the diffusion coefficient of 0.001-M lanthanum chloride?

Solution From Table 6.1-1, the diffusion coefficients of La^{3+} and Cl^- are $0.62 \cdot 10^{-5}cm^2/sec$ and $2.03 \cdot 10^{-5}cm^2/sec$, respectively. In water, the average coefficient can be found either from Eq. 6.1-18 or from Eq. 6.1-21. From Eq. 6.1-21, taking La^{3+} as ion 1 and chloride as ion 2, we get

$$D = \frac{|z_1| + |z_2|}{|z_1|/D_2 + |z_2|/D_1}$$

$$= \left[\frac{|3| + |-1|}{|3|/2.03 \cdot 10^{-5} + |-1|/0.62 \cdot 10^{-5}}\right] cm^2/sec$$

$$= 1.29 \cdot 10^{-5}cm^2/sec$$

From Eq. 6.1-18, because $c_1 = 0.001$ M and $c_2 = 0.003$ M, we can find the same result.

Example 6.1-3: Diffusion of lanthanum chloride in excess sodium chloride How will the result of the previous example be changed if the lanthanum chloride diffuses through 1 M NaCl?

    Solution Answering this question requires the assumption that there are no ternary diffusion effects in this system. These effects may arise because the diffusion of sodium ion couples with the diffusion of chloride ion, which in turn affects the diffusion of La^{3+}. However, these effects vanish for any solute present in high dilution, as $LaCl_3$ is in this case (see Section 7.4).

    Because of the added sodium chloride we cannot use Eq. 6.1-21, which is valid only for a single non-1-1 electrolyte. We can use Eq. 6.1-18. If we again label lanthanum as ion 1 and chloride as ion 2, we recognize that c_1 equals 0.001 M, but c_2 is about 1 M. These unequal concentrations mean that Eq. 6.1-18 becomes

$$-j_1 = \frac{D_1 D_2 (z_2^2 c_2)}{D_1 z_1^2 c_1 + D_2 z_2^2 c_2} \nabla c_1$$

$$= D_1 \nabla c_1$$

In other words, the diffusion of the lanthanum chloride is $0.62 \cdot 10^{-5} \text{cm}^2/\text{sec}$, which is the same as the solitary ion. Thus the diffusion of dilute $LaCl_3$ in concentrated NaCl is dominated by the diffusion of the uncommon ion, La^{3+}.

6.1.4 Diffusion versus Conductance

    Although diffusion is a very common process, diffusion coefficients can be difficult to measure. This is true for most of the systems discussed in this book, including solutions of electrolytes. However, for electrolyte solutions, the electrical resistance and its reciprocal, the electrical conductivity, are very easy to measure. Nothing in my experimental experience is as satisfying as a conductance experiment: I get fantastically accurate results with embarrassingly little effort. Because diffusion and conductance give similar information about the system, it is worth comparing the two processes in some detail.

    The conductance of a single electrolyte in solution is most easily measured in cells like those shown in Fig. 6.1-3. The electrical resistance of the stirred solution is measured with a rapidly oscillating AC field of fixed maximum voltage, so that the solution remains homogeneous throughout the experiment. The resistance is inversely proportional to the current through the cell, but the current, in turn, is proportional to the ion fluxes:

$$(\text{resistance})^{-1} = K_{\text{cell}} \, i = K_{\text{cell}} (z_1 j_1 + z_2 j_2) \tag{6.1-22}$$

The proportionality constant K_{cell} in this relation is a function of the electrode area, the electrode separation, and the cell shape. It is found by calibration of the cell, most commonly with a potassium chloride solution.

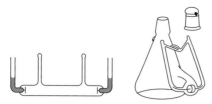

Fig. 6.1-3. Conductance cells. These cells are used to measure with extremely high accuracy the resistance of an electrolyte solution. This information is related to the diffusion coefficient of the electrolyte. As a result, a conductance experiment sometimes is a superior method of studying diffusion.

The ion fluxes in the cell are described by equations analogous to those used for ion diffusion. First, we assume that the ion flux is proportional to the ion concentration:

$$\mathbf{j}_i = c_i \mathbf{v}_i \tag{6.1-23}$$

We also assume that the ion velocity is proportional to the electrical force acting on the ion:

$$\mathbf{v}_i = -u_i z_i \mathcal{F} \nabla \psi \tag{6.1-24}$$

where, as in Eq. 6.1-2, u_i is the ion mobility and ψ is the electrostatic potential acting on the ions. Because in this case the solution is homogeneous, the concentration gradient is zero. The only flux comes from the electrostatic potential applied by the electrodes.

We now can combine Eqs. 6.1-22 through 6.1-24 to find an expression for the resistance in terms of the ion mobilities:

$$(\text{resistance})^{-1} = K_{\text{cell}}(z_1^2 c_1 u_1 + z_2^2 c_2 u_2)\mathcal{F}\nabla\psi \tag{6.1-25}$$

The ion concentrations are related to the total concentration c_T by

$$c_T = c_1/|z_2| = c_2/|z_1| \tag{6.1-26}$$

Equations 6.1-25 and 6.1-26 can now be combined and simplified to define the most convenient measure of conductivity, the equivalent conductance:

$$\Lambda = |z_1|u_1 + |z_2|u_2$$

$$= \{(\text{resistance})[K_{\text{cell}}\mathcal{F}\nabla\psi]|z_1 z_2|c_T^{-1}\} \tag{6.1-27}$$

The quantity Λ is most frequently reported in studies of conductance. It can be measured by determining each of the quantities in the braces. Because the gradient is fixed, the entire quantity in brackets can be treated as a cell constant.

The equivalent conductance Λ can be extremely accurately measured, often to accuracies of 0.01%. It is known to vary slightly with concentration, as shown in Fig. 6.1-4. This variation follows the equation

$$\Lambda = \Lambda_0 - S\sqrt{c_T} + Ec_T \ln c_T + Jc_T + J'c_T^{3/2} \tag{6.1-28}$$

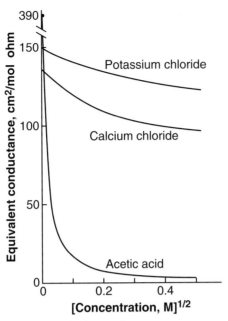

Fig. 6.1-4. Equivalent conductance versus concentration. Conductance varies with concentration, especially at high dilution. For strong electrolytes like KCl and CaCl$_2$, these variations are chemically interesting but practically unimportant. For weak electrolytes like acetic acid, the variation is larger (see Section 6.2).

where Λ_0, S, E, J, and J' are all constants. The limiting equivalent conductance Λ_0 is a property of the ions and is not well understood theoretically. The limiting slope S, first calculated by Onsager, is a function only of the charges on the ions and is thus characteristic of electrostatic interactions between the ions. The higher constants, E, J, and J', include more electrostatic interactions, ion–solvent interactions, and the ion associations more commonly encountered with weak electrolytes.

In many practical problems, the ion transport is well described by assuming that Λ is a constant. After all, the concentration variations are less than twenty percent for aqueous solutions of most strong electrolytes. Some solution chemists who attack this assertion ignore the ion properties implicit in Λ_0 and instead extol those contained in E, J, and J'. If your purpose is knowledge of ion properties, listen to the chemists. If your purpose is knowledge of mass transfer, assume that Λ is a constant.

We now want to relate the equivalent conductance Λ to ion properties and, more specifically, to ion diffusion coefficients. First, because the ions migrate independently in a dilute-solution conductance experiment, we can define, from Eq. 6.1-27,

$$\Lambda = \lambda_1 + \lambda_2 \tag{6.1-29}$$

where

$$\lambda_i = |z_i| u_i \tag{6.1-30}$$

The λ_i, called equivalent ionic conductances, cannot be found from measurements of Λ alone, but require other independent determinations, most commonly the transference numbers given in Eqs. 6.1-13 and 6.1-14. The λ_i depend not only on the ion mobility but also on the charge. More specifically, if two cations have the same size but not the same charge, they will have the same mobility, though not the same equivalent ionic conductance. Note also that Λ is related to the sum of the ionic properties λ_i and hence is an arithmetic average of the ionic properties. In contrast, diffusion is a harmonic average, as shown in Equations 6.1-12 and 6.1-21.

The equivalent ionic conductances are closely related to the ionic diffusion coefficients through the mobilities:

$$D_i = k_B T u_i$$

$$= \left[\frac{k_B T}{|z_i|}\right] \lambda_i \tag{6.1-31}$$

This result is not often used, even though it is simple and valuable. Part of the reason for this neglect is the λ_i are most commonly expressed in "conductance units," which are mercilessly square centimeters per mole ohm. The conversion at 25 °C is

$$D_i \left([=]cm^2/sec\right) = \frac{2.662 \cdot 10^{-7}}{|z_i|} \lambda_i \left([=]cm^2/mol\,ohm\right) \tag{6.1-32}$$

This relation was used to find some of the values in Table 6.1-1.

Equation 6.1-32 suggests that conductance measurements might be a substitute for those of diffusion and other aspects of mass transfer. This would be appealing, because conductance is much easier to measure. Why not measure conductance and forget diffusion?

This idea has both merit and risk. The merit is the simplicity; the two methods do give closely related information. The risk is that the solutes must ionize completely. This effectively restricts these measurements to water, and that is why easily measured conductance is less often reported than difficultly determined diffusion.

Example 6.1-5: Calcium chloride diffusion from conductance Estimate the diffusion coefficient of $CaCl_2$ from conductance measurements. The equivalent ionic conductance at infinite dilution is 59.5 for Ca^{2+} and 76.4 for chloride. The experimental value of the diffusion coefficient is about $1.32 \cdot 10^{-5} cm^2/sec$.

Solution From Eq. 6.1-32 we can find the ionic diffusion coefficients

$$D_{Ca} = 0.79 \cdot 10^{-5} cm^2/sec$$

$$D_{Cl} = 2.03 \cdot 10^{-5} cm^2/sec$$

The diffusion coefficient can be found from these ionic values by using Eq. 6.1-21:

$$D_{CaCl_2} = \left[\frac{2+1}{(2/2.03)+(1/0.79)}\right] \cdot 10^{-5}$$

$$= 1.33 \cdot 10^{-5} cm^2/sec$$

This result is accurate in very dilute solution. At higher concentrations, the diffusion coefficient drops to about $1.1 \cdot 10^{-5}$ cm²/sec at 0.2 M and then rises slightly.

6.2 Associating Solutes

We now switch from solutes that dissociate completely to form ions to solutes that associate to form aggregates. We again want to find the diffusion coefficient averaged over the various species present.

The analysis of these systems began when Arrhenius (1884) suggested that materials like acetic acid partially dissociate in water. Many who study diffusion vaguely remember this variation but ignore it in their experiments. Interestingly, the diffusion of such solutes can lead to curious and dramatic results. These results have been scattered through different academic disciplines and so have tended to be ignored. As an example, consider diffusion of potassium chloride across two thin membranes. The first membrane is just a thin layer of water. The steady-state flux across this membrane is given by

$$j_{KCl} = j_K = -D \frac{dc_K}{dz} \tag{6.2-1}$$

where $c_K = c_{Cl} = c_{KCl}$; and the diffusion coefficient D is the average of the ionic values (cf. Eq. 6.1-12):

$$D = \frac{2}{\frac{1}{D_K} + \frac{1}{D_{Cl}}} \tag{6.2-2}$$

The flux equation is subject to the constraints

$$z = 0, \quad c_{KCl} = C_{KCl,0} \tag{6.2-3}$$

$$z = l, \quad c_{KCl} = 0 \tag{6.2-4}$$

where the $C_{KCl,0}$ is the concentration adjacent to but outside the membrane, and integrating, we find the usual result:

$$j_{KCl} = \frac{D}{l} C_{KCl,0} \tag{6.2-5}$$

In other words, if we double the KCl concentration, we double the flux across this water-filled membrane.

The results for the second thin membrane are different. This membrane consists of a chloroform solution of a macrocyclic polyether, again separating two aqueous solutions. Because the dielectric constant of this second membrane is low, the potassium and chloride ions are largely associated as ion pairs: The ions are stuck together with electrostatic glue. The solute that is diffusing is now actually KCl, and not K^+ and Cl^-. To analyze diffusion in this case, we again begin with the flux equation:

$$j_{KCl} = -D \frac{dc_{KCl}}{dz} \tag{6.2-6}$$

where the diffusion coefficient D is now that of the ion pairs, not an average of the ions. The boundary considerations on this flux equation are:

$$z = 0, \quad c_{KCl} = KC_K C_{Cl} = KC_{KCl}^2 \tag{6.2-7}$$

$$z = l, \quad c_{KCl} = 0 \tag{6.2-8}$$

where the uppercase variables are outside the membrane, and K is a combined partition coefficient and association constant across this membrane's interface. Notice how Eq. 6.2-7 implicitly assumes the fast reaction:

$$\begin{bmatrix} \text{K}^+ \text{ ion in water adjacent} \\ \text{to the membrane} \end{bmatrix} + \begin{bmatrix} \text{Cl}^- \text{ ion in water adjacent} \\ \text{to the membrane} \end{bmatrix}$$

$$\underset{\rightleftharpoons}{\overset{K}{}} \begin{bmatrix} \text{KCl ion pairs at membrane boundary} \\ \text{but within the membrane} \end{bmatrix} \tag{6.2-9}$$

It is just as if a chemical dimerization converted the ions into a new chemical species. As before, we integrate Eq. 6.2-6 to find

$$j_{KCl} = \left[\frac{DK}{l} c_{KCl,0}^2 \right] \tag{6.2-10}$$

The flux is now proportional to the square of the potassium chloride concentration. This square dependence is verified experimentally, as shown in Fig. 6.2-1.

In some cases, we may not be sufficiently astute to realize that the diffusing solutes are associating. For example, if we still thought that the ions – not the ion pairs – were diffusing, then we might analyze our data with the equation

$$j_{KCl} = \frac{D_{apparent}}{l} c_{KCl,0} \tag{6.2-11}$$

When we plotted our results, we would discover that this apparent coefficient varied strongly with concentration. In the example given here, we easily see why this variation occurs:

$$D_{apparent} = DK c_{KCl,0} \tag{6.2-12}$$

In other cases, we may not have the chemical insight to understand why the diffusion coefficient varies with concentration. This section analyzes how concentration-dependent

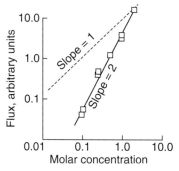

Fig. 6.2-1. Potassium chloride flux across an organic membrane. In these experiments, a concentrated solution of KCl diffuses across a polyether–chloroform membrane into pure water. The flux observed is not proportional to the salt's concentration but to this concentration squared. This effect occurs because potassium and chloride ions associate within the membrane to form ion pairs. [Data from Reusch and Cussler (1973).]

diffusion may result from solute association. Three cases are important: weak electrolytes, detergents, and dyes. Each is discussed below.

6.2.1 Weak Electrolytes

Weak electrolytes will produce solutions of a cation, an anion and a molecule in equilibrium with each other. For example, an aqueous solution of acetic acid contains hydrated protons, acetate ions, and acetic acid molecules, all in local equilibrium as the result of fast association.

We want to describe steady-state diffusion in this associating system. To do so, we write mass balance on the acetate ions (species 1) and on the acetic acid molecules (species 2):

$$0 = -\frac{dj_1}{dz} - r \qquad (6.2\text{-}13)$$

$$0 = -\frac{dj_2}{dz} + r \qquad (6.2\text{-}14)$$

where r is the rate of formation of the molecules (the "dimers"). We add Eq. 6.2-13 to Eq. 6.2-14, and integrate to find the total flux j_T:

$$-j_T = -j_1 - j_2 = D_1\frac{dc_1}{dz} + D_2\frac{dc_2}{dz} \qquad (6.2\text{-}15)$$

where D_1 is the average diffusion coefficient of the ions. For example, for protons and acetate, it is

$$D_1 = \frac{2}{\left[\dfrac{1}{D_H} + \dfrac{1}{D_{CH_3COO}}\right]} \qquad (6.2\text{-}16)$$

We assume Eq. 6.2-15 is subject to boundary conditions like those of a thin membrane:

$$z = 0, \quad c_1 = C_{10}, \quad c_2 = C_{20} \qquad (6.2\text{-}17)$$

$$z = l, \quad c_1 = 0, \quad c_2 = 0 \qquad (6.2\text{-}18)$$

Integrating again, we find

$$j_T = \frac{D_1 C_{10}}{l} + \frac{D_2 C_{20}}{l} \qquad (6.2\text{-}19)$$

In general, we do not know the species concentrations C_{10} and C_{20}. We do know that these are related to the total acetic acid concentration

$$C_T = C_{10} + C_{20} \qquad (6.2\text{-}20)$$

We also know that they are interdependent:

$$C_{20} = KC_{10}^2 = KC_{H^+}C_{CH_3COO^-} \qquad (6.2\text{-}21)$$

where K is the association constant for the diffusing species. We then can rewrite the flux in terms of this constant to find

$$j_T = \left\{ \frac{D_1}{2KC_T}\left(-1 + \sqrt{1 + 4KC_T}\right) + \frac{D_2}{4\,KC_T}\left(-1 + \sqrt{1 + 4KC_T}\right)^2 \right\} \frac{C_T}{l} \quad (6.2\text{-}22)$$

The quantity in braces is the apparent diffusion coefficient of the weak electrolyte.

The apparent diffusion coefficient of the weak electrolyte is concentration depen-dent, the result of the solute–solute association. The physical significance of this con-centration dependence may be clearer if we consider two limits. First, in dilute solutions ($4KC_T \ll 1$), the apparent coefficient equals D_1, the ionic value. This makes sense because dilute solutions will show complete ionization. Second, in concentrated solution ($4KC_T \gg 1$), the apparent coefficient reduces to D_2: molecular diffusion is paramount. Thus concentration-dependent diffusion of weak electrolytes shown by Fig. 6.2-2 reflects association.

Example 6.2-1: Diffusion of acetic acid What is the diffusion coefficient of the acetic acid molecule if the apparent diffusion coefficient of acetic acid is $1.80 \cdot 10^{-5}$ cm^2/sec at 25 °C and 10 M? The pK_a of acetic acid is 4.756.

Solution The pK_a of a weak acid HA is defined as

$$pK_a = -\log_{10} \frac{[H^+][A^-]}{[HA]}$$

In this case, the $[H^+]$ and $[A^-]$ concentrations are equal. Comparing this with Eq. 6.2-21 we see that

$$K = 10^{pK_a} = 5.70 \cdot 10^4 \, 1/\text{mol}$$

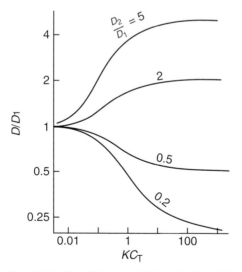

Fig. 6.2-2. The diffusion coefficient of a dimerizing solute. As a solute dimerizes, its average diffusion coefficient changes from that of the monomer to that of the dimer. The concentration C_T at which this occurs is roughly the reciprocal of the association constant K.

If we insert this into Eq. 6.2-22, we find that the term containing D_2 dominates completely, and

$$D_2 \doteq 1.80 \cdot 10^{-5} \mathrm{cm}^2/\sec$$

In passing, note that the diffusion coefficient of the fully ionized acid found from Eq.6.1-12 and Table 6.1-1 is $1.95 \cdot 10^{-5}$ cm^2/sec.

6.2.2 Micelle Formation

We now want to calculate the average diffusion coefficient for solutes that aggregate much more than the simple weak electrolytes discussed earlier. Three cases of this aggregation are shown in Fig. 6.2-3. The one dramatic case is the detergent sodium dodecylsulfate (SDS). Molecules of this detergent remain separate at low concentration but then suddenly aggregate. The resulting aggregates, called "micelles," are most commonly visualized as an ionic hydrophilic skin surrounding an oily hydrophobic core (Fig. 6.2-4(a)). In fact, detergents clean in this way: they capture oil-bearing particles in their cores.

In contrast, molecules of the dye Orange II aggregate gently, resulting in a slow and steady deviation from the unaggregated limit. Such aggregation results from a stacking of dye molecules, like that shown schematically in Fig. 6.2-4(b). When the ease of stacking is the same for all sizes in the stack, this aggregation is called "isodesmic." The third case involving the bile salt taurodeoxycholate is intermediate between the other two.

The two situations of micelle formation and isodesmic stacking represent two limiting forms of solute aggregation. These two limits are discussed in the following paragraphs.

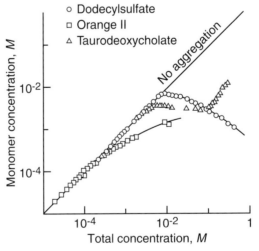

Fig. 6.2-3. Types of solute aggregation. The detergent sodium dodecylsulfate aggregates abruptly to form micelles, and the dye Orange II has its isodesmic aggregates (see Fig. 6.2-4). The bile salt sodium taurodeoxycholate falls between these two limits. These results were obtained using ion-selective electrodes. [Data from Kale, Cussler, and Evans (1980).]

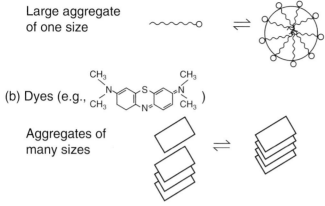

(a) Long-chain surfactants (e.g., $CH_3(CH_2)_{11}SO_4^-$)

Large aggregate
of one size

(b) Dyes (e.g.,)

Aggregates of
many sizes

Fig. 6.2-4. Micelle formation and isodesmic aggregation. In the type of micelle formation discussed here, n monomers combine to form an n-mer. No other sizes are present. In isodesmic association, monomers add with equal facility to monomers or aggregates of any size.

The diffusion coefficient measured in a detergent solution represents an average over the monomer and micelle present in solution. Steady-state diffusion in such a system of monomer and micelle obeys the continuity equations.

$$0 = D_1 \frac{d^2 c_1}{dz^2} - n r_m \tag{6.2-23}$$

$$0 = D_m \frac{d^2 c_m}{dz^2} + r_m \tag{6.2-24}$$

where the subscripts 1 and m refer to the monomer and the micelle, respectively and r_m represents the rate of formation of micelles. Equation 6.2-24 is multiplied by n, added to Eq. 6.2-23 and integrated to give

$$-j_T = D_1 \frac{dc_1}{dz} + n D_m \frac{dc_m}{dz} \tag{6.2-25}$$

The integration constant j_T is the total flux of the solute.

Equation 6.2-25 is not useful because it is written in terms of the unknown gradients of c_1 and c_m, rather than in terms of the known total solute gradient c_T. To remove these unknowns, we could assume that micelle formation is fast, so that

$$c_m = K c_1^n \tag{6.2-26}$$

where K is the equilibrium constant for the fast micelle-forming reaction. We would also need the mass balance:

$$c_T = c_1 + n c_m \tag{6.2-27}$$

We would like to combine Eqs. 6.2-25 through 6.2-27 to get the answer we want. However, while we did this easily in the case of ionic association, we now have an nth-order equation for micelle formation. We can't solve this easily.

To get an approximate solution, we first recognize that detergent solutions typically have physical properties like conductance and surface tension that suddenly change at a critical concentration at which micelles start to form in significant numbers. Above this "critical micelle concentration" c_{CMC}, the monomer concentration c_1 is approximately equal that at the critical micelle concentration, so from Eq. 6.2-27,

$$c_m = \frac{1}{n}(c_T - c_{CMC}) \tag{6.2-28}$$

An estimate of c_1 can now be found from Eq. 6.2-26:

$$c_1 = \left[\frac{1}{nK}(c_T - c_{CMC})\right]^{1/1n} \tag{6.2-29}$$

Inserting these results into Eq. 6.2-25 we find

$$-j_T = \left[D_m + \frac{D_1}{n}\frac{(nK)^{1/n}}{(c_T - c_{CMC})^{1-(1/n)}}\right]\frac{dc_T}{dz} \tag{6.2-30}$$

or, because n is large,

$$-j_T = \left[D_m + \frac{D_1(nK)^{1/n}}{n(c_T - c_{CMC})}\right]\frac{dc_T}{dz} \tag{6.2-31}$$

which is the desired result. The quantity in square brackets is the apparent diffusion coefficient found experimentally.

To my surprise, this analysis works for nonionic detergents. The apparent diffusion coefficient does vary inversely with $(c_T - c_{CMC})$, as shown in Fig. 6.2-5. The intercept on

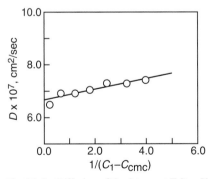

Fig. 6.2-5. Diffusion of the detergent Triton X-100 at 25 °C. The variation with concentration is predicted by Eq. 6.2-31. The intercept is the micelle's diffusion coefficient, and the slope is related to the monomer's diffusion coefficient. [From Weinheimer *et al.* (1981), with permission.]

this plot agrees closely with the micelle's diffusion coefficient estimated in other ways. The slope is consistent with independent measurements of K and n.

However, this analysis does not work for ionic detergents at low ionic strength. For example, the diffusion coefficient of sodium dodecylsulfate increases significantly at concentrations above the critical micelle concentration, as shown in Fig. 6.2-6. This increase is of electrostatic origin, due to small relatively mobile counter ions. At high ionic strength, these electrostatic effects are less important, and Eq. 6.2-31 is again verified.

6.2.3 *Isodesmic Association*

As the next topic in this section, we want to calculate the average diffusion coefficient for systems in which aggregation occurs one molecule at a time. The simplest case is called the isodesmic model. It assumes that

$$c_i = Kc_{i-1}c_1 \tag{6.2-32}$$

where K is an equilibrium constant that is independent of the size of the aggregate. Note that the equilibrium constant for forming dimers from two monomers is assumed to be the same as that for forming heptamers from hexamers and monomers.

Equations 6.2-26 and 6.2-32 show why the isodesmic model and micelle formation represent two extreme limits of solute aggregation. In the isodesmic case, aggregates of any size form with equal facility because all the steps are equal. In the micelle case, aggregates form only of that special micelle of n monomers; the equilibrium constants are zero for all but that special size.

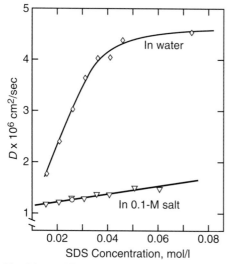

Fig. 6.2-6. Diffusion of sodium dodecylsulfate (SDS) at 25 °C. The diffusion coefficients in this case increase as SDS concentration and solution viscosity rise. This increase is the result of aggregation and electrostatic interaction. [Data from Weinheimer *et al.* (1981).]

To find the apparent diffusion coefficient of a solute associating isodesmically, we again start with the steady-state continuity equations:

$$0 = D_1 \frac{d^2 c_1}{dz^2} - 2r_2 - r_3 - r_4 - \cdots \tag{6.2-33}$$

$$0 = D_2 \frac{d^2 c_2}{dz^2} + r_2 - r_3 \tag{6.2-34}$$

$$0 = D_3 \frac{d^2 c_3}{dz^2} + r_3 - r_4 \tag{6.2-35}$$

$$\vdots$$

Again, these equations can be added together to eliminate reaction terms:

$$0 = \sum_{i=1}^{\infty} i D_i \frac{d^2 c_i}{dz^2} \tag{6.2-36}$$

Integrating this result gives

$$-j_T = \sum_{i=1}^{\infty} i D_i \frac{dc_i}{dz} \tag{6.2-37}$$

where j_T is again an integration constant physically equal to the total solute flux in both aggregated and monomer forms.

As earlier in this section, we now rewrite the unknown concentrations $\{c_i\}$ in terms of the known total concentration of solute. Doing this requires two constraints. One of these is that of isodesmic equilibria (Eq. 6.2-32). The other is a mass balance:

$$c_T = c_1 + 2c_2 + 3c_3 + \cdots$$

$$= \sum_{i=1}^{\infty} i c_i \tag{6.2-38}$$

When these constraints are combined, we find

$$c_T = \frac{c_1}{(1 - Kc_1)^2} \tag{6.2-39}$$

This quadratic can be solved for c_1 as a function of c_T, and the result combined with Eqs. 6.2-32 and 6.2-37 gives the total flux j_T as a function of total solute concentration c_T. This solution is an algebraic mess. A more useful form is the power series

$$-j_T = \left\{ D_1 - Kc_T(4D_1 - 4D_2) + K^2 c_T^2 (15D_1 - 24D_2 + 9D_3) \right.$$

$$\left. -K^3 c_T^3 (56D_1 - 112D_2 + 72D_3 + 16D_4 + \cdots) \right\} \frac{dc_T}{dz} \tag{6.2-40}$$

Note that the apparent diffusion coefficient given in braces does not vary with concentration if the diffusion coefficients are all equal (i.e., if $D_1 = D_2 = D_3 = \cdots$).

6.2.4 Solvation

So far, we have been discussing the effects of solute–solute interactions on the diffusion coefficient. These interactions can be electrostatic, like the case of strong electrolytes. There, our goal was to find the apparent diffusion coefficient that averaged the coefficients of the species present. These solute–solute interactions can also reflect solute association. The association might form a dimer, as for weak electrolytes. The association might produce larger aggregates, as for dyes and detergents. In that case, our goal was to understand the apparent diffusion coefficient.

We now switch from solute–solute interactions to solute–solvent interactions. The first of these occurs when solute and solvent combine to form a new species, which is that actually diffusing. This combination is most carefully studied for water, where it is called hydration. We will discuss other forms of solute–solute interactions in Section 6.3.

The idea of hydration is based on the following flux equation:

$$-j_1 = D_0\left(1 + \frac{\partial \ln \gamma_1}{\partial \ln c_1}\right)\nabla c_1 = \frac{k_B T}{6\pi\mu R_0}\left(1 + \frac{\partial \ln \gamma_1}{\partial \ln c_1}\right)\nabla c_1 \tag{6.2-41}$$

in which D_0 is a new diffusion coefficient, μ is the solvent viscosity, R_0 is the solute radius, and γ_1 is an activity coefficient. This equation makes two implicit assumptions: that the solute's flux is proportional to chemical potential gradient and that the diffusion coefficient in dilute solution is given by the Stokes–Einstein equation.

Hydration can affect this equation in two ways. First, the solute radius R_0 must be that of the hydrated species. This can be related to the true solute radius R'_0 by the equation

$$\tfrac{4}{3}\pi R_0^3 = \tfrac{4}{3}\pi (R'_0)^3 + n\left(\frac{\bar{V}_{H_2O}}{\tilde{N}}\right) \tag{6.2-42}$$

in which $\bar{V}_{H_2O}$ is the molar volume of water, n is the "hydration number," the number of water molecules bound to a solute, and $\tilde{N}$ is Avagadro's number. If the diffusion coefficient at infinite dilution is known, R_0 can be calculated, R'_0 can be estimated from crystallographic data, and n can be calculated. This kind of hydration *decreases* diffusion.

Hydration can also be calculated from the concentration dependence of diffusion by assuming that this concentration dependence is the result of hydration. Ideas like this were first used by Scatchard (1921) to rationalize the activity coefficient of sucrose. To do this, one assumes that the solute activity $c_1\gamma_1$ equals the solute's true mole fraction corrected for hydration:

$$c_1\gamma_1 = \frac{\text{number of hydrated solute molecules}}{\left(\begin{array}{c}\text{number of hydrated solute}\\\text{molecules}\end{array}\right) + \left(\begin{array}{c}\text{number of "free"}\\\text{water molecules}\end{array}\right)}$$

$$= \frac{\text{number of solute molecules}}{(1-n)\left(\begin{array}{c}\text{number of solute}\\\text{molecules}\end{array}\right) + \left(\begin{array}{c}\text{total number of}\\\text{water molecules}\end{array}\right)}$$

$$= \frac{c_1}{(1-n)c_1 + c_2} \tag{6.2-43}$$

The two concentrations are related through the partial molar volumes:

$$c_1 \bar{V}_1 + c_2 \bar{V}_2 = 1 \tag{6.2-44}$$

Combining Eqs. 6.2-42, 6.2-43, and 6.2-44, we obtain

$$D = D_0 \left[1 - \frac{(1 - n - \bar{V}_1/\bar{V}_2)c_1}{1/\bar{V}_2 + (1 - n - \bar{V}_1/\bar{V}_2)c_1} \right] \tag{6.2-45}$$

For dilute solutions, c_1 is small; for solutions of constant density $\bar{V}_1/\bar{V}_2$ is unity, and Eq. 6.2-45 becomes

$$D = D_0 [1 + n\bar{V}_2 c_1 + \cdots] \tag{6.2-46}$$

This result is often decorated with viscosity and electrostatic corrections. However, the basic message remains: hydration tends to *increase* diffusion.

These ideas are frequently qualitatively useful, but they are rarely quantitatively applicable. The data in Table 6.2-1 illustrate this by comparing hydration numbers found from diffusion, from activity coefficients, and from transference methods. Qualitatively, these values supply insights. For example, the diffusion of lithium is slower than that of sodium, which is slower than that of potassium, etc. This suggests that the radii of the diffusing solutes are in the order $Li^+ > Na^+ > K^+ > Cs^+$, exactly the reverse of the ionic radii found in the solid state. Such inverted behavior seems to be the result of hydration.

However, the hydration numbers make little quantitative sense. The values found from Eq. 6.2-42 are shown in the third column of Table 6.2-1. Although these values are often negative, we could force them to be positive by replacing the factor 6π in the

Table 6.2-1 *Hydration numbers found by various methods*

Ion	Observed diffusion coefficient at infinite dilution[a]	Hydration numbers from diffusion at infinite dilution	Hydration numbers from diffusion's concentration dependence[b]	Hydration numbers from activity coefficients[c]	Hydration numbers from transference methods[c]
H^+	9.33	−1.3	–	4	1
Li^+	1.03	1.3	2.8	4	14
Na^+	1.34	0.5	1.2	3	8
K^+	1.96	−0.1	0.9	1	5
Cs^+	2.06	−0.5	0.5	0	5
Cl^-	2.03	−0.7	0	1	4
Br^-	2.08	−0.9	0.2	1	5
I^-	2.04	−1.2	0.7	2	2

Notes: [a] $\times 10^{-5}$ cm^2/sec.
[b] Data of Robinson and Stokes (1960).
[c] Data of Hinton and Amis (1971).

Stokes–Einstein equation with some other theoretically rationalized value. The values calculated from Eq. 6.2-45, shown in the fourth column, do have the courtesy to remain positive, but they are far from being integers. I am always unsure how a cesium ion can react with half a molecule of water. In addition, the hydration numbers found from diffusion show little relation with those calculated from values from the other types of experiments shown in Table 6.2-1. These ideas have only qualitative value.

6.3 Solute–Solvent Interactions

In every case, diffusion is about mixing. In almost every case in this book, we are interested in what happens when two miscible solutions are placed next to each other and then allowed to mix without flow as the result of molecular motion. The speed of this spontaneous mixing is described by diffusion. This diffusion is a consequence of free energy decreases, of the second law of thermodynamics.

In some cases, diffusion occurs much more slowly than expected. This most commonly occurs near a phase boundary where the solution is supersaturated. An example is diffusion in supersaturated solutions of sugar in water, shown in Fig. 6.3-1. Slow diffusion also occurs in solutions near to a consolute point where two liquids first become miscible. Examples of diffusion near consolute points are shown in Fig. 6.3-2. Other related cases occur when an initially homogeneous solution is suddenly quenched to cause a phase separation. This quenching is commonly effected by abruptly lowering the temperature. The phase separation then occurs very rapidly, at a rate proportional to the diffusion.

Each of these cases involves mass transfer driven by changes in free energy, or more exactly, by gradients in chemical potential. Their description requires major changes in

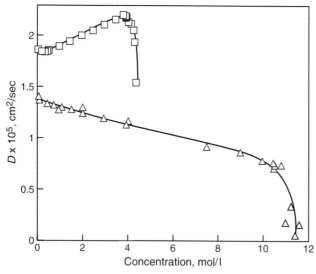

Fig. 6.3-1. Diffusion of sucrose ($\square$) and urea ($\triangle$) in aqueous solutions at 25 °C. The sudden drops occur in supersaturated solutions as the concentration approaches the spinodal limit.

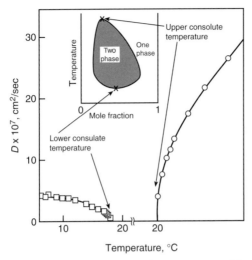

Fig. 6.3-2. Diffusion near consolute temperatures. The squares are for triethylamine–water and the circles represent hexane–nitrobenzene. At the consolute or critical-solution point the binary diffusion coefficient is zero. (Data from Claersson and Sundelöf (1957) and Haase and Siry (1968).)

Fick's law, including the effects of higher order terms in the gradients. In particular, the form postulated for the flux j_1 is

$$-j_1 = \frac{D_0 c_1}{k_B T} \left[\nabla \mu_1 - 2\gamma \nabla^3 x_1 \right]$$

(6.3-1)

where D_0 is a diffusion coefficient, due to Brownian motion and closely related to the diffusion coefficients in normal solutions, γ is an "interfacial influence," which is a characteristic of a phase separation, and x_1 is the mole fraction of species 1. We discuss this more general form of Fick's law in the following paragraphs.

6.3.1 Diffusion Near Spinodal Limits

We begin our discussions by considering spinodal limits, including consolute points. To focus the discussion, imagine we have two partially miscible liquids. When we dissolve a trace of one "solute" liquid in the other "solvent" liquid, we get a true solution. As we increase the amount of solute, we will saturate the solution. This saturation limit is called the "binodal." If we are careful, the solution will remain one supersaturated phase. If we continue to increase the solute concentration, we will reach a new limit of thermodynamic stability called the "spinodal." At a specific temperature, the spinodal will equal the binodal at a concentration called the "consolute point." This point is for liquid–liquid mixtures what the critical point is for gas–liquid phase behavior.

Diffusion coefficients in solutions near spinodal limits and consolute points drop from normal values to near zero, as shown in Figures 6.3-1 and 6.3-2. However, the concentration profiles still relax proportionally to the square root of time, as shown in

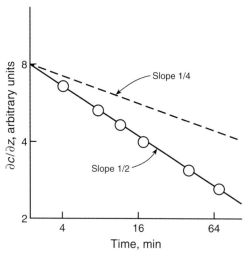

Fig. 6.3-3. Concentration gradient versus time near a consolute point. If Fick's law is valid, the data should fall along a line of slope one-half. If a new diffusion law is involved, the data should fall along a line of one-fourth. [Data from Brunel and Breuer (1971).]

Fig. 6.3-3. They do not relax proportionally to the fourth root of time, which would be significant if the interfacial influence were important. Thus only the first term in the brackets of Equation 6.3-1 is required to explain diffusion near a consolute point.

We can put these ideas on a more quantitive basis by rewriting concentration gradient

$$- j_1 = \left[\frac{D_0 c_1}{k_B T}\left(\frac{\partial \mu_1}{\partial c_1}\right)\right]\nabla c_1 \tag{6.3-2}$$

where the quantity in square brackets is an effective diffusion coefficient. At any spinodal limit, $(\partial \mu_1/\partial c_1)$ is zero, and so the effective diffusion coefficient is zero. This explains the limits of Figs. 6.3-1 and 6.3-2. Note that below any spinodal limit, $(\partial \mu_1/\partial c_1)$ becomes negative and the apparent diffusion coefficient is also negative. This indicates not mixing but phase separation. We will return to this point when we discuss spinodal decomposition.

Alternatively, if we assume

$$\mu_1 = \mu_1^0 + k_B T \ln c_1 \gamma_1 \tag{6.3-3}$$

we may rewrite Equation 6.3-2 as

$$- j_1 = \left[D_0\left(1 + \frac{\partial \ln \gamma_1}{\partial \ln c_1}\right)\right]\nabla c_1 \tag{6.3-4}$$

where again the quantity in square brackets is an effective diffusion coefficient, now written in terms of the activity coefficient γ_1. At a spinodal limit, $(\partial \ln \gamma_1/\partial \ln c_1)$ equals minus one, so the apparent diffusion coefficient is zero. The flux equation using the activity coefficient is commonly mentioned though it does not seem to have been carefully checked.

However, neither Equation 6.3-2 nor 6.3-4 correctly predicts all aspects of diffusion near the consolute point. To illustrate this, consider the system hexane–nitrobenzene, for which results are given in Fig. 6.3-2. The chemical potential of this system is found by experiment to fit the equation

$$\mu_1 = \mu_1^0 + k_B T \ln x_1 + \omega x_2^2 \tag{6.3-5}$$

where ω is a measure of interaction between solute and solvent. This type of chemical potential is sometimes called a "regular solution." At the consolute point of such a solution,

$$\frac{\partial \mu_1}{\partial x_1} = \frac{\partial^2 \mu_1}{\partial x_1^2} = 0 \tag{6.3-6}$$

As a result, $x_1 = x_2 = 0.5$ and $\omega = 2k_B T_C$, where T_C is the consolute temperature. Thus, we find that at the consolute composition,

$$\frac{x_1}{k_B T} \frac{\partial \mu_1}{\partial x_1} = 1 + \frac{\partial \ln \gamma_1}{\partial \ln x_1} = 1 - \frac{4x_1 x_2 T_C}{T} = 1 - \frac{T_C}{T} \tag{6.3-7}$$

Combining this with Eq. 6.2-4 we obtain

$$D = D_0 \left[\frac{T - T_C}{T} \right] \tag{6.3-8}$$

Thus the diffusion coefficient is expected to drop as the temperature is cooled to the consolute point. The coefficient is zero at the consolute point, consistent with Figure 6.3-2. However, the linear temperature variation in brackets is not observed in Fig. 6.3-2 so that Eqs. 6.3-2 and 6.3-8 are inconsistent with experiment.

The reason for this inconsistency is that long-range fluctuations dominate behavior near any spinodal limit, including a consolute point. When fluctuations of concentration and of fluid velocity couple, diffusion occurs. Under ordinary conditions, the concentration fluctuations are dominated by motion of single molecules, but near the critical point, these fluctuations exist even when the average fluid velocity is zero. The result is like a turbulent dispersion coefficient but without flow.

When the details of these coupled fluctuations are considered, the diffusion coefficient is found to be

$$D = \frac{k_B T}{2\pi \mu \xi} \tag{6.3-9}$$

where the correlation length ξ is approximately the average size of a cluster. The approach retains the same temperature and viscosity dependence as the Stokes–Einstein equation. The factor 2π in place of 6π is not a major change. However, both the diffusion coefficient D and the length ξ vary dramatically with the thermodynamic factor $(1 + \partial \ln \gamma_1 / \partial \ln x_1)$.

The calculation of ξ as a function of temperature and composition can proceed in two different ways. The best way is to depend on scaling laws developed for phase transitions that in turn are based most frequently on the Ising model. Such calculations give the temperature dependence at the critical composition:

$$D \propto \left(\frac{T}{T_C} - 1 \right)^{0.62} \tag{6.3-10}$$

Unfortunately, these calculations are not so complete as to give the concentration dependence of ξ and D. The alternative, less accurate route in the calculation of ξ is to use simple models of the chemical potential to find:

$$D = D_0 \left[1 + \frac{A}{x_1 x_2} \left(\frac{\partial \ln x_1}{\partial \ln \gamma_1 x_1} - 1 \right) \right]^{-1/2} \qquad (6.3\text{-}11)$$

in which A is a constant of the order of one-half.

The predictions based on coupled fluctuations are compared with those of the more traditional result in Fig. 6.3-4. This figure includes data on four different systems obtained in five different laboratories using four different experimental methods. The data all appear consistent. They fall very close to the predictions of Eq. 6.3-10, which is based on the coupled fluctuations as described by scaling laws. They are in reasonable agreement with the predications of Eq. 6.3-11, which uses simple statistical models for chemical potential. These results support the explanation of diffusion near the consolute point in terms of coupled fluctuations of concentration and velocity.

6.3.2 Spinodal Decomposition

The strong solute–solvent interactions that cause the diffusion coefficient to drop so sharply near critical points are also central to spinodal decomposition. In many phase separations, a homogeneous solution is cooled so that its equilibrium condition is a two-phase mixture. As in the case discussed above, separation into these two phases can begin as soon as the solution is cooled below its phase boundary or its "binodal." The region just below this phase boundary is metastable, waiting for events that cause the phase separation. The phase separation begins with nucleation of small droplets of the new phase; these droplets grow with time.

In the case of spinodal decomposition, the original solution is rapidly quenched, so that the equilibrium condition drops suddenly through the binodal and below the

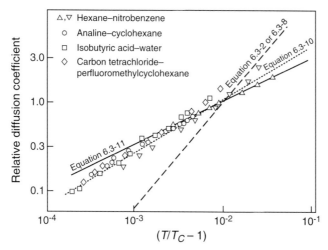

Fig. 6.3-4. Diffusion versus temperature near a consolute point. The classic theories shown by the broken line are much less successful than the predictions of scaling laws (dotted line) or of cluster diffusion (solid line). [From Cussler (1980), with permission.]

spinodal curve. This spinodal curve, which lies within the two-phase region, is the lower boundary of the metastable region. Below this curve, the solution is unstable and phase separation is immediate. No small dust particles are needed to nucleate the phase separation; instead, the separation is spontaneous and fast.

We want to estimate the speed of this separation. Now, however, we do not begin with an imposed chemical potential gradient as is the case in any diffusion experiment. Now, we begin with a quenched system, which is initially homogeneous, of a constant but unstable chemical potential. To describe this unstable system, we use the extended form of Fick's law given in Equation 6.3-1.

We can combine this with a mass balance to obtain

$$\frac{\partial c_1}{\partial t} = -\nabla \cdot j_1$$

$$\frac{\partial x_1}{\partial t} = \frac{D_0}{k_B T} \nabla \cdot \left[\frac{\partial \mu_1}{\partial \ln x_1} \nabla x_1 - 2\gamma x_1 \nabla^3 x_1 \right]$$

(6.3-12)

where γ is the interfacial influence characterizing the phase separation. Note that we have divided both sides of this second equation by the total concentration c, which is assumed constant. Next, we define a perturbation x from the original solution concentration x_{10}

$$x = x_1 - x_{10}$$

(6.3-13)

Inserting this into the previous relation, we find

$$\frac{\partial x}{\partial t} = \frac{D_0}{k_B T} \left[\left(\frac{\partial \mu_1}{\partial \ln x_1} \right)_{x_{10}} \nabla^2 x - 2\gamma x_{10} \nabla^4 x \right]$$

(6.3-14)

The first term in square brackets is the effect of diffusion; the second is the interfacial influence.

The general solution of this equation for the concentration fluctuation is most easily given as a Fourier series

$$x = \sum_{i=1}^{\infty} [\text{amplitude}]_i \cos\left(\frac{2\pi r}{\Lambda_i} - \beta_i\right) e^{-t/\tau_i}$$

(6.3-15)

where r is the distance from some point, Λ_i is a characteristic distance, and τ_i is a characteristic parameter with the dimensions of time. The characteristic times are the key to this problem. If a particular time is positive, the fluctuations in the concentration decay and the solution stays homogeneous. If it is negative, then the concentration fluctuations grow over time, and the phase separation proceeds. Thus our real interest is in the sign of these characteristic times because they will govern whether spinodal decomposition does occur. From the general characteristics of Fourier series, we may show that these times are given by

$$\tau_i = \frac{k_B T}{D_0} \left[\left(\frac{\partial \mu_1}{\partial \ln x_1} \right)_{x_{10}} \left(\frac{2\pi}{\Lambda_i} \right)^2 + 2\gamma x_{10} \left(\frac{2\pi}{\Lambda_i} \right)^4 \right]^{-1}$$

(6.3-16)

These times may be negative and the phase separation immediate if the derivative $(\partial \mu_1 / \partial \ln x_1)_{x_{10}}$ is negative. This derivative is positive at temperatures above the spinodal, even in the metastable region between the spinodal and the solubility limit (i.e. the binodal). It is negative below the spinodal.

We are most interested in the shortest negative time suggested by Eq. 6.3-16 because this time will correspond to the fluctuation that grows fastest. By setting $\partial \tau_i / \partial \Lambda_i$ equal to zero, we can show that the largest characteristic length is

$$\Lambda_{max} = 4\pi \left[- \frac{\gamma x_{10}}{(\partial \mu_1 / \partial \ln x_1)_{x_{10}}} \right]^{1/2} \tag{6.3-17}$$

and that

$$\tau_{min} = - \frac{8 \gamma k_B T}{D_0 x_{10}} \left(\frac{\partial x_1}{\partial \mu_1} \right)^2_{x_{10}} \tag{6.3-18}$$

While we don't know the interfacial influence γ, we do know the characteristic length Λ_{max} in the original solution: it is nothing more than the size of the molecules or clusters. Thus we can combine these last relations to obtain

$$\tau_{min} = \frac{k_B T \Lambda_{max}^2}{2\pi^2 D_0 (\partial \mu_1 / \partial \ln x_1)} \tag{6.3-19}$$

This gives the time characteristic of the fast spinodal decomposition.

The physical significance of this result may be clearer if we return to the case of a regular solution defined by Eq. 6.3-6. Using the chemical potential derivative in Eq. 6.3-7, we find from Eq. 6.3-19 that

$$\tau_{min} = \frac{\Lambda_{max}^2}{2\pi^2 D_0 \left(1 - \frac{4T_c}{T} x_1 x_2 \right)} \tag{6.3-20}$$

Typical values might be Λ_{max} equal to 10^{-7} cm, D_0 equal to 10^{-5} cm^2/sec, T_C equal to 300 K and T of 299 K. In this case, τ_{min} will be $[-10^{-8}$sec$]$: the phase separation will effectively be immediate. This mechanism is rare in gases and ordinary liquids but common in solid alloys and glasses.

Example 6.3-1: Diffusion through a consolute point Imagine a diaphragm cell of two well-stirred compartments (see Example 2.2-4). One compartment contains water, and the other contains triethylamine. Diffusion occurs across the diaphragm between the two compartments. However, this experiment will be made at the consolute temperature 18.6 °C. As a result, somewhere within the diaphragm, the concentration must be that at the consolute point, and the diffusion coefficient at that point will approach zero, as shown in Fig. 6.3-2. What will happen in this experiment?

   Solution If we make a mass between balance on a thin slice of the diaphragm, we find that

$$0 = \frac{-dn_1}{dz} = - \frac{dj_1}{dz}$$

This means that there will be a steady-state flux across the diaphragm. When we combine this result with Fick's law we find

$$-j_i = D(c_1)\frac{dc_1}{dz} = \text{constant}$$

At the consolute concentrations, $D(c_1)$ approaches zero, so dc_1/dz must approach infinity. Thus, in this experiment, the flux behaves normally but the concentration gradient reflects the unusual properties of the consulate point.

6.4 Solute–Boundary Interactions

When a solute diffuses through small pores, its speed may be affected by the size and the chemistry of the pores. For example, a solute will diffuse faster through a large straight pore than through a small crooked one. It may diffuse differently if it adsorbs on the pore's wall and then scoots along the wall at a faster rate than it moves in the bulk.

In this section, we explore these effects in more detail. In some cases our simple goal is to organize experimental results. In other cases we may have more ambitious goals, especially where the altered diffusion is the result only of a different geometry. These cases involve a wide variety of possible mechanisms. Some of these are shown in Figure 6.4-1 for the special case of a cylindrical pore. In the simplest case, shown at the top of the figure, a pressure drop along the pores causes a convective flow. In the second case, where there is no pressure drop but a concentration difference, transport occurs by diffusion. In these two cases, the properties of the fluid in the pores are the same as those of the fluid in the bulk.

In other cases shown in Figure 6.4-1, the basic mechanism of transport changes. For example, it may involve gas diffusion where the gas molecules collide more often with the

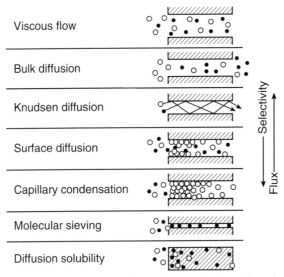

Fig. 6.4-1. Pore diffusion effects, The pore size drops from the top of the figure to the bottom. The selectivity is often larger for smaller pores.

pore walls than with other gas molecules ("Knudsen diffusion"). In still other cases, gas molecules may absorb on the walls and then diffuse ("surface diffusion"), or condense within the pores and move as a liquid ("capillary condensation"). When the pores are of molecular dimensions, one solute may dissolve in solvent held in liquid-filled pores and then diffuse by a "diffusion-solubility mechanism." Such a variety of effects is hard to discuss as anything but a long series of examples.

I have tried to force an organization on these examples as follows. In Section 6.4.1, I have discussed the simplest empirical methods of organizing experimental results. In Section 6.4.2, I have reviewed theories for solute diffusion in a solvent trapped within cylindrical pores in an impermeable solid. In this case, solute–solvent interactions still control diffusion; and the solid only imposes boundary conditions. Cases where the interactions are between the diffusion solute and the pores' boundaries are covered in Section 6.4.3. Finally, cases not of cylindrical pores but of other composite structures are described in Section 6.4.4.

6.4.1 Empirical Descriptions

Imagine a solute diffusing through the fluid-filled pores of the porous solid shown schematically in Fig. 6.4-2. Because the solid itself is impermeable, diffusion takes place only through the cramped and tortuous pores of the composite. Because the pores are not straight, the diffusion effectively takes place over a longer distance than it would in a homogeneous material. Because the solid is impermeable, diffusion occurs over a smaller cross-sectional area than that available in a homogeneous material.

The effects of longer pores and smaller areas are often lumped together in the definition of a new, effective diffusion coefficient D_{eff}

$$D_{\text{eff}} = \varepsilon \frac{D}{\tau} \tag{6.4-1}$$

in which D is the diffusion coefficient in the bulk fluid, ε is the void fraction, and τ is the tortuosity. The tortuosity attempts to account for the longer distance traversed in the pores. Tortuosities usually range between two and six, averaging about three. These values can be rationalized because solutes diffuse in three directions instead of one, so they diffuse about three times as far. Such rationalization is suspect. I have measured tortuosities as high as ten, which I find hard to justify on geometrical arguments alone. Moreover, the tortuosity measured for diffusion may not correlate closely with the

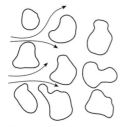

Fig. 6.4-2. Diffusion in a composite. When the particles are impermeable, a diffusing particle must travel a longer path through a reduced cross-sectional area.

tortuosity measured for flow. Still, the great advantage of the tortuosity is its simplicity: it does give a simple number showing how much diffusion will be retarded in a porous solid.

6.4.2 Diffusion in Large Cylindrical Pores

We next turn to diffusion of a solution held within an array of cylindrical pores. Normally, these large pores are assumed to span a thin film, and to all be perpendicular to the surfaces of the film. By "large pores," we imply that the solvent acts as a continuum and that the solute diameter is much smaller than the pore diameter. Not surprisingly, this idealized geometry has been the focus of considerable theoretical effort. In spite of its idealizations, it does provide physical insight.

We first consider results for a gas. For large pores, gas transport through the pores will be described by the Hagen–Poiseville law

$$v = \frac{\varepsilon d^2 \Delta p}{32 \mu l} \tag{6.4-2}$$

in which v is the superficial velocity, ε is the void fraction, μ is the gas viscosity, Δp is the pressure drop across the pore, and d and l are the pore's diameter and length, respectively. If we multiply each side of this equation by the concentration of species 1, we find

$$n_1 = \varepsilon c_1 v = \left[\frac{\varepsilon c_1 d^2}{32 \mu l} \right] \Delta p = \left[\frac{\varepsilon c_1 d^2 RT}{32 \mu} \right] \frac{\Delta c_1}{l} \tag{6.4-3}$$

where $\Delta c_1 \; (= \Delta p / RT)$ is the concentration difference of the ideal gas along the pore. If instead, we have no overall pressure difference along the pores but only a partial pressure difference, we have

$$n_1 = j_1 = [\varepsilon D] \frac{\Delta c_1}{l} \tag{6.4-4}$$

Finally, if the pores are extremely short, we have

$$n_1 = j_1 = \left[\frac{\varepsilon D l}{d} \right] \frac{\Delta c_1}{l} \tag{6.4-5}$$

where d is the pore diameter. In this case, the limitations to diffusion are not actually in the pore itself but in necking down to enter the pore. Details of this case are given in Example 3.5-2.

In Equations 6.4-3 to 6.4-5, the quantity in square brackets is an apparent diffusion coefficient. However, in Equation 6.4-3, it is actually due to convective flow, and in Equation 6.4-5, it represents diffusion *to* the pore, not diffusion *in* the pore. Only Equation 6.4-4 actually describes diffusion *in* the pore. These differences seem obvious in this theoretically based discussion. However, when we have experimental data, we may find that it is more difficult to decide which of the three cases is actually occurring.

Similar subtleties occur for the transport of liquids in large pores. There is no exact parallel to Equation 6.4-3 because a difference in pressure does not cause a difference in concentration. There is a complete parallel to Equation 6.4-5, where the key is diffusion to a very short pore. However, there are new complexities to diffusion within a pore.

To understand these complexities, we consider a solvent at concentration c_2 and a solute at concentration c_1. The solvent is much smaller than the pore's diameter and so can always be treated as a continuum, but the solute's diameter $2R$ is a significant fraction of the pore's diameter d, i.e., $\lambda\ (= 2R/d)$ is less than one but of order one. In this case, the solvent's flux is

$$n_2 = \varepsilon c_2 v_2 = \left[\frac{\varepsilon_2 c_2 d^2}{32\mu}\right]\frac{\Delta p}{l} \tag{6.4-6}$$

This is much like the results for a gas, given in Equation 6.4-3.

However, the solute velocity v_1 is different than v_2 for two reasons. First, the solute will not fit into the entire pore diameter but only into a smaller equivalent pore of diameter $(d - 2R)$. This implies that the void fraction for the solute will be smaller than that for the solvent. Second, because the solute is forced to be more towards the center of the pore, it will encounter average velocities somewhat higher than those averaged over the entire diameter of the pore. One theory typical of efforts on this subject gives for the solute velocity

$$\frac{v_1}{v_2} = (1-\lambda)^2\left[2-(1-\lambda)^2\right]e^{-0.71\lambda^2} \tag{6.4-7}$$

where λ is the ratio of the diameter of the solute to that of the pore. Remember that in this case, transport is by convective flow. This result is important in ultrafiltration.

The results for diffusion in large liquid-filled pores are different. In this case, which is sometimes called hindered diffusion, the solute is modeled as a rigid sphere in a solvent continuum that fills the pore. The solute's transport is retarded by the viscous drag of the solvent, which is affected by the proximity of the pore walls. The diffusion coefficient D is given by:

$$\frac{D}{D_0} = 1 + \frac{9}{8}\lambda\ln\lambda - 1.54\lambda + O(\lambda^2) \tag{6.4-8}$$

where $D_0\ (= k_B T/6\pi\mu\,R_0)$ is the Stokes–Einstein diffusion coefficient. When $\lambda = 0.1$, the diffusion coefficient D is roughly half that in bulk solution; when $\lambda = 0.2$, Eq. 6.4-8 is accurate to within about two percent.

6.4.3 Diffusion in Small Cylindrical Pores

Our next cases occur for small pores. By "small" we mean that the pore diameter is of the same order of magnitude as the molecular size of both solvent and solute. Thus we can no longer approximate the solvent as a continuum. Under this heading, we consider four topics: Knudsen diffusion, surface diffusion, capillary condensation, and sieving.

Knudsen Diffusion

In Knudsen diffusion, diffusing molecules collide with the walls of the pores much more frequently than they collide with other molecules. This type of transport is dominant whenever the distance between molecular collisions is greater than the pore diameter. This ratio of distances is defined as a dimensionless group, the Knudsen number Kn:

$$Kn = \frac{l}{d} \tag{6.4-9}$$

in which l is now the mean free path and d is the pore diameter. If the Knudsen number is small, diffusion has the same characteristics as it does outside of the pores, and it is analyzed with the effective coefficients and tortuosities given earlier. If the Knudsen number is large, diffusion is dominated by collisions with the boundaries; this requires a different description.

For liquids, the mean free path is commonly a few angstroms, so the Knudsen number is almost always small, and Knudsen diffusion is not important. In gases, the mean free path l can be estimated from

$$l = \frac{4k_B T}{\pi \sigma^2 p} \tag{6.4-10}$$

in which σ is the collision diameter of the diffusing species. This mean free path can be large. For example, for air at room temperature and pressure, it is over 60 nm; for hydrogen at 300 °C and 1 atm, it is over 200 nm. Because pores smaller than these values often exist, for example, in porous catalysts, Knudsen diffusion can be a significant effect in gases.

When the mean free path and Knudsen number are large, the diffusion coefficient can be quickly estimated by arguments that parallel those for the kinetic theory of rigid spheres. This theory predicts that

$$D_{Kn} = \frac{1}{3} dv \tag{6.4-11}$$

where v is the molecular velocity. This prediction is the same as that in Eq. 5.1-5 but with the mean free path l replaced by the pore diameter d. Because a molecule's kinetic energy $(\frac{1}{2}mv^2)$ must equal $k_B T$, we expect

$$v = \sqrt{\frac{2k_B T}{m}} \tag{6.4-12}$$

where m is the molecular mass. Thus the Knudsen diffusion coefficient is given by

$$D_{Kn} = \frac{d}{3} \left[\frac{2k_B T}{m} \right]^{\frac{1}{2}} \tag{6.4-13}$$

Unlike gas diffusion outside of the pores, the Knudsen diffusion coefficient is independent of pressure and of the molecular weight of any solvent species.

Surface Diffusion

Like Knudsen diffusion, surface diffusion is much more important for gases than for liquids. In surface diffusion, gas molecules adsorb on the solid pore walls. When the adsorption is physical, the adsorption energy is less than $k_B T$, and the adsorbed solutes are highly mobile. When the adsorption involves more specific chemical interactions (chemisorption), the adsorption energy is greater than $k_B T$ and the adsorbed species tend to be more tightly bound to specific sites. Such tightly bound species are much less mobile than in physical adsorption, but instead are said to "hop" from one site to the next.

Surface diffusion is most commonly measured in a form of diaphragm cell (cf. Section 5.5), with a sample of the porous solid serving as the diaphragm. This cell is used in two ways. In the first way, we place a pure gas on one side of the diaphragm, and a vacuum on the other side. We then measure concentration versus time to find the flux across the membrane. In the second way, we place two binary gas mixtures of different composition but the same pressure on the opposite sides of the diaphragm. Again, we measure concentration changes and use these to find the fluxes of each gas. Note that these fluxes will usually not be of equal magnitude.

From these results, we calculate the surface diffusion flux as follows. As a standard, we measure the flux of a gas that we expect will not adsorb. This gas is most commonly helium. Then, expecting that nonsurface diffusion will occur by the Knudsen mechanism, we calculate the expected flux for the test gas from:

$$j_1(\text{nonsurface}) = j_{\text{He}} \sqrt{\frac{\tilde{M}_{\text{He}}}{\tilde{M}_1}} \qquad (6.4\text{-}14)$$

where $\tilde{M}_1$ is the molecular weight of species i. We now estimate the surface diffusion flux as the difference between the experimental measurement and the nonsurface estimate:

$$j_1(\text{surface}) = j_1(\text{experimental}) - j_1(\text{nonsurface}) \qquad (6.4\text{-}15)$$

Typically, the flux inferred for surface diffusion is less than half of that measured experimentally.

Measurements of surface diffusion can be correlated in terms of random walk, surface mobility, and surface diffusion coefficients. Correlations in terms of surface diffusion coefficients are most similar to the type of analysis used in this book. The surface diffusion coefficient D_s is defined with the equation

$$j_1(\text{surface}) = -\frac{l D_s}{A} \frac{dc_1}{dz} \qquad (6.4\text{-}16)$$

where l and A are the diaphragm thickness and cross-sectional area, respectively, and c_1 is the surface concentration, in units of moles per area. Thus the surface diffusion coefficient has dimensions of $(\text{length})^2/\text{time}$, just like other diffusion coefficients.

Values of surface diffusion coefficients cluster around 10^{-5} cm^2/sec, and so are similar to values in liquids. However, these values vary widely. At room temperature, hydrogen on tungsten has a value of 10^{-7} cm^2/sec and propane on silica has a value of 10^{-3} cm^2/sec.

Surface diffusion coefficients are strong functions of temperature, a characteristic of solids rather than liquids or gases. They are also strong functions of surface concentration: Typically, the surface diffusion coefficient increases sharply as surface coverage increases.

Surface diffusion is often viewed as a step in gas–solid catalytic reactions. This diffusion-based step is often fast relative to other, more selective chemical changes, and hence it does not control the catalytic rate. As a result, surface diffusion is less industrially important than the bulk diffusion described in the remainder of this book.

Capillary Condensation

While Knudsen diffusion and surface diffusion normally involve only gases, capillary condensation involves the conversion from a gas into a liquid. It results from the altered vapor pressure of a liquid inside a pore. This increased vapor pressure p is given by the Kelvin equation

$$RT \ln \frac{p}{p_0} = \frac{2\gamma \tilde{V}}{r} \tag{6.4-17}$$

where p_0 is the bulk vapor pressure, γ is the surface tension, $\tilde{V}$ is the solute's molar volume, and r is the radius of curvature of the liquid inside the pore. Once condensation occurs, transport is by a combination of diffusion and convection across the pore. If the pore completely fills with liquid, its apparent diffusion coefficient is

$$D_{cap} = \frac{\rho_L RT}{\tilde{M}} \left(\frac{d^2}{32\mu_L} \right) \left[1 + \frac{\rho_L RT}{\tilde{M}p} \right] \tag{6.4-18}$$

where ρ_L and μ_L are the density and viscosity of the condensed *liquid*, and p is the mean pressure across the pore. This diffusion coefficient is roughly parallel to that for Pouiseville flow of a vapor given in Equation 6.4-3 but with the liquid properties replacing those of a gas and with the added factor in square brackets. When the pore separates a gas at high pressure from one at low pressure, the result is more complex.

This transport can be dramatically faster than that due to diffusion alone and is an unexpected delight where it occurs experimentally. However, capillary condensation is rarely important for two reasons. First, it exists only when a surface tension exists and hence will not work for gases above their critical temperatures. For example, capillary condensation will not work for separating air at room temperature; it will work only below 155 K, the critical temperature of oxygen. Capillary condensation might be used to separate carbon dioxide and methane at room temperature, for the critical temperature of carbon dioxide is 304 K. Second, capillary condensation is rarely important because it is a small effect. For example, water vapor in 100 nanometer pores showing a surface tension of 72 dynes will condense at 100.2 °C, not 100 °C. This means that only very small pores will be effective.

Molecular Seiving

The final way in which pores can be physically selective also hinges on the relative sizes of the diffusing species and the pore. Now, both solute and solvent are of sizes comparable to the pore. Small molecules pass through the pores but big molecules are retained.

This sieving mechanism is rare but may be highly selective. One case where it is definitely involved is in transport of linear and branched alkanes into zeolites. The linear alkanes diffuse into the zeolites about fifteen times faster than the branched ones do. The reason is that the linear alkanes have a smaller cross-section which can fit into the pores in the zeolite crystal. This difference in diffusion is exploited in some forms of pressure swing adsorption. If thin zeolite layers can be produced commercially, they will have considerable value.

More frequently sieving is postulated as an alternative to what almost certainly is a diffusion–solubility mechanism. For example, sieving is sometimes asserted to be responsible for the selectivity of cellulose acetate reverse osmosis membranes. Sieving is consistent with the slow transport of larger salt ions compared to smaller water molecules. It seems inconsistent with the fast transport of larger phenol molecules compared to smaller water molecules. I would view any new claims of a sieving with skepticism unless the pores have dimensions like those of the solute and the solvent.

6.4.4 Periodic Composites

The cases of diffusion in solids discussed in the previous paragraphs involve a simple geometry – cylindrical pores – and a spectrum of chemistry – surface diffusion, capillary condensation, etc. In this last subsection, we want to discuss cases with a more complex geometry but with simpler chemistry. We will assume that diffusion in each phase will be the same as if it were the only phase present.

We quickly recognize that specifying the geometry only in terms of a void fraction and a tortuosity will not always be sufficient. To illustrate this, we consider two cases of diffusion in a composite membrane containing permeable flakes as shown in Figure 6.4-3. In the two cases, the flakes are present at the same volume fraction ϕ. However, in the case in Figure 6.4-3(a), the flakes are aligned perpendicularly to the membrane's surface, while in Figure 6.4-3(b), they are parallel. In the first case, the resistances to diffusion are approximately in parallel so that the flux will be

$$j_1 = [(1 - \phi)D + \phi D_F]\frac{\Delta c_1}{l} \qquad (6.4\text{-}19)$$

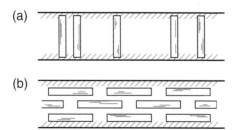

Fig. 6.4-3. Diffusion through a suspension of impermeable flakes. The effective diffusion coefficient varies sharply with orientations shown in (b), but is affected little by the orientation in (a).

where D and D_F are the diffusion coefficients in the continuous phase and in the flakes, respectively. In the second case, the resistances are more nearly in series

$$j_1 = \left[\cfrac{1}{\cfrac{1-\phi}{D} + \cfrac{\phi}{D_F}} \right] \frac{\Delta c_1}{l} \tag{6.4-20}$$

In both cases the quantity in brackets is an effective diffusion coefficient for the composite. The results are very different, showing that volume fraction alone is not enough to describe the composite system.

More exact descriptions of these effects require more exact geometries. One such case occurs when the composite consists of periodically spaced spheres like those shown in Fig. 6.4-4. In this case, we assume that diffusion can take place both in the interstitial region between the spheres and through the spheres themselves. The effective diffusion coefficient D_{eff} can be calculated from

$$\frac{D_{eff}}{D} = \cfrac{\cfrac{2}{D_s} + \cfrac{1}{D} - 2\phi\left(\cfrac{1}{D_s} - \cfrac{1}{D}\right)}{\cfrac{2}{D_s} + \cfrac{1}{D} + \phi\left(\cfrac{1}{D_s} - \cfrac{1}{D}\right)} \tag{6.4-21}$$

in which D is the diffusion coefficient in the interstitial pores, D_s is the diffusion coefficient through the spheres, and ϕ is the volume fraction of the spheres in the composite material (Maxwell, 1873). Strictly speaking, this equation is valid only for dilute suspensions. However, it is routinely applied to experiments with ϕ equal to as high as 0.5 with reasonable agreement. Why this should be so is unclear.

Equation 6.4-21 is a fascinating result. It says that diffusion does not depend on the size of the spheres but only on their volume fraction. It does not matter if the spheres are birdshot or basketballs – the diffusion is the same if the volume fraction is the same.

A second interesting consequence of Eq. 6.4-21 is that the properties of the continuous phase dominate the diffusion process. To demonstrate this, we imagine that the spheres are impenetrable, so that D_s is zero. Then Eq. 6.4-21 becomes

$$\frac{D_{eff}}{D} = \frac{1-\phi}{1+\phi/2} \tag{6.4-22}$$

If ϕ is 0.1, then $D_{eff}/D = 0.86$. The diffusion is eighty-six percent of what it would be without the spheres.

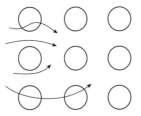

Fig. 6.4-4. Diffusion through a periodic array of spheres. In this case, the spheres and the surrounding continuum have different, nonzero permeabilities.

Now consider the other limit in which diffusion through the spheres is extremely rapid, so that $D_s \rightarrow \infty$. In this case,

$$\frac{D_{\text{eff}}}{D} = \frac{1 + 2\phi}{1 - \phi} \qquad (6.4\text{-}23)$$

If ϕ is still 0.1, then D_{eff}/D is 1.33. Thus changing the diffusion coefficient in the spheres from zero to infinity changes D_{eff}/D only by a factor of 1.6.

The results for impermeable spheres have been extended to impermeable cylinders and impermeable flakes. For impermeable cylinders aligned periodically and parallel to the membrane's surfaces, the result is

$$\frac{D_{\text{eff}}}{D} = \frac{1 - \phi}{1 + \phi} \qquad (6.4\text{-}24)$$

Again, the relative diffusion coefficient (D_{eff}/D) is independent of D and of the cylinders' size. As in the case of the spheres, it doesn't matter whether the cylinders are carbon nanotubes or millimeter-sized glass fibers: the only variable is their volume fraction. The change in diffusion for cylinders is smaller than that for spheres but only slightly. The continuous phase still dominates diffusion.

The result for flakes is more complicated because it includes two limits. In both limits, the flakes are aligned parallel to the membrane surfaces. In the first limit, the flakes are so dilute that they do not overlap. The result is similar to that for spheres or cylinders

$$\frac{D_{\text{eff}}}{D} = \frac{1}{1 + \alpha\phi} \qquad (6.4\text{-}25)$$

where α is the aspect ratio of the flakes. This ratio, equal to the flakes' intermediate dimension divided by the shortest dimension, characterizes the flakes' shape.

The second limit occurs when the flakes overlap, even though they may still be dilute. In this limit, ϕ may still be much less than one, but $\alpha\phi$ is greater than one. In this case, the effective diffusion coefficient is given by

$$\frac{D_{\text{eff}}}{D} = \frac{1}{1 + \alpha^2\phi^2/(1 - \phi)} \qquad (6.4\text{-}26)$$

Like the results for spheres and cylinders, Eq. 6.4-26 does not depend on flake size: 10 mm clay flakes will give the same result as 10 cm dead leaves when the volume fraction and the aspect ratio are equal. However, unlike the previous results, the effective diffusion coefficient depends not on the first power of the volume fraction but on the square. This is because the flakes both increase the tortuosity and reduce the cross-sectional area available for diffusion. Moreover, these effects can be significantly larger than those for other shapes: a factor of ten is not uncommon.

6.4.5 Graham's Law

To conclude this section, we turn to Graham's law which states that at constant pressure the ratio of fluxes of two diffusing gases in a porous medium is proportional to the inverse square root of their molecular weights. This law, based on Graham's original experiment described in Section 2.1, has limited application, but the inverse square root dependence is characteristic of a wide range of situations. Thus this discussion gives us a chance to compare and contrast the wide range of solute–boundary interactions. It provides a summary of diffusion in composite media.

Before we begin our discussion, we review three characteristics of the kinetic theory of gases. First, the actual velocity of gas molecules u_i is of course proportional to the temperature

$$\frac{1}{2} m_i u_i^2 = k_B T \tag{6.4-27}$$

where m_i is the molecular mass and k_B is Boltzman's constant. Rearranging

$$u_i = \sqrt{\frac{2 k_B T}{m_i}} \tag{6.4-28}$$

Second, the mean-free path l_i must be related to the total concentration c_i, which is in turn related to the pressure

$$\left(\frac{\pi}{4} d_i^2\right) l_i = \frac{1}{c_i} = \frac{k_B T}{p_i} \tag{6.4-29}$$

where d is the molecular diameter. Note that c_i is the number of molecules per volume. Third, the number of molecular collisions per area per time ζ_i is given by

$$\zeta_i = c_i u_i \tag{6.4-30}$$

We will use these three results to discuss five special cases.

The five cases of interest are shown schematically in Figure 6.4-5. In these cases, two volumes are separated by some sort of porous medium. The volumes are bounded by pistons which are normally fixed. Each volume normally contains a different gas. However, the transport mechanisms between the volumes may be very different.

We first consider diffusion at constant volume, illustrated schematically in Figure 6.4-5(a). When the diameter of the pores is much larger than the mean free path, this is the common case discussed in detail throughout this book. If one volume initially contains only nitrogen and the other initially contains only hydrogen, diffusion will occur between the two volumes until their concentrations are equal. During this time,

$$n_1 = n_2 \tag{6.4-31}$$

where "1" and "2" represent nitrogen and hydrogen, respectively. Each flux depends on the same gas phase diffusion coefficient which is proportional to the square root of the harmonic average of molecular weights. Second, we consider the familiar case of convection shown in Figure 6.4-5(b). Now a mixed gas flows from left to right. Because the convective velocity of both gases is the same, the ratio of the fluxes is just the ratio of the concentrations

$$v_1 = \frac{n_1}{c_1} = v_2 = \frac{n_2}{c_2} \tag{6.4-32}$$

Diffusion is not involved in this second case.

The third case, in Figure 6.4-5(c), is Knudsen diffusion where the capillary diameter is less than the mean free path. This is the case where a diffusing molecule collides much

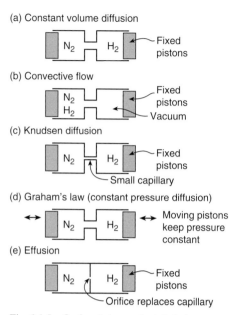

(a) Constant volume diffusion

(b) Convective flow

(c) Knudsen diffusion

(d) Graham's law (constant pressure diffusion)

(e) Effusion

Fig. 6.4-5. Graham's law and related phenomena. All these cases involve gases. Many show a flux proportional to the inverse square root of the gas's molecular weight.

more frequently with the capillary walls than with other diffusing molecules. The diffusion coefficient for each species is proportional to the inverse square root of its molecular weight. Thus the ratio of fluxes is given by

$$\frac{n_1}{n_2} = \sqrt{\frac{m_2}{m_1}} \tag{6.4-33}$$

While this result does give the expected form, it actually is not Graham's law.

To explore Graham's law, we turn to the case shown in Figure 6.4-5(d). In this case, the capillary is again large so that collisions are intermolecular and not with the capillary walls. Now, however, the pistons bounding the volumes are mobile, moved to insure that the pressure on both sides of the capillary is equal. In this particular case, because the hydrogen is more mobile than the nitrogen, the pistons must both be moved to the left. Thus in this case, there *is* a convective velocity, and this case is very different from the conventional diffusion analysis in Figure 6.4-5(a).

To analyze this fourth case, we recognize that because the pressure is constant, there is no net momentum transfer between the left and right volumes. Thus

$$\begin{bmatrix} \text{momentum of} \\ \text{nitrogen} \\ \hline \text{time} \end{bmatrix} = \begin{bmatrix} \text{momentum of} \\ \text{hydrogen} \\ \hline \text{time} \end{bmatrix}$$

$$\begin{bmatrix} \text{momentum of} \\ \text{nitrogen} \\ \hline \text{collision} \end{bmatrix} \begin{bmatrix} \text{collisions of} \\ \text{nitrogen} \\ \hline \text{time} \end{bmatrix} = \begin{bmatrix} \text{momentum of} \\ \text{hydrogen} \\ \hline \text{collision} \end{bmatrix} \begin{bmatrix} \text{collisions of} \\ \text{hydrogen} \\ \hline \text{time} \end{bmatrix}$$

$$m_1 v_1 \zeta_1 = m_2 v_2 \zeta_2 \tag{6.4-34}$$

Combining with Equation 6.4-30

$$(m_1 u_1) c_1 v_1 = (m_2 u_2) c_2 v_2 \tag{6.4-35}$$

Note that the velocities u_i and v_i are very different. The molecular velocity u_i is sonic, perhaps around 10^4 cm/sec. The diffusion velocity v_i is much slower, perhaps around 1 cm/sec. Rearranging the above, we find

$$\frac{n_1}{n_2} = \frac{c_1 v_1}{c_2 v_2} = \frac{m_2 u_2}{m_1 u_1} \tag{6.4-36}$$

Inserting the molecular velocities from Equation 6.4-28, we find

$$\frac{n_1}{n_2} = \sqrt{\frac{m_2}{m_1}} \tag{6.4-37}$$

This inverse square root relation, suggested in 1829, is Graham's law.

Interestingly, there is still another, closely related mechanism called "effusion" which gives the same result for different reasons. In this case, shown in Figure 6.4-5(e), the capillary is replaced by an orifice of zero thickness. Molecules now don't diffuse through a capillary, but just fly through the orifice. They don't collide either with other molecules or with the capillary walls. In this case, we can show that for a circular hole,

$$n_1 = \frac{1}{4} c_1 u_1 = \frac{1}{4} \left(\frac{p_1}{k_B T} \right) \left[\frac{2 k_B T}{m_1} \right]^{\frac{1}{2}} \tag{6.4-38}$$

A similar relation will exist for species "2." If the pressures on both sides of the orifice are equal, we see that

$$\frac{n_1}{n_2} = \sqrt{\frac{m_2}{m_1}} \tag{6.4-39}$$

This relation, called "Graham's law of effusion," has still another physical basis than the cases discussed earlier.

At this point, you can be pretty confused by the nuances of these cases. To help you to keep them distinct, consider the summary shown in Table 6.4-1. As the table shows, many mechanisms can give similar results for different reasons, showing the subtlety of the apparently simple mechanism of diffusion.

Example 6.4-1: Diffusion in a porous catalyst Imagine a catalyst sphere with 30 percent voids to be used for the dehydrogenation reaction

$$C_2H_6 \rightarrow C_2H_4 + H_2$$

At 300 °C and 1 atmosphere, the effective diffusion coefficient of ethane in a 0.5-cm sphere is 0.06 cm^2/sec. What is the tortuosity?

Solution The chemical reaction produces a ternary mixture of ethane, ethylene, and hydrogen. Such a mixture may require consideration of the multicomponent diffusion equations in Chapter 7. However, if conversion is low, the diffusion coefficient

Table 6.4-1 *Graham's Law and Related Phenomena*

Effect	Key idea	Flux Ratio n_1 / n_2	Remarks
Diffusion (Constant Volume Diffusion)	No Flow	-1	Not Graham's Law
Convection	Flow	c_1 / c_2	Not Graham's Law
Knudsen Diffusion	Pore Smaller Than Mean-Free Path	$\sqrt{\dfrac{m_2}{m_1}}$	Not Graham's Law
Diffusion (Constant Pressure Diffusion)	Flow	$\sqrt{\dfrac{m_2}{m_1}}$	Graham's Law!
Effusion	Across Orifice	$\sqrt{\dfrac{m_2}{m_1}}$	Not Graham's Law

can be estimated with the same precision as for a mixture of ethane and ethylene. From Eq. 5.1-1, we find

$$D = \frac{1.86 \cdot 10^{-3} T^{3/2} (1/\tilde{M}_1 + 1/\tilde{M}_2)^{1/2}}{p \sigma_{12}^2 \Omega}$$

$$= \frac{1.86 \cdot 10^{-3} (573)^{3/2} (1/28 + 1/26)^{1/2}}{(1)[(4.23 + 4.16)/2]^2 (0.99)} = 0.40 \, \text{cm}^2/\text{sec}$$

The tortuosity is then (cf. Eq. 6.4-1)

$$\tau = \frac{\varepsilon D}{D_{\text{eff}}} = \frac{(0.3)0.4 \, \text{cm}^2/\text{sec}}{0.06 \, \text{cm}^2/\text{sec}} = 2.0$$

This value is typical.

Example 6.4-2: Pores in cell walls Some experiments on living cells suggests that there are pores 3 nm in diameter in the cell wall. Estimate the diffusion coefficient at 37 °C through such a pore for a solute 0.5 nm in diameter.

Solution To find the solute's diffusion coefficient in bulk solution, we use the Stokes–Einstein equation to find D_0. Combining this with Eq. 6.4-7, we find

$$D = \frac{k_B T}{6\pi\mu R_0} \left[1 + \frac{9}{8} \left(\frac{2R_0}{d} \right) \ln\left(\frac{2R_0}{d} \right) - 1.54 \left(\frac{2R_0}{d} \right) + \cdots \right]$$

$$= \frac{1.38 \cdot 10^{-16} \text{g cm}^2/\text{sec}^2 \, \text{K} \, (310 \, \text{K})}{6\pi(0.01 \text{g/cm sec})(2.5 \cdot 10^{-8} \, \text{cm})}$$

$$\cdot \left[1 + \frac{9}{8} \left(\frac{5}{30} \right) \ln\left(\frac{5}{30} \right) - 1.54 \left(\frac{5}{30} \right) + \cdots \right]$$

$$= (9.1 \cdot 10^{-6} \text{cm}^2/\text{sec})(1 - 0.33 - 0.26 + \cdots)$$

$$= 3.7 \cdot 10^{-6} \text{cm}^2/\text{sec}$$

Note that we have implicitly assumed that the pore is filled with water by using the viscosity of water in this estimate.

Example 6.4-3: Diffusion of hydrogen in small pores   Find the steady diffusion flux at 100 °C and 1 atm for hydrogen diffusing into nitrogen through a plug effectively 0.6 cm thick with 13-nm pores. Then estimate the flux through 18.3 μm pores.

   Solution   The mean free path for hydrogen can be found from Eq. 6.4-10:

$$l = \frac{4k_B T}{\pi \sigma^2 p}$$

$$= \frac{4(1.38 \cdot 10^{-16} \mathrm{g\, cm^2/sec^2\, K})(373\, \mathrm{K})}{\pi(2.83 \cdot 10^{-8}\, \mathrm{cm})^2 (1.01 \cdot 10^6\, \mathrm{g/cm\, sec^2})} = 800\, \mathrm{nm}$$

This mean free path is greater than the pore diameter, so the Knudsen number is large. Thus diffusion takes place in the Knudsen regime. For steady-state transport, the flux is found by applying Equation 6.4-13

$$n_1 = [D_{Kn}] \frac{\Delta c_1}{l}$$

$$= \left[\frac{13 \cdot 10^{-7} \mathrm{cm}}{3} \left(\frac{2 \left(8.31 \cdot 10^7 \frac{\mathrm{g\, cm^2}}{\mathrm{sec^2\, mol}}\right) 373\, \mathrm{K}}{2\mathrm{g/sec}} \right)^{\frac{1}{2}} \right] \left\{ \frac{1\, \mathrm{mol}}{22.4 \cdot 10^3 \mathrm{cm^3}} \left(\frac{273\, \mathrm{K}}{373\, \mathrm{K}}\right) \right\}$$

$$= 0.42 \cdot 10^{-5} \mathrm{mol/cm^2\, sec}$$

There are two interesting features of this result. First, hydrogen molecules spend their time colliding with pore walls, not with nitrogen molecules. Consequently, the properties of nitrogen do not appear in the calculation. Second, we have assumed that the pores are as long as the plug is thick, so the pores are implicitly taken to be straight. Any tortuosity would reduce the flux.

   For the 18.3 μm pores, the mean free path is much less than the pore diameter, and the Knudsen number is small. In this case, the flux equation contains the usual diffusion coefficient calculated from Eq. 5.1-1:

$$n_1 = j_1 = [D] \frac{\Delta c_1}{l}$$

$$= \frac{1.86 \cdot 10^{-3} (373)^{3/2} \left(\frac{1}{2.01} + \frac{1}{28.0}\right)^{1/2}}{1\, \mathrm{atm} \left(\frac{2.92 + 3.68}{2}\right)^2 (0.80)} \left\{ \frac{1\, \mathrm{mol}}{22.4 \cdot 10^3 \mathrm{cm^3}} \left(\frac{273\, \mathrm{K}}{273\, \mathrm{K}}\right) \right\}$$

$$= 6.1 \cdot 10^{-5} \mathrm{mol/cm^2\, sec}$$

This flux is greater than that in the Knudsen limit.

Example 6.4-4: Effective diffusion in an inhomogeneous gel The diffusion coefficient of KCl through a protein gel is $6 \cdot 10^{-7}$ cm^2/sec. However, the gel is not homogeneous, because it contains water droplets about 10^{-2} cm in diameter that are separated by only $2 \cdot 10^{-2}$ cm. The diffusion in these water droplets is about $2 \cdot 10^{-5}$ cm^2/sec. What is the diffusion in the homogeneous gel?

Solution The volume fraction of water can be found by considering a unit cell, $2 \cdot 10^{-2}$ cm on a side s, drawn around each 10^{-2} cm droplet:

$$\phi = \frac{\frac{4}{3}\pi r^3}{s^3} = \frac{\frac{4}{3}\pi \left(\dfrac{10^{-2}\text{cm}}{2}\right)^3}{(2 \cdot 10^{-2}\text{ cm})^3} = 0.065$$

The diffusion in the gel is found from Eq. 6.4-21.

$$\frac{D_{\text{eff}}}{D} = \frac{\dfrac{2}{D_s} + \dfrac{1}{D} - 2\phi\left(\dfrac{1}{D_s} - \dfrac{1}{D}\right)}{\dfrac{2}{D_s} + \dfrac{1}{D} + \phi\left(\dfrac{1}{D_s} - \dfrac{1}{D}\right)}$$

$$\frac{6 \cdot 10^{-7}\text{cm}^2/\text{sec}}{D} = \frac{\dfrac{2}{2 \cdot 10^{-5}\text{ cm}^2/\text{sec}} + \dfrac{1}{D} - 2(0.065)\left(\dfrac{1}{2 \cdot 10^{-5}\text{cm}^2/\text{sec}} - \dfrac{1}{D}\right)}{\dfrac{2}{2 \cdot 10^{-5}\text{ cm}^2/\text{sec}} + \dfrac{1}{D} + 0.065\left(\dfrac{1}{2 \cdot 10^{-5}\text{cm}^2/\text{sec}} - \dfrac{1}{D}\right)}$$

Solving, we find that D equals about $5 \cdot 10^{-7}$ cm^2/sec.

6.5 A Final Perspective

At the start of this book, we argued that the simplest way to look at diffusion was as a dilute solution of a particular solute moving through a homogeneous solvent. Such an argument led to the idea of a diffusion coefficient, a particular property of solute and solvent.

In this chapter, we have discussed the effects on the diffusion coefficient of the solute's interaction with other parts of the system. Sometimes the solute's flux is coupled with that of other solutes. Sometimes, the solute combines with solvent molecules, but near consolute points it avoids them. In a porous medium, the solute's diffusion may be slowed or accelerated; it may collide with pore walls during Knudsen diffusion, or be adsorbed in surface diffusion. In every case, the changes in diffusion can be major.

In describing these effects, scientists have used many methods. For example, the mathematics leading to the equation for hindered diffusion poses, for me, a truly formidable exercise. Obtaining the results of diffusion in composite media required the genius of Clerk Maxwell. However, these descriptions are limited by the particular models of the diffusion process. For example, the ideas of hydration are certainly inexact. The hindered diffusion equation depends on the model of a rigid solute sphere in a solvent continuum. Almost all of these estimates mean making major approximations.

As a result, I believe that the results in this chapter are best applied when using your scientific judgement. I do not think any of the ideas are gospel. Instead, they are approximations, subject to corrections found in future research. I wish you luck in finding these corrections.

Questions for Discussion

1. For a 1-1 electrolyte, does the faster ion control diffusion?
2. For a 1-1 electrolyte, does the faster ion control conduction?
3. Why is proton diffusion so rapid?
4. Why does the diffusion of a soap suddenly drop as the soap concentration is increased?
5. What is "isodermic association"? How does it affect diffusion?
6. Compare diffusion vs. solute concentration of HCl in excess NaCl and of sodium acetate in excess NaCl.
7. Why does the diffusion coefficient go to zero at a consolute point?
8. Does Brownian motion change near a consolute point?
9. What is spinodal decomposition?
10. What is the Knudsen number?
11. Compare the variation of the diffusion coefficient with pore diameter for small and for large pores.
12. Compare this variation with pressure in small and large pores.
13. What is the change in diffusion caused by ten volume percent impermeable spheres?
14. What is it by the same concentrations of cylinders?
15. What is it for the same concentration of flakes whose aspect ratio is 30?
16. What is Graham's law?

Problems

1. You are studying a thin film of 310 stainless steel that apparently is without pores. You clamp this film in a diaphragm cell, put a hydrogen pressure of 0.43 atmospheres on one side, and measure the much smaller hydrogen pressure on the other side. You find the data shown below [N. R. Quick and H. H. Johnson, *Metal Trans. A.*, **10A**, 67 (1979)]. These data show that the flux depends on the square root of hydrogen pressure, a dependence known as Sievert's law. This is believed to occur because molecular hydrogen dissociates into atomic hydrogen within the film. Use your knowledge of diffusion to justify this conclusion.

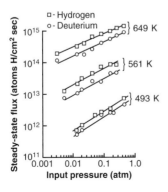

2. The ion diffusion coefficients at 25 °C of Na^+, K^+, Ca^{2+}, and Cl^- are 1.33, 1.9, 0.79, and 2.0 (all times 10^{-5} cm^2/sec). Find the diffusion coefficients and transference numbers for NaCl, KCl, and $CaCl_2$ in water and in excess KCl. *Answer:* In water, the diffusion coefficients are 1.60, 1.95, and 1.32 respectively, all in 10^{-5} cm^2/sec.

3. Calculate the diffusion coefficient at 25 °C of NH_4OH versus concentration. The relevant ionic diffusion coefficients are $D_{NH4} = 1.96$ and $D_{OH} = 5.28$ cm^2/sec. The pK_a of the NH_4^+ is 9.245. Estimate the diffusion coefficient of the NH_4OH molecule from the Wilke–Chang correlation.

4. The uptake of drugs from the intestinal lumen is often strongly influenced by diffusion. For example, consider a water-insoluble steroid for birth control that is solubilized in detergent micelles. These micelles, aggregates of steroid and soap, have a molecular weight of 24,000, an aggregation number of 80, a diameter of 26 angstroms, and a charge of –27. The counter ion is Na^+. (a) What is the diffusion coefficient of the micelle in water at 37 °C? *Answer:* $1.10 \cdot 10^{-5}$ cm^2/sec. (b) What is it in 0.1-M NaCl? Assume the micelle concentration is relatively low. *Answer:* $2.5 \cdot 10^{-6}$ cm^2/sec.

5. Electrolyte solutions can be highly nonideal. In these solutions, the flux equation for a 1–1 univalent electrolyte is often written as

$$-j_T = \frac{D_0 c_T}{RT} \nabla \mu_T$$

where the chemical potential μ_T is given by

$$\mu_T = \mu_T^0 + RT \ln c_T \gamma_T$$

and the activity coefficient γ_T in water at 25 °C is estimated from the Debye–Hückel theory:

$$\ln \gamma_T = -1.02 c_T^{1/2}$$

where c_T is in moles per liter. Using the values in Table 6.1-1, estimate the variation with concentration of the diffusion coefficient of potassium chloride and compare it with the experimental values (J. Zasadzinski).

6. The analytical ultracentrifuge takes a homogeneous solution of a large solute – like a buffered protein – and subjects it to a strong centrifugal field, at perhaps 100,000 rpm. The force exerted on a protein molecule is approximately $m\omega^2 r$, where m is the molecular mass, corrected for buoyancy; ω is the centrifuge's angular velocity; and r is its radius. (a) Parallel the development of the Nernst–Plank equations (Eq. 6.1-5) to derive an extended form of Fick's law that includes centrifugal force. (b) Calculate the steady-state concentration profile that would exist in very long ultracentrifuge experiments when diffusion and centrifugal force are balanced. You may not be able to complete the integration involved; go as far as you can.

7. The following data have been reported for ϵ-caprolactam diffusing in water at 25 °C [E. L. Cussler and P. J. Dunlop, *Austral. J. Chem.*, **19**, 1661 (1966)]:

$c(\text{mol/dm}^3)$	$D(10^{-5}\text{cm}^2/\text{sec})$
0.0514	0.8671
0.0515	0.8669
0.500	0.6978
0.991	0.5254
1.998	0.4160
3.003	0.3311

This solute is believed to dimerize by forming hydrogen bonds. Estimate the equilibrium constant K for this reaction. (G. Jerauld) *Answer:* about 0.5 M^{-1}.

8. Each molecule of sucrose in dilute aqueous solution is believed to combine with about four molecules of water. Such hydration has two effects. First, it increases the size of the sucrose solute and thus retards diffusion. Second, it increases the mole fraction of sucrose and hence may accelerate diffusion. (a) Estimate how the measured diffusion coefficient of sucrose differs from that of the unhydrated sucrose. In this estimate, take the hydrated sucrose diffusion coefficient at infinite dilution as $5.21 \cdot 10^{-6}$ cm^2/sec, its molecular weight as 342, and its solid density as 1.59 g/cm^3. *Answer:* about $5.7 \cdot 10^{-6}$ cm^2/sec. (b) Assume that the diffusion coefficient is given by Eq. 6.2-46 times a viscosity correction. Find how the coefficient varies with concentration.

9. A. Vignes [*Ind. Engr. Chem. Fund.*, 5, 189 (1966)] suggested that the concentration dependence of many liquid diffusion coefficients can be predicted with the equation

$$D = D_0 \left(1 + \frac{\partial \ln \gamma_1}{\partial \ln x_1}\right)$$

$$D_0 = D_1^{x_2} D_2^{x_1}$$

where D_1 is the diffusion coefficient of a trace of species 1 in excess species 2 and D_2 is that of a trace of species 2 in excess species 1. (a) Test the Vignes equation using the following data for ethanol (1) and water (2) at 25 °C [B. R. Hammond and R. H. Stokes, *Trans. Faraday Soc.*, 49, 890 (1953)]:

x_1	$D(10^{-5}\text{cm}_2/\text{sec})$	$1 + \dfrac{\partial \ln \gamma_1}{\partial \ln x_1}$
0.0	1.24	1.00
0.1	0.66	0.76
0.2	0.41	0.41
0.4	0.42	0.355
0.6	0.64	0.53
0.8	0.94	0.77
1.0	1.31	1.00

(b) Using these same data, calculate D_0 from Eq. 6.3-11. Compare how these quantities vary with concentration.

10. At the consolute temperature of a regular solution, the diffusion coefficient is approximately given by

$$D = D_0(1 - 4x_1x_2)^\gamma$$

where D_0 is a constant and x_i is the mole fraction of species i. Assume that the exponent γ equals 0.5. Note that D is zero when $x_1 = x_2 = 0.5$. The volume average velocity is zero, and the total concentration c is constant. Imagine that you are letting pure solute ($x_1 = 1$) diffuse through a long thin capillary into an equally large volume of pure solvent ($x_1 = 0$). You analyze your data as

$$j_1 = \frac{\bar{D}}{l}(c_{10} - c_{1l})$$

Show that $\bar{D}$ equals $D_0/2$.

11. You are separating globular proteins by gel permeation chromatography. One protein has a diameter of 34 angstroms. How much will its diffusion coefficient be reduced by diffusion in 417-angstrom pores?

12. The diffusion coefficients in water at 20 °C of hemoglobin and of catalase are $6.9 \cdot 10^{-7}$ cm^2/sec and $4.1 \cdot 10^{-7}$ cm^2/sec, respectively. They are $4.3 \cdot 10^{-10}$ cm^2/sec and $1.8 \cdot 10^{-10}$ cm^2/sec across a porous membrane. Estimate the pore size in this membrane. *Answer:* 30 nm.

13. Porous catalyst particles are often made by compressing the powdered catalyst into a particle, the pore structure of which can be controlled by the compression process. You are the engineer in charge of quality control at a catalyst manufacturing facility. You are making catalyst with pore sizes around 3 micrometers in diameter. Unfortunately, electron micrographs show one batch of product with pore sizes that are much smaller – about 550 angstroms in diameter. Paradoxically, the catalyst still has the same surface area per volume. The bad batch of particles was to be used in a diffusion-controlled oxidation at 400 °C and 1 atmosphere total pressure. As an estimate of the extent to which these particles will perform off-standard, calculate the diffusion coefficient of O_2 in the two different cylindrical pores. (S. Balloge) *Answer:* 0.13 cm^2/sec in small pores.

14. To estimate the pore size of a porous membrane, you plan to study the flux through the membrane caused by a single gas on the feed side and a vacuum on the permeate side. The pores in the membrane are near-circular cylinders formed by etching radiation tracks. At high feed pressure, you expect normal laminar flow following the Hagen–Poiseuille equation. At low feed pressure, you expect Knudsen diffusion. How do you expect the flux to vary with pressure in each of these cases? (Remember that the viscosity of dilute gases is independent of pressure.)

Further Reading

Arrhenius, S. A. (1884). *Göteborgs K. Vetenskaps-Och Vitterhets-Samhälles Bihang Till Handlingar*, **8**, No. 13–14.

Bard, A. J. and Faulkner, L. R. (2000). *Electrochemical Methods: Fundamentals and Applications*, 2nd ed. New York: Wiley.

Brenner, H. and Edwards, D. A. (1993). *Maromolecular Transport Processes*. Boston: Butterworth-Heinemann.

Brunel, M. E. and Breuer, M. M. (1971). *Diffusion Processes,* Vol. 1, ed. J.N. Sherwood *et al.,* p. 119. London: Gordon and Breach.

Claersson, S. and Sundelöf, L. O. (1957). *Journal de Chimie Physique,* **54,** 914.

Cussler, E. L. (1980). *American Institute of Chemical Engineers Journal,* **26,** 43.

Haase, R. and Siry, M. (1968). *Zeitschrift für Physicalische Chemie (Neue Folge),* **57,** 56.

Hinton, J. F. and Amis, E. S., (1971). *Chemical Reviews,* **71,** 621.

Kale, K., Cussler, E. L., and Evans, D. F. (1980). *Journal of Physical Chemistry,* **84,** 593.

Maxwell, J. C. (1873). *A Treatise on Electricity and Magnetism,* Vol. l, p. 365. Oxford: Clarendon Press.

Mazurin, O. V. (1984). *Phase Separation in Glass.* Amsterdam: North Holland; esp. Chapter 2.

Reusch, C. F., and Cussler, E. L. (1973). *American Institute of Chemical Engineers Journal,* **19,** 736.

Robinson, R. G. and Stokes, R. H. (1960). *Electrolyte Solutions.* London: Butterworth.

Scatchard, G. (1921). *Journal of the American Chemical Society,* **43,** 2406.

Strutt, W. (Lord Rayleigh) (1892). *Philosophical Magazine,* **34,** 481.

Weinheimer, R. M., Evans, D. F., and Cussler, E. L. (1981). *Journal of Colloid and Interface Science,* **80,** 357.

Yang, R. T. (2003). *Adsorbants: Fundamentals and Applications.* New York: Academic Press.

CHAPTER 7

Multicomponent Diffusion

Throughout this book, we have routinely assumed that diffusion takes place in binary systems. We have described these systems as containing a solute and a solvent, although such specific labels are arbitrary. We often have further assumed that the solute is present at low concentration, so that the solutions are always dilute. Such dilute systems can be analyzed much more easily than concentrated ones.

In addition to these binary systems, other diffusion processes include the transport of many solutes. One group of these processes occurs in the human body. Simultaneous diffusion of oxygen, sugars, and proteins takes place in the blood. Mass transfer of bile salts, fats, and amino acids occurs in the small intestine. Sodium and potassium ions cross many cell membranes by means of active transport. All these physiological processes involve simultaneous diffusion of many solutes.

This chapter describes diffusion for these and other multicomponent systems. The formalism of multicomponent diffusion, however, is of limited value. The more elaborate flux equations and the slick methods used to solve them are often unnecessary for an accurate description. There are two reasons for this. First, multicomponent effects are minor in dilute solutions, and most solutions are dilute. For example, the diffusion of sugars in blood is accurately described with the binary form of Fick's law. Second, some multicomponent effects are often more lucid if described without the cumbersome equations splattered through this chapter. For example, the diffusion of oxygen and carbon dioxide in blood is better described by considering explicitly the chemical reactions with hemoglobin.

Nonetheless, some concentrated systems are best described using multicomponent diffusion equations. Examples of these systems, which commonly involve unusual chemical interactions, are listed in Table 7.0-1. They are best described using the equations derived in Section 7.1. These equations can be rationalized using the theory of irreversible thermodynamics, a synopsis of which is given Section 7.2. In most cases, the solution to multicomponent diffusion problems is automatically available if the binary solution is available; the reasons for this are given in Section 7.3. Some values of ternary diffusion coefficients are given in Section 7.4 as an indication of the magnitude of the effects involved. Finally, tracer diffusion is detailed as an example of ternary diffusion in Section 7.5.

7.1 Flux Equations for Multicomponent Diffusion

Binary diffusion is often most simply described by Fick's law relative to the volume average velocity v^0:

$$-j_i = c_i(v^0 - v_i) = D\nabla c_i \tag{7.1-1}$$

Table 7.0-1 *Systems with large multicomponent effects*

Type of System	Examples
Solutes of very different sizes	Hydrogen–methane–argon
	Polystyrene–cyclohexane–toluene
Solutes in highly nonideal	Mannitol–sucrose–water
solutions	Acetic acid–chloroform–water
Concentrated electrolytes	Sodium sulfate–sulfuric acid–water
	Hydrogen chloride–polyacrylic acid–water
Concentrated alloys	Zinc–cadmium–silver
	Chromium–nickel–cobalt

In many cases, multicomponent diffusion is described by generalizing this equation to an n-component system:

$$-j_i = c_i(v^0 - v_i) = \sum_{j=1}^{n-1} D_{ij} \nabla c_j \tag{7.1-2}$$

in which the D_{ij} are multicomponent diffusion coefficients. The relation between these coefficients and the binary values is not known except for the dilute-gas limit, given for ternary diffusion in Table 7.1-1. In general, the diffusion coefficients are not symmetric ($D_{ij} \neq D_{ji}$). The diagonal terms (the D_{ii}) are called the "main term" diffusion coefficients, because they are commonly large and similar in magnitude to the binary values. The off-diagonal terms (the $D_{ij,i \neq j}$), called the "cross-term" diffusion coefficients, are often ten percent or less of the main terms. Each cross term gives a measure of the flux of one solute that is engendered by the concentration gradient of a second solute.

For an n-component system, this equation contains $(n-1)^2$ diffusion coefficients. This implies that one component must be arbitrarily designated as the solvent n. Because of the Onsager reciprocal relations discussed in Section 7.2, the coefficients are not all independent but instead are subject to certain restraints:

$$\sum_{j=1}^{n-1}\sum_{l=1}^{n-1} \left(\frac{\partial \mu_l}{\partial c_i} \right)_{c_{k \neq i,n}} \alpha_{lj} D_{jk} = \sum_{j=1}^{n-1}\sum_{l=1}^{n-1} \left(\frac{\partial \mu_l}{\partial c_k} \right)_{c_{i \neq k,n}} \alpha_{lj} D_{ji} \tag{7.1-3}$$

where

$$\alpha_{lj} = \left(\delta_{lj} + \frac{c_j \bar{V}_l}{c_n \bar{V}_n} \right) \tag{7.1-4}$$

where $\bar{V}_i$ is either a partial molar or partial specific volume, depending on whether the concentration is in moles per volume or mass per volume. These restraints reduce the number of diffusion coefficients required to describe diffusion to $\left(\frac{1}{2}\right)[n(n-1)]$ for an n-component system. However, because application of these restraints requires detailed thermodynamic information that is rarely available, the restraints are frequently impossible to apply, and by default the system is treated as having $(n-1)^2$ independent diffusion coefficients.

Equation 7.1-2 is the most useful form of the multicomponent flux equations. Because of an excess of theoretical zeal, many who work in this area have nurtured a glut of alternatives. These zealots most commonly use different driving forces or reference

Table 7.1-1 *Ternary diffusion coefficients: known functions of binary values for ideal gases*

$$D_{11} = \left[\frac{\dfrac{y_1}{\mathcal{D}_{12}} + \dfrac{y_2 + y_3}{\mathcal{D}_{23}}}{\dfrac{y_1}{\mathcal{D}_{12}\mathcal{D}_{13}} + \dfrac{y_2}{\mathcal{D}_{12}\mathcal{D}_{23}} + \dfrac{y_3}{\mathcal{D}_{13}\mathcal{D}_{23}}} \right]$$

$$D_{12} = \left[\frac{y_1 \left(\dfrac{1}{\mathcal{D}_{12}} - \dfrac{1}{\mathcal{D}_{13}} \right)}{\dfrac{y_1}{\mathcal{D}_{12}\mathcal{D}_{13}} + \dfrac{y_2}{\mathcal{D}_{12}\mathcal{D}_{23}} + \dfrac{y_3}{\mathcal{D}_{13}\mathcal{D}_{23}}} \right]$$

$$D_{21} = \left[\frac{y_2 \left(\dfrac{1}{\mathcal{D}_{12}} - \dfrac{1}{\mathcal{D}_{23}} \right)}{\dfrac{y_1}{\mathcal{D}_{12}\mathcal{D}_{13}} + \dfrac{y_2}{\mathcal{D}_{12}\mathcal{D}_{23}} + \dfrac{y_3}{\mathcal{D}_{13}\mathcal{D}_{23}}} \right]$$

$$D_{22} = \left[\frac{\dfrac{y_1 + y_3}{\mathcal{D}_{13}} + \dfrac{y_2}{\mathcal{D}_{12}}}{\dfrac{y_1}{\mathcal{D}_{12}\mathcal{D}_{13}} + \dfrac{y_2}{\mathcal{D}_{12}\mathcal{D}_{23}} + \dfrac{y_3}{\mathcal{D}_{13}\mathcal{D}_{23}}} \right]$$

velocities. Unfortunately, most of their answers are of limited value. The exception is for some metal alloys.

The best alternative to Eq. 7.1-2 is the Maxwell–Stefan equation for dilute gases:

$$\nabla y_i = \sum_{j=1}^{n-1} \frac{y_i y_j}{\mathcal{D}_{ij}} (v_j - v_i) \tag{7.1-5}$$

This equation has two major advantages over Eq. 7.1-2. First, these diffusion coefficients are the binary values found from binary experiments or calculated from the Chapman–Enskog theory given in Section 5.1. Second, the Stefan–Maxwell equations do not re-quire designating one species as solvent, which is sometimes an inconvenience when using Eq. 7.1-2.

These advantages can be compromised for multicomponent liquid mixtures. There, the nonideal solutions require a somewhat different form

$$\frac{\nabla \mu_i}{RT} = \sum_{j=1}^{n-1} \frac{x_j}{\mathcal{D}'_{ij}} (v_j - v_i) \tag{7.1-6}$$

For an ideal solution in which

$$\mu_i = \mu_i^0 + RT \ln x_i \tag{7.1-7}$$

this reduces to the ideal gas form. The new $\mathcal{D}'_{ij}$ are a new set of diffusion coefficients often believed to be more closely related to the binary form. This belief seems to me to rest more on faith than on data. Still, some researchers believe that this Maxwell–Stefan formulation is superior to Equation 7.1-2 because it does not require designating a solvent.

At the same time, the Maxwell–Stefan form has a serious disadvantage. It is difficult to combine with mass balances without designating one of the species as a solvent. Moreover, in many cases we benefit from identifying transport in one direction as occurring by diffusion and in the other direction as dominated by convection. When I use the Maxwell–Stefan form, I can lose this physical insight. Thus in practice, the advantage of this form is often lost. As a result, I feel Eq. 7.1-2 remains the most useful form of flux equation. We next examine the origins of these equations more carefully using irreversible thermodynamics.

7.2 Irreversible Thermodynamics

The multicomponent flux equations given in Eq. 7.1-2 are empirical generalizations of Fick's law that define a set of multicomponent diffusion coefficients. Because such definitions are initially intimidating, many have felt the urge to rationalize the origin of these equations and buttress this rationale with "more fundamental principles." This emotional need is often met with derivations based on irreversible thermodynamics.

Because the derivation of irreversible thermodynamics is straightforward, it seems on initial reading to be extremely valuable. After all those years of laboring under the restraint of equilibrium, the treatment of departures from equilibrium seems like a new freedom. Eventually one realizes that although irreversible thermodynamics does give the proper form of the flux equations and clarifies the number of truly independent coefficients, this information is of little value because it is already known from experiment. Irreversible thermodynamics tells us nothing about the nature and magnitude of the coefficients in the multicomponent equations, nor the resulting size and nature of the multicomponent effects. These are the topics in which we are interested. As a result, irreversible thermodynamics has enjoyed an overoptimistic vogue, first in chemical physics, next in engineering, and then in biophysics. Subsequently, it has been deemphasized as its limitations have become recognized. Because irreversible thermodynamics is of limited utility in describing multicomponent diffusion, only the barest outline will be given here.

7.2.1 The Entropy Production Equation

Three basic postulates are involved in the derivation of Eq. 7.1-2 (Fitts, 1962). The first postulate states that thermodynamic variables such as entropy, chemical potential, and temperature can in fact be correctly defined in a *differential volume* of a system that is not at equilibrium. This is an excellent approximation, except for systems that are very far from equilibrium, such as explosions. In the simple derivation given here, we assume a system of constant density, temperature, and pressure, with no net flow or chemical reaction. More complete equations without these assumptions are derived elsewhere (e.g., Haase, 1969).

The mass balance for each species in this type of system is given by

$$\frac{\partial c_i}{\partial t} = -\nabla \cdot \boldsymbol{n}_i = -\nabla \cdot \boldsymbol{j}_i \tag{7.2-1}$$

In this continuity equation, we use the fact that at no net flow and constant density, n_i equals j_i, the flux relative to the volume or mass average velocity. We also imply that the concentration is expressed in mass per unit volume. The left-hand side of this equation represents solute accumulation, and the right-hand side represents the solute diffusing in minus that diffusing out. The energy equation is similar:

$$\rho \frac{\partial \hat{H}}{\partial t} = -\nabla \cdot \boldsymbol{q} - \nabla \cdot \sum_{i=1}^{n} \bar{H}_i j_i \tag{7.2-2}$$

where $\boldsymbol{q}$ is the conductive heat flux, and $\bar{H}_i$ is the partial specific enthalpy. The left-hand side of this relation is the accumulation, the first term on the right-hand side is the energy conducted in minus that conducted out, and the second term is the energy diffusing in minus that diffusing out. Because we are assuming an isothermal system, $\boldsymbol{q}$ is presumably zero; we include it here so that the equation will look more familiar.

By parallel arguments, we can write a similar equation for entropy:

$$\rho \frac{\partial \hat{S}}{\partial t} = -\nabla \cdot \boldsymbol{J}_s + \sigma \tag{7.2-3}$$

By analogy, the term on the left must be the entropy accumulation. The first term on the right includes $\boldsymbol{J}_s$, which is entropy in minus entropy out by both convection and diffusion. The second term on the right, σ, gives the entropy produced in the process. This entropy production, which must be positive, is the quantitative measure of irreversibility in the system and represents a novel contribution of irreversible thermodynamics.

To find the entropy production, we first recognize that in this isothermal system,

$$d\hat{G} = d\hat{H} - Td\hat{S} = \frac{1}{\rho} \sum_{i=1}^{n} \mu_i dc_i \tag{7.2-4}$$

in which μ_i is the partial Gibbs free energy per unit mass, not the usual form of chemical potential; and ρ is the total mass density. This equation suggests that

$$\rho T \frac{\partial \hat{S}}{\partial t} = \rho \frac{\partial \hat{H}}{\partial t} - \sum_{i=1}^{n} \mu_i \frac{\partial c_i}{\partial t} \tag{7.2-5}$$

Combining with Eqs. 7.2-1 and 7.2-2

$$\rho T \frac{\partial \hat{S}}{\partial t} = -\nabla \cdot \boldsymbol{q} - \nabla \cdot \sum_{i=1}^{n} \bar{H}_i j_i - \sum_{i=1}^{n} \mu_i (\nabla \cdot j_i) \tag{7.2-6}$$

However,

$$\mu_i (\nabla \cdot j_i) = \nabla \cdot (\bar{H}_i - T\bar{S}_i) j_i - (j_i \cdot \nabla \mu_i) \tag{7.2-7}$$

Combining Eqs. 7.2-6 and 7.2-7,

$$\rho \frac{\partial \hat{S}}{\partial t} = -\nabla \cdot \left[\frac{\boldsymbol{q}}{T} + \sum_{i=1}^{n} \bar{S}_i j_i \right] - \frac{1}{T} \sum_{i=1}^{n} j_i \cdot \nabla \mu_i \tag{7.2-8}$$

By comparison with the entropy balance, Eq. 7.2-3, we see that the entropy flux is

$$J_s = \frac{q}{T} + \sum_{i=1}^{n} \hat{S}_i j_i \tag{7.2-9}$$

The first and second terms on the right-hand side are the entropy flux by conduction and by diffusion, respectively.

The entropy production can also be found by comparing Eqs. 7.2-3 and 7.2-8:

$$\sigma = -\frac{1}{T} \sum_{i=1}^{n} j_i \cdot \nabla \mu_i \tag{7.2-10}$$

The terms in this equation have units of energy per volume per time per temperature. Not all the fluxes and gradients in Eq. 7.2-1 are independent, because

$$\sum_{i=1}^{n} j_i = 0 \tag{7.2-11}$$

and, because the pressure and temperature are constant,

$$\sum_{i=1}^{n} c_i \nabla \mu_i = 0 \tag{7.2-12}$$

Using these restraints, we can rewrite Eq. 7.2-10 in terms of $n - 1$ fluxes and gradients relative to any reference velocity. In particular, for the mass average velocity, we can show that

$$\sigma = -\frac{1}{T} \sum_{i=1}^{n-1} j_i \cdot X_i \tag{7.2-13}$$

with the more general driving forces X_i given by

$$X_i = \sum_{j=1}^{n-1} \left(\delta_{ij} + \frac{c_j}{c_n} \right) \nabla \mu_j \tag{7.2-14}$$

Strictly speaking, Eqs. 7.3-13 and 7.3-14 apply only to the mass average reference velocity and j_i should be the flux relative to this velocity. Other reference velocities can also be used with other general forces. For example, for the volume average velocity, we may show that

$$\sigma = -\frac{1}{T} \sum_{i=1}^{n-1} j_i \cdot X_i^0 \tag{7.2-15}$$

where j_i is now relative to the volume average velocity, where

$$X_i^0 = \sum_{j=1}^{n-1} \alpha_{ij} \nabla \mu_j \tag{7.2-16}$$

and where the α_{ij} are given by Eq. 7.1-4. Eq. 7.2-15 is identical with Eq. 7.2-13 for a system of constant density, when the partial specific volumes all equal the reciprocal of the density, and volume and mass fractions are identical. We will use the volume average velocity and the associated fluxes and forces in the remainder of this chapter because these forms are those commonly used for fluids.

7.2.2 The Linear Laws

The second postulate in the derivation of irreversible thermodynamics is that a linear relation exists between the forces and fluxes in Eq. 7.2-15

$$-j_i = \sum_{j=1}^{n-1} L_{ij} X_j^0, \qquad (7.2\text{-}17)$$

where the L_{ij} have the mind-bending name of "Onsager phenomenological coefficients." These L_{ij} are strong functions of concentration, especially in dilute solution, where they approach zero as $c_i \to 0$. The linear law can be derived mathematically by use of a Taylor series in which all but the first terms are neglected, but because I am unsure when this neglect is justified, I prefer to regard the linear relation as a postulate.

7.2.3 The Onsager Relations

The third and final postulate is that the L_{ij} are symmetric, that is,

$$L_{ij} = L_{ji} \qquad (7.2\text{-}18)$$

These symmetry conditions, called the Onsager reciprocal relations (Onsager, 1931), can be derived by means of perturbation theory if "microscopic reversibility" is valid. The physical significance of microscopic reversibility is best visualized for a binary collision in which two molecules start in some initial positions, collide, and wind up in some new positions. If the velocities of these molecules are reversed and if microscopic reversibility is valid, the two molecules will move backward, retracing their paths through the collision to regain their original initial positions, just like a movie running backward. Those unfamiliar with the temperament of molecules running backward may be mollified by recalling that the symmetry suggested by Eq. 7.2-18 has been verified experimentally. Thus we can accept Eq. 7.2-18 as a theoretical result or as an experimentally verified postulate.

7.2.4 The Flux Equations

Using these three postulates, we can easily complete the derivation of the multicomponent flux equations from irreversible thermodynamics. We first rewrite Eq. 7.2-17 in terms of concentration gradients. Because the $\bar{V}_i$ are partial extensive quantities,

$$\sum_{i=1}^{n} \bar{V}_i \nabla c_i = 0 \qquad (7.2\text{-}19)$$

Those less well versed in thermodynamics can get the same result by assuming that the partial molar volumes are constant. As a result, only $n - 1$ concentration gradients are independent:

$$\nabla \mu_i = \sum_{j=1}^{n-1} \left(\frac{\partial \mu_i}{\partial c_j}\right)_{c_{k \neq j,n}} \nabla c_j \tag{7.2-20}$$

Note that the concentrations that are held constant in this differentiation differ from those that are commonly held constant in partial differentiation. If we combine Eqs. 7.2-16, 7.2-17, and 7.2-20, we obtain

$$-j_i = \sum_{j=1}^{n-1} D_{ij} \nabla c_j \tag{7.2-21}$$

where

$$D_{ij} = \sum_{k=1}^{n-1} \sum_{l=1}^{n-1} L_{ik} \alpha_{kl} \left(\frac{\partial \mu_l}{\partial c_j}\right)_{c_{m \neq j,n}} \tag{7.2-22}$$

where the α_{kl} are those given by Eq. 7.1-4. Thus, by starting our argument with conservation equations plus an equation for entropy production, we have derived multicomponent diffusion equations using only three postulates.

We still know nothing from this theory about the diffusion coefficients D_{ij}; we must evaluate these from experiment. Finding these coefficients commonly requires solving the flux equations with the techniques developed in the next section.

7.3 Solving the Multicomponent Flux Equations

In general, solving the multicomponent diffusion problems is not necessary if the analogous binary problem has already been solved (Toor, 1964; Stewart and Prober, 1964). We can mathematically convert the multicomponent problem into a binary problem, look up the binary solution, and then convert this solution back into the multicomponent one. In other words, multicomponent problems usually can be solved using a cookbook approach; little additional work is needed. Some use this cookbook to convert fairly comprehensible binary problems into multicomponent goulash that is harder to understand than necessary.

In this section, we first give the results for ternary diffusion and then for the general approach. By starting with the ternary results, we hope to help those who need to solve simple problems. They should not have to dig through the matrix algebra unless they decide to do so.

7.3.1 The Ternary Solutions

A binary diffusion problem has a solution that can be written as

$$\Delta c_1 = \Delta c_{10} F(D) \tag{7.3-1}$$

In this, Δc_1 is a concentration difference that generally varies with position and time, Δc_{10} is some reference concentration difference containing initial and boundary conditions, and $F(D)$ is the explicit function of position and time. For example, for the diaphragm cell, the binary solution is (see Example 2.2-4)

$$(c_{1B} - c_{1A}) = (c_{1B}^0 - c_{1A}^0)e^{-\beta Dt} \tag{7.3-2}$$

where c_{1i}^0 and c_{1i} are the concentrations in the diaphragm-cell compartment i at times zero and t, respectively, β is the cell calibration constant, and D is the diffusion coefficient. By comparison of Eqs. 7.3-1 and 7.3-2, we see that Δc_1 is $c_{1B} - c_{1A}$, Δc_{10} is $c_{1B}^0 - c_{1A}^0$, and $F(D)$ is $e^{-\beta Dt}$.

Every binary diffusion problem has an analogous ternary diffusion problem that is described by similar differential equations and similar initial and boundary conditions. The differential equations differ only in the form of Fick's law that is used. The conditions are also parallel. For example, in a binary problem the solute concentration may be fixed at a particular boundary, so in the corresponding ternary problem, solute concentrations will also be fixed at the corresponding boundary. When this is true, the ternary diffusion problems have the solutions

$$\Delta c_1 = P_{11}F(\sigma_1) + P_{12}F(\sigma_2) \tag{7.3-3}$$

and

$$\Delta c_2 = P_{21}F(\sigma_1) + P_{22}F(\sigma_2) \tag{7.3-4}$$

in which the concentration differences Δc_1 and Δc_{10} are the dependent and independent values in the binary problem, $F(D)$ is again the solution to the binary problem, and the values of σ_i and P_{ij} are given in Table 7.3-1 (Cussler, 1976). The σ_i are the eigenvalues (with relative weighting factors ρ) of the diffusion-coefficient matrix and hence are a type of pseudobinary diffusion coefficient.

The calculation of the ternary diffusion profile is now routine. For example, the result for solute 1 in the diaphragm cell will be

$$c_{1B} - c_{1A} = \frac{(D_{11} - \sigma_2)(c_{1B}^0 - c_{1A}^0) + D_{12}(c_{2B}^0 - c_{2A}^0)}{\sigma_1 - \sigma_2}e^{-\sigma_1 dt}$$

$$+ \frac{(D_{11} - \sigma_1)(c_{1B}^0 - c_{1A}^0) + D_{12}(c_{2B}^0 - c_{2A}^0)}{\sigma_2 - \sigma_1}e^{-\sigma_2 dt} \tag{7.3-5}$$

The results for the second solute can be found from Eq. 7.3-4 or by rotating the indices in Eq. 7.3-5.

Example 7.3-1: Fluxes for ternary free diffusion Find the fluxes and the concentration profiles in a dilute ternary free-diffusion experiment. In such an experiment, one ternary solution is suddenly brought into contact with a different composition of the same ternary solution. Find the flux and the concentrations versus position and time at small times.

Table 7.3-1 *Factors for solution of ternary diffusion problems*

Eigenvalues

$$\sigma_1 = \tfrac{1}{2}[D_{11} + D_{22} + \sqrt{(D_{11} - D_{22})^2 + 4D_{12}D_{21}}]$$

$$\sigma_2 = \tfrac{1}{2}[D_{11} + D_{22} - \sqrt{(D_{11} - D_{22})^2 + 4D_{12}D_{21}}]$$

Weighting factors

$$P_{11} = \left(\frac{D_{11} - \sigma_2}{\sigma_1 - \sigma_2}\right)\Delta c_{10} + \left(\frac{D_{12}}{\sigma_1 - \sigma_2}\right)\Delta c_{20}$$

$$P_{12} = \left(\frac{D_{11} - \sigma_1}{\sigma_2 - \sigma_1}\right)\Delta c_{10} + \left(\frac{D_{12}}{\sigma_2 - \sigma_1}\right)\Delta c_{20}$$

$$P_{21} = \left(\frac{D_{21}}{\sigma_1 - \sigma_2}\right)\Delta c_{10} + \left(\frac{D_{22} - \sigma_2}{\sigma_1 - \sigma_2}\right)\Delta c_{20}$$

$$P_{22} = \left(\frac{D_{21}}{\sigma_2 - \sigma_1}\right)\Delta c_{10} + \left(\frac{D_{22} - \sigma_1}{\sigma_2 - \sigma_1}\right)\Delta c_{20}$$

Note: For further definitions, see Eqs. 7.1-2, 7.3-1, 7.3-3, and 7.3-4.

Solution When the two solutions come in contact for only a short time, they are effectively infinitely thick. The binary solution of this problem is (see Eq. 2.3-15)

$$\frac{c_1 - c_{10}}{c_{1\infty} - c_{10}} = \mathrm{erf}\,\frac{z}{\sqrt{4Dt}}$$

in which c_{10} and $c_{1\infty}$ are the concentrations where the solutions are contacted (at $z = 0$) and far into one solution (at $z = \infty$), respectively, z and t are the position and time, and D is the binary diffusion coefficient. By comparison with Eq. 7.3-1, we see that Δc_1 is $c_1 - c_{10}$, Δc_{10} is $c_{1\infty} - c_{10}$ and $F(D)$ equals the error function of $z/\sqrt{4Dt}$. As a result the concentration profile for solute 1 will be

$$c_1 - c_{10} = \left[\frac{(D_{11} - \sigma_2)(c_{1\infty} - c_{10}) + D_{12}(c_{2\infty} - c_{20})}{\sigma_1 - \sigma_2}\right]\mathrm{erf}\,\frac{z}{\sqrt{4\sigma_1 t}}$$

$$+ \left[\frac{(D_{11} - \sigma_1)(c_{1\infty} - c_{10}) + D_{12}(c_{2\infty} - c_{20})}{\sigma_2 - \sigma_1}\right]\mathrm{erf}\,\frac{z}{\sqrt{4\sigma_2 t}}$$

The close similarity between this result and that for the diaphragm cell is obvious.

The fluxes can be found in the same manner as the concentration profile. Because the solutions are dilute, there is negligible convection induced by diffusion, so

$$-n_1 \doteq -j_1 = D_{11}\frac{\partial c_1}{\partial z} + D_{12}\frac{\partial c_2}{\partial z}$$

Combining this with Eqs. 2.3-17, 7.3-3, and 7.3-4,

$$-j_1 = (D_{11}P_{11}+D_{12}P_{21})\frac{e^{-z^2/4\sigma_1 t}}{\sqrt{\pi\sigma_1 t}}$$

$$+(D_{11}P_{12}+D_{12}P_{22})\frac{e^{-z^2/4\sigma_2 t}}{\sqrt{\pi\sigma_2 t}}$$

where again the P_{ij} are given in Table 7.3-1. These results are complex algebraically but straightforward conceptually.

7.3.2 The General Solution

We now turn from the detail of ternary diffusion to the more general solution of the multicomponent problems. The general solution of these equations is most easily presented in terms of linear algebra, a notation that is not used elsewhere in this book. In this presentation, we consider the species concentrations as a vector of $\underline{c}$ and the multicomponent diffusion coefficients as a matrix $\underline{\underline{D}}$.

In matrix notation, the multicomponent flux equations are

$$-j = \underline{\underline{D}} \cdot \nabla\underline{c} \tag{7.3-6}$$

The continuity equations for this case are

$$\frac{\partial\underline{c}}{\partial t} + (\nabla \cdot v^0\underline{c}) = -\nabla \cdot j \tag{7.3-7}$$

These are subject to the initial and boundary conditions

$$\Delta\underline{c}(x, y, z, t = 0) = \Delta\underline{c}_0 \tag{7.3-8}$$

$$\Delta\underline{c}(B, t) = 0 \tag{7.3-9}$$

$$\frac{\partial\underline{c}}{\partial z}(b, t) = 0 \tag{7.3-10}$$

where B and b represent two boundaries of the system. Note that the boundary conditions on all concentrations must have the same functional form. This is a serious restriction only for the case of simultaneous diffusion and chemical reaction.

We now assume that there exists a nonsingular matrix $\underline{\underline{t}}$ that can diagonalize $\underline{\underline{D}}$:

$$\underline{\underline{t}}^{-1} \cdot \underline{\underline{D}} \cdot \underline{\underline{t}} = \underline{\underline{\sigma}} = \begin{bmatrix} \sigma_1 & 0 & 0 & \cdots \\ 0 & \sigma_2 & 0 & \cdots \\ 0 & 0 & \sigma_3 & \cdots \\ \vdots & \vdots & \vdots & \end{bmatrix} \tag{7.3-11}$$

where $\underline{\underline{t}}^{-1}$ is the inverse of $\underline{\underline{t}}$ and $\underline{\underline{\sigma}}$ is the diagonal matrix of the eigenvalues of the diffusion coefficient matrix $\underline{\underline{D}}$. The assumption that $\underline{\underline{D}}$ can be put into diagonal form is not necessary for a general mathematical solution, but because this assumption is valid for all cases encountered in practice, it is used here. For the case of ternary diffusion,

$$\underset{=}{t} = \begin{bmatrix} t_{11} & t_{12} \\ t_{21} & t_{22} \end{bmatrix} = \frac{\begin{bmatrix} 1 & \dfrac{D_{12}}{D_{22} - \sigma_1} \\ \dfrac{D_{22} - \sigma_2}{D_{12}} & 1 \end{bmatrix}}{\left[1 - \left(\dfrac{\sigma_2 - D_{22}}{\sigma_1 - D_{22}} \right) \right]} = \frac{\begin{bmatrix} 1 & \dfrac{D_{11} - \sigma_1}{D_{21}} \\ \dfrac{D_{21}}{D_{11} - \sigma_2} & 1 \end{bmatrix}}{\left[1 - \left(\dfrac{\sigma_1 - D_{11}}{\sigma_2 - D_{11}} \right) \right]} \tag{7.3-12}$$

Correspondingly,

$$\underset{=}{t}^{-1} = \begin{bmatrix} t_{11}^{-1} & t_{12}^{-1} \\ t_{21}^{-1} & t_{22}^{-1} \end{bmatrix} = \frac{\begin{bmatrix} 1 & \dfrac{D_{12}}{\sigma_1 - D_{22}} \\ \dfrac{\sigma_2 - D_{22}}{D_{12}} & 1 \end{bmatrix}}{\det\left(\underset{=}{t} \right)} = \frac{\begin{bmatrix} 1 & \dfrac{\sigma_1 - D_{11}}{D_{21}} \\ \dfrac{D_{21}}{\sigma_2 - D_{11}} & 1 \end{bmatrix}}{\det\left(\underset{=}{t} \right)} \tag{7.3-13}$$

where

$$\det\left(\underset{=}{t} \right) = \frac{\sigma_1 - \sigma_2}{\sigma_1 - D_{22}} = \frac{\sigma_2 - \sigma_1}{\sigma_2 - D_{11}} \tag{7.3-14}$$

Remember that the product of $\underset{=}{t}$ and its inverse $\underset{=}{t}^{-1}$ is the unit matrix.

We now use this new matrix $\underset{=}{t}$ to define a new combined concentration $\underline{\Psi}$

$$\underline{c} = \underset{=}{t} \cdot \underline{\Psi} \tag{7.3-15}$$

We combine Eqs. 7.3-6, 7.3-7, and 7.3-15 and premultiply the equation by $\underset{=}{t}^{-1}$ to obtain

$$\frac{\partial \underline{\Psi}}{\partial t} + \nabla \cdot \nu^0 \underline{\Psi} = \underset{=}{\sigma} \cdot \nabla^2 \underline{\Psi} \tag{7.3-16}$$

which represents a set of scalar equations

$$\frac{\partial \Psi_i}{\partial t} + \nabla \cdot \nu^0 \Psi_i = \sigma_i \cdot \nabla^2 \Psi_i \tag{7.3-17}$$

In this operation, we have made the assumption that $\underset{=}{D}$ and hence both $\underset{=}{t}$ and $\underset{=}{\sigma}$ are not functions of composition.

The initial and boundary conditions can also be written in terms of the new combined concentration $\underline{\Psi}$:

$$\Delta \underline{\Psi}(x, y, z, 0) = \Delta \underline{\Psi}_0 = \underset{=}{t}^{-1} \cdot \Delta \underline{c}_0 \tag{7.3-18}$$

$$\Delta \underline{\Psi}(B, t) = 0 \tag{7.3-19}$$

$$\frac{\partial \underline{\Psi}}{\partial z}(b, t) = 0 \tag{7.3-20}$$

Thus a set of coupled differential equations has been separated into uncoupled equations written in terms of the new concentration Ψ.

Equations 7.3-16 through 7.3-20 have exactly the same form as the associate *binary* diffusion problem:

$$\frac{\partial c_1}{\partial t} + \nabla \cdot v^0 c_1 = D\nabla^2 c_1 \tag{7.3-21}$$

which has the same initial and boundary conditions for each species as those given in Eqs. 7.3-8 through 7.3-10. If this binary problem has the solution

$$\Delta c_1 = F(D)\Delta c_{10} \tag{7.3-22}$$

then Eqs. 7.3-17 through 7.3-20 must have the solution

$$\Delta\Psi_i = F(\sigma_i)\Delta\Psi_{i0} \tag{7.3-23}$$

where the eigenvalue σ_i is substituted everywhere that the binary diffusion coefficient occurs in the binary solution. If we rewrite our solution in terms of the actual concentrations, we find that

$$\Delta \underline{c} = \underline{t} \cdot \underline{F}(\underline{\sigma}) \cdot \underline{t}^{-1} \cdot \Delta \underline{c}_0 \tag{7.3-24}$$

Thus we know the concentration profiles in the multicomponent system in terms of its binary analogue. The results for the ternary case are given in Eqs. 7.3-3 and 7.3-4.

Many find this derivation difficult to grasp, even after they apparently understand every step. Their trouble usually stems from a mathematical, not physical, problem. They do not see why the derivation is more than a trick, a slick invention. The reason is that Eq. 7.3-17 and its associated conditions are shown to be mathematically the same as the binary solution. If we change the symbol Ψ_i to c_1, Eq. 7.3-17 and Eq. 7.3-21 are exactly the same. The physical circumstances in the multicomponent problem may be more elaborate, but the identity of the differential equations signals that the mathematical solutions are identical.

Example 7.3-2: Steady-state multicomponent diffusion across a thin film In steady-state binary diffusion, we found that the solute's concentration varied linearly across a thin film. Will solute concentrations vary linearly in the multicomponent case? What will the flux be?

    Solution By comparison with Eq. 2.2-9, we see that

$$(c_1 - c_{10}) = \left(\frac{z}{l}\right)(c_{1l} - c_{10})$$

By comparing this with Eq. 7.3-22, we see that $F(D)$ equals (z/l). From Eq. 7.3-24, for the multicomponent case,

$$\Delta \underline{c} = \left(\underline{t} \cdot \frac{z}{l}\underline{\delta} \cdot \underline{t}^{-1}\right) \cdot \Delta \underline{c}_0 = \left(\frac{z}{l}\right)\Delta \underline{c}_0$$

Thus the concentration profile of each solute remains linear. The flux is

$$-\underline{j} = \underline{\underline{D}} \cdot \nabla \underline{c}$$

$$= \underline{\underline{D}} \cdot \frac{\Delta \underline{c}_0}{l}$$

or

$$j_i = \sum_{j=1}^{n-1} \frac{D_{ij}}{l}(c_{j0} - c_{jl})$$

Note that a solute's flux can be in the opposite direction to that expected if other gradients exist in the system.

7.4 Ternary Diffusion Coefficients

In this section, we report a variety of values for ternary diffusion coefficients. These coefficients support the generalizations given at the beginning of this chapter that multicomponent effects were significant when the system was concentrated and contained interacting species. These interactions can originate from chemical reactions, from electrostatic coupling, or from major differences in molecular weights.

Typical diffusion coefficients for gases are shown in Table 7.4-1. These values are not experimental, but are calculated from the Chapman–Enskog theory (see Section 5.1) and from Table 7.1-1. The first two rows in the table show how the values of D_{12} and D_{21} are larger as the solution becomes concentrated. The second and third rows refer to the same solution but with a different species chosen as the solute. The difference in the diffusion coefficients illustrates why ternary diffusion coefficients can be difficult to interpret. The final three rows are other characteristic situations.

Table 7.4-1 *Ternary diffusion coefficients in gases at 25°C*

System	D_{11}	D_{12}	D_{21}	D_{22}
Hydrogen ($x_1 = 0.05$) Methane ($x_2 = 0.05$) Argon ($x_3 = 0.90$)	0.78	−0.00	0.03	0.22
Hydrogen ($x_1 = 0.2$) Methane ($x_2 = 0.2$) Argon ($x_3 = 0.6$)	0.76	−0.01	0.12	0.25
Argon ($x_1 = 0.6$) Methane ($x_2 = 0.2$) Hydrogen ($x_3 = 0.2$)	0.64	−0.39	−0.12	0.37
Carbon dioxide ($x_1 = 0.2$) Oxygen ($x_2 = 0.2$) Nitrogen ($x_3 = 0.6$)	0.15	−0.00	−0.01	0.19
Hydrogen ($x_1 = 0.2$) Ethylene ($x_2 = 0.2$) Ethane ($x_3 = 0.6$)	0.56	0.00	0.11	0.13
Benzene ($x_1 = 0.2$) Cyclohexane ($x_2 = 0.2$) Hexane ($x_3 = 0.6$)	0.028	0.000	0.001	0.026

Note: All coefficients have units of square centimeters per second and are calculated from the equations in Table 7.1-1.

Ternary diffusion coefficients in liquids and solids cannot be found from binary values, but only from experiments. When experiments are not available, which is usually the case, one can make estimates by assuming that the Onsager phenomenological coefficients are a diagonal matrix; that is,

$$L_{ij,\ i \neq j} = 0 \tag{7.4-1}$$

In addition, we can assume that the main-term coefficients are related to the binary values given by

$$L_{ii} = \left(\frac{D_i c_i}{RT} \right) \tag{7.4-2}$$

where D_i is the coefficient of species i in the solvent. These assumptions can be combined with Eq. 7.1-4 and Eq. 7.2-20 to give

$$D_{ij} = \left(\frac{D_i c_i}{RT} \right) \sum_{l=1}^{n-1} \left(\delta_{il} + \frac{c_l \bar{V}_i}{c_n \bar{V}_n} \right) \left(\frac{\partial \mu_l}{\partial c_j} \right)_{c_{k \neq j,n}} \tag{7.4-3}$$

This is equivalent to saying that ternary effects result from activity coefficients. I routinely use this equation for making initial estimates.

Experimental values of ternary diffusion coefficients characteristic of liquids are shown in Table 7.4-2. In cases like KCl–NaCl–water, KCl–sucrose–water, and toluene–chlorobenzene–bromobenzene, the cross-term diffusion coefficients are small, less than ten percent of the main diffusion coefficients. In these cases, we can safely treat the diffusion as a binary process.

The cross-term diffusion coefficients are much more significant for interacting solutes. In cases like HBr–KBr–water and H_2SO_4–Na_2SO_4–water, this interaction is ionic; in other cases, it may involve hydrogen-bond formation. Cross-term diffusion coefficients and the resulting ternary effects should be especially large in partially miscible systems, where few measurements have been made.

The ternary diffusion coefficients in metals shown in Table 7.4-3 have the largest cross-term values. As a result, the flux of one component in an alloy can be against its concentration gradient, from low concentration into higher concentration. These effects are especially interesting when they are superimposed on the elaborate phase diagrams characteristic of alloys because they can lead to local phase separations that dramatically alter the material's properties. As in gases and liquids, the methods of estimating ternary diffusion coefficients are risky. One must either rely on relations like Eq. 7.4-3 or undertake the difficult experiments involved. As a result, many avoid ternary diffusion even when they suspect it is important.

7.5 Tracer Diffusion

Imagine we want to study the diffusion of steroids like progesterone through human blood. The amounts of these steroids will be very small, making direct chemical analysis difficult. As a result, we synthesize steroids that contain carbon 14 as

Table 7.4-2 *Ternary diffusion coefficients in liquids at 25 °C*[a]

System	D_{11}	D_{12}	D_{21}	D_{22}
1.5-M KCl (1) 1.5-M NaCl (2) H_2O (3)[b]	1.80	0.33	0.10	1.39
0.10-M HBr (1) 0.25-M KBr (2) H_2O (3)[c]	5.75	0.05	−2.20	1.85
1-M H_2SO_4 (1) 1-M Na_2SO_4 (2) H_2O (3)[d]	2.61	−0.04	−0.51	0.91
0.06 g/cm^3 KCl (1) 0.03 g/cm^3 sucrose (2) H_2O (3)[e]	1.78	0.02	0.07	0.50
2-M urea (1) O-M^{14} C-tagged urea (2) H_2O (3)[f]	1.24	0.01	0.00	1.23
32 mol% hexadecane (1) 35 mol% dodecane (2) 33 mol% hexane (3)[g]	1.03	0.23	0.27	0.97
25 mol% toluene (1) 50 mol% chlorobenzene (2) 25 mol% bromobenzene (3)[h]	1.85	−0.06	−0.05	1.80
0.326 g/cm^3 benzene (1) 0.265 g/cm^3 propanol (2) Carbon tetrachloride (3)[i]	1.64	0.78	0.17	1.33
5 wt% cyclohexane (1) 5 wt% polystyrene (2) 90 wt% toluene (3)[j]	2.03	−0.09	−0.02	0.09

Notes: [a]All values × 10^{-5} square centimeters per second. [b]P. J. Dunlop, *J. Phys. Chem.*, **63**, 612 (1959). [c]A. Reojin, *J. Phys. Chem.*, **76**, 3419 (1972). [d]R. P. Wendt, *J. Phys. Chem.*, **66**, 1279 (1962). [e]E. L. Cussler and P. J. Dunlop, *J. Phys. Chem.*, **70**, 1880 (1966). [f]J. G. Albright and R. Mills, *J. Phys. Chem.*, **69**, 3120 (1966). [g]T. K. Kett and D. K. Anderson, *J. Phys. Chem.*, **73**, 1268 (1969). [h]J. K. Burchard and H. L. Toor, *J. Phys. Chem.*, **66**, 2015 (1962). [i]R. A. Graff and T. B. Drew, *IEC Fund.*, **7**, 490 (1968) (data at 200 °C). [j]E. L. Cussler and E. N. Lightfoot, *J. Phys. Chem.*, **69**, 1135 (1965).

a radioactive label. We then measure the steroid concentration, and calculate diffusion coefficients from these concentration measurements.

This measurement of tracer diffusion in dilute solution is a good strategy. Such a use of radioactive tracers provides a near-unique opportunity for a specific chemical analysis in highly dilute solution. Such analysis is especially important in biological systems, where complex chemistry may compromise analysis. Moreover, in dilute solution, the diffusion coefficients found with radioactive tracers are almost always indistinguishable from those measured in other ways. Exceptions occur in those systems in which the solute moves by a jump mechanism like that for protons (see Fig. 6.1.-1) or in which the solute's molecular weight is significantly altered by the isotopic mass.

Table 7.4-3 Ternary interdiffusion coefficients in solids[a]

Ternary system 1-2-3	Composition (mol%) and structure at temperature studied	Temperature (°C)	D_{11}^3	D_{12}^3	D_{21}^3	D_{22}^3
C-Si-Fe[b]	0.46C-1.97Si-97.57Fe (FCC)	1,050	4.8×10^{-7}	0.3×10^{-7}	$\simeq 0$	2.3×10^{-9}
Al-Ni-Fe	47Al-18Ni-35Fe (BCC)[c]	1,004	4.4×10^{-11}	-0.2×10^{-11}	0.3×10^{-11}	1.6×10^{-11}
	43Al-8.5Ni-48.5Fe (BCC)[c]	1,004	16.4×10^{-11}	-5.9×10^{-11}	0.3×10^{-11}	2.5×10^{-11}
	8Al-44.5Ni-48.5Fe (BCC)[d]	1,000	23.3×10^{-12}	-9.2×10^{-12}	-9.9×10^{-12}	16.4×10^{-12}
Cr-Ni-Co[e]	9.5Cr-20.4Ni-70Co	1,300	0.6×10^{-9}	-0.13×10^{-11}	-0.12×10^{-10}	0.23×10^{-9}
	8.8Cr-40Ni-51Co		1.09×10^{-9}	-0.6×10^{-11}	-2.6×10^{-10}	0.47×10^{-9}
	9.2Cr-79Ni-11.8Co (FCC)		1.25×10^{-9}	-2.5×10^{-11}	-5.1×10^{-10}	0.74×10^{-9}
Zn-Cd-Ag[f]	13.1Zn-3.5Cd-83.4Ag	600	1.2×10^{-10}	1.3×10^{-10}	0.13×10^{-10}	1.2×10^{-10}
	18.1Zn-4.4Cd-77.5Ag		3.3×10^{-10}	2.4×10^{-10}	0.6×10^{-10}	2.1×10^{-10}
	16.4Zn-8.5Cd-75.1Ag (FCC)		4.4×10^{-10}	4.5×10^{-10}	1.4×10^{-10}	5.5×10^{-10}
Zn-Mn-Cu[g]	10.3Zn-1.8Mn-87.9Cu (FCC)	850	1.82×10^{-9}	0.11×10^{-9}	-0.02×10^{-9}	1.46×10^{-9}
Zn-Ni-Cu[h]	19Zn-43Ni-38Cu (FCC)	775	5.1×10^{-11}	-0.8×10^{-11}	-1.7×10^{-11}	1.2×10^{-11}
V-Zr-Ti[i]	9V-9Zr-82Ti	800	2.3×10^{-10}	0.1×10^{-10}	1.0×10^{-10}	4.4×10^{-10}
	17.5V-19.5Zr-63Ti		2.9×10^{-10}	1.5×10^{-10}	0.7×10^{-10}	1.8×10^{-10}
	37.5V-7.5Zr-55Ti		0.16×10^{-10}	0.18×10^{-10}	0.1×10^{-10}	0.23×10^{-10}
	5.0V-77.5Zr-17.5Ti (FCC)		12.4×10^{-10}	2.6×10^{-10}	-0.8×10^{-10}	2.8×10^{-10}
Cu-Ag-Au[j]	13.1Cu-34.0Ag-52.9Au	725	1.0×10^{-10}	0.1×10^{-10}	1.7×10^{-10}	1.3×10^{-10}
	60.3Cu-12.9Ag-26.8Au (FCC)		2.3×10^{-10}	1.1×10^{-10}	1.8×10^{-10}	3.1×10^{-10}
Co-Ni-Fe[k]	10.3Co-31.4Ni-58.3Fe	1,315	4×10^{-10}	0.9×10^{-10}	3×10^{-10}	7.1×10^{-10}
	35.5Co-35.4Ni-29.1Fe		6.5×10^{-10}	2.7×10^{-10}	3.2×10^{-10}	7.3×10^{-10}
	31.1Co-65.6Ni-3.3Fe (FCC)		6.1×10^{-10}	0.2×10^{-10}	4.0×10^{-10}	8.8×10^{-10}

Note: [a]All diffusion coefficients are in square centimeters per second and are based on a solvent-fixed reference frame. [b]J. Kirkaldy, Can. J. Phys., 35, 435 (1957). [c]T. D. Moyer and M. A. Dayananda, Met. Trans., 7A, 1035 (1976). [d]G. H. Cheng and M. A. Dayananda, Met. Trans., 10A, 1415 (1979). [e]G. Guy and V. Leroy, The Electron Microprobe, eds. McKinley, T.D., Heinrich, K.F.J., and Wittry, D.B., New York: John Wiley and Sons (1966). [f]P. T. Carlson, M. A. Dayananda, and R. E. Grace, Met. Trans., 3, 819 (1972). [g]M. A. Dayananda and R. E. Grace, Trans. Met. Soc. AMIE, 233, 1287 (1965). [h]R. D. Sisson, Jr. and M. A. Dayananda, Met. Trans., 8A, 1849 (1977). [i]A. Brunsch and S. Steel, Zeit. Metallkunde, 65, 765 (1974). [j]T. O. Ziebold and R. E. Ogilvie, Trans. Met. Soc. AMIE, 239, 942 (1967). [k]A. Vignes and J. P. Sabatier, Trans. Met. Soc. AMIE, 245, 1795 (1969).

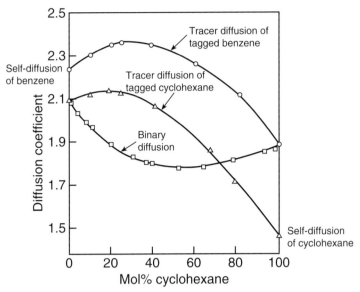

Fig. 7.5-1. Binary and tracer diffusion at 25 °C. The tracer diffusion coefficient equals the binary coefficient only in certain special cases. All coefficients are $\times 10^{-5}$ square centimeters per second. [From Mills (1965), with permission.]

In concentrated solution, tracer diffusion is a much more complex process which may not provide coefficients identical with those in the binary system. This is illustrated by the data in Fig. 7.5-1. In this figure, we see that the diffusion coefficients using different radioactive isotopes can differ from each other and from the binary diffusion coefficient. On reflection, we realize that this is not surprising; the diffusion of radioactively tagged benzene in untagged benzene is obviously a different process than the diffusion of tagged cyclohexane in benzene.

Explaining these differences requires more careful definitions (Albright and Mills, 1965). *Binary diffusion* occurs with two chemically distinct species. In contrast, *intra-diffusion* occurs with three distinguishable species. One of these species is chemically different. The other two species are very similar, for they have the same chemical formula, the same boiling point, the same viscosity, and so forth. They differ only in their isotopic composition or their assymetrical structure. Nonetheless, this means that intra-diffusion involves three species.

There are two important special cases of intradiffusion. The first, *tracer diffusion*, is the limit when the concentration of one similar species is small. This is the usual situation when one uses radioactive isotopes, for high concentrations of radioactive material are expensive, risky, and unnecessary. The second special case, *self-diffusion*, occurs when the system contains a radioactively tagged solute in an untagged but otherwise chemically identical solvent. This system may also contain traces of other solutes and so still may have more than two components. These different definitions are identified in Fig. 7.5-1.

The best available description of these various forms of diffusion is supplied by the multicomponent equations developed earlier in this chapter. Indeed, tracer diffusion is a simple example by which you can test your understanding of these ideas. To begin this description, we define the tracer as species 1, the identical unlabeled compound as species 2, and the different species as the solvent 3. The flux equations for this system are then

$$-j_1 = D_{11}\nabla c_1 \tag{7.5-1}$$

$$-j_2 = D_{21}\nabla c_1 + D_{22}\nabla c_2 \tag{7.5-2}$$

The coefficient D_{12} is zero because the tracer concentration c_1 is always near zero. When c_1 and c_2 are both very small, D_{11} is the tracer diffusion coefficient of species 1 in species 3. When c_3 is very small, D_{11} is the self-diffusion coefficient of species 1 in species 2. We will imitate the literature and relabel the coefficient D_{11} as D^*, a reminder that it is often radioactively tagged.

We can also reach conclusions about the coefficients D_{22} and D_{21}. Since species 1 is always present at vanishingly small concentrations, D_{22} must be the binary diffusion coefficient D of species 2 in solvent 3. This has other implications. The total flux of species 1 and 2 must be the sum of the fluxes above

$$- (j_1 + j_2) = (D_{11} + D_{21})\nabla c_1 + D_{22}\nabla c_2$$

$$= (D^* + D_{21})\nabla c_1 + D\nabla c_2 \tag{7.5-3}$$

But now imagine that our radiation detector is broken, so we can't measure c_1; we can only measure $(c_1 + c_2)$. We can still measure the binary diffusion coefficient D using the relation

$$- (j_1 + j_2) = D\nabla(c_1 + c_2) \tag{7.5-4}$$

By comparing Eqs. 7.5-3 and 7.5-4, we see that

$$D_{21} = D - D^* \tag{7.5-5}$$

Thus in this special case of ternary diffusion, the four diffusion coefficients can be written in terms of two: the tracer and the binary. This reduction to two coefficients is a consequence of the chemical identity of the solutes 1 and 2.

The physical reasons why the tracer and the binary coefficients are different can most easily by seen for the case of a dilute gas mixture of a tagged solute 1, an untagged solute 2, and a solvent 3. Diffusion in this system is described in terms of solute–solvent collisions and solute–solute collisions. Solute–solvent collisions are characterized by collision diameters σ_{13} and ε_{13}. Solute–solute collisions are described by σ_{12} and ε_{12}. With these diameters and energies, the binary diffusion coefficient can be shown from Table 7.1-1 to be a function only of solute–solvent collisions:

$$D_{22} = D = D(\sigma_{23}, \sigma_{23}) = D(\sigma_{13}, \sigma_{13}) \tag{7.5-6}$$

On the other hand, the intradiffusion coefficient D^* is seen from this table to be a weighted harmonic average of solute–solvent and solute–solute collisions:

$$D_{11} = D^* = \cfrac{1}{\cfrac{y_3}{D(\sigma_{23}, \sigma_{23})} + \cfrac{y_1 + y_2}{D(\sigma_{12}, \sigma_{12})}} \tag{7.5-7}$$

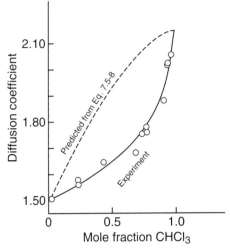

Fig. 7.5-2. Binary diffusion predicted from tracer diffusion. In general, binary diffusion cannot be predicted from tracer diffusion and activity data using empirical relations like Eq. 7.5-8. The data, for chloroform–carbon tetrachloride at 25 °C, are square centimeters per second. [From Kelly, Wirth, and Anderson (1971), with permission.]

Note that when $(y_1 + y_2)$ is nonzero, D^* is not equal to D. In the limit of infinite dilution, both y_1 and y_2 approach zero, and D^* equals D.

Many investigators have tried to discover empirical connections between binary diffusion and intradiffusion. The most common is the assertion that

$$D = D^* \left(1 + \frac{\partial \ln \gamma_1}{\partial \ln c_1} \right) \tag{7.5-8}$$

in which D is the binary diffusion coefficient, D^* is the intradiffusion coefficient measured with a radioactive tracer, and the quantity in parentheses is the increasingly familiar activity correction for diffusion. This empirical assertion is often buttressed by theoretical arguments, especially those based on the irreversible thermodynamics described in Section 7.2. Equation 7.5-8 does not always work experimentally, as shown by the results in Fig. 7.5-2.

Why Eq. 7.5-8 sometimes fails is illustrated by the case of dilute gases. Binary diffusion involves only solute–solvent interactions. Intradiffusion and tracer diffusion are the result not only of solute–solvent interactions but also of solute–solute interactions. Thus D^* contains different information than D, information characteristic of dynamic collisions as well as equilibrium activities. This difference means in general that D^* cannot be found only from D and activity coefficients.

Example 7.5-1: Tracer and binary diffusion of hydrogen and benzene Find the tracer diffusion coefficient of ^{14}C-tagged benzene in gas mixtures of hydrogen and benzene. At 25 °C, the binary diffusion coefficient is 0.40 cm^2/sec, and the self-diffusion coefficient of benzene is 0.03 cm^2/sec.

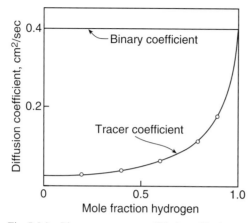

Fig. 7.5-3. Binary versus tracer diffusion of hydrogen gas and benzene vapor. The benzene is the species tagged with radioactivity. The differences between the binary and tracer values are unusually large in this case.

Solution To be consistent with the preceding development, let species 1, 2, and 3 be radioactively tagged benzene, untagged benzene, and hydrogen, respectively. Then, from Eq. 7.5-6, we see that the binary coefficient is

$$D = 0.40 \, \text{cm}^2/\text{sec}$$

This coefficient is independent of concentration. The tracer diffusion coefficient is found from Eq. 7.5-7

$$D^* = \cfrac{1}{\cfrac{y_3}{0.40} + \cfrac{1-y_3}{0.03}}$$

This result is shown versus hydrogen concentration in Fig. 7.5-3. In this case, the binary and tracer values differ by an unusually large amount, a consequence of the exceptional mobility of hydrogen.

7.6 Conclusions

Diffusion frequently occurs in multicomponent systems. When these systems are dilute, the diffusion of each solute can be treated with a binary form of Fick's law. In concentrated solutions, the fluxes and concentration profiles deviate significantly from binary expectations only in exceptional cases. These exceptions include mixed gases containing hydrogen, mixed weak electrolytes, partially miscible species, and some alloys.

When multicomponent diffusion is significant, it is best described with a generalized form of Fick's law containing $(n - 1)^2$ diffusion coefficients in an n-component system. This form of diffusion equation can be rationalized using irreversible thermodynamics. Concentration profiles in these multicomponent cases can be directly inferred from the binary results. However, multicomponent diffusion coefficients are difficult to estimate, and experimental values are fragmentary. As a result, you should make very sure that you need the more complicated theory before you attempt to use it.

Questions for Discussion

1. In what multicomponent mixture can diffusion be accurately described with the binary form of Fick's law?
2. When are multicomponent diffusion coefficients different from the binary values?
3. What is a cross-term diffusion coefficient?
4. When can diffusion in one phase go from low concentration to higher concentration?
5. What is an advantage and a disadvantage of the Fick's law form of ternary diffusion equations (Eq. 7.1-2)?
6. What is an advantage and a disadvantage of the Maxwell–Stefan form of ternary diffusion equations (Eq. 7.1-5)?
7. What are the Onsager reciprocal relations?
8. Will multicomponent effects be greatest in gases, liquids, or solids?
9. How can you find a ternary concentration profile if you know the solution of an analogous binary one?
10. When will tracer diffusion equal binary diffusion?
11. Imagine a system of equimolar amounts of hydrogen and ethylene containing a trace of tritium. The diffusion coefficient of tritium would not equal the diffusion coefficient of hydrogen. Explain why without using equations.
12. Diffusion of two gases in a porous medium can be treated as ternary diffusion, where the third species is the porous medium itself. Write and simplify equations for this case.

Problems

1. Imagine a thin membrane separating two large volumes of aqueous solution. The membrane is 0.014 cm thick and has a void fraction of 0.32. One solution contains 2-M H_2SO_4 and the other 2-M Na_2SO_4. As a result, there is no gradient of sulfate across the membrane. Ternary diffusion coefficients for this system are given in Table 7.4-2. What is the sulfate flux? *Answer:* $5.6 \cdot 10^{-7}$ mol/cm^2 sec.

2. A solution of 12 mol% hexadecane (1), 55 mol% dodecane (2), and 33 mol % hexane (3) is diffusing at 25 °C in a diaphragm cell into a solution of 52 mol% hexadecane (1), 15 mol% dodecane (2), and 33 mol% hexane (3). The cell constant of the cell is 3.62 cm^{-2}, and the ternary diffusion coefficients are

$$D_{11} = 1.03, \quad D_{12} = 0.23,$$
$$D_{21} = 0.27, \quad D_{22} = 0.97$$

all times 10^{-5} cm^2/sec. Plot the concentration differences Δc_1 and Δc_2 versus time.

3. In a two-bulb capillary diffusion apparatus like that in Fig. 3.1-2, one bulb contains 75% H_2 and 25% C_6H_6, and the other contains 65% H_2, 34.9% C_6H_6, and 0.1% radioactively tagged C_6H_6. The system is at 0°C. We can measure diffusion in one of two ways. First, we can measure the concentration change of all the benzene using a gas chromatograph. Second, we can measure the concentration difference of the radioactive isotopes. How different are these results? To answer this problem, let 1 be

tracer, 2 be untagged benzene, and 3 be hydrogen solvent. (a) Find the ternary diffusion coefficients assuming that the radioactive concentration is much less than the nonradioactive. (b) Using the binary solution, write out the ternary one. (c) Combine parts (a) and (b) to find Δc_1 and Δc_2 versus βt, where β is the cell constant of this apparatus.

4. An iron bar containing 0.86 mol% carbon is joined with a bar containing 3.94 mol% silicon. The two bars are then heated to 1,050 °C for 13 days; under these conditions, there is only one equilibrium phase, FCC austenite. Calculate the carbon concentration profile under these conditions using the values in Table 7.4-3. Remember that these coefficients are relative to the *solvent average* velocity.

5. In practical work, air is often treated as if it is a pure species. This problem tests the accuracy of this assumption for diffusion. Imagine a large slab of an isotropic porous solid centered at $z=0$. To the left, at $z<0$, the solid's pores initially contain pure hydrogen; to the right, at $z>0$, they initially contain pure air. If air were really a single component, then the mole fraction of hydrogen y_1 would vary as follows (see Section 2.3, assuming that the total molar concentration c is a constant.):

$$y_1 = \frac{1}{2}\left(1 - \text{erf}\frac{z}{\sqrt{4Dt}}\right)$$

Because air is really a mixture, the exact solution involves ternary diffusion coefficients that can be calculated from Table 7.1-1. Calculate the ternary concentration profile and compare it with the binary one (S. Gehrke).

6. You are using the diaphragm cell to study diffusion in the ternary system sucrose(1)–KCl(2)–water(3). Instead of measuring the concentration differences of each species in these experiments, you find it convenient to measure the overall density and refractive-index differences, defined as

$$\Delta\rho = H_1\Delta\rho_1 + H_2\Delta\rho_2$$
$$\Delta n = R_1\Delta\rho_1 + R_2\Delta\rho_2$$

In separate experiments, you find $H_1=0.379$, $H_2=0.602$, $R_1=0.1414$, and $R_2=0.1255$. You find the calibration constant of the cell to be 0.462 cm^{-2}. Other relevant data are in the following table: [E. L. Cussler and P. J. Dunlop, *J. Phys. Chem.*, **70**, 1880 (1966)]:

	Exp. 20	Exp. 26	Exp. 24	Exp. 22
$\Delta\rho^{10}$	0.0000	0.00277	0.01111	0.01500
$\Delta\rho^{20}$	0.0150	0.01250	0.00313	0.00000
Δn^0	86.33	89.88	89.96	97.21
$\Delta\rho^0$	0.00904	0.00856	0.00609	0.00569
Δn	28.24	33.56	46.34	55.38
$\Delta\rho$	0.00293	0.00299	0.00279	0.00315
$10^{-5}\beta t$	0.627	0.620	0.9526	1.0598

Use these data to calculate the four ternary diffusion coefficients, and compare them with the following values found with the Gouy interferometer: $D_{11}=0.497$, $D_{12}=0.021$,

$D_{21} = 0.069$, $D_{22} = 1.775$ (all times 10^{-5} cm^2/sec). *Answer:* $D_{11} = 0.498$, $D_{12} = 0.022$, $D_{21} = 0.071$, $D_{22} = 1.776$ (all times 10^{-5} cm^2/sec).

7.   Ternary diffusion effects are expected to be common in the molten silicates that occur in the center of the Earth. In a study of one such melt, Spera and Trial [*Science* **259**, 204 (1993)] report for 40 mol% CaO (1)–20 mol% Al$_2$O$_3$–40 mol% SiO$_2$ at 1500 K that

$$D_{11} = (10.0 \pm 0.10) \cdot 10^{-7} \text{cm}^2/\text{sec}; \quad D_{12} = (-2.8 \pm 0.8) \cdot 10^{-7} \text{cm}^2/\text{sec};$$

$$D_{21} = (-4.2 \pm 0.8) \cdot 10^{-7} \text{cm}^2/\text{sec}; \quad D_{22} = (7.3 \pm 0.4) \cdot 10^{-7} \text{cm}^2/\text{sec};$$

Large coefficients like these provide a good chance to check the Onsager reciprocal relations (cf. Eq. 7.2-18):

$$L_{12} = L_{21}$$

This is equivalent to

$$D_{11}\alpha_{12} + D_{21}\alpha_{22} = D_{12}\alpha_{11} + D_{22}\alpha_{21}$$

These authors also estimate that

$$\alpha_{11} = 8.15 \cdot 10^6 \text{ J/kg}; \quad \alpha_{12} = 4.25 \cdot 10^6 \text{ J/kg};$$

$$\alpha_{21} = 4.25 \cdot 10^6 \text{ J/kg}; \quad \alpha_{22} = 4.00 \cdot 10^6 \text{ J/kg};$$

Do the Onsager relations hold?

Further Reading

Albright, J.G., and Mills, R. (1965). *Journal of Physical Chemistry*, **69**, 3120.

Cussler, E.L. (1976). *Multicomponent Diffusion*. Amsterdam: Elsevier.

deGroot, S.R., and Mazur, P. (1962). *Non-Equilibrium Thermodynamics*. Amsterdam: North Holland.

Fitts, D.D. (1962). *Non-Equilibrium Thermodynamics*. New York: McGraw-Hill.

Haase, R. (1969). *Thermodynamics of Irreversible Processes*. London: Addison-Wesley.

Katchalsky, A., and Curran, P.F. (1967). *Non-Equilibrium Thermodynamics in Biophysics*. Cambridge, MA: Harvard University Press.

Kelly, C.M., Wirth, G.B., and Anderson, D.K. (1971). *Journal of Physical Chemistry*, **75**, 3293.

Lamm, O. (1944). *Arkiv for Kemi, Minerologi och Geologi*, **2**, 1813.

Mills, R. (1965). *Journal of Physical Chemistry*, **69**, 3116.

Onsager, L. (1931). *Physical Review*, **37**, 405; (1931)**38**, 2265.

Onsager, L. (1945). *New York Academy of Sciences Annals*, **46**, 241.

Stewart, W.E., and Prober, R. (1964). *Industrial and Engineering Chemistry Fundamentals*, **3**, 224.

Taylor, R., and Krishna, R. (1993). *Multicomponent Mass Transfer*. New York: Wiley.

Toor, H.L. (1964). *American Institute of Chemical Engineers Journal*, **10**, 448; 460.

PART III

Mass Transfer

CHAPTER 8

Fundamentals of Mass Transfer

Diffusion is the process by which molecules, ions, or other small particles spontaneously mix, moving from regions of relatively high concentration into regions of lower concentration. This process can be analyzed in two ways. First, it can be described with Fick's law and a diffusion coefficient, a fundamental and scientific description used in the first two parts of this book. Second, it can be explained in terms of a mass transfer coefficient, an approximate engineering idea that often gives a simpler description. It is this simpler idea that is emphasized in this part of this book.

Analyzing diffusion with mass transfer coefficients requires assuming that changes in concentration are limited to that small part of the system's volume near its boundaries. For example, in the absorption of one gas into a liquid, we assume that gases and liquids are well mixed, except near the gas–liquid interface. In the leaching of metal by pouring acid over ore, we assume that the acid is homogeneous, except in a thin layer next to the solid ore particles. In studies of digestion, we assume that the contents of the small intestine are well mixed, except near the villi at the intestine's wall. Such an analysis is sometimes called a "lumped-parameter model" to distinguish it from the "distributed-parameter model" using diffusion coefficients. Both models are much simpler for dilute solutions.

If you are beginning a study of diffusion, you may have trouble deciding whether to organize your results as mass transfer coefficients or as diffusion coefficients. I have this trouble too. The cliché is that you should use the mass transfer coefficient approach if the diffusion occurs across an interface, but this cliché has many exceptions. Instead of depending on the cliché, I believe you should always try both approaches to see which is better for your own needs. In my own work, I have found that I often switch from one to the other as the work proceeds and my objectives evolve.

This chapter discusses mass transfer coefficients for dilute solutions; extensions to concentrated solutions are deferred to Section 9.5. In Section 8.1, we give a basic definition for a mass transfer coefficient and show how this coefficient can be used experimentally. In Section 8.2, we present other common definitions that represent a thicket of prickly alternatives rivaled only by standard states for chemical potentials. These various definitions are why mass transfer often has a reputation with students of being a difficult subject. In Section 8.3, we list existing correlations of mass transfer coefficients; and in Section 8.4, we explain how these correlations can be developed with dimensional analysis. Finally, in Section 8.5, we discuss processes involving diffusion across interfaces, a topic that leads to overall mass transfer coefficients found as averages of more local processes. This last idea is commonly called mass transfer resistances in series.

8.1 A Definition of a Mass Transfer Coefficient

The definition of mass transfer is based on empirical arguments like those used in developing Fick's law in Chapter 2. Imagine we are interested in the transfer of mass from some interface into a well-mixed solution. We expect that the amount transferred is proportional to the concentration difference and the interfacial area:

$$\left(\begin{array}{c}\text{rate of mass}\\\text{transferred}\end{array}\right) = k \left(\begin{array}{c}\text{interfacial}\\\text{area}\end{array}\right)\left(\begin{array}{c}\text{concentration}\\\text{difference}\end{array}\right) \qquad (8.1\text{-}1)$$

where the proportionality is summarized by k, called a mass transfer coefficient. If we divide both sides of this equation by the area, we can write the equation in more familiar symbols:

$$N_1 = k\,(c_{1i} - c_1) \qquad (8.1\text{-}2)$$

where N_1 is the flux at the interface and c_{1i} and c_1 are the concentrations at the interface and in the bulk solution, respectively. The flux N_1 includes both diffusion and convection; it is like the total flux n_1 except that it is located at the interface. The concentration c_{1i} is at the interface but in the same fluid as the bulk concentration c_1. It is often in equilibrium with the concentration across the interface in a second, adjacent phase; we will defer discussion of transport across this interface until Section 8.5.

The physical meaning of the mass transfer coefficient is clear: it is the rate constant for moving one species from the boundary into the bulk of the phase. A large value of k implies fast mass transfer, and a small one means slow mass transfer. The mass tranfer coefficient is like the rate constant of a chemical reation, but written per area, not per volume. As a result, its dimensions are of velocity, not of reciprocal time. Those learning about this subject sometimes call the mass transfer coefficient the "velocity of diffusion."

The flux equation in Eq. 8.1-2 makes practical sense. It says that if the concentration difference is doubled, the flux will double. It also suggests that if the area is doubled, the total amount of mass transferred will double but the flux per area will not change. In other words, this definition suggests an easy way of organizing our thinking around a simple constant, the mass transfer coefficient k.

Unfortunately, this simple scheme conceals a variety of approximations and ambiguities. Before introducing these complexities, we shall go over some easy examples. These examples are important. Study them carefully before you go on to the harder material that follows.

Example 8.1-1: Humidification   Imagine that water is evaporating into initially dry air in the closed vessel shown schematically in Fig. 8.1-1(a). The vessel is isothermal at 25 °C, so the water's vapor pressure is 3.2 kPa. This vessel has 0.8 l of water with 150 cm^2 of surface area in a total volume of 19.2 l. After 3 min, the air is five percent saturated. What is the mass transfer coefficient? How long will it take to reach ninety percent saturation?

   Solution   The flux at 3 min can be found directly from the values given:

$$N_1 = \frac{\left(\begin{array}{c}\text{vapor}\\\text{concentration}\end{array}\right)\left(\begin{array}{c}\text{air}\\\text{volume}\end{array}\right)}{\left(\begin{array}{c}\text{liquid}\\\text{area}\end{array}\right)(\text{time})}$$

$$= \frac{0.05\left(\dfrac{3.2}{101}\right)\left(\dfrac{1\ \text{mol}}{22.4\ \text{liters}}\right)\left(\dfrac{273}{298}\right)(18.4\ \text{liters})}{\left(150\ \text{cm}^2\right)(180\ \text{sec})} = 4.4\cdot10^{-8}\ \text{mol/cm}^2\ \text{sec}$$

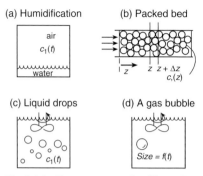

Fig. 8.1-1. Four easy examples. We analyze each of the physical situations shown in terms of mass transfer coefficients. In (a), we assume that the air is at constant humidity, except near the air–water interface. In (b), we assume that water flowing through the packed bed is well mixed, except very close to the solid spheres. In (c) and (d), we assume that the liquid solution, which is the continuous phase, is at constant composition, except near the droplet or bubble surfaces.

The concentration difference is that at the water's surface minus that in the bulk solution. That at the water's surface is the value at saturation; that in bulk at short times is essentially zero. Thus, from Eq. 8.1-2, we have

$$4.4 \cdot 10^{-8} \ \mathrm{mol/cm^2\ sec} = k \left(\frac{3.2}{101} \frac{1\ \mathrm{mol}}{22.4 \cdot 10^3\ \mathrm{cm^3}} \frac{273}{298} - 0 \right)$$

$$k = 3.4 \cdot 10^{-2} \mathrm{cm/sec}$$

This value is lower than that commonly found for transfer in gases. The time required for ninety percent saturation can be found from a mass balance:

$$\left(\begin{array}{c} \mathrm{accumulation} \\ \mathrm{in\ gas\ phase} \end{array} \right) = \left(\begin{array}{c} \mathrm{evaporation} \\ \mathrm{rate} \end{array} \right)$$

$$\frac{d}{dt} Vc_1 = -AN_1 = -kA \left[c_1(\mathrm{sat}) - c_1 \right]$$

The air is initially dry, so

$$t = 0, \quad c_1 = 0$$

We use this condition to integrate the mass balance:

$$\frac{c_1}{c_1(\mathrm{sat})} = 1 - e^{-(kA/V)t}$$

Rearranging the equation and inserting the values given, we find

$$t = -\frac{V}{kA} \ln \left(1 - \frac{c_1}{c_1(\mathrm{sat})} \right)$$

$$= -\frac{18.4 \cdot 10^3\ \mathrm{cm^3}}{\left(3.4 \cdot 10^{-2}\ \mathrm{cm/sec} \right) \cdot \left(150\ \mathrm{cm^2} \right)} \ln \left(1 - 0.9 \right)$$

$$= 8.3 \cdot 10^3\ \mathrm{sec} = 2.3\ \mathrm{hr}$$

It takes over two hours to saturate the air this much.

Example 8.1-2: Mass transfer in a packed bed Imagine that 0.2-cm diameter spheres of benzoic acid are packed into a bed like that shown schematically in Fig. 8.1-1(b). The spheres have 23 cm^2 surface per 1 cm^3 of bed. Pure water flowing at a superficial velocity of 5 cm/sec into the bed is 62% saturated with benzoic acid after it has passed through 100 cm of bed. What is the mass transfer coefficient?

  Solution The answer to this problem depends on the concentration difference used in the definition of the mass transfer coefficient. In every definition, we choose this difference as the value at the sphere's surface minus that in the solution. However, we can define different mass transfer coefficients by choosing the concentration difference at various positions in the bed. For example, we can choose the concentration difference at the bed's entrance and so obtain

$$N_1 = k \left[c_1(\text{sat}) - 0 \right]$$

$$\frac{0.62 \, c_1(\text{sat}) \, (5 \text{ cm/sec}) \, A}{\left(23 \text{ cm}^2/\text{cm}^3 \right) (100 \text{ cm}) \, A} = k c_1(\text{sat})$$

where A is the bed's cross-section. Thus

$$k = 1.3 \cdot 10^{-3} \text{cm/sec}$$

This definition for the mass transfer coefficient is infrequently used.

  Alternatively, we can choose as our concentration difference that at a position z in the bed and write a mass balance on a differential volume $A\Delta z$ at this position:

$$(\text{accumulation}) = \left(\begin{array}{c} \text{flow in} \\ \text{minus flow out} \end{array} \right) + \left(\begin{array}{c} \text{amount of} \\ \text{dissolution} \end{array} \right)$$

$$0 = A \left(c_1 v^0 \big|_z - c_1 v^0 \big|_{z+\Delta z} \right) + (A\Delta z) a N_1$$

where a is the sphere surface area per bed volume. Substituting for N_1 from Eq. 8.1-2, dividing by $A\Delta z$, and taking the limit as Δz goes to zero, we find

$$\frac{d c_1}{d z} = \frac{k a}{v^0} \left[c_1(\text{sat}) - c_1 \right]$$

This is subject to the initial condition that

$$z = 0, \ c_1 = 0$$

Integrating, we obtain an exponential of the same form as in the first example:

$$\frac{c_1}{c_1(\text{sat})} = 1 - e^{-(ka/v^0)z}$$

Rearranging the equation and inserting the values given, we find

$$k = \left(\frac{v^0}{az}\right) \ln\left(1 - \frac{c_1}{c_1(\text{sat})}\right)$$

$$= -\frac{5 \text{ cm/sec}}{\left(23 \text{ cm}^2/\text{cm}^3\right)(100 \text{ cm})} \ln(1 - 0.62)$$

$$= 2.1 \cdot 10^{-3} \text{ cm/sec}$$

This value is typical of those found in liquids. This type of mass transfer coefficient definition is preferable to that used first, a point explored further in Section 8.2.

A tangential point worth discussing is the specific chemical system of benzoic acid dissolving in water. This system is academically ubiquitous, showing up again and again in problems of mass transfer. Indeed, if you read the literature, you can get the impression that it is a system where mass transfer is very important, which is not true. Why is it used so much?

Benzoic acid is studied thoroughly for three distinct reasons. First, its concentration is relatively easily measured, for the amount present can be determined by titration with base, by ultraviolet spectrophotometry of the benzene ring, or by radioactively tagging either the carbon or the hydrogen. Second, the dissolution of benzoic acid is accurately described by one mass transfer coefficient. This is not true of all dissolutions. For example, the dissolution of aspirin is essentially independent of events in solution. Third, and most subtle, benzoic acid is solid, so mass transfer takes place across a solid–fluid interface. Such interfaces are an exception in mass transfer problems; fluid–fluid interfaces are much more common. However, solid–fluid interfaces are the rule for heat transfer, the intellectual precursor of mass transfer. Experiments with benzoic acid dissolving in water can be compared directly with heat transfer experiments. These three reasons make this chemical system popular.

Example 8.1-3: Mass transfer in an emulsion   Bromine is being rapidly dissolved in water, as shown schematically in Fig. 8.1-1(c). Its concentration is about half saturated in 3 minutes. What is the mass transfer coefficient?

Solution   Again, we begin with a mass balance:

$$\frac{d}{dt} V c_1 = A N_1 = A k \left[c_1(\text{sat}) - c_1\right]$$

$$\frac{dc_1}{dt} = ka\left[c_1(\text{sat}) - c_1\right]$$

where $a (= A/V)$ is the surface area of the bromine droplets divided by the volume of aqueous solution. If the water initially contains no bromine,

$$t = 0, \quad c_1 = 0$$

Using this in our integration, we find

$$\frac{c_1}{c_1(\text{sat})} = 1 - c^{-kat}$$

Rearranging,

$$ka = -\frac{1}{t} \ln \left(1 - \frac{c_1}{c_1(\text{sat})} \right)$$

$$= -\frac{1}{3 \text{ min}} \ln (1 - 0.5)$$

$$= 3.9 \cdot 10^{-3} \text{ sec}^{-1}$$

This is as far as we can go; we cannot find the mass transfer coefficient, only its product with a.

Such a product occurs often and is a fixture of many mass transfer correlations. The quantity ka is very similar to the rate constant of a first-order reversible reaction with an equilibrium constant equal to unity. This particular problem is similar to the calculation of a half-life for radioactive decay.

Example 8.1-4: Mass transfer from an oxygen bubble   A bubble of oxygen originally 0.1 cm in diameter is injected into excess stirred water, as shown schematically in Fig. 8.1-1(d). After 7 min, the bubble is 0.054 cm in diameter. What is the mass transfer coefficient?

  Solution   This time, we write a mass balance not on the surrounding solution but on the bubble itself:

$$\frac{d}{dt} \left(c_1 \frac{4}{3} \pi r^3 \right) = A N_1$$

$$= -4\pi r^2 k [c_1(\text{sat}) - 0]$$

This equation is tricky; c_1 refers to the oxygen concentration in the bubble, 1 mol/22.4 l at standard conditions, but $c_1(\text{sat})$ refers to the oxygen concentration at saturation in water, about $1.5 \cdot 10^{-3}$ mol/l under similar conditions. Thus

$$\frac{dr}{dt} = -k \frac{c_1(\text{sat})}{c_1}$$

$$= -0.034k$$

This is subject to the condition

$$t = 0, \quad r = 0.05 \text{ cm}$$

so integration gives

$$r = 0.05 \text{ cm} - 0.034\, kt$$

Inserting the numerical values given, we find

$$0.027 \text{ cm} = 0.05 \text{ cm} - 0.034k \,(420 \text{ sec})$$

$$k = 1.6 \cdot 10^{-3} \text{cm/sec}$$

Table 8.2-1 *Mass transfer coefficient compared with other rate coefficients*

Effect	Basic equation	Rate	Force	Coefficient
Mass transfer	$N_1 = k\Delta c_1$	Flux per area relative to an interface	Difference of concentration	The mass transfer coefficient k ($[=]L/t$) is a function of flow
Diffusion	$-j_1 = D\nabla c_1$	Flux per area relative to the volume average velocity	Gradient of concentration	The diffusion coefficient D ($[=]L^2/t$) is a physical property independent of flow
Dispersion	$-\overline{c_1' v_1'} = E\nabla \bar{c}_1$	Flux per area relative to the mass average velocity	Gradient of time averaged concentration	The dispersion coefficient E ($[=]L^2/t$) depends on the flow
Homogeneous chemical reaction	$r_1 = \kappa_1 c_1$	Rate per volume	Concentration	The rate constant κ_1 ($[=]1/t$) is a chemical property independent of flow
Heterogeneous chemical reaction	$r_1 = \kappa_1 c_1$	Flux per interfacial area	Concentration	The rate constant κ_1 ($[=]L/t$) is a chemical surface property often defined in terms of a bulk concentation

Remember that this coefficient is defined in terms of the concentration in the liquid. It would be numerically different if it were defined in terms of the gas-phase concentration.

8.2 Other Definitions of Mass Transfer Coefficients

We now want to return to some of the problems we glossed over in the simple definition of a mass transfer coefficient given in the previous section. We introduced this definition with the implication that it provides a simple way of analyzing complex problems. We implied that the mass transfer coefficient will be like the density or the viscosity, a physical quantity that is well defined for a specific situation.

In fact, the mass transfer coefficient is often an ambiguous concept, reflecting nuances of its basic definition. To begin our discussion of these nuances, we first compare the mass transfer coefficient with the other rate constants given in Table 8.2-1. The mass transfer coefficient seems a curious contrast, a combination of diffusion and dispersion. Because it involves a concentration difference, it has different dimensions than the diffusion and dispersion coefficients. It is a rate constant for an interfacial physical reaction, most similar to the rate constant of an interfacial chemical reaction.

Unfortunately, the definition of the mass transfer coefficient in Table 8.2-1 is not so well accepted that the coefficient's dimensions are always the same. This is not true for the other processes in this table. For example, the dimensions of the diffusion coefficient are always taken as L^2/t. If the concentration is expressed in terms of mole fraction or

Table 8.2-2 *Common definitions of mass transfer coefficients*

Basic equation	Typical units of k^a	Remarks
$N_1 = k\Delta c_1$	cm/sec	Common in the older literature; used here because of its simple physical significance
$N_1 = k_p\Delta p_1$	mol/cm^2 sec Pa	Common for a gas adsorption; equivalent forms occur in medical problems
$N_1 = k_x\Delta x_1$	mol/cm^2 sec	Preferred for practical calculations, especially in gases
$N_1 = k\Delta c_1 + c_1 v^0$	cm/sec	Rarely used; an effort to include diffusion-induced convection (cf. k in Eq. 9.5-2 et seq.)

Notes: ^{a} In this table, N_1 is defined as moles/$L^2 t$, and c_1 as moles/L^3. Parallel definitions where N_1 is in terms of $M/L^2 t$ and c_1 is M/L^3 are easily developed. Definitions mixing moles and mass are infrequently used.

partial pressure, then appropriate unit conversions are made to ensure that the diffusion coefficient keeps the same dimensions.

This is not the case for mass transfer coefficients, where a variety of definitions are accepted. Four of the more common of these are shown in Table 8.2-2. This variety is largely an experimental artifact, arising because the concentration can be measured in so many different units, including partial pressure, mole and mass fractions, and molarity.

In this book, we will frequently use the first definition in Table 8.2-2, implying that mass transfer coefficients have dimensions of length per time. If the flux is expressed in moles per area per time we will express the concentration in moles per volume. If the flux is expressed in mass per area per time, we will give the concentration in mass per volume. This choice is the simplest for correlations of mass transfer coefficients reviewed in this chapter and for predictions of these coefficients given in Chapter 9. Expressing the mass transfer coefficient in dimensions of velocity is also simplest in the cases of chemical reaction and simultaneous heat and mass transfer described in Chapters 16, 17, and 21.

However, in some other cases, alternative forms of the mass transfer coefficients lead to simpler final equations. This is especially true for gas adsorption, distillation, and extraction described in Chapters 10–14. There, we will frequently use k_x, the third form in Table 8.2-2, which expresses concentrations in mole fractions. In some cases of gas absorption, we will find it convenient to respect seventy years of tradition and use k_p, with concentrations expressed as partial pressures. In the membrane separations in Chapter 18, we will mention forms like k_x but will carry out our discussion in terms of forms equivalent to k.

The mass transfer coefficients defined in Table 8.2-2 are also complicated by the choice of a concentration difference, by the interfacial area for mass transfer, and by the treatment of convection. The basic definitions given in Eq. 8.1-2 or Table 8.2-1 are ambiguous, for the concentration difference involved is incompletely defined. To explore the ambiguity more carefully, consider the packed tower shown schematically in Fig. 8.2-1. This tower is basically a piece of pipe standing on its end and filled with crushed inert material like broken glass. Air containing ammonia flows upward through the column. Water trickles down through the column and absorbs the ammonia: ammonia is scrubbed out of the gas mixture with water. The flux of ammonia into the water is proportional to the

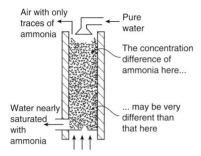

Fig. 8.2-1. Ammonia scrubbing. In this example, ammonia is separated by washing a gas mixture with water. As explained in the text, the example illustrates ambiguities in the definition of mass transfer coefficients. The ambiguities occur because the concentration difference causing the mass transfer changes and because the interfacial area between gas and liquid is unknown.

ammonia concentration at the air–water interface minus the ammonia concentration in the bulk water. The proportionality constant is the mass transfer coefficient. The concentration difference between interface and bulk is not constant but can vary along the height of the column. Which value of concentration difference should we use?

In this book, we always choose to use the local concentration difference at a particular position in the column. Such a choice implies a "local mass transfer coefficient" to distinguish it from an "average mass transfer coefficient." Use of a local coefficient means that we often must make a few extra mathematical calculations. However, the local coefficient is more nearly constant, a smooth function of changes in other process variables. This definition was implicitly used in Examples 8.1-1, 8.1-3, and 8.1-4 in the previous section. It was used in parallel with a type of average coefficient in Example 8.1-2.

Another potential source of ambiguity in the definition of the mass transfer coefficient is the interfacial area. As an example, we again consider the packed tower in Fig. 8.2-1. The surface area between water and gas is often experimentally unknown, so that the flux per area is unknown as well. Thus the mass transfer coefficient cannot be easily found. This problem is dodged by lumping the area into the mass transfer coefficient and experimentally determining the product of the two. We just measure the flux per column volume. This may seem like cheating, but it works like a charm.

Finally, mass transfer coefficients can be complicated by diffusion-induced convection normal to the interface. This complication does not exist in dilute solution, just as it does not exist for the dilute diffusion described in Chapter 2. For concentrated solutions, there may be a larger convective flux normal to the interface that disrupts the concentration profiles near the interface. The consequence of this convection, which is like the concentrated diffusion problems in Section 3.3, is that the flux may not double when the concentration difference is doubled. This diffusion-induced convection is the motivation for the last definition in Table 8.2-2, where the interfacial velocity is explicitly included. Fortunately, many transfer-in processes like distillation often approximate equimolar counterdiffusion, so there is little diffusion-induced convection. Also fortunately, many other solutions are dilute, so diffusion induced convection is minor. We will discuss the few cases where it is not minor in Section 9.5.

I find these points difficult, hard to understand without careful thought. To spur this thought, try solving the examples that follow.

Example 8.2-1: The mass transfer coefficient in a blood oxygenator   Blood oxygenators are used to replace the human lungs during open-heart surgery. To improve oxygenator design, you are studying mass transfer of oxygen into water in one specific blood oxygenator. From published correlations of mass transfer coefficients, you expect that the mass transfer coefficient based on the oxygen concentration difference in the water is $3.3 \cdot 10^{-3}$ cm/sec. You want to use this coefficient in an equation given by the oxygenator manufacturer

$$N_1 = k_p \left(p_{O_2} - p_{O_2}^* \right)$$

where p_{O_2} is the actual oxygen partial pressure in the gas, and $p_{O_2}^*$ is the hypothetical oxygen partial pressure (the "oxygen tension") that would be in equilibrium with water under the experimental conditions. The manufacturer expressed both pressures in mm Hg. You also know the Henry's law constant of oxygen in water at your experimental conditions:

$$p_{O_2} = 44,000 \text{ atm } x_{O_2}$$

where x_{O_2} is the mole fraction of the total oxygen in the water.

Find the mass transfer coefficient k_p.

  Solution Because the correlations are based on the concentrations in the liquid, the flux equation must be

$$N_1 = 3.3 \cdot 10^{-3} \text{ cm/sec} \left(c_{O_{2i}} - c_{O_2} \right)$$

where $c_{O_{2i}}$ and c_{O_2} refer to concentrations in the water at the interface and in the bulk aqueous solution, respectively. We can convert these concentrations to the oxygen tensions as follows:

$$c_{O_2} = c \, x_{O_2} = \left(\frac{\rho}{\tilde{M}} \frac{p_{O_2}}{H} \right)$$

where c is the total concentration in the liquid water, ρ is the liquid's density, H is the Henry's law constant, and $\tilde{M}$ is its average molecular weight. Because the solution is dilute,

$$c_{O_2} = \left[\frac{1 \text{ g/cm}^3}{18 \text{ g/mol}} \frac{1}{4.4 \cdot 10^4 \text{ atm}} \frac{\text{atm}}{760 \text{ mm Hg}} \right] p_{O_2}$$

Combining with the earlier definitions, we see

$$N_1 = 3.3 \cdot 10^{-3} \frac{\text{cm}}{\text{sec}} \left[\frac{1.67 \cdot 10^{-9} \text{ mol}}{\text{cm}^3 \text{ mm Hg}} \right] \left(p_{O_2} - p_{O_2}^* \right)$$

$$= 5.5 \cdot 10^{-12} \left[\frac{\text{mol}}{\text{cm}^2 \text{ sec mm Hg}} \right] \left(p_{O_2} - p_{O_2}^* \right)$$

The new coefficient k_p equals $5.5 \cdot 10^{-12}$ in the units given.

Example 8.2-2: Converting units of ammonia mass transfer coefficients A packed tower is being used to study ammonia scrubbing with 25 °C water. The mass transfer coefficients reported for this tower are 1.18 lb mol NH_3/hr ft^2 for the liquid and 1.09 lb mol NH_3/hr ft^2 atm for the gas. What are these coefficients in centimeters per second?

Solution From Table 8.2-2, we see that the units of the liquid-phase coefficient correspond to k_x. Thus

$$k = \frac{\tilde{M}_2}{\rho} k_x$$

$$= \left(\frac{18 \text{ lb/lb mol}}{62.4 \text{ lb/ft}^3}\right) \left(\frac{1.18 \text{ lb mol } NH_3}{\text{ft}^2 \text{ hr}}\right) \left(\frac{30.5 \text{ cm}}{\text{ft}}\right) \left(\frac{\text{hr}}{3{,}600 \text{ sec}}\right)$$

$$= 2.9 \cdot 10^{-3} \text{cm / sec}$$

For the gas phase, we see from Table 8.2-2 that the coefficient has the units of k_p. Thus

$$k = RTk_p$$

$$= \left(\frac{1.314 \text{ atm ft}^3}{\text{lb mol K}}\right) \left(\frac{1.09 \text{ lb mol}}{\text{hr ft}^2 \text{ atm}}\right) \left(\frac{30.5 \text{ cm}}{\text{ft}}\right) \left(\frac{\text{hr}}{3{,}600 \text{ sec}}\right) (298 \text{ K})$$

$$= 3.6 \text{ cm/sec}$$

These conversions take time and thought, but are not difficult.

Example 8.2-3: Averaging a mass transfer coefficient Imagine two porous solids whose pores contain different concentrations of a particular dilute solution. If these solids are placed together, the flux N_1 from one to the other will be (see Section 2.3)

$$N_1 = \sqrt{D/\pi t}\, \Delta c_1$$

By comparison with Eq. 8.1-2, we see that the local mass transfer coefficient is

$$k = \sqrt{D/\pi t}$$

Note that this coefficient is initially infinite.

We want to correlate our results not in terms of this local value but in terms of a total experimental time t_0. This implies an average coefficient, $\bar{k}$, defined by

$$\bar{N}_1 = \bar{k}\, \Delta c_1$$

where $\bar{N}_1$ is the total solute transferred per area divided by t_0. How is $\bar{k}$ related to k?

Solution From the problem statement, we see that

$$\bar{N}_1 = \frac{\int_0^{t_0} N_1 dt}{\int_0^{t_0} dt} = \frac{\int_0^{t_0} \sqrt{D/\pi t}\, \Delta c_1\, dt}{t_0} = 2\sqrt{D/\pi t_0}\, \Delta c_1$$

Thus

$$\bar{k} = 2\sqrt{D/\pi t_0}$$

which is twice the value of k evaluated at t_0. Note that "local" refers here to a particular time rather than a particular position.

Example 8.2-4: Log mean mass transfer coefficients   Consider again the packed bed of benzoic acid spheres shown in Fig. 8.1-1(b) that was basic to Example 8.1-2. Mass transfer coefficients in a bed like this are sometimes reported in terms of a log mean driving force:

$$N_1 = k_{\log} \left(\frac{\Delta c_{1,\text{inlet}} - \Delta c_{1,\text{outlet}}}{\ln \left(\dfrac{\Delta c_{1,\text{inlet}}}{\Delta c_{1,\text{outlet}}} \right)} \right)$$

For this specific case, N_1 is the total benzoic acid leaving the bed per time divided by the total surface area in the bed. The bed is fed with pure water, and the benzoic acid concentration at the sphere surfaces is at saturation; that is, it equals $c_1(\text{sat})$. Thus

$$N_1 = k_{\log} \frac{[c_1(\text{sat}) - 0] - [c_1(\text{sat}) - c_1(\text{out})]}{\ln \left(\dfrac{c_1(\text{sat}) - 0}{c_1(\text{sat}) - c_1(\text{out})} \right)}$$

Show how $k_{\log}$ is related to the local coefficient k used in the earlier problem.

 Solution   By integrating a mass balance on a differential length of bed, we showed in Example 8.1-2 that for a bed of length L,

$$\frac{c_1(\text{out})}{c_1(\text{sat})} = 1 - e^{-kaL/v^0}$$

Rearranging, we find

$$\frac{c_1(\text{sat}) - c_1(\text{out})}{c_1(\text{sat}) - 0} = e^{-kaL/v^0}$$

Taking the logarithm of both sides and rearranging,

$$v^0 = \frac{kaL}{\ln \left(\dfrac{c_1(\text{sat}) - 0}{c_1(\text{sat}) - c_1(\text{out})} \right)}$$

Multiplying both sides by $c_1(\text{out})$,

$$c_1(\text{out}) v^0 = kaL \left(\frac{[c_1(\text{sat}) - 0] - [c_1(\text{sat}) - c_1(\text{out})]}{\ln \left(\dfrac{c_1(\text{sat}) - 0}{c_1(\text{sat}) - c_1(\text{out})} \right)} \right)$$

By definition,

$$N_1 = \frac{c_1(\text{out}) v^0 A}{a(AL)}$$

where A is the bed's cross-section and AL is its volume. Thus

$$N_1 = k \, \frac{[c_1(\text{sat}) - 0] - [c_1(\text{sat}) - c_1(\text{out})]}{\ln \left(\dfrac{c_1(\text{sat}) - 0}{c_1(\text{sat}) - c_1(\text{out})} \right)}$$

and

$$k_{\log} = k$$

The coefficients are identical.

Many argue that the log mean mass transfer coefficient is superior to the local value used mostly in this book. Their reasons are that the coefficients are the same or (at worst) closely related and that $k_{\log}$ is macroscopic and hence easier to measure. After all, these critics assert, you implicitly repeat this derivation every time you make a mass balance. Why bother? Why not use $k_{\log}$ and be done with it?

This argument has merit, but it makes me uneasy. I find that I need to think through the approximations of mass transfer coefficients every time I use them and that this review is easily accomplished by making a mass balance and integrating. I find that most students share this need. My advice is to avoid log mean coefficients until your calculations are routine.

8.3 Correlations of Mass Transfer Coefficients

In the previous two sections we have presented definitions of mass transfer coefficients and have shown how these coefficients can be found from experiment. Thus we have a method for analyzing the results of mass transfer experiments. This method can be more convenient than diffusion when the experiments involve mass transfer across interfaces. Experiments of this sort include liquid–liquid extraction, gas absorption, and distillation.

However, we often want to predict how one of these complex situations will behave. We do not want to correlate experiments; we want to avoid experiments if possible. This avoidance is like that in our studies of diffusion, where we often looked up diffusion coefficients so that we could calculate a flux or a concentration profile. We wanted to use someone else's measurements rather than painfully make our own.

8.3.1 Dimensionless Numbers

In the same way, we want to look up mass transfer coefficients whenever possible. These coefficients are rarely reported as individual values, but as correlations of dimensionless numbers. These numbers are often named, and they are major weapons that engineers use to confuse scientists. These weapons are effective because the names sound so scientific, like close relatives of nineteenth-century organic chemists.

The characteristics of the common dimensionless groups frequently used in mass transfer correlations are given in Table 8.3-1. Sherwood and Stanton numbers involve the mass transfer coefficient itself. The Schmidt, Lewis, and Prandtl numbers

Table 8.3-1 *Significance of common dimensionless groups*

Group[a]	Physical meaning	Used in
Sherwood number $\dfrac{kl}{D}$	$\dfrac{\text{mass transfer velocity}}{\text{diffusion velocity}}$	Usual dependent variable
Stanton number $\dfrac{k}{v^0}$	$\dfrac{\text{mass transfer velocity}}{\text{flow velocity}}$	Occasional dependent variable
Schmidt number $\dfrac{\nu}{D}$	$\dfrac{\text{diffusivity of momentum}}{\text{diffusivity of mass}}$	Correlations of gas or liquid data
Lewis number $\dfrac{\alpha}{D}$	$\dfrac{\text{diffusivity of energy}}{\text{diffusivity of mass}}$	Simultaneous heat and mass transfer
Prandtl number $\dfrac{\nu}{\alpha}$	$\dfrac{\text{diffusivity of momentum}}{\text{diffusivity of mass}}$	Heat transfer; included here for completeness
Reynolds number $\dfrac{lv}{\nu}$	$\dfrac{\text{inertial forces}}{\text{viscous forces}}$ or $\dfrac{\text{flow velocity}}{\text{"momentum velocity"}}$	Forced convection
Grashof number $\dfrac{l^3\, g\, \Delta\rho\,/\,\rho}{\nu^2}$	$\dfrac{\text{buoyancy forces}}{\text{viscous forces}}$	Free convection
Péclet number $\dfrac{v^0 l}{D}$	$\dfrac{\text{flow velocity}}{\text{diffusion velocity}}$	Correlations of gas or liquid data
Second Damköhler number or (Thiele modulus)^2 $\dfrac{\kappa l^2}{D}$	$\dfrac{\text{reaction velocity}}{\text{diffusion velocity}}$	Correlations involving reactions (see Chapters 16–17)

Note: [a] The symbols and their dimensions are as follows: D diffusion coefficient (L^2/t); g acceleration due to gravity (L/t^2); k mass transfer coefficient (L/t); l characteristic length (L); v^0 fluid velocity (L/t); α thermal diffusivity (L^2/t); κ first-order reaction rate constant (t^{-1}); ν kinematic viscosity (L^2/t); $\Delta\rho/\rho$ fractional density change.

involve different comparisons of diffusion, and the Reynolds, Grashof, and Peclet numbers describe flow. The second Damköhler number, which certainly is the most imposing name, is one of many groups used for diffusion with chemical reaction.

A key point about each of these groups is that its exact definition implies a specific physical system. For example, the characteristic length l in the Sherwood number

kl/D will be the membrane thickness for membrane transport, but the sphere diameter for a dissolving sphere. A good analogy is the dimensionless group "efficiency." An efficiency of thirty percent has very different implications for a turbine and for a running deer. In the same way, a Sherwood number of 2 means different things for a membrane and for a dissolving sphere. This flexibility is central to the correlations that follow.

8.3.2 Frequently Used Correlations

Correlations of mass transfer coefficients are conveniently divided into those for fluid–fluid interfaces and those for fluid–solid interfaces. The correlations for fluid–fluid interfaces are by far the more important, for they are basic to absorption, extraction, and distillation. These correlations of mass transfer coefficients are also important for aeration and water cooling. These correlations usually have no parallel correlations in heat transfer, where fluid–fluid interfaces are not common.

Some of the more useful correlations for fluid–fluid interfaces are given in Table 8.3-2. The accuracy of these correlations is typically of the order of thirty percent, but larger errors are not uncommon. Raw data can look more like the result of a shotgun blast than any sort of coherent experiment because the data include wide ranges of chemical and physical properties. For example, the Reynolds number, that characteristic parameter of forced convection, can vary 10,000 times. The Schmidt number, the ratio (ν/D), is about 1 for gases but about 1000 for liquids. Over a more moderate range, the correlations can be more reliable. Still, while the correlations are useful for the preliminary design of small pilot plants, they should not be used for the design of full-scale equipment without experimental checks for the specific chemical systems involved.

Many of the correlations in Table 8.3-2 have the same general form. They typically involve a Sherwood number, which contains the mass transfer coefficient, the quantity of interest. This Sherwood number varies with Schmidt number, a characteristic of diffusion. The variation of Sherwood number with flow is more complex because the flow has two different physical origins. In most cases, the flow is caused by external stirring or pumping. For example, the liquids used in extraction are rapidly stirred; the gas in ammonia scrubbing is pumped through the packed tower; the blood in an artificial kidney is pumped through the dialysis unit. This type of externally driven flow is called "forced convection." In other cases, the fluid velocity is a result of the mass transfer itself. The mass transfer causes density gradients in the surrounding solution; these in turn cause flow. This type of internally generated flow is called "free convection." For example, the dispersal of pollutants and the dissolution of drugs are often accelerated by free convection.

The dimensionless form of the correlations for fluid–fluid interfaces may disguise the very real quantitative similarities between them. To explore these similarities, we consider the variations of the mass transfer coefficient with fluid velocity and with diffusion coefficient. These variations are surprisingly uniform. The mass transfer coefficient varies with about the 0.7 power of the fluid velocity in four of the five correlations for packed towers in Table 8.3-2. It varies with the diffusion coefficient to the

Table 8.3-2 *Selected mass transfer correlations for fluid–fluid interfaces*[a]

Physical situation	Basic equation[b]	Key variables	Remarks
Liquid in a packed tower	$k\left(\dfrac{1}{\nu g}\right)^{1/3} = 0.0051\left(\dfrac{\nu^0}{a\nu}\right)^{0.67}\left(\dfrac{D}{\nu}\right)^{0.50}(ad)^{0.4}$	a = packing area per bed volume; d = nominal packing size	Probably the best available correlation for liquids; tends to give lower value than other correlations
	$\dfrac{kd}{D} = 25\left(\dfrac{d\nu^0}{\nu}\right)^{0.45}\left(\dfrac{\nu}{D}\right)^{0.5}$	d = nominal packing size	The classical result, widely quoted; probably less successful than above
	$\dfrac{k}{\nu^0} = a\left(\dfrac{d\nu^0}{\nu}\right)^{-0.3}\left(\dfrac{D}{\nu}\right)^{0.5}$	d = nominal packing size	Based on older measurements of height of transfer units (HTUs); a is of order one
Gas in a packed tower	$\dfrac{k}{aD} = 3.6\left(\dfrac{\nu^0}{a\nu}\right)^{0.70}\left(\dfrac{\nu}{D}\right)^{1/3}(ad)^{-2.0}$	a = packing area per bed volume; d = nominal packing size	Probably the best available correlation for gases
Pure gas bubbles in a stirred tank	$\dfrac{kd}{D} = 1.2(1-\varepsilon)^{0.36}\left(\dfrac{d\nu^0}{\nu}\right)^{0.64}\left(\dfrac{\nu}{D}\right)^{1/3}$	d = nominal packing size; ε = bed void fraction	Again, the most widely quoted classical result
	$\dfrac{kd}{D} = 0.13\left(\dfrac{(P/V)d^4}{\rho\nu^3}\right)^{1/4}\left(\dfrac{\nu}{D}\right)^{1/3}$	d = bubble diameter; P/V = stirrer power per volume	Note that k does not depend on bubble size
Pure gas bubbles in an unstirred tank	$\dfrac{kd}{D} = 0.31\left(\dfrac{d^3 g\,\Delta\rho/\rho}{\nu^2}\right)^{1/3}\left(\dfrac{\nu}{D}\right)^{1/3}$	d = bubble diameter; $\Delta\rho$ = density difference between bubble and surrounding fluid	Drops 0.3-cm diameter or larger
Small liquid drops rising in unstirred solution	$\dfrac{kd}{D} = 1.13\left(\dfrac{d\nu^0}{D}\right)^{0.8}$	d = drop diameter; ν^0 = drop velocity	These small drops behave like rigid spheres
Falling films	$\dfrac{kz}{D} = 0.69\left(\dfrac{z\nu^0}{D}\right)^{0.5}$	z = position along film; ν^0 = average film velocity	Frequently embroidered and embellished

Notes: [a] The symbols used include the following: D is the diffusion coefficient; g is the acceleration due to gravity; k is the local mass transfer coefficient; ν^0 is the superficial fluid velocity; and ν is the kinematic viscosity.

[b] Dimensionless groups are as follows: $\dfrac{d\nu^0}{\nu}$ and $\nu^0/a\nu$ are Reynolds numbers; ν/D is the Schmidt number; $d^3 g(\Delta\rho/\rho)/\nu^2$ is the Grashof number; kd/D is the Sherwood number, and $k/(\nu g)^{1/3}$ is an unusual form of the Stanton number.

0.5 to 0.7 power in every one of the correlations. Thus any theory that we derive for mass transfer across fluid–fluid interfaces should imply variations with velocity and diffusion coefficient like those shown here.

Some frequently quoted correlations for fluid–solid interfaces are given in Table 8.3-3. These correlations are rarely important in common separation processes like absorption and extraction. They can be important in leaching, in membrane separations, and in adsorption. However, the chief reason that these correlations are quoted in undergraduate and graduate courses is that they are close analogues to heat transfer. Heat transfer is an older subject, with a strong theoretical basis and more familiar nuances. This analogy lets lazy lecturers merely mumble, "Mass transfer is just like heat transfer" and quickly compare the correlations in Table 8.3-3 with the heat transfer parallels.

The correlations for solid–fluid interfaces in Table 8.3-3 are much like their heat transfer equivalents. More significantly, these less important, fluid–solid correlations are analogous but more accurate than the important fluid–fluid correlations in Table 8.3-2. Accuracies for solid–fluid interfaces are typically average $\pm$ 10%; for some correlations like laminar flow in a single tube, accuracies can be $\pm$ 1%. Such precision, which is truly rare for mass transfer measurements, reflects the simpler geometry and more stable flows in these cases. Laminar flow of one fluid in a tube is much better understood than turbulent flow of gas and liquid in a packed tower.

The correlations for fluid–solid interfaces often show mathematical forms like those for fluid–fluid interfaces. The mass transfer coefficient is most often written as a Sherwood number, though occasionally as a Stanton number. The effect of diffusion coefficient is most often expressed as a Schmidt number. The effect of flow is most often expressed as a Reynolds number for forced convection, and as a Grashof number for free convection.

These fluid–solid dimensionless correlations can conceal how the mass transfer coefficient varies with fluid flow v^0 and diffusion coefficient D, just as those for fluid–fluid interfaces obscured these variations. Often k varies with the square root of v^0. The variation is lower for some laminar flows and higher for some turbulent flows. Usually, k is said to vary with $D^{2/3}$, though this variation is rarely checked carefully by those who develop the correlations. Variation of k with $D^{2/3}$ does have some theoretical basis, a point explored further in Chapter 9.

Example 8.3-1: Gas scrubbing with a wetted-wall column Air containing a water-soluble vapor is flowing up and water is flowing down in the experimental column shown in Fig. 8.3-1. The water flow in the 0.07-cm-thick film is 3 cm/sec, the column diameter is 10 cm, and the air is essentially well mixed right up to the interface. The diffusion coefficient in water of the absorbed vapor is $1.8 \cdot 10^{-5}$ cm^2/sec. How long a column is needed to reach a gas concentration in water that is 10% of saturation?

 Solution The first step is to write a mass balance on the water in a differential column height Δz:

$$(\text{accumulation}) = (\text{flow in minus flow out}) + (\text{absorption})$$

$$0 = \left[\pi d l v^0 c_1\right]_z - \left[\pi d l v^0 c_1\right]_{z+\Delta z} + \pi d \Delta z k [c_1(\text{sat}) - c_1]$$

Table 8.3-3 *Selected mass transfer correlations for fluid–solid interfaces*[a]

Physical situation	Basic equation[b]	Key variables	Remarks
Membrane	$\dfrac{kl}{D} = 1$	l = membrane thickness	Often applied even where membrane is hypothetical
Laminar flow along flat plate[c]	$\dfrac{kL}{D} = 0.646 \left(\dfrac{Lv^0}{\nu}\right)^{1/3} \left(\dfrac{\nu}{D}\right)^{1/3}$	L = plate length v^0 = bulk velocity	Solid theoretical foundation, which is unusual
Turbulent flow through horizontal slit	$\dfrac{kd}{D} = 0.026 \left(\dfrac{dv^0}{\nu}\right)^{0.8} \left(\dfrac{\nu}{D}\right)^{1/3}$	v^0 = average velocity in slit $d = [2/\pi]$ (slit width)	Mass transfer here is identical with that in a pipe of equal wetted perimeter
Turbulent flow through circular pipe	$\dfrac{kd}{D} = 0.026 \left(\dfrac{dv^0}{\nu}\right)^{0.8} \left(\dfrac{\nu}{D}\right)^{1/3}$	v^0 = average velocity in slit d = pipe diameter	Same as slit, because only wall regime is involved
Laminar flow through circular tube	$\dfrac{kd}{D} = 1.62 \left(\dfrac{d^2 v^0}{LD}\right)^{1/3}$	d = pipe diameter L = pipe length v^0 = average velocity in tube	Very strong theoretical and experimental basis
Flow outside and parallel to a capillary bed	$\dfrac{kd}{D} = 1.25 \left(\dfrac{d^2 v^0}{\nu l}\right)^{0.93} \left(\dfrac{\nu}{D}\right)^{1/3}$	$d = 4$ cross-sectional area/(wetted perimeter) v^0 = superficial velocity	Not reliable because of channeling in bed
Flow outside and perpendicular to a capillary bed	$\dfrac{kd}{D} = 0.80 \left(\dfrac{dv^0}{\nu}\right)^{0.47} \left(\dfrac{\nu}{D}\right)^{1/3}$	d = capillary diameter v^0 = velocity approaching bed	Reliable if capillaries evenly spaced
Forced convection around a solid sphere	$\dfrac{kd}{D} = 2.0 + 0.6 \left(\dfrac{dv^0}{\nu}\right)^{1/2} \left(\dfrac{\nu}{D}\right)^{1/3}$	d = sphere diameter v^0 = velocity of sphere	Very difficult to reach $(kd/D) = 2$ experimentally; no sudden laminar-turbulent transition
Free convection around a solid sphere	$\dfrac{kd}{D} = 2.0 + 0.6 \left(\dfrac{d^3 \Delta \rho g}{\rho \nu^2}\right)^{1/4} \left(\dfrac{\nu}{D}\right)^{1/3}$	d = sphere diameter g = gravitational acceleration	For a 1-cm sphere in water, free convection is important when $\Delta \rho = 10^{-9}$ g/cm³
Packed beds	$\dfrac{k}{v^0} = 1.17 \left(\dfrac{dv^0}{\nu}\right)^{-0.42} \left(\dfrac{D}{\nu}\right)^{2/3}$	d = particle diameter v^0 = superficial velocity	The superficial velocity is that which would exist without packing
Spinning disc	$\dfrac{kd}{D} = 0.62 \left(\dfrac{d^2 \omega}{\nu}\right)^{1/2} \left(\dfrac{\nu}{D}\right)^{1/3}$	d = disc diameter ω = disc rotation (radians/time)	Valid for Reynolds numbers between 100 and 20,000

Notes: [a] The symbols used include the following: D is the diffusion coefficient of the material being transferred; k is the local mass transfer coefficient; ρ is the fluid density; ν is the kinetmatic viscosity. Other symbols are defined for the specific situation.
[b] The dimensionless groups are defined as follows: (dv^0/ν) and $(d^2\omega/\nu)$ are the Reynolds number; (dv^0/ν) is the Schmidt number; $(d^3\Delta\rho g/\rho\nu^2)$ is the Grashof number, kd/D is the Sherwood number; k/v^0 is the Stanton number.
[c] The mass transfer coefficient given here is the value averaged over the length L.

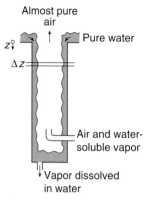

Fig. 8.3-1. Gas scrubbing in a wetted-wall column. A water-soluble gas is being dissolved in a falling film of water. The problem is to calculate the length of the column necessary to reach a liquid concentration equal to ten percent saturation.

in which d is the column diameter, l is the film thickness, v^0 is the flow, and c_1 is the vapor concentration in the water. This balance leads to

$$0 = -lv^0 \frac{dc_1}{dz} + k[c_1(\text{sat}) - c_1]$$

From Table 8.3-2, we have

$$k = 0.69 \left(\frac{Dv^0}{z} \right)^{1/2}$$

We also know that the entering water is pure; that is, when

$$z = 0, \; c_1 = 0$$

Combining these results and integrating, we find

$$\frac{c_1}{c_1(\text{sat})} = 1 - e^{-1.38(Dz/l^2v^0)^{1/2}}$$

Inserting the numerical values given,

$$z = \left(\frac{l^2 v^0}{1.38\,D} \right) \left[\ln \left(1 - \frac{c_1}{c_1(\text{sat})} \right) \right]^2$$

$$= \left(\frac{(0.07\text{ cm})^2\,(3\text{ cm/sec})}{(1.38)\,1.8 \cdot 10^{-5}\text{cm}^2/\text{sec}} \right) [\ln(1 - 0.1)]^2$$

$$= 6.6\,\text{cm}$$

This type of system has been studied extensively, though its practical value is small.

Example 8.3-2: Dissolution rate of a spinning disc   A solid disc of benzoic acid 2.5 cm in diameter is spinning at 20 rpm and 25 °C. How fast will it dissolve in a large volume of water? How fast will it dissolve in a large volume of air? The diffusion coefficients are $1.00 \cdot 10^{-5}$ cm^2/sec in water and 0.233 cm^2/sec in air. The solubility of benzoic acid in water is 0.003 g/cm^3; its equilibrium vapor pressure is 0.30 mm Hg.

Solution   Before starting this problem, try to guess the answer. Will the mass transfer be higher in water or in air?

In each case, the dissolution rate is

$$N_1 = kc_1(\text{sat})$$

where $c_1(\text{sat})$ is the concentration at equilibrium. We can find k from Table 8.3-3:

$$k = 0.62 D \left(\frac{\omega}{\nu}\right)^{1/2} \left(\frac{\nu}{D}\right)^{1/3}$$

For water, the mass transfer coefficient is

$$k = 0.62 \left(1.00 \cdot 10^{-5} \text{cm}^2/\text{sec}\right) \left(\frac{(20/60)\,(2\pi/\text{sec})}{0.01 \text{cm}^2/\text{sec}}\right)^{1/2} \left(\frac{0.01 \text{cm}^2/\text{sec}}{1.00 \cdot 10^{-5} \text{cm}^2/\text{sec}}\right)^{1/3}$$

$$= 0.90 \cdot 10^{-3} \text{cm}/\text{sec}$$

Thus the flux is

$$N_1 = (0.90 \cdot 10^{-3} \text{ cm}/\text{sec})(0.003 \text{ g/cm}^3)$$

$$= 2.7 \cdot 10^{-6} \text{ g/cm}^2 \text{ sec}$$

For air, the values are very different:

$$k = 0.62 \left(0.233 \text{ cm}^2/\text{sec}\right) \left(\frac{(20/60)\,(2\pi/\text{sec})}{0.15 \text{ cm}^2/\text{sec}}\right)^{1/2} \left(\frac{0.15 \text{ cm}^2/\text{sec}}{0.23 \text{ cm}^2/\text{sec}}\right)^{1/3}$$

$$= 0.47 \text{ cm}/\text{sec}$$

which is much larger than before. However, the flux is

$$N_1 = (0.47 \text{ cm}/\text{sec}) \left[\left(\frac{0.3 \text{ mm Hg}}{760 \text{ mm Hg}}\right) \left(\frac{1 \text{ mol}}{22.4 \cdot 10^3 \text{ cm}^3}\right) \left(\frac{273}{298}\right) \left(\frac{122 \text{ g}}{\text{mol}}\right)\right]$$

$$= 0.9 \cdot 10^{-6} \text{g/cm}^2 \text{ sec}$$

The flux in air is about one-third of that in water, even though the mass transfer coefficient in air is about 500 times larger than that in water. Did you guess this?

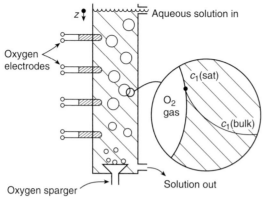

Fig. 8.4-1. An experimental apparatus for the study of aeration. Oxygen bubbles from the sparger at the bottom of the tower partially dissolve in the aqueous solution. The concentration in this solution is measured with electrodes that are specific for dissolved oxygen. The concentrations found in this way are interpreted in terms of mass transfer coefficients; this interpretation assumes that the solution is well mixed, except very near the bubble walls.

8.4 Dimensional Analysis: The Route to Correlations

The correlations in the previous section provide a useful and compact way of presenting experimental information. Use of these correlations quickly gives reasonable estimates of mass transfer coefficients. However, when we find the correlations inadequate, we will be forced to make our own experiments and develop our own correlations. How can we do this?

Mass transfer correlations are developed using a method called dimensional analysis. This method can be learned via the two specific examples that follow. Before embarking on this description, I want to emphasize that most people go through three mental states concerning this method. At first they believe it is a route to all knowledge, a simple technique by which any set of experimental data can be greatly simplified. Next they become disillusioned when they have difficulties in the use of the technique. These difficulties commonly result from efforts to be too complete. Finally, they learn to use the method with skill and caution, benefiting both from their past successes and from their frequent failures. I mention these three stages because I am afraid many may give up at the second stage and miss the real benefits involved. We now turn to the examples.

8.4.1 Aeration

Aeration is a common industrial process and yet one in which there is often serious disagreement about correlations. This is especially true for deep-bed fermentors and for sewage treatment, where the rising bubbles can be the chief means of stirring. We want to study this process using the equipment shown schematically in Fig. 8.4-1. We plan to inject pure oxygen into a variety of aqueous solutions and measure the oxygen concentration in the bulk solution with oxygen selective electrodes. We expect to vary the average bubble velocity v, the solution's density ρ and viscosity μ, the entering bubble diameter d, and the depth of the bed L. Of course, we also expect that mass transfer will vary with the diffusion coefficient, but because the solute is always oxygen, we ignore this constant coefficient here.

We measure the steady-state oxygen concentration as a function of position in the bed. These data can be summarized as a mass transfer coefficient in the following way. From a mass balance, we see that

$$0 = -v\frac{dc_1}{dz} + ka\left[c_1(\text{sat}) - c_1\right] \tag{8.4-1}$$

where a is the total bubble area per column volume. This equation, a close parallel to the many mass balances in Section 8.1, is subject to the initial condition

$$z = 0, \; c_1 = 0 \tag{8.4-2}$$

Thus

$$ka = \frac{v}{z}\ln\left(\frac{c_1(\text{sat})}{c_1(\text{sat}) - c_1(z)}\right) \tag{8.4-3}$$

Ideally, we would like to measure k and a independently, separating the effects of mass transfer and geometry. This would be difficult here, so we report only the product ka.

Our experimental results now consist of the following:

$$ka = ka(v, \rho, \mu, d, z) \tag{8.4-4}$$

We assume that this function has the form

$$ka = [\text{constant}] \, v^\alpha \rho^\beta \mu^\gamma d^\delta z^\varepsilon] \tag{8.4-5}$$

where both the constant in the square brackets and the exponents are dimensionless. Now the dimensions or units on the left-hand side of this equation must equal the dimensions or units on the right-hand side. We cannot have centimeters per second on the left-hand side equal to grams on the right. Because ka has dimensions of the reciprocal of time $(1/t)$, v has dimensions of length/time (L/t), ρ has dimensions of mass per length cubed (M/L^3), and so forth, we find

$$\frac{1}{t} \; [=] \; \left(\frac{L}{t}\right)^\alpha \left(\frac{M}{L^3}\right)^\beta \left(\frac{M}{Lt}\right)^\gamma (L)^\delta \, (L)^\varepsilon \tag{8.4-6}$$

The only way this equation can be dimensionally consistent is if the exponent on time on the left-hand side of the equation equals the sum of the exponents on time on the right-hand side:

$$-1 = -\alpha - \gamma \tag{8.4-7}$$

Similar equations hold for the mass:

$$0 = \beta + \gamma \tag{8.4-8}$$

and for the length:

$$0 = \alpha - 3\beta - r + \delta + \varepsilon \tag{8.4-9}$$

Equations 8.4-7 to 8.4-9 give three equations for the five unknown exponents.

We can solve these equations in terms of the two key exponents and thus simplify Eq. 8.4-5. We choose the two key exponents arbitrarily. For example, if we choose the exponent on the viscosity γ and that on column height ε, we obtain

$$\alpha = 1 - \gamma \tag{8.4-10}$$

$$\beta = -\gamma \tag{8.4-11}$$

$$\gamma = \gamma \tag{8.4-12}$$

$$\delta = -\gamma - \varepsilon - 1 \tag{8.4-13}$$

$$\varepsilon = \varepsilon \tag{8.4-14}$$

Inserting these results into Eq. 8.4-5 and rearranging, we find

$$\left(\frac{kad}{v}\right) = [\text{constant}] \left(\frac{dv\rho}{\mu}\right)^{-\gamma} \left(\frac{z}{d}\right)^{\varepsilon} \tag{8.4-15}$$

The left-hand side of this equation is a type of Stanton number. The first term in parentheses on the right-hand side is the Reynolds number, and the second such term is a measure of the tank's depth.

This analysis suggests how we should plan our experiments. We expect to plot our measurements of Stanton number versus two independent variables: Reynolds number and z/d. We want to cover the widest possible range of independent variables. Our resulting correlation will be a convenient and compact way of presenting our results, and everyone will live happily ever after.

Unfortunately, it is not always that simple for a variety of reasons. First, we had to assume that the bulk liquid was well mixed, and it may not be. If it is not, we shall be averaging our values in some unknown fashion, and we may find that our correlation extrapolates unreliably. Second, we may find that our data do not fit an exponential form like Eq. 8.4-5. This can happen if the oxygen transferred is consumed in some sort of chemical reaction, which is true in aeration. Third, we do not know which independent variables are important. We might suspect that ka varies with tank diameter, or sparger shape, or surface tension, or the phases of the moon. Such variations can be included in our analysis, but they make it complex.

Still, this strategy has produced a simple method of correlating our results. The foregoing objections are important only if they are shown to be so by experiment. Until then, we should use this easy strategy.

8.4.2 Drug Dissolution

A standard test for drug dissolution uses a cylindrical basket made of 40 mesh 316 stainless steel and immersed in a larger vessel of degassed water at a fixed pH and at 37 °C. Tablets of drug are placed in the basket. The basket is then rotated at a fixed speed. Samples are drawn from the water and analyzed. If the amount of drug dissolved is reproduceable, these tablets may be acceptable for marketing.

To impove reproduceability, we are interested in analyzing this dissolution test in more detail. We want a dimensional analysis suitable for organizing our results. However, this problem is sufficiently broad that many variables could be involved. For example, we would normally expect that water's properties like viscosity and pH will be important. However, if we include all these properties, we will get such an elaborate correlation that we really won't be able to test it much.

Thus we want to guess an answer which implies neglecting less important factors. In the case of water, we will at least temporarily assume that its viscosity doesn't change much and so can't be that big of a factor. We also assume that the major effect of pH is a change in drug solubility and not a change in mass transfer coefficient.

We then try to guess the answer. Right away, we can see two possibilities. First, dissolution may be limited by mass transfer from the tablets to the edge of the basket. This says the basket radius R will be important. Second, dissolution may be limited by mass transfer from the basket to the surrounding solution. Then the basket rotation will be important.

To put these ideas on a more quantitative basis, we postulate that the mass transfer coefficient for drug dissolution k is

$$k = k(R, \omega, d, D) \tag{8.4-16}$$

where R is the basket radius, ω is the angular rotation of the basket, d is the tablet diameter, and D is the drug's diffusion coefficient in water. We also assume that

$$k = [\text{constant}] R^\alpha \omega^\beta d^\gamma D^\delta \tag{8.4-17}$$

On dimensional grounds, this says

$$\frac{L}{t} [=] (L)^\alpha \left(\frac{1}{t}\right)^\beta (L)^\gamma \left(\frac{L^2}{t}\right)^\delta$$

As before, the dimensions on the left-hand side must equal those on the right. For length L,

$$1 = \alpha + \gamma + 2\delta$$

For time t

$$-1 = -\beta - \delta$$

There is no equation for mass M, because it appears in none of the variables. If we choose as key exponents β and γ, we have

$$\alpha = -1 + 2\beta - \gamma$$

$$\beta = \beta$$

$$\gamma = \gamma$$

$$\delta = 1 - \beta$$

Inserting these values into the above and rearranging, we have

$$\frac{kR}{D} = [\text{constant}] \left(\frac{R^2\omega}{D}\right)^{\beta} \left(\frac{d}{R}\right)^{\gamma}$$

The dimensionless group raised to the β power is a Péclet number; that raised to the γ power is a ratio of lengths.

This correlation will be successful only if we have guessed the right key variables. It may work if the tablets remain intact. It probably won't work if they disintegrate. Still, the analysis gives us a good first step in organizing our results.

8.5 Mass Transfer Across Interfaces

We now turn to mass transfer across interfaces, from one fluid phase to the other. This is a tricky subject, one of the main reasons that mass transfer is felt to be a difficult subject. In the previous sections, we used mass transfer coefficients as an easy way of describing diffusion occurring from an interface into a relatively homogeneous solution. These coefficients involved approximations and sparked the explosion of definitions exemplified by Table 8.2-2. Still, they are an easy way to correlate experimental results or to make estimates using the published relations summarized in Tables 8.3-2 and 8.3-3.

In this section, we extend these definitions to transfer across an interface, from one well-mixed bulk phase into another different one. This case occurs more frequently than does transfer from an interface into one bulk phase; indeed, I had trouble dreaming up the examples earlier in this chapter. Transfer across an interface again sparks potentially major problems of unit conversion, but these problems are often simplified in special cases.

8.5.1 The Basic Flux Equation

Presumably, we can describe mass transfer across an interface in terms of the same type of flux equation as before:

$$N_1 = K\Delta c_1 \tag{8.5-1}$$

where N_1 is the solute flux relative to the interface, K is called an "overall mass transfer coefficient," and Δc_1 is some appropriate concentration difference. But what is Δc_1?

Choosing an appropriate value of Δc_1 turns out to be difficult. To illustrate this, consider the three examples shown in Fig. 8.5-1. In the first example in Figure 8.5-1(a), hot benzene is placed on top of cold water; the benzene cools and the water warms until they reach the same temperature. Equal temperature is the criterion for equilibrium, and the amount of energy transferred per time turns out to be proportional to the temperature difference between the liquids. Everything seems secure.

As a second example, shown in Fig. 8.5-1(b), imagine that a benzene solution of bromine is placed on top of water containing the same concentration of bromine. After a while, we find that the initially equal concentrations have changed, that the bromine

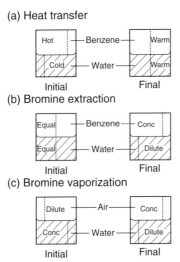

(a) Heat transfer

(b) Bromine extraction

(c) Bromine vaporization

Fig. 8.5-1. Driving forces across interfaces. In heat transfer, the amount of heat transferred depends on the temperature difference between the two liquids, as shown in (a). In mass transfer, the amount of solute that diffuses depends on the solute's "solubility" or, more exactly, on its chemical potential. Two cases are shown. In (b), bromine diffuses from water into benzene because it is much more soluble in benzene; in (c), bromine evaporates until its chemical potentials in the solutions are equal. This behavior complicates analysis of mass transfer.

concentration in the benzene is much higher than that in water. This is because the bromine is more soluble in benzene, so that its concentration in the final solution is higher.

This result suggests which concentration difference we can use in Eq. 8.5-1. We should not use the concentration in benzene minus the concentration in water; that is initially zero, and yet there is a flux. Instead, we can use the concentration actually in benzene minus the concentration that would be in benzene that was in equilibrium with the actual concentration in water. Symbolically,

$$N_1 = K[c_1(\text{in benzene}) - Hc_1(\text{in water})] \tag{8.5-2}$$

where H is a partition coefficient, the ratio at equilibrium of the concentration in benzene to that in water. Note that this does predict a zero flux at equilibrium.

A better understanding of this phenomenon may come from the third example, shown in Fig. 8.5-1(c). Here, bromine is vaporized from water into air. Initially, the bromine's concentration in water is higher than that in air; afterward, it is lower. Of course, this reversal of the concentration in the liquid might be expressed in moles per liter and that in gas as a partial pressure in atmospheres, so it is not surprising that strange things happen.

As you think about this more carefully, you will realize that the units of pressure or concentration cloud a deeper truth: Mass transfer should be described in terms of the more fundamental chemical potentials. If this were done, the peculiar concentration differences would disappear. However, chemical potentials turn out to be difficult to use in practice, and so the concentration differences for mass transfer across interfaces will remain complicated by units.

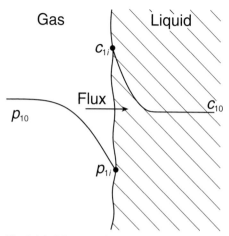

Fig. 8.5-2. Mass transfer across a gas–liquid interface. In this example, a solute vapor is diffusing from the gas on the left into the liquid on the right. Because the solute concentration changes both in the gas and in the liquid, the solute's flux must depend on a mass transfer coefficient in each phase. These coefficients are combined into an overall flux equation in the text.

8.5.2 The Overall Mass Transfer Coefficient

We want to extend these qualitative observations in more exact equations. To do this, we consider the example of the gas–liquid interface in Fig. 8.5-2. In this case, gas on the left is being transferred into the liquid on the right. The flux in the gas is

$$N_1 = k_p(p_1 - p_{1i}) \qquad (8.5\text{-}3)$$

where k_p is the gas-phase mass transfer coefficient (in, for example, mol/cm^2 sec Pa), p_1 is the bulk pressure, and p_{1i} is the interfacial pressure. Because the interfacial region is thin, the flux across it will be in steady state, and the flux in the gas will equal that in the liquid. Thus,

$$N_1 = k_x(x_{1i} - x_1) \qquad (8.5\text{-}4)$$

where the liquid-phase mass transfer coefficient k_x is, for example, in mol/cm^2 sec, and x_{1i} and x_1 are the interfacial and bulk mole fractions, respectively.

We now need to eliminate the unknown interfacial concentrations from these equations. In almost all cases, equilibrium exists across the interface:

$$p_{1i} = Hx_{1i} \qquad (8.5\text{-}5)$$

where H is a type of Henry's law or partition constant (here in units of pressure). Combining Eqs. 8.5-3 through 8.5-5, we can find the interfacial concentrations

$$x_{1i} = \frac{p_{1i}}{H} = \frac{k_p p_1 + k_x x_1}{k_p H + k_x} \qquad (8.5\text{-}6)$$

and the flux

$$N_1 = \frac{1}{1/k_p + H/k_x} (p_1 - Hx_1) \tag{8.5-7}$$

You should check the derivations of these results because they are important.

Before proceeding further, we make a quick analogy. This result is often compared to an electric circuit containing two resistances in series. The flux corresponds to the current, and the concentration difference $p_1 - Hx_1$ corresponds to the voltage. The resistance is then $1/k_p + H/k_x$ which is roughly a sum of two resistances in series. This is a good way of thinking about these effects. You must remember, however, that the resistances $1/k_p$ and $1/k_x$ are not directly added, but always weighted by partition coefficients like H.

We now want to write Eq. 8.5-7 in the form of Eq. 8.5-1. We can do this in two ways. First, we can write

$$N_1 = K_x \left(x_1^* - x_1 \right) \tag{8.5-8}$$

where

$$K_x = \frac{1}{1/k_x + 1/k_p H} \tag{8.5-9}$$

and

$$x_1^* = \frac{p_1}{H} \tag{8.5-10}$$

K_x is called an "overall liquid-side mass transfer coefficient," and x_1^* is the hypothetical liquid mole fraction that would be in equilibrium with the bulk gas. Alternatively,

$$N_1 = K_p \left(p_1 - p_1^* \right) \tag{8.5-11}$$

where

$$K_p = \frac{1}{1/k_p + H/k_x} \tag{8.5-12}$$

and

$$p_1^* = Hc_{10} \tag{8.5-13}$$

K_p is an "overall gas-side mass transfer coefficient," and p_1^* is the hypothetical partial pressure that would be in equilibrium with the bulk liquid.

8.5.3 Details of the Partition Coefficient

The analysis above is standard, a fixture of textbooks on transport phenomena and unit operations. It can be reproduced by almost everyone who studies the subject.

However, I know that it is unevenly understood in physical terms, both by students and by experienced professionals. Indeed, explaining this repeatedly is a large part of my consulting practice. As a result, I want to pause in the standard development and go over the two tricky points again, in more detail.

The first of the tricky points comes from the basic flux equation, given by (for example)

$$N_1 = K_L \left(c_1^* - c_1 \right) \tag{8.5-14}$$

In this equation, the flux across the interface is proportional to a concentration difference. The concentration difference is confusing. Clearly, c_1 is the bulk concentration of species "1," for example, in the liquid phase. The other concentration c_1^* is harder to understand. Saying that it is "the concentration that would exist in the liquid if the other phase were in equilibrium" is true but may not help much. Alternatively, we can ask what c_1^* will be when the flux is zero. Then, c_1^* equals c_1 and we are in equilibrium.

The second tricky point in the analysis comes from the partition coefficient H. I have casually described H as a partition coefficient or a Henry's law coefficient, neglecting the complex units which can be involved. To illustrate these, we can compare three common forms of partition coefficients, defined at equilibrium

$$p_1 = Hx_1 \tag{8.5-15}$$

$$y_1 = mx_1 \tag{8.5-16}$$

$$c_{1G} = H' c_{1L} \tag{8.5-17}$$

Here, p_1 is the partial pressure of species "1" in the gas; x_1 and y_1 are the corresponding mole fractions in the liquid and gas, respectively; and c_{1L} and c_{1G} are the molar concentrations of species "1" in the liquid and gas, respectively. The quantities $H, m,$ and H' are all partition coefficients, and all can describe the same equilibrium between gas and liquid. However, they don't always have the same dimensions; and they are not numerically equal even when they do have the same dimensions.

To illustrate these differences, we consider the system of oxygen gas which is partly dissolved in water. From equilibrium experiments, we find

$$p_1 = [43000 \text{ atm}] \, x_1 \tag{8.5-18}$$

where p_1 is the partial pressure of oxygen, and x_1 is the mole fraction of oxygen dissolved in water. To convert this into a ratio of mole fractions, we remember that

$$\frac{p_1}{p} = y_1 = \left(\frac{H}{p} \right) x_1 \tag{8.5-19}$$

Thus at atmospheric pressure, m is 42000. We can also reform Henry's law in terms of molar concentrations

$$c_{1G} = \frac{p_1}{RT} = \left(\frac{H}{c_L RT} \right) c_{1L} \tag{8.5-20}$$

where c_L is the total molar concentration in the liquid, about 55 M for water. Because H' equals the quantity in parentheses, at 298 K, H' is 31.

We now turn to a variety of examples illustrating mass transfer across an interface. These examples have the annoying characteristic that they are initially difficult to do, but they are trivial after you understand them. Remember that most of the difficulty comes from that ancient but common curse: unit conversion.

Example 8.5-1: Oxygen mass transfer   Use Equation 8.5-9 to estimate the overall liquid-side mass transfer coefficient K_x at 25 °C for oxygen from water into air. In this estimate, assume that the film thickness is 10^{-2} cm in liquids but 10^{-1} cm in gases.

   Solution   For oxygen in air, the diffusion coefficient is 0.23 cm^2/sec; for oxygen in water, the diffusion coefficient is $2.1 \cdot 10^{-5}$ cm^2/sec. The Henry's law constant in this case is $4.2 \cdot 10^4$ atmospheres. We need only calculate k_x and k_p and plug these values into Eq. 8.5-9. Finding k_x is easy:

$$k_x = k_L c_L = \frac{D_L c_L}{0.01 \text{ cm}} = \frac{2.1 \cdot 10^{-5} \text{ cm}^2/\text{sec}}{0.01 \text{ cm}} \left(\frac{\text{mol}}{18 \text{ cm}^3} \right)$$

$$= 1.2 \cdot 10^{-4} \text{ mol/cm}^2 \text{ sec}$$

Finding k_p involves the unit conversions given in Table 8.2-2:

$$k_p = \frac{k_G}{RT} = \frac{D_G}{(0.1 \text{ cm}) (RT)}$$

$$= \frac{0.23 \text{ cm}^2/\text{ sec}}{(0.1 \text{ cm}) \left(82 \text{ cm}^3 \text{ atm/mol K} \right) (298 \text{ K})}$$

$$= 9.4 \cdot 10^{-5} \text{mol/cm}^2 \text{ sec atm}$$

Inserting these results into Eq. 8.5-9, we find

$$K_x = \frac{1}{1/k_x + 1/k_p H}$$

$$= \frac{1}{\left(\dfrac{\text{cm}^2 \text{ sec}}{1.2 \cdot 10^{-4} \text{ mol}} \right) + \dfrac{\text{cm}^2 \text{ sec atm}}{9.4 \cdot 10^{-5} \text{ mol } (43000 \text{ atm})}}$$

$$= 1.2 \cdot 10^{-4} \text{mol/cm}^2 \text{sec}$$

The mass transfer is completely dominated by the liquid-side resistance. This would also be true if we calculated the overall gas-side mass transfer coefficient, K_p. Gas absorption is commonly controlled by mass transfer in the liquid and is one reason that reactive liquids are effective.

Example 8.5-2: Benzene mass transfer Estimate the overall liquid-side mass transfer coefficient in the distillation of benzene and toluene. At the concentrations used, you expect a temperature of 90 °C and (at equilibrium)

$$y_1^* = 0.70x_1 + 0.39$$

The molar volume of the liquid is about 97 cm^3/mol. As in the previous problem, assume that the film thickness in the liquid is 0.01 cm and in the gas is 0.1 cm.

   Solution   In the vapor, the diffusion coefficient is about 0.090 cm^2/sec; in the liquid, it is $1.9 \cdot 10^{-5}$ cm^2/sec. As in the previous example,

$$k_x = k_L c_L = \frac{D_L c_L}{0.01 \text{ cm}}$$

$$= \frac{1.9 \cdot 10^{-5} \text{ cm}^2/\text{sec} \left(\dfrac{1 \text{ mol}}{97 \text{ cm}^3} \right)}{0.01 \text{ cm}}$$

$$= 2.0 \cdot 10^{-5} \frac{\text{mol}}{\text{cm}^2 \text{ sec}}$$

If the distillation is at atmospheric pressure, the Henry's law constant is 0.70 atm (cf. Equations 8.5-18 and 8.5-19). Then

$$k_p = \frac{k_G}{RT} = \frac{D_G}{0.1 \text{ cm} \, (RT)}$$

$$= \frac{0.090 \text{ cm}^2/\text{sec}}{0.1 \text{ cm} \left(82 \text{ cm}^3 \text{ atm}/\text{mol K} \right) (363 \text{ K})}$$

$$= 3.0 \cdot 10^{-5} \text{mol}/\text{cm}^2 \text{ sec atm}$$

Inserting these results into Equation 8.5-9 gives

$$K_x = \frac{1}{1/k_x + 1/k_p H}$$

$$= \frac{1}{\left(\dfrac{\text{cm}^2 \text{ sec}}{2.0 \cdot 10^{-5} \text{ mol}} \right) + \dfrac{\text{cm}^2 \text{ sec atm}}{3.0 \cdot 10^{-4} \text{ mol} \, (0.70 \text{ atm})}}$$

$$= 1.8 \cdot 10^{-5} \text{mol}/\text{cm}^2 \text{ sec}$$

The distillation is about equally controlled by the liquid and the vapor. This is typical.

Example 8.5-3: Perfume extraction Jasmone ($C_{11}H_{16}O$) is a valuable material in the perfume industry, used in many soaps and cosmetics. Suppose we are recovering this material from a water suspension of jasmine flowers by an extraction with toluene. The aqueous phase is continuous, with suspended flowers and toluene droplets. The mass transfer coefficient in the toluene droplets is $3.0 \cdot 10^{-4}$ cm^2/sec; the mass transfer coefficient in the aqueous phase is $2.4 \cdot 10^{-3}$ cm^2/sec. Jasmone is about 150 times more soluble in toluene than in the suspension. What is the overall mass transfer coefficient?

Solution For convenience, we designate all concentrations in the toluene phase with a prime and all those in the water without a prime. The flux is

$$N_1 = k(c_{10} - c_{1i}) = k'(c'_{1i} - c'_{10})$$

The interfacial concentrations are in equilibrium:

$$c'_{1i} = Hc_{1i}$$

Eliminating these interfacial concentrations, we find

$$N_1 = \left[\frac{1}{1/k' + H/k}\right](Hc_{10} - c'_{10})$$

The quantity in square brackets is the overall coefficient K' that we seek. This coefficient is based on a driving force in toluene. Inserting the values,

$$K' = \frac{1}{\dfrac{1}{3.0 \cdot 10^{-4} \text{ cm/sec}} + \dfrac{150}{2.4 \cdot 10^{-3} \text{ cm/sec}}}$$

$$= 1.5 \cdot 10^{-5} \text{cm/sec}$$

Similar results for the overall coefficient based on a driving force in water are easily found.

Two points about this problem deserve mention. First, the result is a complete parallel to Eq. 8.5-12, but for a liquid–liquid interface instead of a gas–liquid interface. Second, mass transfer in the water dominates the process even though the mass transfer coefficient in water is larger because jasmone is so much more soluble in toluene.

Example 8.5-4: Overall mass transfer coefficients in a packed tower We are studying gas absorption into water at 2.2 atm total pressure in a packed tower containing Berl saddles. From earlier experiments with ammonia and methane, we believe that for both gases the mass transfer coefficient times the packing area per tower volume is 18 lb mol/hr ft^3 for the gas side and 530 lb mol/hr ft^3 for the liquid side. The values for these two gases may be similar because methane and ammonia have similar molecular weights. However, their Henry's law constants are different: 75 atm for ammonia and 41,000 atm for methane. What is the overall gas-side mass transfer coefficient for each gas?

Solution This is essentially a problem in unit conversion. Although you can extract the appropriate equations from the text, I always feel more confident if I repeat parts of the derivation.

The quantity we seek, the overall gas-side transfer coefficient K_y, is defined by

$$N_1 a = K_y a (y_1 - y_1^*)$$
$$= k_y a (y_1 - y_{1i})$$
$$= k_x a (x_{1i} - x_1)$$

where y_1 and x_1 are the gas and liquid mole fractions.

The interfacial concentrations are related by Henry's law:

$$p_{1i} = py_{1i} = Hx_{1i}$$

When these interfacial concentrations are eliminated, we find that

$$\frac{1}{K_y a} = \frac{1}{k_y a} + \frac{H/p}{k_x a}$$

In passing, we recognize that y_1^* must equal Hx_{10}/p.

We can now find the overall coefficient for each gas. For ammonia,

$$\frac{1}{K_y a} = \frac{1}{18 \text{ lb mol/hr ft}^3} + \frac{75 \text{ atm}/2.2 \text{ atm}}{530 \text{ lb mol/hr ft}^3}$$

$$K_y a = 8.3 \text{ lb mol/hr ft}^3$$

The overall resistance is affected by resistances in both the gas and the liquid. For methane,

$$\frac{1}{K_y a} = \frac{1}{18 \text{ lb mol / hr ft}^3} + \frac{41,000 \text{ atm} / 2.2 \text{ atm}}{530 \text{ lb mol / hr ft}^3}$$

$$K_y a = 0.03 \text{ lb mol / hr ft}^3$$

The overall mass transfer coefficient for methane is smaller and is dominated by the liquid-side mass transfer coefficient.

8.6 Conclusions

This chapter presents an alternative model for diffusion, one using mass transfer coefficients rather than diffusion coefficients. The model is most useful for mass transfer across phase boundaries. It assumes that large changes in the concentration occur only very near these boundaries and that the solutions far from the boundaries are well mixed. Such a description is called a lumped-parameter model.

Mass transfer coefficients provide especially useful descriptions of diffusion in complex multiphase systems. They are basic to the analysis and design of industrial processes like absorption, extraction, and distillation. Mass transfer coefficients are not useful in chemistry when the focus is on chemical kinetics or chemical change. They are not useful in studies of the solid state, where concentrations vary with both position and time, and lumped-parameter models do not help much.

In this chapter, we have shown how experimental results can be analyzed in terms of mass transfer coefficients. We have also shown how values of these coefficients can be efficiently organized as dimensionless correlations, and we have cataloged published correlations that are commonly useful. These correlations are complicated by problems with units that come out of a plethora of closely related definitions. These complications are most confusing for mass transfer across fluid–fluid interfaces, from one well-mixed phase into another.

All in all, the material in this chapter is a solid alternative for analyzing diffusion near interfaces. It is basic stuff for chemical engineers, but it is an unexplored method for many others. It repays careful study.

Questions for Discussion

1. What are the dimensions of a mass transfer coefficient?
2. What are the dimensions of a diffusion coefficient?
3. How can you convert from a mass transfer coefficient based on a concentration difference to one based on a partial pressure difference?
4. Which is typically bigger, a mass transfer coefficient in a liquid or one in a gas?
5. Why do overall mass transfer coefficients vary with partition coefficients, but overall heat transfer coefficients do not?
6. If the flow doubles, how much will the mass transfer coefficient typically change?
7. If the diffusion coefficient doubles, how much does the mass transfer coefficient change?
8. If you have a system with the same solute concentration everywhere, will mass transfer occur?
9. When will a bulk concentration equal an interfacial concentration? When won't it?
10. What are typical units for partition coefficients?

Problems

1. A wet t-shirt hung on a hanger has a total surface area of about 0.6 m². It loses water as follows:

time (pm)	Weight (g)
3:15	661
3:20	640
3:48	580
4:00	553

If the saturation vapor pressure of the water of 20 mm Hg gives a water concentration of 20 g/m³ and the room has a relative humidity of 30%, estimate the mass transfer coefficient from the t-shirt.

2. Water flows through a thin tube, the walls of which are lightly coated with benzoic acid. The benzoic acid is dissolved very rapidly and so is saturated at the pipe's wall. The water flows slowly, at room temperature and 0.1 cm/sec. The pipe is 1 cm in diameter. Under these conditions, the mass transfer coefficient k varies along the pipe:

$$\frac{kd}{D} = 1.62 \left(\frac{d^2 v}{DL} \right)^{1/3}$$

where d and L are the diameter and length of the pipe and v is the average velocity in the pipe. What is the average concentration of benzoic acid in the water after 2 m of pipe?

3. Water containing 0.1-M benzoic acid flows at 0.1 cm/sec through a 1-cm-diameter rigid tube of cellulose acetate, the walls of which are permeable to small electrolytes. These walls are 0.01 cm thick; solutes within the walls diffuse as through water. The tube is immersed in a large well-stirred water bath. Under these conditions, the flux of benzoic acid from the bulk to the walls can be described by the correlation in Problem 8.2. After 50 cm of tube, what fraction of a 0.1-M benzoic acid solution has been removed? Remember that there is more than one resistance to mass transfer in this system.

4. How much is the previous answer changed if the benzoic acid solution in the tube is in benzene, not water?

5. A disk of radioactively tagged benzoic acid 1 cm in diameter is spinning at 20 rpm in 94 cm^3 of initially pure water. We find that the solution contains benzoic acid at $7.3 \cdot 10^{-4}$ g/cm^3 after 10 hr 4 min and $3.43 \cdot 10^{-3}$ g/cm^3 after a long time (i.e., at saturation). (a) What is the mass transfer coefficient? *Answer:* $8 \cdot 10^{-4}$ cm/sec. (b) How long will it take to reach 14% saturation? (c) How closely does this mass transfer coefficient agree with that expected from the theory in Example 3.4-3?

6. As part of the manufacture of microelectronic circuits, silicon wafers are partially coated with a 5,400-Å film of a polymerized organic film called a photoresist. The density of this polymer is 0.96 g/cm^3. After the wafers are etched, this photoresist must be removed. To do so, the wafers are placed in groups of twenty in an inert "boat," which in turn is immersed in strong organic solvent. The solubility of the photoresist in the solvent is $2.23 \cdot 10^{-3}$ g/cm^3. If the photoresist dissolves in 10 minutes, what is its mass transfer coefficient? (S. Balloge) *Answer:* $4 \cdot 10^{-5}$ cm/sec.

7. You are studying mass transfer of a solute from a gas across a gas–liquid interface into a reactive liquid. The mole fraction in the bulk gas is 0.01; that in the bulk liquid is 0.00; and the equilibrium across the interface is

$$y_i^* = \frac{x_i}{3}$$

The individual mass transfer coefficient k_y and k_x are 0.50 and 0.60 mol/m^2 sec, respectively. (a) What is the interfacial concentration in the vapor? (b) Sketch, to scale, the mole fractions in both phases across the interface.

8. Calculate the fraction of the resistance to SO_2 transport in the gas and liquid membrane phases for an SO_2 scrubber operating at 100 °C. The membrane liquid, largely ethylene glycol, is $5 \cdot 10^{-3}$ cm thick. In it, the SO_2 has a diffusion coefficient of about $0.85 \cdot 10^{-5}$ cm^2/sec and a solubility of

$$\frac{0.026 \text{ mol } SO_2/1}{\text{mm Hg of } SO_2}$$

In the stack gas, there is an unstirred film adjacent to the membrane 0.01 cm thick, and the SO_2 has a diffusion coefficient of 0.13 cm^2/sec. (W. J. Ward)

9. Estimate the average mass transfer coefficient for water evaporating from a film falling at 0.82 cm/sec into air. The air is at 25 °C and 2 atm, and the film is

186 cm long. Express your result in cubic centimeters of H_2O vapor at STP/ (hr cm^2 atm).

10. Find the dissolution rate of a cholesterol gallstone 1 cm in diameter immersed in a solution of bile salts. The solubility of cholesterol in this solution is about $3.5 \cdot 10^{-3}$ g/cm^3. The density difference between the bile saturated with cholesterol and that containing no cholesterol is about $3 \cdot 10^{-3}$ g/cm^3; the kinematic viscosity of this solution is about 0.06 cm^2/sec; the diffusion coefficient of cholesterol is $1.8 \cdot 10^{-6}$ cm^2/sec. *Answer: 0.2 g per month.*

11. Air at 100 °C and 2 atm is passed through a bed 1 cm in diameter composed of iodine spheres 0.07 cm in diameter. The air flows at a rate of 2 cm/sec, based on the empty cross-section of bed. The area per volume of the spheres is 80 cm^2/cm^3, and the vapor pressure of the iodine is 45 mm Hg. How much iodine will evaporate from a bed 13 cm long, assuming a bed porosity of 40%? *Answer: 5.6 g/hr.*

12. The largest liquid–liquid extraction process is probably the dewaxing of lubricants. After they are separated by distillation, crude lubricant stocks still contain significant quantities of wax. In the past, these waxes were precipitated by cooling and separated by filtration; now, they are extracted with mixed organic solvents. For example, one such process uses a mixture of propane and cresylic acid. You are evaluating a new mixed solvent for dewaxing that has physical processes like those of catechol. You are using a model lubricant with properties characteristic of hydrocarbons. Waxes are 26.3 times more soluble in the extracting solvent than they are in the lubricant. You know from pilot-plant studies that the mass transfer coefficient based on a lubricant-side driving force is

$$K_L a = 16,200 \text{ lb/ft}^3 \text{ hr}$$

What will it be (per second) if the driving force is changed to that on the solvent side?

13. You need to estimate an overall mass transfer coefficient for solute adsorption from an aqueous solution of density 1.3 g/cm^3 into hydrogel beads 0.03 cm in diameter. The coefficient sought K_y is defined by

$$N_1 = K_y(y - y^*)$$

where N_1 has the units of g/cm^2 sec, and the y's have units of solute mass fraction in the water. The mass transfer coefficient k_S in the solution is 10^{-3} cm/sec; that within the beads is given by

$$k_B = \frac{6D}{d}$$

where d is the particle diameter and D is the diffusion coefficient, equal here to $3 \cdot 10^{-6}$ cm^2/sec. Because the beads are of hydrogel, the partition coefficient is one. Estimate K_y in the units given.

14. A horizontal pipe 10 inches in diameter is covered with an inch of insulation that is 36% voids. The insulation has been soaked with water. The pipe is now drying slowly and hence almost isothermally in 80 °F air that has a relative humidity of about 55%. Estimate how long it will take the pipe to dry, assuming that capillarity always brings any liquid water to the pipe's surface. (H. A. Beesley)

Further Reading

Bird, R. B., Stewart, W. E., and Lightfoot, E. N. (2002). *Transport Phenomena*, 2nd ed. New York: Wiley.

McCabe, W. L., Smith, J. C., and Harriot, P. (2005). *Unit Operations of Chemical Engineering*, 7th ed. New York: McGraw-Hill.

Mills, A. F. (2001). *Mass Transfer*. Englewood Cliffs: Prentice Hall.

Seader, J. D. and Henley, E. J. (2006). *Separation Process Principles*, 2nd ed. New York: Wiley.

CHAPTER 9

Theories of Mass Transfer

In this chapter we want to connect mass transfer coefficients, diffusion coefficients, and fluid flow. In seeking these connections, we are combining the previous chapter, which deals with mass transfer, with the first two sections of the book, which dealt with diffusion.

To find these connections, we will develop theories of mass transfer. These theories are rarely predictive, but they clarify the chemistry and physics which are involved. They are less predictive because they are most often for fluid–fluid interfaces whose geometry is not well known. They are much more successful for solid–fluid interfaces, which are much better defined. Unfortunately, fluid–fluid interfaces are much more important for mass transfer than fluid–solid interfaces are.

Before reviewing the common theories, we should identify exactly what we want to predict. Almost always, we want to predict the mass transfer coefficient k as a function of the diffusion coefficient D and the fluid velocity v. In many cases, convection will be forced, i.e., the velocity will be caused by mechanical forces like pressure drop imposed from outside the system. In occasional cases, convection will be free, the consequence of gravity driven flows often caused by the mass transfer itself. While we will discuss both cases, we will stress forced convection because it is more important and more common in chemical processing.

We can infer what we most want to predict by looking at the mass transfer correlations for fluid–fluid interfaces given in the previous chapter. The most important are those for gas treating, both by absorption and by stripping. Those correlations are in Table 8.3-2. For liquids, k most commonly varies with D to the 0.5 power. For gases, k varies with D to the (2/3) power. For liquids, k varies with v to the 0.45, 0.67, and 0.70 power. For gases, k varies with v to the 0.50 to 0.8 power, most often around the (2/3) power. Thus we would be pleased with a theory which predicted that

$$k \propto v^{2/3} D^{1/2} \qquad (9.0\text{-}1)$$

We will seek such a prediction in this chapter.

We can also seek predictions based on solid–fluid interfaces, although these have less practical value. We will also consider mass transfer within a short tube, which is important for artificial kidneys and blood oxygenators; and mass transfer on a flat plate, which is important because of strong parallels with heat transfer. However, while these solid–fluid interfaces are often detailed pedagogically, they are less significant than the results for fluid–fluid interfaces.

The sections of this chapter are different attempts to give a prediction of Equation 9.0-1. In Sections 9.1 and 9.2, we discuss the film theory, based on diffusion across a thin film; and the penetration and surface-renewal theories, based on diffusion into a semi-infinite slab. In Section 9.3, we discuss why these theories do not predict Equation 9.0-1, and how this disagreement may be resolved. In Section 9.4, we talk about mass transfer in a short tube and from a flat plate. In Section 9.5, we discuss mass transfer with diffusion-generated convection. The result is an overview of theories of mass transfer.

Before we begin this exploration, we should sound a caution. Mass transfer theories are rarely, if ever, a substitute for experiment. I know that many are tempted to use them as such; and I recognize that they are often helpful for inter- polation and extrapolation. However, don't let professors' pride in their calculations obscure the fact that these theories make poor predictions. Their value is not as an alternative to experiment; their value is an increased appreciation for what is happening.

9.1 The Film Theory

The simplest theory for interfacial mass transfer, shown schematically in Fig. 9.1-1, assumes that a stagnant film exists near every interface. This film, also called an unstirred layer, is almost always hypothetical, for fluid motions commonly occur right up to even a solid interface. Nonetheless, such a hypothetical film, suggested first by Nernst in 1904, gives the simplest model of the interfacial region.

We now imagine that a solute present at high dilution is slowly diffusing across this film. The restriction to high dilution allows us to neglect the diffusion-induced convec- tion perpendicular to the interface, so we can use the simple results in Chapter 2, rather than the more complex ones in Chapter 3. The steady-state flux across this thin film can be written in terms of the mass transfer coefficient:

$$N_1 = k(c_{1i} - c_1) \tag{9.1-1}$$

in which N_1 is the flux relative to the interface, k is the mass transfer coefficient, and c_{1i} and c_1 are the interfacial and bulk concentrations in the fluid to the right of the interface.

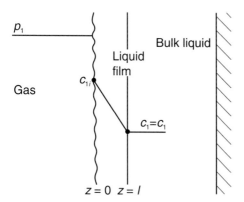

Fig. 9.1-1. The film theory for mass transfer. In this model, the interfacial region is idealized as a hypothetical film or "unstirred layer." Mass transfer involves diffusion across this thin film. Note that the constant value c_{10} implies no resistance to mass transfer in the gas.

The flux across this film can also be calculated in terms of the diffusion coefficient (see Section 2.2):

$$N_1 = n_1|_{z=0} \doteq j_1|_{z=0} = \frac{D}{l}(c_{1i} - c_1) \qquad (9.1\text{-}2)$$

The approximation that the total flux n_1 equals the diffusion flux j_1 reflects the assumption that the solution is dilute. If we compare Eq. 9.1-1 and Eq. 9.1-2, we see that

$$k = \frac{D}{l} \qquad (9.1\text{-}3)$$

This result can be dignified as

$$\begin{pmatrix} \text{Sherwood} \\ \text{number} \end{pmatrix} = \frac{kl}{D} = 1 \qquad (9.1\text{-}4)$$

Such dignity seems silly now but will be useful later.

This simplest theory says that the mass transfer coefficient k is proportional to the diffusion coefficient D and independent of the fluid velocity v. Doubling diffusion doubles mass transfer; doubling flow has no effect. This is not at all what we set out to predict, given in Equation 9.0-1. Of course, the variation of k with v has been lumped into the unknown film thickness l. This thickness is almost never known a priori, but must be found from measurements of k and D. But if we cannot predict k from the film theory, what value has this theory?

The film theory is valuable for two reasons. First, it provides simple physical insight into mass transfer, for it shows in very simple terms how resistance to mass transfer might occur near an interface. Second, it often accurately predicts changes in mass transfer caused by other factors, like chemical reaction or concentrated solution. As a result, the film theory is the picture around which most people assemble their ideas. In fact, we have already implicitly used it in the correlations of mass transfer coefficients in Chapter 8. These correlations are almost always written in terms of the Sherwood number:

$$\begin{pmatrix} \text{Sherwood} \\ \text{number} \end{pmatrix} = \frac{\begin{pmatrix} \text{mass transfer} \\ \text{coefficient} \end{pmatrix} \begin{pmatrix} \text{a characteristic} \\ \text{length} \end{pmatrix}}{\begin{pmatrix} \text{diffusion} \\ \text{coefficient} \end{pmatrix}}$$

$$= F\begin{pmatrix} \text{other system} \\ \text{variables} \end{pmatrix} \qquad (9.1\text{-}5)$$

By using a characteristic length, we imply a form equivalent to Eq. 9.1-3:

$$\begin{pmatrix} \text{mass transfer} \\ \text{coefficient} \end{pmatrix} = \frac{(\text{diffusion coefficient})}{(\text{characteristic length})} \begin{pmatrix} \text{some correction} \\ \text{factor} \end{pmatrix} \qquad (9.1\text{-}6)$$

In some theories, we predict a mass transfer coefficient divided by the fluid velocity, so that a Stanton number seems to be the natural variable. Still, we religiously rewrite our results in terms of a Sherwood number, genuflecting toward the film theory.

Example 9.1-1: Finding the film thickness Carbon dioxide is being scrubbed out of a gas using water flowing through a packed bed of 1-cm Berl saddles. The carbon dioxide is absorbed at a rate of $2.3 \cdot 10^{-6}$ mol/cm^2 sec. The carbon dioxide is present at a partial pressure of 10 atm, the Henry's law coefficient H is 600 atms, and the diffusion coefficient of carbon dioxide in water is $1.9 \cdot 10^{-5}$ cm^2/sec. Find the film thickness.

     Solution   We first find the interfacial concentration of carbon dioxide:

$$p_1 = Hx_1 = H\left(\frac{c_{1i}}{c}\right)$$

$$10 \text{ atm} = 600 \text{ atm} \left(\frac{c_{1i}}{\text{mol} / \left(18 \text{ cm}^3\right)}\right)$$

Thus c_{1i} equals $9.3 \cdot 10^{-4}$ mol/cm^3. We use Eq. 9.1-1 to find the mass transfer coefficient:

$$N_1 = k(c_{1i} - c_1)$$

$$2.3 \cdot 10^{-6} \text{ mol/cm}^2 \text{ sec} = k[(9.3 \cdot 10^{-4} \text{ mol/cm}^3) - 0]$$

As a result, k is $2.5 \cdot 10^{-3}$ cm/sec. Finally, we use Eq. 9.1-3 to find the film thickness:

$$l = \frac{D}{k} = \frac{1.9 \cdot 10^{-5} \text{ cm}^2/\text{sec}}{2.5 \cdot 10^{-3} \text{ cm/sec}} = 0.0076 \text{ cm}$$

Values around 10^{-2} cm are typical of many mass transfer processes in liquids.

9.2 Penetration and Surface-Renewal Theories

     These theories provide a better physical picture of mass transfer than the film theory in return for a modest increase in mathematics. The net gain in understanding is often worth the price. Moreover, although the physical picture is still limited, similar equations can be derived from other, more realistic physical pictures.

9.2.1 Penetration Theory

     The model basic to this theory, suggested by Higbie in 1935, is shown schematically in Fig. 9.2-1. As before, we define the mass transfer coefficient into this film as

$$N_1 = k(c_{1i} - c_1) \tag{9.2-1}$$

where N_1 is the flux across the interface; c_{1i} is the interfacial solute concentration in the liquid in equilibrium with the well-stirred gas; and c_1 is the bulk solute concentration, far into the liquid. The diffusion flux shown in Fig. 9.2-1 can be calculated using the arguments in Section 2.5. The key assumption is that the falling film is thick. Other important assumptions are that in the z direction, diffusion is much more important than convection, and in the x direction, diffusion is much less important than convection.

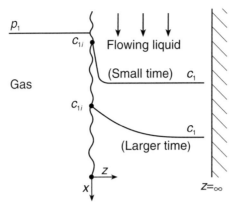

Fig. 9.2-1. The penetration theory for mass transfer. Here, the interfacial region is imagined to be a very thick film continuously generated by flow. Mass transfer now involves diffusion into this film. In this and other theories, the interfacial concentration in the liquid is assumed to be in equilibrium with that in the gas.

These assumptions lead to an equation for the interfacial flux:

$$N_1 = n_1|_{z=0} \doteq j_1|_{z=0} = \sqrt{Dv_{max}/\pi x}(c_{1i} - c_1) \tag{9.2-2}$$

where v_{max} is the interfacial velocity of the water. One should remember that here N_1 is the flux at the interface and that the flux will have smaller values within the fluid. Moreover, it is a point value for some specific x. The interfacial flux averaged over x is given by

$$N_1 = \frac{1}{WL} \int_0^L \int_0^W n_1|_{z=0} \, dy \, dx \tag{9.2-3}$$

where W and L are the width and length of the exposed film shown in Fig. 9.2-1. Combining this with the foregoing, we see that the average flux is

$$N_1 = 2\sqrt{Dv_{max}/\pi L}(c_{1i} - c_1) \tag{9.2-4}$$

Comparing this with Eq. 9.2-1, we see that the mass transfer coefficient is

$$k = 2\sqrt{Dv_{max}/\pi L} \tag{9.2-5}$$

The quantity L/v_{max}, sometimes called the contact time, will not be known a priori in complicated situations, just as the film thickness l was unknown in film theory.

Equation 9.2-5 is often written in terms of dimensionless groups using the fact that the average velocity in the film v^0 is two-thirds the maximum velocity v_{max}. The result (where v is the kinematic viscosity of the fluid) is

$$\frac{kL}{D} = \left(\frac{6}{\pi}\right)^{1/2} \left(\frac{Lv^0}{D}\right)^{1/2} = \left(\frac{6}{\pi}\right)^{1/2} \left(\frac{Lv^0}{v}\right)^{1/2} \left(\frac{v}{D}\right)^{1/2} \tag{9.2-6}$$

or

$$
\begin{pmatrix} \text{Sherwood} \\ \text{number} \end{pmatrix} = \left(\frac{6}{\pi}\right)^{1/2} \begin{pmatrix} \text{Péclet} \\ \text{number} \end{pmatrix}^{1/2}
$$
$$
= \left(\frac{6}{\pi}\right)^{1/2} \begin{pmatrix} \text{Reynolds} \\ \text{number} \end{pmatrix}^{1/2} \begin{pmatrix} \text{Schmidt} \\ \text{number} \end{pmatrix}^{1/2} \tag{9.2-7}
$$

The use of the Sherwood number suggests a film theory, even though nothing like a film exists here.

We should expect that the penetration theory result in Equation 9.2-5 and the film theory result in Equation 9.1-3 should bracket all observed mass transfer conditions. After all, the penetration theory is based on transfer into a semi-infinite fluid, and the film theory is based on transfer across a thin film. All of nature should fall between these two limits of semi-infinite slab and thin film.

However, in practice, our success is limited. As we explained at the start of the chapter, we expect that the mass transfer coefficient k should vary with the square root of the diffusion coefficient D (cf. Equation 9.0-1). This is consistent with the penetration theory. We also expect that k should vary with the two-thirds power of the fluid velocity v. This is larger than that expected by both film and penetration theories. This shortcoming will be explored more in Section 9.3.

9.2.2 Surface-Renewal Theory

Before we explore why experiments show a larger velocity variation than these theories predict, we want to discuss the physical picture of the penetration theory in more depth. This theory does successfully predict the square root dependence of k on D, but on the basis of the naïve and unrealistic physical picture in Figure 9.2-1. As a result, we can sensibly ask if there isn't a better physical picture which gives the same square root dependence.

Such a picture, suggested by Dankwerts in 1951, is the surface-renewal theory. The model used for the surface-renewal theory is shown schematically in Fig. 9.2-2. The specific geometry used in the film and penetration theories is replaced with the vaguer picture of two regions. In one "interfacial" region, mass transfer occurs by means of the penetration theory. However, small volumes or elements of this interfacial region are not static, but are constantly exchanged with new elements from a second "bulk" region. This idea of replacement or "surface renewal" makes the penetration theory a part of a more believable process.

The mathematical description of this surface renewal depends on the length of time that small fluid elements spend in the interfacial region. The concept suggests the definition

$$
E(t)dt = \begin{pmatrix} \text{the probability that a given} \\ \text{surface element will be at the surface} \\ \text{for time } t \end{pmatrix} \tag{9.2-8}
$$

The quantity $E(t)$ is the residence-time distribution used so often in the description of the chemical kinetics of stirred reactors. Obviously, the sum of these probabilities is unity:

$$
\int_0^\infty E(t)dt = 1 \tag{9.2-9}
$$

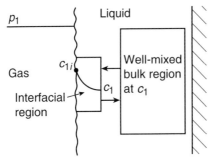

Fig. 9.2-2. The surface-renewal theory for mass transfer. This approach tries to apply the mathematics of the penetration theory to a more plausible physical picture. The liquid is pictured as two regions, a large well-mixed bulk and an interfacial region that is renewed so fast that it behaves as a thick film. The surface renewal is caused by liquid flow.

We now assume that the transfer of different interfacial elements into the bulk region is random. Stated another way, we assume that the interfacial region is uniformly accessible, so that any surface element is equally likely to be withdrawn. In this case, the fraction of surface elements θ remaining at time t must be

$$\theta = e^{-t/\tau} \tag{9.2-10}$$

in which τ is a characteristic constant. This fraction θ must also be the sum of the probabilities from time t to infinity

$$\theta = \int_t^\infty E(t)dt \tag{9.2-11}$$

Thus the residence time distribution of surface elements is

$$E(t) = \frac{e^{-t/\tau}}{\tau} \tag{9.2-12}$$

The physical significance of τ in these equations is an average residence time for an element in the interfacial region.

Armed with these probabilities, we can average the mass transfer coefficient over time. In the interfacial region, the flux is that for diffusion into an infinite slab:

$$n_1|_{z=0} \doteq \sqrt{D/\pi t}(c_{1i} - c_1) \tag{9.2-13}$$

Of course, the interfacial region is anything but infinite, but when the surface is quickly renewed and τ is small, it momentarily behaves as if it were. The average flux is then

$$N_1 = \int_0^\infty E(t)n_1|_{z=0}dt = \sqrt{D/\tau}(c_{1i} - c_1) \tag{9.2-14}$$

By comparison with Eq. 9.2-1, we see that

$$k = \sqrt{D/\tau} \tag{9.2-15}$$

As in the penetration theory, doubling the diffusion coefficient increases the mass transfer coefficient by $\sqrt{2}$.

At this point, some conclude that the surface-renewal theory is much ado about nothing. It predicts the same variation with diffusion as the penetration theory, and the new residence time τ is as unknown as the film thickness l introduced in the film theory. This new theory may at first seem of small value.

The value of the surface-renewal theory is that the simple math basic to the penetration theory is extended to a more realistic physical situation. Although the result is less exact than we might wish, the surface-renewal theory does suggest reasonable ways to think about mass transfer in complex situations. Such thoughts can lead to more effective correlations and to better models.

Example 9.2-1: Finding the adjustable parameters of the penetration and surface-renewal theories What are the contact time L/v_{max} and the surface residence time τ for the carbon dioxide scrubber described in Example 9.1-1?

Solution In this earlier example, we were given that D was $1.9 \cdot 10^{-5}$ cm^2/sec, and we calculated that k was $2.5 \cdot 10^{-3}$ cm/sec. Thus, from Eq. 9.2-5,

$$2.5 \cdot 10^{-3} \text{ cm/sec} = 2\sqrt{[(1.9 \cdot 10^{-5} \text{ cm}^2/\text{sec})/\pi](v_{max}/L)}$$

$$\frac{L}{v_{max}} = 3.9 \text{ sec}$$

Similarly, from Eq. 9.2-15,

$$2.5 \cdot 10^{-3} \text{ cm/sec} = \sqrt{(1.9 \cdot 10^{-5} \text{ cm}^2/\text{sec})/\tau}$$

$$\tau = 3.0 \text{ sec}$$

Deciding whether these values are physically realistic requires additional information.

9.3 Why Theories Fail

The mass transfer theories developed in the previous sections of this chapter are not especially successful. To be sure, the penetration and surface-renewal theories do predict that mass transfer does vary with the square root of the diffusion coefficient, consistent with many correlations. However, neither the film theory nor the surface-renewal theory predicts how mass transfer varies with flow. The penetration theory predicts variation with the square root of flow, less than that indicated by most correlations. This failure to predict the variation of mass transfer with flow is especially disquieting: the film and penetration theories should bracket all behavior because a thin film and a semi-infinite slab bracket all possible geometries.

Something is seriously wrong. To see what, imagine that we are removing traces of ammonia from a gas with aqueous sulfuric acid. The absorption takes place in a packed tower where the gas flows upwards and the acid trickles countercurrently downwards.

In our case, the absorption is effectively irreversible and controlled by diffusion in the gas phase. We can describe this absorption by a mass balance on the ammonia in a differential volume of gas in the tower

$$\text{Accumulation} = (\text{ammonia flow in} - \text{out}) - (\text{ammonia absorbed})$$

$$= -v\frac{dc_1}{dz} - ka(c_1 - 0) \tag{9.3-1}$$

where a is the interfacial area per tower volume. The concentration difference is $(c_1 - 0)$ because absorption by the acid is irreversible, i.e., c_1^* is zero. Note also that k is the coefficient in the gas, not an overall coefficient. This is because reaction with acid so accelerates ammonia uptake within the liquid that the liquid resistance is small. This mass balance is subject to the boundary condition where the gas enters the tower

$$z = 0, \quad c_1 = c_{10} \tag{9.3-2}$$

Integration leads to the familiar result

$$\frac{c_1}{c_{10}} = e^{-kaz/v} \tag{9.3-3}$$

This ammonia concentration drops exponentially as we move along the tower.

If we were trying to measure the mass transfer coefficients in this tower, we would use a rearranged form of this result.

$$k = \left(\frac{v}{az}\right) \ln \frac{c_{10}}{c_1} \tag{9.3-4}$$

We would then measure the concentration c_1 at different velocities and calculate the mass transfer coefficients. Measurements like these are the basis of the correlations given in Tables 8.3-2 and 8.3-3. At the same time, these measurements are those giving values of k which vary too strongly with velocity.

The reason why, suggested by Schlunder in 1977, is that the gas flow in the packed tower is not as uniform as we are implicitly assuming when we rattle off the analysis above. To illustrate the consequences of an uneven velocity, imagine that the gas flow consists of the two parts suggested in Fig. 9.3-1. Part of this flow – a fraction $(1 - \theta)$ – is so intimately contacted with the acid that all ammonia is removed. The rest of flow – the remaining fraction θ – channels past the acid so quickly that no ammonia is removed. Then the exiting concentration is

$$c_1 = [1 - \theta] \, 0 + [\theta] \, c_{10} \tag{9.3-5}$$

If we insert this result into Equation (9.3-4), we find

$$k = \left(\frac{v}{az}\right) \ln \left(\theta^{-1}\right) \tag{9.3-6}$$

If the fraction θ stays constant with changes in flow, and if we stupidly believe that flow is uniform, then we will find that k varies linearly with v. Many of the correlations in

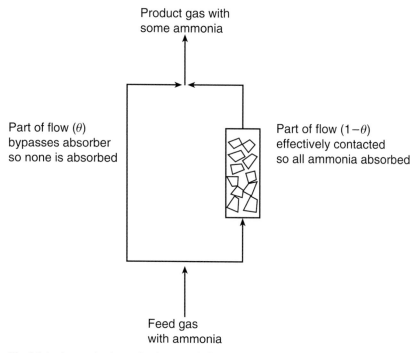

Fig. 9.3-1. Ammonia absorption in a poorly functioning column. The ammonia which bypasses gives a misleading mass transfer correlation.

Chapter 8 predict that k does vary with v more strongly than expected, and hence are presumably compromised by effects like those Schlunder postulates. This suggests we should always view correlations skeptically, as the following example suggests.

Example 9.3-1: Apparent mass transfer coefficients caused by bypassing. This example illustrates how uneven flow can give apparent variations of mass transfer coefficients with flow which are larger than those expected from the theories described earlier. These flow variations are close to those reported in popular correlations.

(a) Imagine we have a system in which the mass transfer coefficient k is 10^{-3} cm/sec, the velocity v is 1 cm/sec, and the product al is 10^{3}. However, half the flow in this system bypasses the region where mass transfer occurs. What will the apparent mass transfer coefficient be?

Solution From the relations above, we expect that

$$\frac{c_l}{c_{10}} = 0 + (1 - 0)e^{-kal/v}$$

$$= 0.5 + (1 - 0.5)\exp^{-10^{-3}\,\text{cm/sec} \times (1000/(1\,\text{cm/sec}))}$$

$$= 0.684$$

Then from Equation 9.3-6.

$$(\text{apparent } k) = \frac{1 \text{ cm/sec}}{1000} \ln\left(\frac{1}{0.684}\right)$$

$$= 0.38 \cdot 10^{-3} \text{ cm/sec}$$

The apparent mass transfer coefficient is 60 percent less than the true value in that part of the system where mass transfer actually occurs.

(b) Now imagine that we increase the flow three times. The true mass transfer coefficient increases according to the penetration theory, that is, with the square root of velocity or by a factor of $\sqrt{3}$. However, if we don't realize that half the flow is bypassing the mass transfer region, we will calculate a new apparent mass transfer coefficient. If we use this new coefficient and that of part (a), how will we conclude that mass transfer varies with velocity?

Solution Parallel to part (a), we find

$$\frac{c_1}{c_{10}} = 0.5 + (1 - 0.5)\exp\left(-\sqrt{3} \cdot 10^{-3}\text{cm/sec } (1000)\right)/(3 \text{ cm/sec})$$

$$= 0.781$$

Thus the apparent coefficient is

$$(\text{apparent } k) = \frac{3 \text{ cm/sec}}{1000} \ln\left(\frac{1}{0.781}\right)$$

$$= 0.74 \cdot 10^{-3} \text{ cm/sec}$$

From these values of the apparent k, we find

$$\text{apparent } k \propto v^{0.65}$$

This apparent velocity variation is larger than the actual square root dependence. It is a consequence of uneven flow causing bypassing in the experiments.

9.4 Theories for Solid–Fluid Interfaces

While mass transfer most commonly takes place across fluid–fluid interfaces, many well-developed theories are based on fluid–solid interfaces. There are two reasons for this. First, fluid–solid interfaces are more easily specified: they normally do not wave or ripple but sit right where they were. Second, fluid–solid interfaces are common for heat transfer, so results calculated and verified for heat transfer can be confidently converted to mass transfer.

In this section we develop two such theories. The first is for mass transfer out of a solution in laminar flow in a short tube. This "Graetz–Nusselt problem" finds some application for blood oxygenators and artificial kidneys. The second example is the "boundary layer theory" for flow across a dissolving, sharp-edged plate. This example has almost no value for mass transfer, but is an analogue of an important heat transfer problem. In sketching the development of this theory, I have abridged the mathematical development more than in other parts of the book. This is because both theories are tangential to mass transfer and discussed in detail in heat transfer references.

9.4.1 The Graetz–Nusselt Problem

This theory calculates how the mass transfer coefficient varies with the fluid's flow and the solute's diffusion. In other words, it finds the mass transfer as a function of quantities like Reynolds and Schmidt numbers.

The problem, shown schematically in Fig. 9.4-1, again assumes a dilute solute, so that the velocity profile is parabolic, as expected for laminar flow. The detailed solution depends on the exact boundary conditions involved. The most important case assumes fixed solute concentration at the wall of a short tube. In this case, a mass balance on the solute gives

$$\begin{pmatrix} \text{solute} \\ \text{accumulation} \end{pmatrix} = \begin{pmatrix} \text{solute in minus} \\ \text{solute out} \\ \text{by diffusion} \end{pmatrix} + \begin{pmatrix} \text{solute in minus} \\ \text{solute out} \\ \text{by convection} \end{pmatrix} \tag{9.4-1}$$

or, in symbolic terms,

$$0 = D \left(\frac{1}{r} \frac{\partial}{\partial r} r \frac{\partial c_1}{\partial r} + \frac{\partial^2 c_1}{\partial z^2} \right) - 2v^0 \left(1 - \left(\frac{r}{R} \right)^2 \right) \frac{\partial c_1}{\partial z} \tag{9.4-2}$$

where r is the radial position and R is the tube radius. This equation can be found either by a mass balance on the washer-shaped region $(2\pi r \Delta r \Delta z)$ or by the appropriate simplification of the general mass balance given in Table 3.4-2.

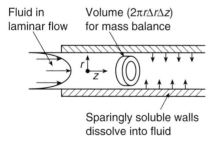

Fluid in   Volume $(2\pi r \Delta r \Delta z)$
laminar flow   for mass balance

Sparingly soluble walls
dissolve into fluid

Fig. 9.4-1. The Graetz–Nusselt problem. In this case, a pure solvent flowing laminarly in a cylindrical tube suddenly enters a section where the tube's walls are dissolving. The problem is to calculate the wall's dissolution rate and hence the mass transfer coefficient. The problem's solutions, based on analogies with heat transfer, are useful for designing artificial kidneys and blood oxygenators.

For short tubes, solute diffusion occurs mainly near the wall and the bulk of the fluid near the tube's axis is pure solvent. As a result, axial diffusion is small and $\partial^2 c_1/\partial z^2$ can be neglected. We then define a new coordinate:

$$s = R - r \tag{9.4-3}$$

Rewriting the mass balance in terms of this variable,

$$0 = D \frac{\partial^2 c_1}{\partial s^2} - \frac{4v^0 s}{R} \frac{\partial c_1}{\partial z} \tag{9.4-4}$$

In this, we have used the fact that s/R is much less than unity near the wall, so the wall curvature and the velocity variation can be ignored. The boundary conditions for this differential equation are the following:

$$z = 0, \quad \text{all } s, \quad c_1 = 0 \tag{9.4-5}$$

$$z > 0, \quad s = 0, \quad c_1 = c_{1i} \tag{9.4-6}$$

$$z > 0, \quad s = \infty, \quad c_1 = 0 \tag{9.4-7}$$

The intriguing condition is, of course, the third one, because this implies that the tube be short.

To solve this problem, we define the new dimensionless variable

$$\xi = s \left(\frac{4v^0}{9DRz} \right)^{1/3} \tag{9.4-8}$$

The differential equation now becomes

$$0 = \frac{d^2 c_1}{d\xi^2} + 3\xi^2 \frac{dc_1}{d\xi} \tag{9.4-9}$$

subject to

$$\xi = 0, \quad c_1 = c_{1i} \tag{9.4-10}$$

$$\xi = \infty, \quad c_1 = 0 \tag{9.4-11}$$

Integration and use of the boundary conditions gives

$$c_1 = c_{1i} \frac{\int_\xi^\infty e^{-\xi^3} d\xi}{\Gamma\left(\frac{4}{3}\right)} \tag{9.4-12}$$

The numerator in this expression is the incomplete gamma function. We now know the concentration profile.

To find the mass transfer coefficient k, we again compare the definition

$$N_1 = k(c_{1i} - 0) \tag{9.4-13}$$

with the value found from Fick's law:

$$N_1 = n_1|_{r=R} \doteq -D\frac{\partial c_1}{\partial r}\Big|_{r=R} = D\frac{\partial c_1}{\partial s}\Big|_{s=0}$$

$$= D\left(\frac{4v^0}{9DRz}\right)^{1/3}\frac{\partial c_1}{\partial \xi}\Big|_{\xi=0} = D\frac{\left(\dfrac{4v^0}{9DRz}\right)^{1/3}}{\Gamma\!\left(\dfrac{4}{3}\right)}(c_{1i}-0) \qquad (9.4\text{-}14)$$

Comparison gives

$$k = \frac{D}{\Gamma\!\left(\frac{4}{3}\right)}\left(\frac{4v^0}{9DRz}\right)^{1/3} \qquad (9.4\text{-}15)$$

an important limit due to Lévique.

As in the penetration theory, this k is a local value located at fixed z. If we average this coefficient over a pipe length L, we find, after rearrangement, the average mass transfer coefficient k_{avg}

$$\frac{k_{avg}d}{D} = \frac{3^{1/3}}{\Gamma\!\left(\frac{4}{3}\right)}\left(\frac{d^2 v^0}{DL}\right)^{1/3}$$

$$= \frac{3^{1/3}}{\Gamma\!\left(\frac{4}{3}\right)}\left(\frac{dv^0}{v}\right)^{1/3}\left(\frac{v}{D}\right)^{1/3}\left(\frac{d}{L}\right)^{1/3} \qquad (9.4\text{-}16)$$

where $d\,(=2R)$ is the diameter. Eq. 9.4-16 may be written as

$$\left(\begin{array}{c}\text{Sherwood}\\ \text{number}\end{array}\right) = 1.62\left(\begin{array}{c}\text{Reynolds}\\ \text{number}\end{array}\right)^{1/3}\left(\begin{array}{c}\text{Schmidt}\\ \text{number}\end{array}\right)^{1/3}\left(\begin{array}{c}\text{diameter}\\ \text{length}\end{array}\right)^{1/3} \qquad (9.4\text{-}17)$$

Mass transfer experiments give a numerical coefficient of 1.64, close to the predicted value.

Example 9.4-1: Mass transfer of benzoic acid   Water is flowing at 6.1 cm/sec through a pipe 2.3 cm in diameter. The walls of a 14-cm section of this pipe are made of benzoic acid, whose diffusion coefficient in water is $1.00 \cdot 10^{-5}$ cm^2/sec. Find the average mass transfer coefficient k_{avg} over this section.

Solution   From Equation 9.4-16, we find

$$k_{avg} = \frac{3^{1/3}}{\Gamma\!\left(\frac{4}{3}\right)}\left(\frac{D}{d}\right)\left(\frac{d^2 v^0}{DL}\right)^{1/3}$$

$$= 1.62\left(\frac{1\cdot 10^{-5}\,\text{cm}^2\text{/sec}}{2.3\,\text{cm}}\right)\left(\frac{(2.3\,\text{cm})^2 6.1\,\text{cm/sec}}{1\cdot 10^{-5}\,\text{cm}^2\text{/sec})\,14\,\text{cm}}\right)^{1/3}$$

$$= 4.3\cdot 10^{-4}\,\text{cm/sec}$$

This will be correct if the flow is laminar and the tube is short. The Reynolds number is

$$\frac{d v^0 \rho}{\mu} = \frac{(2.3\,\text{cm})(6.1\,\text{cm/sec})(1\,\text{g/cm}^3)}{0.01\,\text{g/cm sec}} = 1,400$$

That is less than 2,100, the transition to turbulent flow. If the pipe is short

$$s \ll R$$

(i.e., diffusion never penetrates very far from the wall). By dimensional arguments or by analogies with Chapter 2, we guess a characteristic value of s:

$$s = \sqrt{4DL/v^0} \ll R$$

$$= \sqrt{4(1.00 \cdot 10^{-5}\text{cm}^2\text{/sec})(14\,\text{cm})/(6.1\,\text{cm/sec})} = 0.01\,\text{cm} \ll 1.15\,\text{cm}$$

Thus the benzoic acid dissolution does take place in a short pipe, where short is defined in terms of diffusion.

9.4.2 Mass Transfer From a Plate

We now turn to a second example of mass transfer at solid–fluid interfaces: that of dissolution of a sharp-edged plate. In the Graetz–Nusselt example given directly above, we assumed the parabolic velocity profile was already established, and we then calculated the concentration profile. Here, we need to calculate both the velocity and the concentration profiles.

In this case, we imagine that a plate, made of a sparingly soluble solute, is immersed in a rapidly flowing solvent. We want to find the rate at which solute dissolves. In more scientific terms, we want to calculate how the Sherwood number varies with the Reynolds and Schmidt numbers.

This physical situation is shown in the oft-quoted schematic in Fig. 9.4-2(a). The flow over the top of the plate is disrupted by the drag caused by the plate; this region of disruption, called a boundary layer, becomes larger as the flow proceeds down the plate. The boundary layer is usually defined as the locus of distances over which 99% of the disruptive effect occurs. Such specificity is obviously arbitrary. While this flow pattern develops, the sparingly soluble solute dissolves off the plate. This dissolution produces the concentration profiles shown in Figure 9.4-2(b). The distance that the solute penetrates defines a new concentration boundary layer, but this layer is not the same as that observed for flow.

Our theoretical development involves the calculation of these two boundary layers. We first calculate that for flow. We begin by assuming a velocity profile for the fluid flowing parallel to the flat plate:

$$v_x = a_0 + a_1 y + a_2 y^2 + a_3 y^3 + \cdots \tag{9.4-18}$$

(a) Fluid drag on a flat
    plate generates a
    "boundary layer"

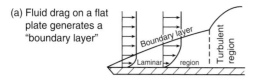

(b) Slow dissolution of the
    plate also affects a
    thinner region near
    the plate

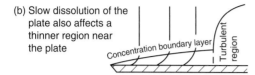

(c) To calculate the effect
    of flow on dissolution,
    use the control volume
    shown

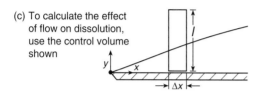

Fig. 9.4-2. The boundary layer theory for mass transfer. In this theory, both the flow and the diffusion are analyzed for specific geometries like that of a flat plate. The results are accurate for the specific case, but are purchased with considerable mathematical effort.

where y is the distance normal to the plate and the a_i are constants. These constants are independent of y, but may vary with x. The velocity described by this equation is subject to four boundary conditions:

$$y = 0, \quad v_x = 0 \tag{9.4-19}$$

$$\frac{\partial^2 v_x}{\partial y^2} = 0 \tag{9.4-20}$$

$$y = \infty, \quad v_x = v^0 \tag{9.4-21}$$

$$\frac{\partial v_x}{\partial y} = 0 \tag{9.4-22}$$

The first of these conditions says that the fluid sticks to the plate, and the second says that because the plate is solid, the stress on it is a constant. The other two conditions say that far from the plate, the plate has no effect.

As an approximation, we can replace Eqs. 9.4-21 and 9.4-22 with

$$y = \delta, \quad v_x = v^0 \tag{9.4-23}$$

$$\frac{\partial v_x}{\partial y} = 0 \tag{9.4-24}$$

Now we can find the constants in Eq. 9.4-18: a_0 is zero by Eq. 9.4-19; a_2 is zero by Eq. 9.4-20, and so

$$\frac{v_x}{v^0} = \frac{3}{2}\left(\frac{y}{\delta}\right) - \frac{1}{2}\left(\frac{y}{\delta}\right)^3 \tag{9.4-25}$$

At this point, I always feel that I have gotten something for nothing, until I remember that δ is an unknown function of x. Moreover, we should remember that Eq. 9.4-18 was arbitrarily truncated and that more accurate calculations might require more terms.

We now turn to the overall mass and momentum balances. Because the solute is sparingly soluble; its dissolution has a negligible effect on the overall mass balance. Because the process is in steady state, the accumulation is zero. Thus the mass balance is

$$0 = \left(\int_0^l W\rho v_x dy\right)_x - \left(\int_0^l W\rho v_x dy\right)_{x+\Delta x} - \rho v_y W\Delta x \tag{9.4-26}$$

in which W and l are the width and length of the plate. Dividing by $W\Delta x$, taking the limit as this area goes to zero, and assuming ρ is constant gives

$$v_y = -\frac{d}{dx}\int_0^l v_x dy \tag{9.4-27}$$

This overall mass balance can be supplemented by an x-momentum balance which in symbolic terms becomes

$$0 = \left[\left(\int_0^l W\rho v_x v_x dy\right)_x - \left(\int_0^l W\rho v_x v_x dy\right)_{x+\Delta x}\right]$$
$$+ (0 - \rho v^0 v_y W\Delta x) + (\tau_0 W\Delta x) \tag{9.4-28}$$

Dividing by $W\Delta x$, and taking the limit as this area goes to zero, and using Equation 9.4-27 to eliminate v_y, we find

$$-\tau_0 = \frac{d}{dx}\int_0^l (v^0 - v_x)\rho v_x dy \tag{9.4-29}$$

When y is greater than the boundary layer δ, v^0 equals v_x. Moreover, the stress τ_0 is given by Newton's law of viscosity. Thus

$$\mu \frac{\partial v_x}{\partial y}\bigg|_{y=0} = \frac{d}{dx}\int_0^\delta (v^0 - v_x)\rho v_x dy \tag{9.4-30}$$

The left-hand side of this equation comes from the shear force; the right-hand side represents momentum convection, rewritten with the help of the overall mass balance.

We now combine Eqs. 9.4-25 and 9.4-30 to find a differential equation for the boundary layer:

$$\delta \frac{\partial \delta}{\partial x} = \frac{140}{13} \left(\frac{\mu}{\rho v^0} \right) \tag{9.4-31}$$

This is subject to the boundary condition

$$x = 0, \quad \delta = 0 \tag{9.4-32}$$

Integration is straightforward:

$$\frac{\delta}{x} = \left(\frac{280}{13} \right)^{1/2} \left(\frac{x v^0 \rho}{\mu} \right)^{-1/2} \tag{9.4-33}$$

Because we know δ, we now know v_x from Eq. 9.4-25. Thus we have solved the fluid mechanics part of this example.

Next, we must calculate the smaller boundary layer for the concentration. We begin by assuming this boundary layer is given by:

$$c_1 = a_0 + a_1 y + a_2 y^2 + a_3 y^3 + \cdots \tag{9.4-34}$$

We know this function must be subject to the boundary conditions

$$y = 0, \quad c_1 = c_{1i} \tag{9.4-35}$$

$$\frac{\partial^2 c_1}{\partial y^2} = 0 \tag{9.4-36}$$

$$y = \infty, \quad c_1 = 0 \tag{9.4-37}$$

$$\frac{\partial c_1}{\partial y} = 0 \tag{9.4-38}$$

The first two conditions indicate that both concentration and flux are constant at the plate's surface. The last two conditions apply deep into the fluid, where the plate has no effect; they can be replaced by

$$y = \delta_c, \quad c_1 = 0 \tag{9.4-39}$$

$$\frac{\partial c_1}{\partial y} = 0 \tag{9.4-40}$$

where δ_c is the concentration boundary layer. Note that this distance is less than the boundary layer δ, caused by fluid drag. Using these conditions, we quickly discover

$$\frac{c_1}{c_{1i}} = 1 - \frac{3}{2}\left(\frac{y}{\delta_c}\right) + \frac{1}{2}\left(\frac{y}{\delta_c}\right)^3 \tag{9.4-41}$$

This type of manipulation should seem familiar, for it is completely parallel to that used to find the flow boundary layer. As before, we make a mass balance on the solute in a control volume $lW\Delta x$

$$0 = \left[\left(\int_0^l Wc_1v_xdy\right)_x - \left(\int_0^l Wc_1v_xdy\right)_{x+\Delta x}\right] + (N_1W\Delta x - 0) \tag{9.4-42}$$

Dividing by the area $W\Delta x$ and taking the limit as this area becomes small, we find

$$-N_1 = -n_1|_{y=0} = -\frac{d}{dx}\int_0^l c_1v_xdy \tag{9.4-43}$$

This equation is simplified in two ways: by replacing n_1 with Fick's law and by remembering that c_1 is zero from δ_c to l. The result is

$$D\frac{\partial c_1}{\partial y}\bigg|_{y=0} = -\frac{d}{dx}\int_0^{\delta_c} c_1v_xdy \tag{9.4-44}$$

We combine Eqs. 9.4-25, 9.4-33, 9.4-41, and 9.4-44 to find a differential equation for the concentration boundary layer. In this combination we assume that δ_c is smaller than δ, an approximation whose chief justification must be experimental. The result can be written as

$$\frac{4}{3}x\frac{d}{dx}\left(\frac{\delta_c}{\delta}\right)^3 + \left(\frac{\delta_c}{\delta}\right)^3 = \left(\frac{D\rho}{\mu}\right) \tag{9.4-45}$$

If δ_c is smaller, it must develop more slowly than δ, and the boundary on this equation is

$$x = 0, \quad \frac{\delta_c}{\delta} = 0 \tag{9.4-46}$$

Integration leads to:

$$\left(\frac{\delta_c}{\delta}\right)^3 = \left(\frac{D\rho}{\mu}\right) \tag{9.4-47}$$

Combining this with Eq. 9.4-33 gives δ_c as a function of position.

The final manipulation in this analysis is the reformulation of the boundary layer as the mass transfer coefficient. By definition,

$$N_1 = k(c_{1i} - 0) \tag{9.4-48}$$

and by Fick's law,

$$N_1 = n_1|_{y=0} \doteq -D\frac{\partial c_1}{\partial y}\Big|_{y=0} \tag{9.4-49}$$

We have the concentration profile in terms of δ_c in Eq. 9.4-41, so

$$N_1 = \frac{3Dc_{1i}}{2\delta_c} \tag{9.4-50}$$

Note again how this result parallels the film theory, with $2\delta_c/3$ equivalent to the film thickness. Combining this with Eq. 9.4-33 and 9.4-47, we find

$$N_1 = 0.323\frac{Dc_{1i}}{x}\left(\frac{xv^0\rho}{\mu}\right)^{1/2}\left(\frac{\mu}{\rho D}\right)^{1/3} \tag{9.4-51}$$

Comparing this with Eq. 9.4-48,

$$\frac{kx}{D} = 0.323\left(\frac{xv^0\rho}{\mu}\right)^{1/2}\left(\frac{\mu}{\rho D}\right)^{1/3} \tag{9.4-52}$$

We can average this result at a particular x over a length L to find

$$\frac{k_{\text{avg}}L}{D} = 0.646\left(\frac{Lv^0\rho}{\mu}\right)^{1/2}\left(\frac{\mu}{\rho D}\right)^{1/3} \tag{9.4-53}$$

The prediction agrees with experiments for a flat plate when the boundary layer is laminar, which occurs when the Reynolds number ($xv^0\rho/\mu$) is less than 300,000.

Example 9.4-2: Calculation of mass transfer coefficients from boundary layer theory
Water flows at 10 cm/sec over a sharp-edged plate of benzoic acid. The dissolution of benzoic acid is diffusion-controlled, with a diffusion coefficient of $1.00 \cdot 10^{-5}$ cm^2/sec. Find (a) the distance at which the laminar boundary layer ends, (b) the thickness of the flow and concentration boundary layers at that point, and (c) the local mass transfer coefficients at the leading edge and at the position of transition.
    Solution (a) The length before turbulence begins can be found from

$$\frac{xv^0\rho}{\mu} = \frac{x(10\,\text{cm/sec})(1\,\text{g/cm}^3)}{0.01\,\text{g/cm sec}} = 300{,}000$$

Thus the transition occurs when x is 300 cm.
    (b) The boundary layer for flow can be found from Eq. 9.4-33:

$$\delta = \left(\frac{280}{13}\right)^{1/2}x\left(\frac{\mu}{xv^0\rho}\right)^{1/2}$$

$$= \left(\frac{280}{13}\right)^{1/2}(300\,\text{cm})\left(\frac{0.01\,\text{g/cm sec}}{(300\,\text{cm})(10\,\text{cm/sec})(1\,\text{g/cm}^3)}\right)^{1/2}$$

$$= 2.5\,\text{cm}$$

The boundary layer for concentration is given by Eq. 9.4-47:

$$\delta_c = \left(\frac{D\rho}{\mu}\right)^{1/3}\delta$$

$$= \left(\frac{(1.00 \cdot 10^{-5}\text{cm}^2\text{/sec})(1\,\text{g/cm}^3)}{0.01\,\text{g/cm sec}}\right)^{1/3}(2.5\,\text{cm}) = 0.25\,\text{cm}$$

The concentration boundary layer is thinner, just as was assumed in the derivation given earlier.

(c) The local mass transfer coefficients can be found from Eqs. 9.4-52. At the sharp edge of the plate, the local mass transfer coefficient is infinity. Where the transition to turbulent flow occurs, the value is given by

$$k = 0.323\frac{D}{x}\left(\frac{xv^0\rho}{\mu}\right)^{1/2}\left(\frac{\mu}{\rho D}\right)^{1/3} = 0.323\left(\frac{1.00 \cdot 10^{-5}\text{cm}^2\text{/sec}}{300\,\text{cm}}\right)$$

$$\cdot (300{,}000)^{1/2}\left(\frac{0.01\,\text{g/cm sec}}{(1\,\text{g/cm}^3)(1.00 \cdot 10^{-5}\text{cm}^2\text{/sec})}\right)^{1/3}$$

$$= 5.9 \cdot 10^{-5}\text{cm/sec}$$

Again, this result is for a solid–water interface.

9.5 Theories for Concentrated Solutions

By this time, we should recognize one omnipresent assumption in all of the foregoing theories: the assumption of a dilute solution. Restricting our arguments to dilute solution allows a focus on diffusion and a neglect of the convection that diffusion itself can generate. In terms of this book, the restriction to dilute solution uses the simple ideas in Chapter 2, not the more complex concepts in Chapter 3.

The restriction to dilute solution is less serious than it might first seem. While correlations of mass transfer coefficients like those in Chapter 8 are often based on dilute solution experiments, these correlations can often be successfully used in concentrated solutions as well. For example, in distillation, the concentrations at the vapor–liquid interface may be large, but the large flux of the more volatile component into the vapor will almost exactly equal the large flux of the less volatile component out of the vapor. There is a lot of mass transfer, but not much diffusion-induced convection. Thus constant molar overflow in distillation implies a small volume average velocity normal to the interface, and mass transfer correlations based on dilute solution measurements should still work for these concentrated solutions.

In a few cases, however, these simple ideas of mass transfer fail. This failure is most commonly noticed as a mass transfer coefficient k that depends on the driving force. In other words, we define as before

$$N_1 = k\Delta c_1 = K(c_{1i} - c_1) \tag{9.5-1}$$

where c_{1i} is the concentration at the interface and c_1 is that in the bulk. As expected, we find that k is a function of Reynolds and Schmidt numbers. However, when mass transfer is fast, k may also be a function of Δc_1. Stated another way, if we double Δc_1, we may not double N_1 even when the Reynolds and Schmidt numbers are constant.

These shortcomings lead to alternative definitions of mass transfer coefficients that include the effects of diffusion-induced convection. One such definition is

$$N_1 = k(c_{1i} - c_1) + c_{1i}v^0$$

$$= k(c_{1i} - c_1) + c_{1i}(\bar{V}_1 N_1 + \bar{V}_2 N_2) \tag{9.5-2}$$

in which k is now the mass transfer coefficient for rapid mass transfer, $\bar{V}_i$ is the partial molar volume of species i, and v^0 is the velocity at the interface. Note that the convective term is defined in terms of the interfacial concentration c_{1i}, not in terms of some average value.

We want to calculate this new coefficient, just as we calculated the dilute coefficient in earlier sections of this chapter. In general, we might expect to repeat the whole chapter, producing an entirely new series of equations for the film, penetration, and surface-renewal, and boundary-layer theories. However, these calculations not only would be difficult but also would retain the unknown parameters like film thickness and contact time.

To avoid these unknowns, we adopt a new strategy, one that we shall use more later, especially in the study of coupled mass transfer and chemical reaction. We assume that from experiments we know the mass transfer coefficient in dilute solution. We then calculate the corrections caused by fast mass transfer in concentrated solution, as predicted by the film theory. In other words, we find the ratio of the mass transfer coefficient in concentrated solution to that in dilute solution. The ratio found from the film theory turns out to be close to that found from other theories. Thus the film theory gives a reasonable estimate of the changes engendered by diffusion-induced convection.

To make this calculation, imagine a thin film like that shown schematically in Fig. 9.5-1. A mass balance on a thin shell Δz thick shows that the total flux is a constant:

$$0 = -\frac{dn_1}{dz} \tag{9.5-3}$$

Integrating and combining with Fick's law gives

$$n_1 = -D\frac{dc_1}{dz} + c_1 v^0 = -D\frac{dc_1}{dz} + c_1(\bar{V}_1 n_1 + \bar{V}_2 n_2) \tag{9.5-4}$$

The first term on the right-hand side is the flux due to diffusion, and the second term is the flux due to diffusion-induced convection. This equation is subject to the boundary conditions

$$z = 0, \quad c_1 = c_{1i} \tag{9.5-5}$$

$$z = l, \quad c_1 = c_1 \tag{9.5-6}$$

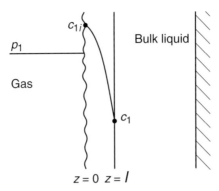

Fig. 9.5-1. The film theory for fast mass transfer. When mass transfer is rapid, the formulations of mass transfer given in earlier parts of this chapter can break down. This is because the diffusion process itself can generate convection normal to the interface. As a result, the simple concentration profile shown in Fig. 9.1-1 for the film theory becomes more complicated. Still, correction factors for fast mass transfer based on this simple theory turn out to be reasonably accurate.

In dilute solution, c_1 is small, and the diffusion-induced convection is negligible, so Eq. 9.5-4 is easily integrated to give Eq. 9.1-2. As a result, the mass transfer coefficient in dilute solution k^0 is D/l, as stated in Eq. 9.1-3.

In concentrated solutions, c_1 is large and no easy simplifications are possible. However, because n_1 and n_2 are constants, v^0 is as well, and Eq. 9.5-4 can be integrated to give

$$\frac{c_1 - n_1/v^0}{c_{1i} - n_1/v^0} = e^{v^0 l/D}$$ (9.5-7)

This equation can be rearranged:

$$N_1 = n_1|_{z=0} = \left(\frac{v^0}{e^{v^0 l/D} - 1}\right)(c_{1i} - c_1) + c_{1i}v^0$$ (9.5-8)

If we compare this result with Eq. 9.5-1, we find

$$k = \frac{v^0}{e^{v^0 l/D} - 1}$$ (9.5-9)

We then can eliminate the unknown l by using the dilute-solution coefficient k^0 given by Eq. 9.1-3:

$$k^0 = \frac{D}{l}$$ (9.5-10)

Combining this with Eq. 9.5-9, we find

$$\frac{k}{k^0} = \frac{v^0/k^0}{e^{v^0/k^0} - 1}$$ (9.5-11)

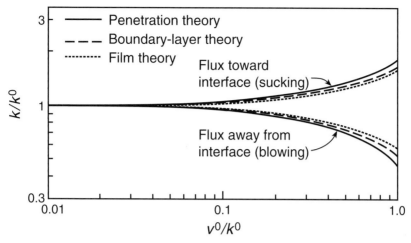

Fig. 9.5-2. Correction factors for rapid mass transfer. This figure gives the mass transfer coefficient k as a function of the interfacial convection v^0. In dilute solution, v^0 is small and k approaches the slow mass transfer limit k^0. In concentrated solution, k may reach a new value, although estimates of this value from different theories are about the same. (The boundary layer theory shown is for a Schmidt number of 1,000.)

or, in terms of a power series,

$$k = k^0 \left(1 - \frac{v^0}{2k^0} + \frac{(v^0)^2}{12(k^0)^2} - \cdots \right) \tag{9.5-12}$$

Note that the mass transfer coefficient k can be either increased or decreased in concentrated solution. If a large convective flux blows from the interface into the bulk, then v^0 is positive and k is less than k^0. If a large flux sucks into the interface from the bulk, then v^0 is negative and k is greater than k^0. These changes are made clearer by the following example.

Before proceeding to this example, we compare these film results with those found from other mass transfer theories (Bird *et al.*, 2002) in Fig. 9.5-2. All theories give similar results. To be sure, more complex theories provide greater detail in the form of a Schmidt-number dependence, but this is rarely a major factor. The corrections given by the film theory are sufficient in many cases.

Example 9.5-1: Fast benzene evaporation Benzene is evaporating from a flat porous plate into pure flowing air. Using the film theory, estimate how much a concentrated solution increases the mass transfer rate beyond that expected for a simple theory. Then calculate the resulting change in the mass transfer coefficient defined by Eq. 9.5-2. In other words, find $N_1/k^0 c_{1i}$ and k/k^0 as a function of the vapor concentration of benzene at the surface of the plate.

  Solution The benzene evaporates off the plate into air flowing parallel to the plate. Thus the air flux n_2 is zero. Moreover, if air and benzene behave as ideal gases, $\bar{V}_1 = \bar{V}_2 = c^{-1}$, and $v^0 = n_1/c$. In addition, $c_1 = 0$. Thus from Eq. 9.5-7,

$$N_1 = n_1|_{z=0} = \frac{Dc}{l} \ln\left(\frac{c}{c - c_{1i}} \right)$$

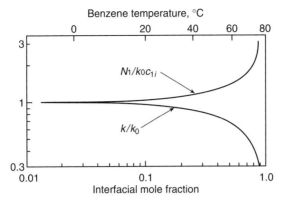

Fig. 9.5-3. Fast benzene evaporation. This figure gives the flux N_1 and mass transfer coefficient k for fast benzene evaporation through stagnant air relative to values expected for slow mass transfer. These estimates are based on the film theory, although other estimates would be similar.

This is equivalent to the result in Equation 3.3-11. In dilute solution, the flux will be Dc_{1i}/l, and k^0 will be D/l; thus

$$\frac{N_1}{k^0 c_{1i}} = -\frac{1}{x_{1i}} \ln(1 - x_{1i})$$

Values for this flux are given in Fig. 9.5-3. The values of mass transfer coefficient can be found by combining these results with Eq. 9.5-11:

$$\frac{k}{k^0} = \frac{N_1/ck^0}{e^{N_1/ck^0} - 1} = -\left(\frac{1 - x_{1i}}{x_{1i}}\right) \ln(1 - x_{1i})$$

Note that k/k^0 is less than unity. Values of this ratio are also given in Fig. 9.5-3.

9.6 Conclusions

By this time, the catalog of theories given in this chapter may cloud perspective. We may understand each step and each equation but still be confused about the arguments used. We can gain insight by stepping back, reviewing our goals, and seeing what we have accomplished.

Our goal was to predict the experimentally based correlations of mass transfer coefficients. This prediction would then clarify the basis of mass transfer, giving us a physical picture relating mass transfer and diffusion coefficients. Such a picture could potentially show us how to make mass transfer faster and more efficient.

The different theoretical efforts to achieve this goal are summarized in Table 9.6-1. These efforts predict mass transfer coefficients vary with diffusion coefficients to powers ranging from 0.0 to 1.0, and clustering around 0.5. This is close to the average of the various correlations. This implies that the physical picture of the penetration and surface-renewal theories is superior to that of the film theory.

The theoretical efforts in Table 9.6-1 are much less reassuring when applied to the variation of the mass transer coefficient with velocity. Theories predict that mass transfer

Table 9.6-1 *Theories for mass transfer coefficients*

Method	Interface	Basic form	$f(\text{flow})$	$f(D)$	Advantages	Disadvantages
Film theory	Fluid–Fluid	$k = \dfrac{D}{l}$	$v^{0.0}$	$D^{1.0}$	Simple; often good base for extensions	Film thickness l is unknown
Penetration theory	Fluid–Fluid	$k = 2\sqrt{Dv_{max}/\pi L}$	$\dfrac{1}{v^{\frac{1}{2}}}$	$D^{\frac{1}{2}}$	Simplest including flow	Contact time (L/v^0) usually unknown
Surface-renewal theory	Fluid–Fluid	$k = \sqrt{D/\tau}$	$v^{0.0}$	$D^{\frac{1}{2}}$	Similar math to penetration theory, but better physical picture	Surface-renewal rate (τ) is unknown
Bypassing theory	Fluid–Fluid	Eq 9.3-6	$v^{1.0}$	$D^{0.0}$	Bypassing explains some data	Effects are not due to mass transfer
Graetz–Nusselt theory	Solid–Fluid	$k = \left(\dfrac{3^{1/3}}{\Gamma\left(\frac{4}{3}\right)}\right) \dfrac{D}{d} \left(\dfrac{d^2 v^0}{DL}\right)^{1/3}$	$\dfrac{1}{v^{\frac{1}{3}}}$	$D^{\frac{2}{3}}$	Exact result for short tubes	Valid only in laminar flow in tubes
Boundary-layer theory	Solid–Fluid	$k = 0.323 \dfrac{D}{L} \left(\dfrac{Lv^0}{v}\right)^{1/2} \left(\dfrac{v}{D}\right)^{\frac{1}{3}}$	$v^{\frac{1}{2}}$	$D^{\frac{2}{3}}$	Good parallels with heat transfer	Unrealistic picture for mass transfer

varies with the 0 to 0.5 power of velocity. Because these theories range from thin films to semi-infinite slabs, we expect experiments should fall somewhere between these limits. They don't: they show a higher velocity dependence. This is likely the result of uneven flow, and so represents an opportunity for improving equipment. Indeed, more modern tower packing, especially structured packing, does show smaller velocity variations than that in older correlations. We will return to this situation in the discussions of industrial operations in subsequent chapters.

Questions for Discussion

1. Theories of mass transfer try to predict the mass transfer coefficient k from a variety of process variables. What velocity dependence do they predict?
2. What do theories predict for the variation of k with diffusion coefficient D?
3. How does the mass transfer coefficient vary with viscosity?
4. What variation of k with D and velocity v does the film theory predict?
5. Both the penetration theory and the surface-renewal theory presume diffusion into a semi-infinite slab. What is the difference between them?
6. Mass transfer data often show larger variations of k with velocity v than expected. Why?
7. Could the mass transfer coefficient k ever decrease as the velocity increases?
8. When will the Leveque limit of the Graetz–Nusselt problem (Equation 9.4-15) be valid?
9. Will the absorption of NH_3 from a concentrated vapor be affected by diffusion-induced convection?
10. Will the mass transfer in differential distillation be affected by diffusion-induced convection?
11. Does mass transfer of a concentrated solute in a solvent decrease or increase the mass transfer coefficient?
12. If mass transfer theories aren't accurate, what good are they?

Problems

1. As part of a study of O_2 absorption in water in a small packed tower, you find that the outlet concentration of O_2 is $1.1 \cdot 10^{-3}$ M. The partial pressure of O_2 in the tower is about 0.21 atm; the total area in the tower is 1.37 m^2; the liquid flow rate is 1.62 l/min. (The diffusion coefficient of oxygen in water is $1.8 \cdot 10^{-5}$ cm^2/sec.) Find (a) the film or unstirred-layer "thickness," (b) the "contact time," and (c) the "average residence time on the surface." Assume no gas-phase resistance. *Answer:* 0.005 cm, 1.6 sec, 1.2 sec.

2. The mass transfer coefficient in gas–solid fluidized beds has been correlated with the dimensionless expression [J. C. Chu, J. Kalil, and W. A. Wetteroth, *Chem. Engr. Prog.*, **49**, 141 (1953)]

$$\frac{k_p}{G} \left(\frac{\mu}{\rho D} \right)^{2/3} = 1.77 \left(\frac{dG\tilde{M}}{\mu(1-\varepsilon)} \right)^{-0.44}$$

where k_p is the mass transfer coefficient (moles per (length)^2 time pressure), G is total convective flux (moles per (length)^2 time pressure), d is particle diameter, and ε is bed porosity. You are studying 0.1-cm particles of coal burning in a bed with $\varepsilon = 0.42$ fluidized by air at 1,250 °C. The air flux at 1.7 atm is 380 lb/ft^2 hr. What is the film or unstirred-layer thickness around these particles? (H. Beesley) *Answer:* 0.03 cm.

3. Copper is adsorbed from a stream flowing at 5.1 l/hr using a countercurrent flow of 23 lb/hr ion-exchange resin. The bed has a volume of 160 l. The resin area per volume is 40 cm^2/cm^3, and its density is 1.1 g/cm^3. The equilibrium concentration of copper in solution varies with that in the resin as follows:

c(solution), (mol/l)	c(beads), (g Cu/g beads)
0.02	0.011
0.10	0.043
0.25	0.116
2.40	0.200

The resin enters without copper. The copper solution flows in at 0.40 M, but leaves at 0.056 M. What is the mass transfer coefficient and the penetration time into the beads? (R. Contraras) *Answer:* $7 \cdot 10^{-3}$ cm/sec and 0.2 sec.

4. Ether and water are contacted in a small stirred tank. An iodine-like solute is originally present in both phases at $3 \cdot 10^{-3}$ M. However, it is 700 times more soluble in ether. Diffusion coefficients in both phases are around 10^{-5} cm^2/sec. Resistance to mass transfer in the ether is across a 10^{-2}-cm film; resistance to mass transfer in the water involves a surface renewal time of 10 sec. What is the solute concentration in the ether after 20 minutes? *Answer:* $5 \cdot 10^{-3}$ mol/l.

5. One handbook gives the following correlation for mass transfer into a flow fluid from a single solid sphere:

$$\frac{kRTd}{Dp} = 0.276 \left\{ \left(\frac{dv\rho}{\mu} \right)^{1/2} \left(\frac{\mu}{D\rho} \right)^{1/3} \right\}$$

In one set of experiments in air for a napthalene sphere 1.26 cm in diameter, the function of the Reynolds and Schmidt numbers in braces equals 980. The diffusion coefficient of napthalene in air is 0.074 cm^2/sec. What is the film thickness for mass transfer? What is the surface-renewal time?

6. Hikita et al. [*Chemical Engr. Sci.*, **45**, 437–442 (1990)] report Sherwood numbers for a laminar wave-free falling film entering a wetted wall column. Their result is

$$\frac{kl}{D} = \frac{2}{\sqrt{\pi}} \left\{ l \left(\frac{g}{v^2} \right)^{1/3} \right\} \left(\frac{v}{Dt^*} \right)^{1/2}$$

where l is the liquid thickness, g is the acceleration due to gravity, and t^* is a complex dimensionless time. Briefly describe how or if this is consistent with your knowledge of mass transfer.

7. To grow embryos, eggs must breathe, a process limited by oxygen diffusion through pores in the shell. Data from a large number of birds, from warblers to ostriches, suggest that

$$\left(\frac{\text{oxygen uptake,}}{\text{cm}^3/\text{day mm Hg}}\right) = 0.02(\text{egg mass,g})$$

and

$$\left(\text{pore area, mm}^2\right) = 0.04\,(\text{egg mass, g})$$

Pore length varies from 0.1 to 1 mm and is correlated by

$$\left(\text{pore length, mm}\right) = 0.03(\text{egg mass, g})^{1/2}$$

[H. Rahn, A. Ar, and C. V. Paganelli, *Sci. Amer.*, **240**(2), 46 (1979)]. Do these results make sense in terms of the film theory of mass transfer?

8. Develop the intermediate steps from which you can use the film theory for fast mass transfer in concentrated solution. More specifically, (a) integrate Eq. 9.5-4 to find Eq. 9.5-7, (b) rearrange this result to give Eq. 9.5-8, and (c) show that Eq. 9.5-10 can be expanded to give Eq. 9.5-11.

9. A large pancake-shaped drop of water loses 18% of its area in two hours of sitting at 24 °C in air at 10% relative humidity. The drop's thickness is about constant. If the same drop were placed in an 80 °C oven containing air at the same absolute humidity, how long would it take to lose the same fraction of its area? (G. Jerauld).

10. In this book, we routinely assume that there is no slip at a solid–fluid interface (i.e., that at a stationary solid interface, the fluid velocity is zero). This leads to expressions for mass transfer in laminar flow in short tubes like

$$k = \frac{D}{\Gamma(\frac{4}{3})}\left(\frac{4v^0}{9DRz}\right)^{1/3}$$

(see Eq. 9.4-15). However, for some polymer solutions, this assumption is not valid, and the solution will slip along a smooth wall [A. M. Kraynik and W. R. Schowalter, *J. Rheology*, **25**, 95 (1981)]. Find the mass transfer coefficient in the case of total slip, and calculate the ratio of this slippery case and the more common one.

11. Methylene chloride, a solvent used in the photographic and drug industries, is a major pollutant and a potent carcinogen. Accordingly, there is currently a national effort aimed at finding substitutes for this solvent. Failing this, there is interest in designing packed towers using excess dry air to strip this solvent out of water and feed the resulting vapor stream to an incinerator. The problem is that the methylene chloride is to be stripped under conditions where large quantities of water evaporate simulta- neously. In other words, the mass transfer of the methylene chloride may be altered by

the mass transfer of the water. (a) Present equations giving the water flux under these conditions. Use these to calculate the diffusion-induced convection. (b) Show how this convection alters the flux of methylene chloride, even though the methylene chloride is dilute. (c) To make this quantitative, assume the maximum water concentration in the vapor is half the total concentration and that both mass transfer coefficients are equal. What is the change in methylene chloride flux?

Further Reading

Bird, R. B., Stewart, W. E., and Lightfoot, E. N. (2002). *Transport Phenomena*, 2nd ed. New York: Wiley.

Dankwerts, P. V. (1951). *Industrial and Engineering Chemistry*, **43**, 1460.

Graetz, L. (1880). *Zeitschrift für Mathematik und Physik*, **25**, 316, 375.

Higbie, R. (1935). *Transactions of the American Institute of Chemical Engineers*, **31**, 365

Lévêque, M. A. (1928). *Annales des Mines*, **13**, 201, 305, 381.

Levich, V. (1962). *Physiochemical Hydrodynamics* Englewood Cliffs, NJ: Prentice-Hall.

Lewis, W. K. and Chang, K. C. (1928). *Transactions of the American Institute of Chemical Engineers*, **21**, 127.

Nernst, W. (1904). *Zeitschrift für Physikalische Chemie*, **47**, 52.

Nusselt, W. (1909). *Zeitschrift des Vereines der Deutscher Ingenieure*, **53**, 1750; (1910) 54, 1154.

Schlichting, H. Gersten, K., Krause, H., and Mayes, C. (2000). *Boundary Layer Theory*, 8th ed. Berlin: Springer.

Schlunder, E. V. (1977). *Chemical Engineering Science*, **32**, 339.

Skelland, A. H. (1974). *Diffusional Mass Transfer*. New York: Wiley Interscience.

CHAPTER 10

Absorption

The most common use of the mass transfer coefficients developed in Chapter 8 is the analytical description of large-scale separation processes like gas absorption and distillation. These mass transfer coefficients can describe the absorption of a solute vapor like SO_2 or NH_3 from air into water. They describe the distillation of olefins and alkanes, the extraction of waxes from lubricating oils, the leaching of copper from low-grade ores, and the speed of drug release.

Mass transfer coefficients are useful because they describe how fast these separations occur. They thus represent a step beyond thermodynamics, which establishes the maximum separations that are possible. They are a step short of analyses using diffusion coefficients, which have a more exact fundamental basis. Mass transfer coefficients are accurate enough to correlate experimental results from industrial separation equipment, and they provide the basis for designing new equipment.

All industrial processes are affected by mass transfer coefficients but to different degrees. Gas absorption, the focus of this chapter, is an example of what is called "differential contacting" and depends directly on mass transfer coefficients. Many mechanical devices, including blood oxygenators and kidney dialyzers, are analyzed similarly, as discussed in the next chapter. Distillation, the most important separation, is idealized in two ways. In the first, it is treated as "differential contracting" and analyzed in a parallel way to absorption, as described in Chapter 12. In the second idealization, distillation is approximated as a cascade of near equilibrium "stages." Such "staged contacting," is detailed in Chapter 13. The efficiency of these stages, included in that chapter, again depends on mass transfer coefficients. Other separation processes, often treated as either differential or staged contacting, are described in later chapters.

Understanding absorption is the key to all these operations. This understanding is usually clouded by presenting the ideas largely in mathematical terms. All chemistry and all simple limits are implied rather than explained. As a result, novices often understand every step of the analysis but have a poor perspective of the overall problem. To avoid this dislocation, we begin in Section 10.1 with a description of the gases to be absorbed and the liquid solvents that absorb them. A few of these liquids depend only on the solubility of the gas; many more liquids react chemically with the components of the gas.

Once this chemical problem is stated, we must decide two dimensions of the column: how fat and how tall. (I know this sounds like an on-line dating service, but these really are the questions.) To decide how fat a column is needed, we turn in Section 10.2 to the physical equipment used. This physical equipment is simple, but it is constrained by the fluid mechanics of the gas and liquid flowing past each other. These flows are complicated, described by largely empirical correlations. The best strategy may be to follow the turnkey procedure to resolve the fluid mechanics.

Once we know how fat the column should be, we turn to the issue of how tall. In Section 10.3 we begin with the simple case of dilute absorption and in Section 10.4 extend this to the parallel case of concentrated absorption. For the dilute solution case, we assume a linear isotherm, that is, that a solute's solubility in liquid is directly proportional to its partial pressure in the gas. This leads to a simple analytical solution. For the concentrated case, we expect that the isotherm is nonlinear and we recognize that the liquid and gas flows change within the equipment. Now the solution requires numerical integration. These two sections mirror our earlier discussion of diffusion, where the simple case of dilute diffusion in Chapter 2 gave way to the concentrated and more general results in Chapter 3. Understanding the dilute case is the key to the concentrated case.

10.1 The Basic Problem

When you drive by a chemical plant at night, the most impressive part is the lights. They outline every piece of equipment. When you look more closely, you can often see three types of silhouettes. The tallest are the thin distillation columns, which are described in Chapter 12 and 13. The next tallest are the fat gas absorption columns, which are the subject of this chapter. Ironically, the shortest silhouettes are the chemical reactors, charged with reagents to make the desired products.

This relative size has a moral: while the chemical plant would not exist without the chemical reactors, the biggest expense – the biggest equipment – will often be in the separation equipment. This separation equipment centers on distillation and gas absorption, the two most important unit operations. The analysis and the design of these operations is central to the entire chemical industry.

I have found that distillation is better understood than gas absorption. I believe that this is because everyone knows that distillation is how you concentrate ethanol from water: Distillation is how you turn wine into brandy. In contrast, few know what gas absorption is for. What specific gases are absorbed, anyway? What liquids absorb the gases? What happens to the liquids afterwards? I find this ignorance ironic because of increased environmental concerns. Gas absorption is the chief method for controlling industrial air pollution, yet many with environmental interests remain ignorant of its nature.

In this section, I want to begin to remove this ignorance. I want to list the gases that we most often seek to remove and to give rough limits for the inlet and exit concentrations. I want to explain where these gas mixtures occur. I will do so qualitatively, without equations; there are more than enough equations in later sections. I do want to make one point now about cost. In these systems, the cost of absorption is usually log linear. It costs twice as much to remove 99% as it does to remove 90%, and it costs three times as much to remove 99.9% as it does to remove 90%. This increasing cost should be a key in environmental legislation.

10.1.1 Which Gases are Absorbed

Most gas absorption aims at separation of acidic impurities from mixed gas streams. These acidic impurities include carbon dioxide (CO_2), hydrogen sulfide (H_2S), sulfur dioxide (SO_2), and organic sulfur compounds. The most important of these are CO_2 and H_2S, which occur at concentrations of five to fifty percent. The organic sulfur

Table 10.1-1 *Gas treating in major industrial processes*

Process	Gases to be removed	Common targets (% Acid gas)
Ammonia manufacture	CO_2, NH_3, H_2S	< 10 ppm CO_2
Coal gas:		
High Btu gas	CO_2, H_2S, COS	500 ppm CO_2; 0.01 ppm H_2S
Low Btu gas	H_2S	100 ppm H_2S
Ethylene manufacture	H_2S, CO_2	< 1 ppm H_2S, 1 ppm CO_2
Flue gas desulfurization	SO_2	90% removal
Hydrogen manufacture	CO_2	<0.1% CO_2
Natural gas upgrading	H_2S, CO_2, N_2, RSH	< 4 ppm H_2S; < 1% CO_2
Oil desulfurization	H_2S	100 ppm H_2S
Refinery desulfurization	CO_2, H_2S, COS	10 ppm H_2S
Syn gas for chemicals feedstock	CO_2, H_2S	< 500 ppm CO_2; < 0.01 ppm H_2S

compounds include carbonyl sulfide (COS) and merceptans, which are like alcohols with a sulfur atom in place of the oxygen. Merceptans stink: For example, butyl merceptan is responsible for the stench of skunks.

Other impurities vary widely. One common impurity is water, which can be removed by either absorption or adsorption. Another is ammonia (NH_3), which is basic, rather than acidic. Sulfur trioxide (SO_3), prussic acid (HCN), and nitrogen oxides (NO_x) are of concern because of their high chemical reactivity. Oxygen must be removed from some reagent streams, and nitrogen can be absorbed to upgrade natural gas.

The occurrence of some of these streams and the targets for their removal are summarized in Table 10.1-1. The ubiquitous presence of H_2S reflects the fact that fossil fuels, especially coal and petroleum, contain large amounts of sulfur. Moreover, as the world becomes more industrialized, the targets will decrease. This is particularly true for SO_2 in flue gas, which is the source of acid rain.

10.1.2 Liquids Used as Absorbents

The choice of a liquid absorbent depends on the concentrations in the feed gas mixture and on the percent removal desired. If the impurity concentration in the feed gas is high, perhaps ten to fifty percent, we can often dissolve most of the impurity in a nonvolatile, nonreactive liquid. Such a nonreactive liquid is called a physical solvent. If the impurity concentration is lower, around one to ten percent, we will tend to use a liquid capable of fast, reversible chemical reaction with the gas to be removed. Such a reversibly reactive liquid is referred to as a "chemical solvent." If the concentration of the gas to be removed is lower still, we may be forced to use an adsorbent that reacts irreversibly, an expensive alternative that may produce solid waste.

These generalizations may be clearer if we consider the case of H_2S. If we have a concentrated feed stream, we can dissolve the H_2S in liquids like ethylene glycol or propylene carbonate, which are physical solvents. At lower feed concentrations of H_2S, we would commonly use aqueous solutions of alkylamines. One common example is monoethanol amine: $H_2N\,CH_2\,CH_2\,OH$. As you can see, this is like ammonia but with one proton replaced with ethanol. Such species react reversibly with acid gases like H_2S, so their aqueous solutions are chemical solvents. Finally, if the H_2S is present only in traces, we can remove these traces with an aqueous solution of NaOH. This will produce a waste stream of NaHS, which is discarded.

Gas absorption processes produce a liquid containing high concentrations of the impurity. This commonly is removed – stripped – by heating the liquid so that the impurity bubbles out. Often, this removal is accelerated by pumping an inert gas – a sweep stream – through the hot liquid. Recently, chemical companies have been bothered by the high cost of heating the large volumes of absorbing liquids. To avoid these costs, they have begun using absorbents whose chemical reactivity is pressure sensitive. Because swings in pressure can be less expensive than swings in temperature, I expect that the switch to pressure-sensitive absorbents will continue.

Both temperature and pressure swings yield a concentrated impurity requiring disposal. The disposal is again illustrated by the example of H_2S. The H_2S is normally stripped from amines by heating; the concentrated H_2S stream is split into two. The first stream is burned:

$$H_2S + \frac{3}{2} O_2 \rightarrow SO_2 + H_2O$$

This product gas is reacted catalytically with the H_2S remaining in the second stream:

$$SO_2 + 2H_2S \rightarrow 3S + 2H_2O$$

In these two steps the H_2S is converted into solid sulfur which can be sold. This "Claus process" has been the key to sulfur recovery for almost a century.

Thus gas absorption centers on removing an impurity from a gas stream into a liquid. In the rest of this chapter, we will discuss the analysis and design of equipment for this task. Our discussion will be entirely on physical solvents, that is, on nonreactive liquids. Gas absorption in physical solvents is perhaps twenty times less common than absorption in chemical solvents, that is, in reacting liquids. We focus our discussion on physical solvents because they are much simpler; we will explore chemical solvents in Chapter 17.

10.2 Absorption Equipment

Gas absorption at an industrial scale is most commonly practiced in packed towers like that shown in Fig. 10.2-1. A packed tower is essentially a piece of pipe set on its end and filled with inert material or "tower packing." Liquid poured into the top of the tower trickles down through the packing; gas pumped into the bottom of the tower flows countercurrently upward. The intimate contact between gas and liquid achieved in this way effects the gas absorption.

Analyzing a packed tower involves both mass transfer and fluid mechanics. The mass transfer, detailed in Section 10.3, determines the height of the packed tower. In other words, it determines how tall the tower needs to be. This mass transfer is described as molar flows, partly because of the chemical reactions that often occur. The fluid mechanics, described in this section, determines the cross-sectional area of the packed

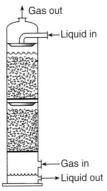

Fig. 10.2-1. A packed tower used for gas absorption. A gas mixture enters the bottom of the tower and flows out the top. Part of this mixture is absorbed by liquid flowing countercurrently, from top to bottom.

tower. The fluid mechanics is described as mass flows, a consequence of the physics that control the process. It controls how fat the tower needs to be. To describe the physics, we discuss the tower packing, the flows themselves, and the estimation of the tower's cross-section.

10.2.1 Tower Packing

The fluid mechanics in the packed tower is dominated by the inert material in the packed tower. This material can be small pieces dumped randomly or larger structures carefully stacked inside the tower. Random packing is cheaper and common; structured packing is more expensive but more efficient. The efficiency is typically improved by around 30%, a significant gain when producing commodity chemicals at low margins.

Typical random packings, shown in Fig. 10.2-2, replace the crushed material used in early chemical processing. The earliest packing, the Raschig ring, was modeled on the necks of broken wine bottles available along the lower Rhine River. Tower packings try to permit both high fluid flow and high interfacial area between the gas and the liquid. These goals are in conflict: High fluid flow implies a few large channels through the tower, and high interfacial area requires many small channels. Thus tower packings are compromises, developed with 80 years of empiricism. The Raschig rings and the Berl saddles are described as first-generation packings, the Intalox saddles and Pall rings are second generation, and the Nutter rings are third generation. All aim at the same goal: fast flow with big area.

Typical structured packings consist of larger assemblies, which often look like louvers and are shown in Fig. 10.2-3. These larger structures are stacked inside the column rather than dumped into it. In some cases, a single large assembly will be lowered into the column as a unit. Structured packing seems to give an interfacial area between gas and liquid which is about the same as that through random packing. However, it gives much more even flows so that both gas and liquid move past each other countercurrently with less bypassing.

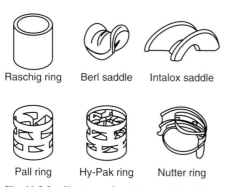

Raschig ring   Berl saddle   Intalox saddle

Pall ring   Hy-Pak ring   Nutter ring

Fig. 10.2-2. Six types of random packing. These packings aim to resolve the conflicting goals of fast flow and large interfacial areas.

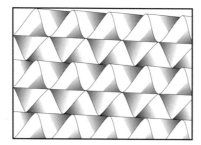

Fig. 10.2-3. Typical structured packing. More expensive packings like this offer excellent mass transfer and greater capacity.

I find that I can appreciate the differences between random and structured packing by looking at boxes of pasta in the supermarket. The Raschig rings are like penne or rigatoni; the Berl saddles are like farfalle and orecchiette; the Pall and Nutter rings are more like rotari and radiatori. All pasta are intended to catch the sauce, to spread it out, and to retard its flow. Random packings try to spread out the liquid, giving a large interfacial area.

Structured packing is like lasagna. Interestingly, a 500 g box of lasagna is smaller than 500 g boxes of the other random types. This is because the spaces between the lasagna noodles are more regular, not random; and so the box is more evenly and efficiently packed. The same regular spacing in structured tower packing is a main cause for its improved absorption efficiency.

10.2.2 Tower Fluid Mechanics

The liquid and gas flows through these tower packings cannot be arbitrarily set but must be adjusted within a narrow, empirically defined range. To see why this range is important, imagine you have a tall glass filled with ice. You blow air into the bottom of the glass through a straw, and you pour cola into the glass at the same time. You watch what happens.

If you pour the cola at a very slow rate, it won't flow evenly downward through the ice but will run down in only a few places. Such a flow, called "channeling," occurs when the gas or liquid flow is much greater at some points than at others. Such channeling is undesirable, for it can substantially reduce interfacial area and hence mass transfer. It is usually minor in crushed solid packing and is minimal in commercially purchased random packing, except at very low liquid flows.

If you now pour the cola faster (remember to keep blowing), you get a case where the cola flows through all the ice more evenly, with your breath bubbling up through it. The conditions where these relatively even flows begin, called "loading," is a requirement for good mass transfer. When loading begins, the flows may slightly decrease, but the dramatic increase in the gas–liquid area means that mass transfer is fast. You almost always want to operate a packed tower in this loaded condition.

However, if you now begin blowing much harder, you will push in so much air that the cola can't flow into the column, but splashes backward, out of the top of the glass. This condition is called "flooding." It not only reduces mass transfer but also decreases the cola that is flowing into the glass.

These same three conditions – channeling, loading, and flooding – can exist inside of any packed tower. You will want to use liquid flows that are high enough to avoid channeling and achieve loading. You will want to use gas flows that are low enough to avoid flooding. But you will also want flows that are large enough for a specific task, for example, large enough to treat 600 m^3/min of flue gas. You must choose the packing and the shape of the packed tower to allow these flows without flooding.

The constraint of flooding is especially important because it governs the maximum flows that can pass through the column. We can see why flooding occurs by considering some special cases. First, imagine that at a given gas flow we increase the liquid flow causing thicker liquid films on the packing and, hence, smaller gaps between the packing. The gas pressure builds so that the liquid may suddenly start to flow backwards, flooding the column. The second special case is the reverse, where at constant liquid flow we increase the gas flow. This increases drag on the liquid, slowing its flow, and increasing the liquid thickness. Again, the liquid plugs the gaps between the packing, and the tower floods.

10.2.3 Tower Cross-Sectional Area

At this point, we should restate our objective. We aim to analyze industrial gas absorption in packed towers. This analysis depends most strongly on mass balances and rate equations given in subsequent sections of this chapter. It also depends on the fluid mechanics within the tower, which is the subject of this section.

In most cases, the absorption process that interests us will have specified flows of gases and liquids. These flows must load but not flood the tower. They normally will cause a pressure drop of around 1.0 in H$_2$O per ft packing. We achieve these conditions by changing the tower's cross-sectional area. This changes the gas and liquid fluxes, that is, the amount of fluid per cross-sectional area per time. By increasing the cross-sectional area at constant flows, we decrease the pressure drop, the fluxes, and the velocities of the gas and liquid flowing past each other.

The correlations often used for estimating tower cross-sections are shown in Fig. 10.2-4. This figure is tricky, a mixture of dimensionless and dimensional quantities. The abscissa, often called the "flow parameter," is dimensionless:

$$\frac{L''}{G''} \sqrt{\frac{\rho_G}{\rho_L}} = \left[\frac{\frac{1}{2}\rho_L v_L^2}{\frac{1}{2}\rho_G v_L^2}\right]^{\frac{1}{2}} \tag{10.2-1}$$

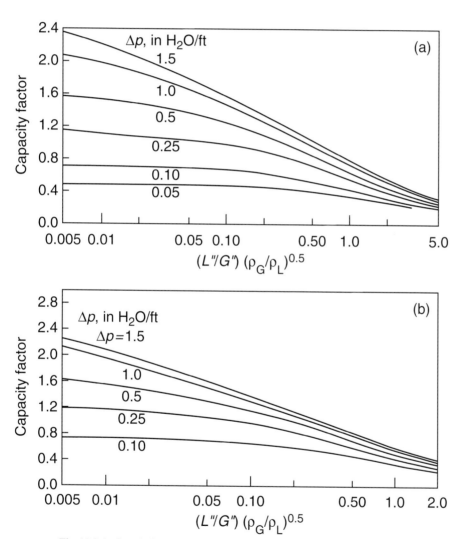

Fig. 10.2-4. Correlation for estimating tower cross-section. The figure plots the capacity factor on the ordinate vs. the flow parameter on the abscissa for random packing (a) and for structured packing (b).

where $L''(= \rho_L v_L)$ and $G''(= \rho_G v_G)$ are the mass fluxes of liquid and gas, respectively; ρ_L and ρ_G are the corresponding densities; and v_L and v_G are the superficial velocities. Two points about this abscissa merit mention. First, while L'' and G'' are mass fluxes and hence depend on the tower's cross-sectional area, their ratio L''/G'' does not depend on this area. Second, the flow parameter is a measure of relative kinetic energy: It is the square root of the ratio of the liquid's kinetic energy to that in the gas.

For the random packing described by Figure 10.2-4(a), the capacity factor is defined as

$$\begin{bmatrix} \text{capacity} \\ \text{factor} \end{bmatrix} = \frac{G''F^{0.5}v^{0.05}}{(\rho_L - \rho_G)^{0.5}} \tag{10.2-2}$$

where G'' is the vapor flux, in $\mathrm{lb/ft^2}$ sec; F is the packing factor, in $\mathrm{ft^{-1}}$; v is the kinematic viscosity, in centistokes, and ρ_L and ρ_G are the liquid and vapor densities, in $\mathrm{lb/ft^3}$. The packing factor F, which is roughly inversely proportional to the packing size, is tabulated for some common packings in Table 10.2-1.

For the structured packing shown in Figure 10.2-4(b), the corresponding definition of the capacity factor is

$$\begin{bmatrix} \text{capacity} \\ \text{factor} \end{bmatrix} = \frac{G''}{(\rho_L - \rho_G)^{0.5}} \tag{10.2-3}$$

where G'', ρ_L and ρ_G have the same definitions and the same units. I cannot stress enough that these definitions and these figures are not dimensionless but must be used with the particular units given. This melange of mixed metric and English units is a historical artifact born in the largely American development of the petrochemical industry.

Table 10.2-1 *Packing factors F and areas per volume (per foot) for random packings*

| | Nominal packing size (in) | | | | | | | |
	$\frac{1}{2}$	$\frac{5}{8}$	$\frac{3}{4}$	1	$1\frac{1}{4}$	$1\frac{1}{2}$	2	3
Raschig rings (ceramic)	580 (111)	380 (100)	255 (80)	179 (58)		93 (38)	65 (28)	37 (19)
Raschig rings (1/32 in metal)	300 (128)	170	155 (84)	115 (63)				
Raschig rings (1/16 in metal)	410 (118)	300	220 (72)	144 (57)		83 (41)	57 (31)	32 (21)
Berl saddles (ceramic)	240 (142)		170 (82)	110 (76)		65 (44)	45 (32)	
Pall rings (metal)		81 (104)		56 (63)		40 (39)	27 (31)	18
Pall rings (plastic)		95 (104)		55 (63)		40 (39)	26 (31)	17 (26)
Intalox saddles (ceramic)	200 (190)		145 (102)	92 (78)		52 (60)	40 (36)	22
Hy-Pak rings (metal)				45 (69)		29 (42)	26 (33)	16 (31)

Note: The areas per volume are given in parentheses. (Abstracted from Strigle, 1987.)

The physical significance of the ordinate is obscure. We recognize that for random packing it is dominated by the ratio

$$\frac{G''F^{0.5}}{\rho_L} \propto \left[\frac{\frac{1}{2}\rho_G v_G^2}{\rho_L g(packing\ size)} \right]^{1/2} \qquad (10.2\text{-}4)$$

Where g is the acceleration due to gravity. This suggests that the ordinate can be regarded as the ratio of the kinetic energy in the gas to the potential energy in the liquid. The other factors, like $v^{0.05}$ in Eq. 10.2-2, are later empiricisms. The history of the correlation is also curious. The original form, due to Sherwood, Shipley, and Holloway, gives only the flooding limit. The curves at constant pressure drop were added by Leva. More recently, many have quarreled over the form, correctly feeling that something with a better-defined physical significance would be preferable. While I agree, I recognize that Fig. 10.2-4 is the starting point for most who are working on absorption design, so I have retained it here. More recent work suggests using a simple expression for random packing at this limit:

$$\Delta p(\text{flooding}) = 0.12\ F^{0.7} \qquad (10.2\text{-}5)$$

where Δp(flooding) is the pressure drop at flooding, again in H_2O/ft packing; and F is the packing factor, in ft^{-1}.

To use Fig. 10.2-4 to find the tower's cross-sectional area, we must first know the gas and liquid flows and hence the flow parameter on the figure's ordinate. After we choose a packing from the myriad available, we want to choose a pressure drop in the tower. In conventional practice, absorbers are designed to operate at pressure drops of 0.2 to 0.6 in H_2O/ft. The lower pressure drop will minimize foaming. Alternatively, we can calculate the column's performance at flooding and arbitrarily choose to operate at a gas flux equal to half the flooding value. In both this method and the previous one, we must make sure to design the tower for the point where the maximum flows of gas and liquid occur. For absorption, this is normally at the tower's bottom; for stripping, it is normally at the top.

Using Fig. 10.2-4 for estimating the tower's cross-section is straightforward though complicated. Remember this figure has two major limitations. First, it implies that at large gas flows, the cross-sectional area should become independent of liquid flux. It is simply flow through a packed bed. Second, this figure is largely based on liquid and gas density differences like those of water and air. These tend to give optimistic predictions for nonaqueous systems (i.e., smaller than optimal tower cross-sections). Thus in non-aqueous systems, like those involving ethylene and propylene, you may need different methods. Again, make early estimates with the methods in this section, and then discuss your case with equipment suppliers.

Example 10.2-1: Estimating a tower cross-section You are planning to reduce the 2% carbon dioxide in 23 lb/sec of a natural gas stream using absorption in aqueous diethyl-amine flowing at 40 lb/sec. You want to use either 1 ½ inch Raschig rings or 1 ½ inch Pall rings. In either case, you want to design for a pressure drop of 0.25 in H_2O/ft, so that foaming is minimized. Under operating conditions, the densities of the gas and the liquid are 2.8 and 63 lb/ft³, respectively; the liquid's viscosity is 2 centistokes.

What should the tower's cross-sectional area be?

Solution This problem illustrates the routine use of Fig. 10.2-4. We first calculate the flow parameter:

$$\frac{L''}{G''}\sqrt{\frac{\rho_G}{\rho_L}} = \frac{40 \text{ lb/sec}}{23 \text{ lb/sec}}\sqrt{\frac{2.8 \text{ lb/ft}^3}{63 \text{ lb/ft}^3}} = 0.37$$

From Fig. 10.2-4, the capacity factor is thus

$$\left[\begin{array}{c}\text{capacity}\\\text{factor}\end{array}\right] = \frac{G'' F^{0.5} \nu^{0.05}}{(\rho_L - \rho_G)^{0.5}}$$

$$0.92 = \frac{G''\left(\dfrac{93}{\text{ft}}\right)^{0.5}(2 \text{ cs})^{0.05}}{\left(63 - 2.8 \text{ lb/ft}^3\right)^{0.5}}$$

$$G'' = 0.72 \text{ lb/ft}^2 \text{ sec}$$

Thus the tower has a cross-section of $(23/0.72) = 32$ ft². This corresponds to a diameter of 6.4 ft². For the Pall rings, the packing factor is 40 ft⁻¹, so the diameter is smaller, about 5.2 ft. However, while there is little difference in the tower diameter for these two packings, there may be a considerable change in the tower height.

10.3 Absorption of a Dilute Vapor

We now return to the analysis of gas absorption in a packed tower. In many cases, we will want to use the analysis to estimate the tower's height. In other cases, we will want to use our analysis to organize experimental results as mass transfer coefficients. In any case, we will build on the fluid mechanics described in the earlier section, a description that allowed estimating the tower's cross-section.

To simplify our analysis, we will begin with the case of a dilute solute vapor absorbed from a gas into a liquid. This focus on the dilute limit makes the physical significance clearest. Because the vapor is dilute, the molar gas flux G and the molar liquid flux L are both constants everywhere within the tower. With this simplification, we then need three key equations:

(1) a solute mole balance on both gas and liquid,
(2) a solute equilibrium between gas and liquid, and
(3) a solute mole balance on either gas or liquid.

These three keys are traditionally called an operating line, an equilibrium line, and a rate equation, respectively.

10.3.1 Analytical Description of Dilute Absorption

We begin with a mole balance on the solute in both gas and liquid. We make this balance on a small tower volume $A\Delta z$ located at position z in the tower:

$$
\begin{pmatrix}
\text{solute entering} \\
\text{minus that leaving} \\
\text{in the gas}
\end{pmatrix}
=
\begin{pmatrix}
\text{solute entering} \\
\text{minus that leaving} \\
\text{in the liquid}
\end{pmatrix}
\tag{10.3-1}
$$

$$
GA(y|_{z+\Delta z} - y|_z) = LA(x|_{z+\Delta z} - x|_z)
$$

where y and x are the mole fractions in the gas and liquid, respectively. When we divide by the volume $A\Delta z$, we find

$$
G\frac{dy}{dz} = L\frac{dx}{dz}
\tag{10.3-2}
$$

Rearranging,

$$
\frac{dy}{dx} = \frac{L}{G}
\tag{10.3-3}
$$

subject to (at $z = 0$),

$$
x = x_0, \quad y = y_0
\tag{10.3-4}
$$

where the subscript 0 indicates the streams at the top of the tower. Remember that the gas is leaving and the liquid is entering at this position. Integrating,

$$
y = y_0 + \frac{L}{G}(x - x_0)
\tag{10.3-5}
$$

This first key equation, which is nothing more than a mole balance, is called the "operating line."

The second key equation for analyzing absorption is an equilibrium relation for the solute in the gas and in the liquid. Because the solute is dilute, this has the form

$$
y^* = mx
\tag{10.3-6}
$$

where m is closely related to a Henry's law constant. This relation, briefly discussed in Section 8.5.3, is a frequent source of error because the units of the concentrations are not carefully considered. Remember also that y^* does not exist at the same tower position as x. In fact, x is the actual liquid mole fraction, y^* is the gas mole fraction which would be in equilibrium with that liquid, and y is the actual gas mole fraction. This second key equation is called the "equilibrium line."

The third key relation, the rate equation, is found by another solute mole balance on the differential volume $A\Delta z$ but on the gas only:

$$
\begin{pmatrix}
\text{solute} \\
\text{accumulation}
\end{pmatrix}
=
\begin{pmatrix}
\text{solute flow} \\
\text{in minus that out}
\end{pmatrix}
-
\begin{pmatrix}
\text{solute lost} \\
\text{by absorption}
\end{pmatrix}
\tag{10.3-7}
$$

In symbolic terms, this can be written as

$$0 = GA(y|_{z+\Delta z} - y|_z) - K_ya(A\Delta z)(y - y^*) \tag{10.3-8}$$

in which a represents the packing area per volume and K_y is the overall gas phase mass transfer coefficient. Values for a are given in parentheses for a variety of common packings in Table 10.2-1. Again, we divide this equation by the volume $A\Delta z$ and take the limit as this volume goes to zero:

$$0 = G\frac{dy}{dz} - K_ya(y - y^*) \tag{10.3-9}$$

This rate equation, a mole balance on the solute in the vapor, is the third key in our analysis.

We now complete our analysis by integrating Eq. 10.3-9. To do so, we first combine it with the equilibrium line in Eq. 10.3-6 and rearrange the result:

$$l = \int_0^l dz = \frac{G}{K_ya}\int_{y_0}^{y_l}\frac{dy}{y - y^*} = \frac{G}{K_ya}\int_{y_0}^{y_l}\frac{dy}{y - mx} \tag{10.3-10}$$

where l is the tower height. We further combine this with the operating line, Eq. 10.3-5:

$$l = \frac{G}{K_ya}\int_{y_0}^{y_l}\frac{dy}{y - m\left[x_0 + \dfrac{G}{L}(y - y_0)\right]} \tag{10.3-11}$$

The important result can be written in a variety of useful forms:

$$l = \frac{G}{K_ya}\left[\left(\frac{1}{1 - \dfrac{mG}{L}}\right)\ln\left(\frac{y_l - y_l^*}{y_0 - y_0^*}\right)\right]$$

$$= \frac{G}{K_ya}\left[\left(\frac{1}{1 - \dfrac{mG}{L}}\right)\ln\left(\frac{y_l - mx_l}{y_0 - mx_0}\right)\right] \tag{10.3-12}$$

Solving for the height l is as easy as plugging in the numbers.

This result merits reflection. First, although the analysis repeatedly exploits the assumption of dilute solution, the extension to concentrated solutions is straightforward. Second, we have implied mass transfer of a solute vapor from a gas into a liquid; such a process is called "gas scrubbing." We can repeat the identical analysis for mass transfer of a vapor from a liquid into a gas; such a reversed process is called "stripping." Third, we have written the preceding equations in terms of gas-phase mole fractions; we could write completely analogous equations for liquid-phase mole fractions:

$$l = \frac{L}{K_xa}\left[\left(\frac{1}{1 - \dfrac{L}{mG}}\right)\ln\left(\frac{x_l - y_l/m}{x_0 - y_0/m}\right)\right] \tag{10.3-13}$$

Note that the overall mass transfer coefficient is different in Eqs. 10.3-12 and 10.3-13. Understanding the difference between the coefficients takes care.

10.3.2 *Alternative Descriptions of Dilute Absorption*

Equations 10.3-12 and 10.3-13 are the basis of most modern analyses of dilute absorption. They are especially appropriate for chemical solvents because reactions of the absorbing solute can be incorporated into the overall mass transfer coefficient. However, the equations above are usually supplemented in two alternative ways.

The first supplement to these equations is to look at the operating line and the equilibrium line in graphical terms, as shown in Figure 10.3-1. The operating line, which is of course the mole balance, is a plot of x vs. y, a straight line with slope (L/G). One end of the line, at the top of the column, is at (x_0, y_0); the other end of the line, at the bottom of the column, is at (x_l, y_l). The equilibrium line, which is an equilibrium energy balance, is a plot y^* vs. x. It starts at the origin and has slope m. At a given value of x, the distance between operating and equilibrium lines $(y - y^*)$ is the driving force for the mass transfer. This type of construction is extremely valuable because it shows visually what the equations describe more abstractly.

Plots of operating and equilibrium lines are especially useful because they illustrate features of the separation which may be harder to see algebraically. For example, imagine that the species being absorbed ionizes when it dissolves in the liquid as, for example, ammonia may do. Then the equilibrium line may be nonlinear, and the distance between operating and equilibrium lines may be much smaller in some ranges. Such a "pinch," which can make the separation more difficult, is easy to see on a plot like that in Fig. 10.3-1.

This type of plot is also useful for anticipating the effect of changing flows in the column. For example, we will often design our process so that the liquid flow is about 1.5 times the minimum liquid flow required. This minimum liquid flow would normally be so slow that it would exit in near equilibrium with the entering gas. Such a flow is

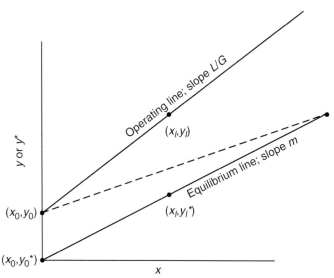

Fig. 10.3-1. Graphical representation of absorption. The vertical difference between the operating and equilibrium line is $(y - y^*)$, the driving force for absorption.

easily seen in Fig. 10.3-1 as the dashed line. Of course, if we did use such a minimum flow, we would need an infinitely tall tower. Such a flow is impractical in practice but often useful as a basis for design. This type of graph is an important supplement to the equations.

The second way to supplement Eqs. 10.3-12 and 10.3-13 breaks them into a column efficiency and a separation difficulty. In particular, Eq. 10.3-12 can be rearranged as

$$l = \text{HTU} \cdot \text{NTU} \tag{10.3-14}$$

where HTU is a height of a transfer unit defined as

$$\text{HTU} = \frac{G}{K_x a} \tag{10.3-15}$$

and NTU is a "number of transfer units" given by

$$\text{NTU} = \frac{1}{1 - \dfrac{mG}{L}} \ln \frac{y_l - mx_l}{y_0 - mx_0} \tag{10.3-16}$$

Other definitions of HTU and NTU can be based on other forms of the overall mass transfer coefficients. The use of "transfer units" is a rough parallel with the use of "stages" in distillation or the term "theoretical plates" in chromatography. As such, it is a historical genuflection by the more recent absorption analyses in the direction of the older equilibrium stage separation analysis.

The use of HTUs and NTUs does have a sound physical interpretation. The NTUs are a measure of the difficulty of the separation, of the distance the final streams will be from equilibrium. If the NTUs are large, the separation is hard. The HTUs, on the other hand, give an idea of the efficiency of the equipment. A small HTU is a sign of a good tower, implying, for example, a large interfacial area per volume or fast mass transfer. Moreover, because the overall mass transfer coefficient often depends on the velocity, the HTU can be largely independent of flow over the practical range: It tends to be between 0.3 m and 1.0 m. Learn to use Eq. 10.3-12 and the idea of an HTU interchangeably.

Example 10.3-1: Carbon dioxide absorption A packed tower uses an organic amine to absorb carbon dioxide. The entering gas, which contains 1.26 mol% CO_2, is to leave with only 0.04 mol% CO_2. The amine enters pure, without CO_2. If the amine left in equilibrium with the entering gas (which it doesn't), it would contain 0.80 mol% CO_2. The gas flow is 2.3 mol/sec, the liquid flow is 4.8 mol/sec, the tower's diameter is 40 cm, and the overall mass transfer coefficient times the area per volume $K_y a$ is $5 \cdot 10^{-5}$ mol/cm^3 sec. How tall should this tower be?

Solution   We first make an overall carbon dioxide balance to find the exiting liquid concentration:

$$GA(y_l - y_0) = LA(x_l - x_0)$$

$$2.3 \text{ mol/sec} (0.0126 - 0.0004) = 4.8 \text{ mol/sec} (x_0 - 0)$$

$$x_0 = 0.00585$$

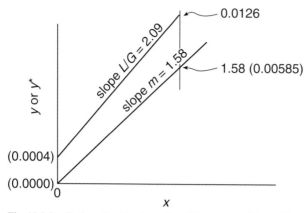

Fig. 10.3-2. Carbon dioxide absorption. The upper and lower lines are again the operating and equilibrium lines.

Next we find the equilibrium constant

$$y_0 = mx_0{}^*$$
$$0.0126 = m(0.0080)$$
$$m = 1.58$$

Now we insert the values given into Eq. 10.3-12:

$$l = \frac{2.3 \text{ mol/sec}}{\left[\frac{\pi}{4}(40 \text{ cm})^2\right] 5 \cdot 10^{-5} \text{ mol/cm}^3\text{sec}} \left[\left(\frac{1}{1 - \frac{1.58\,(2.3)}{4.8}}\right)\right]$$
$$\ln\left(\frac{0.0126 - 1.58\,(.00585)}{0.0004 - 1.58(0)}\right)\right] = 3.2 \text{ m}$$

The meaning of this result is underscored by its graphical representation in Fig. 10.3-2. Note in particular the concentrations at the top and the bottom of the column. Note how the driving force $(y - y^*)$ is smallest at the top, where the CO_2 free solution enters.

Example 10.3-2: Oxygen stripping   You are testing a new packed tower to strip oxygen from water using excess nitrogen. The oxygen-free water is to be used in microelectric manufacture. Your tower is small, about 2 m high and 0.6 m in diameter, filled with 1-in Hy-Pak rings. You expect the value of mG for oxygen is large and the dominant transfer coefficient in the liquid will be $2.2 \cdot 10^{-3}$ cm/sec. The water flow is to be 300 cm^3/sec. How much oxygen can we remove with this tower?

  Solution   To begin, we recognize that because the nitrogen gas flow is in excess, y_0 and y_l are zero, and (L/mG) is much less than one. As a result, the operating line is horizontal, as shown in Figure 10.3-3. Because we are now dealing with stripping, the operating line is below the equilibrium line. This is because the solute oxygen is being transferred from liquid to gas. In the case of gas absorption in Figures 10.3-1 and 10.3-2, the operating line lay above the equilibrium line, showing that solute was transferred from gas to liquid.

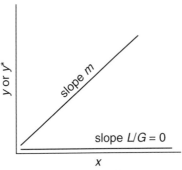

Fig. 10.3-3. Oxygen stripping. Because oxygen is being transferred from liquid to excess gas, the driving force is now $(y^* - y)$, and the operating line is below the equilibrium line.

In this case, Eq. 10.3-13 becomes

$$l = \frac{L}{K_x a}\left[\ln\left(\frac{x_0}{x_l}\right)\right]$$

From Table 8.2-2 and the fact that mass transfer in the liquid is controlling,

$$K_x = k_x = k_L c_L$$

From Table 10.2-1, a is 60 ft^2/ft^3, or 2.26 cm^2/cm^3. Thus

$$200 \text{ cm} = \frac{\left(300 \text{ cm}^3/\text{sec}\right) c_L}{\frac{\pi}{4}(60 \text{ cm})^2 \, 2.2 \cdot 10^{-3} \text{ cm/sec } c_L}\ln\left(\frac{x_0}{x_l}\right)$$

$$\frac{x_l}{x_0} = 8 \cdot 10^{-5}$$

We're removing over 99.9% of the oxygen. An interesting exercise is to check the nitrogen flow implied by this calculation.

Example 10.3-3: Alternative forms of absorption equations Show that Eq. 10.3-12 can be rewritten in the form

$$l = \frac{v_G}{K_G a}\left[\frac{1}{1 - \dfrac{H v_G}{v_L}}\ln\left(\frac{c_{1G,0} - H c_{1L,l}}{c_{1G,0} - H c_{1L,l}}\right)\right]$$

where v_G and v_L are the superficial velocities of gas and liquid, where K_G is defined by

$$N_1 = K_G\left(c_{1G} - c_{1G}^*\right)$$

and where the Henry's law constant H is given by

$$c_{1G}^* = Hc_{1L}$$

Solution We first recognize that

$$G = c_G v_G$$
$$L = c_L v_L$$

where c_G and c_L are the total molar concentrations in gas and liquid, respectively. We then rewrite Eq. 10.3-6 as

$$\frac{c_{1G}^*}{c_G} = m \frac{c_{1L}}{L}$$

so

$$H = \frac{mc_G}{c_L}$$

Finally, from Table 8.2-2,

$$K_y = K_G c_G,$$

Inserting the values of m, K_y, G, and L into Eq. 10.3-12 gives the desired result.

10.4 Absorption of a Concentrated Vapor

In this section, we want to extend the preceding analysis to the case of a concentrated vapor. As before, we plan to accomplish this absorption using a packed tower. As before, we must decide on an appropriate tower packing and on liquid and gas fluxes that will avoid flooding. As before, we depend on a variety of mole balances, though now for concentrated solutions.

Before we develop these new mass balances, we can benefit by looking at our analysis for a dilute vapor in a somewhat different way. This analysis depended on three key equations. The first key equation, the operating line, came from a mole balance for solute in both gas and liquid. The second key equation, the equilibrium line, gave the concentration which would exist if the gas were in equilibrium with the liquid. The third key equation involved the rate of mass transfer and was a mass balance written on only one phase, which in our case was the gas.

We could represent the first and second key equations graphically, as shown in Fig. 10.4-1. The operating line in this figure plots y vs. x; and the equilibrium line plots y^* vs. x. Thus we can read off values of the driving force $(y - y^*)$ and integrate the rate equation. Of course, in the dilute case, all the equations are linear so integration is easily analytical.

For concentrated absorption, the analysis depends on the same three key equations. However, because a lot of solute is transferred, the gas flux G gets smaller as the gas flows up the column, and the liquid flux L gets larger as it flows down the column. As a result,

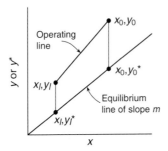

Fig. 10.4-1. Designing an absorption tower for a dilute vapor. The height of the tower is based on the values shown. The straight equilibrium and operating lines reflect the dilute solutions.

the slope of the operating line L/G changes as we go from bottom ($z = l$) to the top of the column ($z = 0$). The operating line is no longer linear.

For concentrated absorption, the equilibrium line reflects complex chemistry and so is normally nonlinear. This complexity may include solute ionization which can cause the plot of y^* vs. x to curve upwards. It may include either positive or negative deviations to Raoult's law, which may cause the equilibrium line to curve in either direction.

Calculating the tower height for concentrated absorption again involves finding values of the driving force ($y - y^*$) and plugging them into the rate equation. The only change is that the driving force now varies in a more complicated way, as suggested by Fig. 10.4-2. Everything else is the same, although the integration of the rate equation must now be numerical.

We can put these ideas on a somewhat more quantitative basis by making a mole balance on both gas and liquid. The result is a parallel to Eq. 10.3-2:

$$0 = - \frac{d}{dz}(Gy) + \frac{d}{dz}(Lx) \tag{10.4-1}$$

Before, the flux of gas G and that of liquid L were nearly constant because the absorbing species was always dilute. Now, however, we expect that

$$G_0 = G(1 - y) \tag{10.4-2}$$

where G_0 is the flux of the nonabsorbing gas. For example, if we are using water to absorb SO_2 out of air, G_0 is the flux of air. Similarly,

$$L_0 = L(1 - x) \tag{10.4-3}$$

where L_0 is the flux of the nonvolatile liquid. When we combine these equations and integrate, we find

$$y = \frac{\dfrac{y_0}{1 - y_0} + \dfrac{L_0}{G_0}\left(\dfrac{x}{1 - x} - \dfrac{x_0}{1 - x_0}\right)}{1 + \left(\dfrac{y_0}{1 - y_0}\right) + \dfrac{L_0}{G_0}\left(\dfrac{x}{1 - x} - \dfrac{x_0}{1 - x_0}\right)} \tag{10.4-4}$$

This mole balance is the operating line for a concentrated vapor, the analog of Eq. 10.3-5. It reduces to this equation as the concentrations become small. However, in general, its shape is more like that in Fig. 10.4-2.

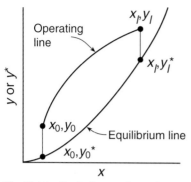

Fig. 10.4-2. Designing an absorption tower for a concentrated vapor. The height of the tower is again based on the values in the figure. The curved equilibrium and operating lines reflect the fact that both gas and liquid are concentrated solutions.

The next step is the specification of a new equilibrium relation analogous to Eq. 10.3-6:

$$y^* = y^*(x) \tag{10.4-5}$$

This relation is often not written in an analytical form, but simply presented as a table or graph of experimental results. The important point is that y^* and x are no longer directly proportional, related by a single, constant coefficient. Instead, they vary nonlinearly, as exemplified by the equilibrium line in Fig. 10.4-2.

The final step is the rate equation, a mole balance on the gas in a differential tower volume:

$$0 = \frac{d}{dz}(Gy) - K_ya(y - y^*) \tag{10.4-6}$$

We combine this result with Eq. 10.4-2 to find

$$0 = -\frac{G_0}{(1-y)^2}\frac{dy}{dz} - K_ya(y - y^*) \tag{10.4-7}$$

where K_y is the overall coefficient based on a mole fraction driving force. Rearranging,

$$l = \int_0^l dz = \frac{G_0}{K_ya}\int_{y_0}^{y_l}\frac{dy}{(1-y)^2(y - y^*)} \tag{10.4-8}$$

This result for concentrated solutions reduces to Eq. 10.3-11 for dilute solutions, where $(1-y)$ is about unity.

The tower height l can be found by integrating Eq. 10.4-8, using values of y and y^* read from a plot like Fig. 10.4-2. The integration is limited by two key assumptions made during the analysis. These subtle assumptions merit review. First, we are assuming absorption of a single vapor from an inert gas into a nonvolatile liquid. The gas is inert in the sense that only negligible amounts dissolve in the liquid; the liquid is nonvolatile in the sense that only negligible amounts evaporate into the gas.

The second key assumption in this analysis is that a single value of mass transfer coefficient can adequately express the mass transfer in a concentrated solution. There are two reasons why this may not be true. Most obviously, this overall coefficient is a function of the equilibrium line

$$\frac{1}{K_y} = \frac{1}{k_y} + \frac{m}{k_x} \tag{10.4-9}$$

where k_y and k_x are the individual mass transfer coefficients in the gas and liquid, respectively; and m is the slope of the equilibrium line. But because this varies, K_y may vary. Second, concentrated absorption involves transport of large amounts of vapor through thin layers of stagnant inert gas. This leads to diffusion-engendered convection normal to the surface, a topic detailed in Sections 3.1 and 9.5. These factors are commonly ignored in practice.

Example 10.4-1: Ammonia scrubbing A gas mixture at 0 °C and 1 atm flowing at 1.20 m^3/sec, and containing 37% NH_3, 16% N_2, and 47% H_2 is to be scrubbed with water containing a little sulfuric acid at 0 °C. The exit gas should contain 1% NH_3 and the exit liquid 23 mol% NH_3.

Design a packed tower to carry out this task. The tower should use 2-in Berl saddles, which have a surface area per volume 105 m^2/m^3 (cf. Table 10.2-1). It should operate with a pressure drop of 0.5 in H_2O per foot of packing. Pilot-plant data suggest that the overall gas-side mass transfer coefficient in this tower will be 0.0032 m/sec; this value is larger than normal because of the chemical reaction of ammonia with water.

In this design, answer the following specific questions: (a) What is the flow of pure water into the top of the tower? (b) What tower diameter should be used? (c) How tall should the tower be?

Solution (a) We first find the total flow AG_0 of the nonabsorbed gases (i.e., of N_2 and H_2):

$$AG_0 = 0.63 \left(\frac{1.20 \ m^3/sec}{22.4 \ m^3/kg \, mol} \right)$$

$$= 0.0338 \ kg \, mol/sec$$

We then find the ammonia transferred:

$$\left(\begin{matrix} NH_3 \\ absorbed \end{matrix} \right) = 0.37 \left(\frac{1.20 \ m^3/sec}{22.4 \ m^3/kg \, mol} \right) - \left(\frac{0.01(0.0338 \ kg \, mol/sec)}{0.99} \right)$$

$$= 0.0195 \ kg \, mol/sec$$

From this, we find the desired water flow AL_0:

$$AL_0 = \left(\frac{0.77}{0.23} \right) (0.0195 \ kg \, mol/sec)$$

$$= 0.0652 \ kg \, mol/sec$$

(b) The risk of flooding is greatest at the bottom of the tower where the flows are greatest. Moreover, because flooding is determined by fluid mechanics, it depends on mass flows, not molar flows. To make this conversion, we first find that the average molecular weight of the gas is 11.7. Then we see that

$$\left(\begin{array}{c}\text{total flow}\\ \text{of gas}\end{array}\right) = \frac{11.7 \text{ kg}}{\text{kg mol}}\left(\frac{0.0338 \text{ kg mol/sec}}{0.63}\right)$$

$$= 0.628 \text{ kg/sec}$$

The average molecular weight of the liquid stream (neglecting any H_2SO_4) is 17.8 kg, so

$$\left(\begin{array}{c}\text{total flow}\\ \text{of liquid}\end{array}\right) = \frac{17.8 \text{ kg}}{\text{kg mol}}\left(\frac{0.0652 \text{ kg mol/sec}}{0.77}\right)$$

$$= 1.51 \text{ kg/sec}$$

Thus

$$\left(\frac{\text{liquid flow}}{\text{gas flow}}\right)\sqrt{\rho_G/\rho_L} = \frac{1.51 \text{ kg/sec}}{0.628 \text{ kg/sec}}\sqrt{0.522 \text{ kg/m}^3/10^3 \text{ kg/m}^3}$$

$$= 0.055$$

Remembering that the values found from Fig. 10.2-2 imply specific dimensions, we find the capacity factor is

$$1.3 = \frac{G''F^{0.5}\nu^{0.05}}{(\rho_L - \rho_G)^{0.5}}$$

$$= \frac{G''(45)^{0.5}1^{0.05}}{(62.4 - 0)^{0.5}}$$

Thus in the units implicit in this figure

$$G'' = 1.53 \text{ lb/ft}^2 \text{ sec}$$

In units appropriate to this problem, this is equivalent to

$$G'' = 0.062 \text{ kg/m}^2 \text{ sec}$$

But the total is so that 0.628 kg/sec, so the cross-sectional area A is about 10 m². In other words, the tower diameter is about 3.6 m.

(c) The calculation of the tower's height can begin with Eq. 10.4-8. Because K_y equals $K_G c_G$

$$\text{HTU} = \frac{G_0}{K_G a c_G}$$

$$= \frac{[0.0338 \text{ kg mol/sec}] / [10 \text{ m}^2]}{(0.0032 \text{ m/sec})\left(105 \text{ m}^2/\text{m}^3\right)\left(1 \text{ kg mol/22.4 m}^3\right)}$$

$$= 0.23 \text{ m}$$

To find the number of transfer units NTU, we first plot values of y versus x using Eq. 10.4-4, shown as the operating line in Fig. 10.4-2. We also plot y^* versus x, shown as the equilibrium line in the figure. We then read off values of y versus y^* at fixed x, and integrate Eq. 10.4-8 from $y_0 = 0.37$ to $y_l = 0.01$. The result is

$$l = 10 \text{ m}$$

Problems of stripping gases are very similar except that the operating line falls below the equilibrium line.

10.5 Conclusions

This chapter analyzes gas absorption, an important separation process in chemical manufacture and pollution control. Gas absorption commonly is effected in packed towers filled with inert packing that gives a large interfacial area between gas and liquid. The gas rises through the tower; the liquid trickles countercurrently downward. The liquid is often chemically reactive, binding the solutes being absorbed. For example, acid gases like H_2S are absorbed into aqueous solutions of amines. However, the analysis in this chapter implies nonreactive liquids; reactive liquids are discussed in Chapter 17.

The analysis of gas absorption depends on fluid mechanics and on mass transfer. The fluid mechanics determines the acceptable range of gas and liquid fluxes, which are adjusted by changing the cross-sectional area of the tower. The mass transfer coefficients determine the rate of absorption and hence the height of the packed tower. This height can be estimated by either algebraic or geometric methods. The algebraic formulation is simple for the common case of a dilute solute, a case detailed in Section 10.3. This case depends on three key relations: an overall mole balance, a thermodynamic equilibrium, and a rate equation. This dilute case is the easiest way to learn about absorption.

The geometric analysis of absorption is suitable for either dilute or concentrated systems. It also depends on the same three key relations. Almost perversely, the overall mole balance is now called the operating line and the thermodynamic equilibrium is called the equilibrium line. The rate equation sometimes has the mass transfer coefficients rewritten in terms of new quantities called HTUs, height of transfer units, which are measures of the efficiency of the packed tower. These new terms provide occasional physical insight; simultaneously, they are effective at discouraging the inexperienced from trying to learn about gas absorption. If you are inexperienced, don't give up. Work hard on the dilute limit; be encouraged by the fact that the concentrated limit and the geometric analysis are more complicated, but involve no new ideas.

Questions for Discussion

1. In absorption, what are two gases which are commonly absorbed?
2. What are two species which are stripped?

3. What is the flow parameter? What is the capacity factor?
4. How can you avoid flooding in existing equipment?
5. What is an advantage and a disadvantage of using a smaller size of random packing?
6. What is the basis of the operating line?
7. When will the assumption of constant liquid and gas fluxes be accurate?
8. How can the slope of the operating line be changed?
9. What is the basis of the equilibrium line?
10. How can the slope of the equilibrium line be changed?
11. What does the number of transfer units NTU signify?
12. Do you want to have a small height of transfer unit HTU?
13. How can you reduce the HTU?
14. What is the effect on tower operation of the heat of absorption?
15. You are given an overall mass transfer coefficient, based on gas side driving force of 0.08 cm/sec. You use this in a tower where the area/volume is 6.6 cm^{-1} and the gas flux $7 \cdot 10^{-4}$ mol/cm^2 sec. What is the HTU (based on the gas side driving force)?

Problems

1. A packed tower is being used to scrub ammonia out of a stream containing only 3% of that gas. The tower contains 1-cm Raschig rings; it is 50 cm in diameter and 4.3 m high. The gas flow of 0.93 kg/sec is at 30 °C and is largely air at 100% relative humidity and 1,100 mm Hg; it leaves the tower with only $2.2 \cdot 10^{-6}$% NH$_3$. The liquid flow of 6.7 kg/sec is also at 30 °C. The Henry's law constant under these conditions is $y_{NH_3} = 0.85 x_{NH_3}$. What is the mass transfer coefficient K_G in this tower?

2. A process gas containing 4% chlorine (average molecular weight 30) is being scrubbed at a rate of 14 kg/min in a 13.2-m packed tower 60 cm in diameter with aqueous sodium carbonate at 850 kg/min. Ninety-four percent of the chlorine is removed. The Henry's law constant (y_{Cl_2}/x_{Cl_2}) for this case is 94; the temperature is a constant 10 °C, and the packing has a surface area of 82 m^2/m^3. (a) Find the overall mass transfer coefficient K_G. (b) Assume that this coefficient results from two thin films of equal thickness, one on the gas side and one on the liquid. Assuming that the diffusion coefficients in the gas and in the liquid are 0.1 cm^2/sec and 10^{-5} cm^2/sec, respectively, find this thickness. (c) Which phase controls mass transfer?

3. Find the height of a packed tower that uses air to strip hydrogen sulfide out of a water stream containing only 0.2% H$_2$S. In this design, assume that the temperature is 25 °C, the liquid flow is 58 kg/sec, the liquid out contains only 0.017 mol% H$_2$S, the air enters with 9.3% H$_2$S, and the entire tower operates at 90 °C. The tower diameter and the packing are 50-cm and 1.0-cm Raschig rings, respectively, and the air flow should be 50% of the value at flooding. The value of $K_L a$ is 0.23 sec^{-1}, and the Henry's law constant (y_{H_2S}/x_{H_2S}) is 1,440.

4. Chlorinating drinking water kills microbes but produces trace amounts of chloroform. You want to remove this chloroform by air stripping, that is, by blowing air through

the water to remove the chloroform as vapor. Such a process is the opposite of gas absorption. You know the equilibrium line is

$$y^* = 170x$$

You know that the mass transfer coefficients in the vapor and the liquid in your equipment are 0.16 cm/sec and $8.2 \cdot 10^{-3}$ cm/sec. You also know the gas velocity is 16 cm/sec and the packing has $a = 6.6$ cm^{-1}. (a) Sketch typical equilibrium and operating lines for this process. (b) Find the HTU based on an overall gas-phase driving force.

5. A 2-ft diameter packed column with a packing of 20 ft of 1-in Berl saddles contains air at 1 atm and 20 °C. The tower is apparently close to flooding, and $\Delta p = 24$ in of H_2O. The mass velocity of liquid is 8.5 times that of the gas. (a) If the tower is repacked with $1\frac{1}{2}$ in polypropylene Pall rings, what would the pressure drop be? (b) How much higher a liquid flow rate could be used if the pressure drop were the same as it was with the Berl saddles?

6. You want to remove 90% of the SO_2 in a flue gas stream using a packed tower that is 0.7 m in diameter. The tower has an HTU, based on the gas-side resistance, of 0.26 m. This low value, for structured packing, is partly due to a buffered absorbing liquid for which

$$y = 8.4x$$

This liquid enters without containing SO_2. If you adjust the flow of the liquid so that the driving force $(y - y^*)$ is constant, how tall a tower will you need?

7. A process gas containing 0.18% sulfur oxides is being scrubbed countercurrently and differentially with an aqueous solution of completely unloaded hindered amines in a tower packed with structured packing. The height of a transfer unit, based on an overall gas-side mole fraction driving force, is 41 cm. The amine solution, flowing at twice the minimum rate, has an equilibrium line of

$$y^* = 0.03x$$

If the exit sulfur oxide concentration is 0.001%, what is the concentration of sulfur oxides in the exiting amine solution?

8. You have a gaseous effluent containing a mole fraction $860 \cdot 10^{-6}$ H_2S which you need to process to under a mole fraction of $2 \cdot 10^{-6}$ before you will be permitted to discharge it. You have a caustic solution for an absorption which initially has no H_2S and which has an equilibrium line of

$$y^* = 0.083x$$

You plan to use a liquid flow which is twice the minimum to carry out this task. (At the minimum flow, you would need an infinitely tall tower.) Because you are using a third-generation packing, the area per volume is large, 200 m^2/m^3. The overall mass transfer coefficient K_L is $4.6 \cdot 10^{-6}$ m/sec, based on a liquid side driving force. The total gas and liquid concentrations are 50 mol/m^3 and 17,000 mol/m^3 respectively and the gas flux is 16.8 mol/m^2 sec. How tall should this tower be?

9. You want to use an existing 7.4-m packed column to remove 99% of the methylene chloride from an entering stream containing 114 ppm. The methylene chloride will be absorbed by mineral oil in which it is extremely soluble (i.e., m is very small). The column has a surface area per volume of 5 cm^2/cm^3, and a mass transfer coefficient K_G of $8.1 \cdot 10^{-4}$ cm/sec, based on a gas-side driving force. You plan to use an air flux of $1.4 \cdot 10^{-5}$ mol/cm^2 sec. (a) Will this column work? (b) You have a ten-stage absorption tower with which you may also achieve the same separation. If you still want 99% removal, what absorption factor A ($= L/mG$) should you use?

10. As the result of liquid–liquid extractions for antibiotics, we produce 63,000 mol/hr of water containing 0.2 mol% isobutanol. Because the charge for putting this stream in the sewer is going to be dramatically increased, we want to strip the alcohol out of the water with 62,000 mol/hr of air. We will then use the air in our power plant and thus burn the butanol.

   We have a 25-m packed tower, 0.7 m in diameter, which we can use for the stripping. We expect the overall mass transfer coefficient in this tower will be about

$$K_G a = 1290/\mathrm{hr}$$

or

$$K_y a = 1.60 \cdot 10^{-5}\, \mathrm{mol/cm^3 sec}$$

or

$$\mathrm{HTU}_{0,y} = 2.80\,\mathrm{m}$$

   We expect to run the tower at 1 atm and hot, where the vapor pressure of isobutanol is 400 mm Hg. What percent of the isobutanol can we remove?

11. A fragrance is stripped out of flowers using 5 mol of air. This fragrance is to be recovered with 1 mol of oil, which is initially fragrance free. The equilibrium for the fragrance between the air and oil is given by

$$y^* = 0.2\,x$$

   You want to evaluate two possible separations. (a) First, imagine that you mix air and oil in a stirred tank. At equilibrium, what percentage of fragrance is in the oil? (b) Now imagine that air and oil are contacted countercurrently in a packed tower at 0.5 mol air/hr and 0.1 mol oil/hr. Draw operating and equilibrium lines, and label the slopes. (c) If the tower and the mass transfer coefficient give six transfer units, what percentage of the fragrance is now in the oil?

12. Increasingly, natural gas wells contain significant amounts of nitrogen in the desired methane. These gases could easily be separated by cryogenic distillation because their boiling points are -196 and -164 °C for N_2 and CH_4, respectively. However,

this is too expensive. As an alternative, you are testing kerosene solutions of a new compound which reacts with the nitrogen and so allows its removal by gas absorption. In one experiment, 6 mol/sec feed containing 10% N_2 is scrubbed with 26 mol/sec of absorbent solution in a 3 m tower. Eighty-eight percent of the nitrogen is removed. Equilibrium experiments show that the mole fraction in the gas is four times larger than that in the liquid. What is the HTU in this column (based on gas concentrations)?

13. You have been running experiments in the pilot plant to remove a mercaptan from a gas stream. (Mercaptans are, among other things, responsible for the smells of garlic and skunks.) You find that with a gas feed containing 100 ppm, you can remove 90% of the mercaptan in a tower 18 m tall if you run countercurrently with a pure liquid water feed, using a flow which makes the operating line parallel to the equilibrium line, which is

$$y^* = mx$$

where y^* and x are given in ppm. (a) What is the value of the HTU based on $(y - y^*)$? (b) You are now ready to scale up this process to 100 greater flow and 99% removal. You will handle the greater flow with a bigger tower diameter, so the fluxes G and L are the same. You also plan to add a reactive solute to the water so that the equilibrium line is now

$$y^* = 0$$

If the HTU is expected to be 2 m, how tall should the process tower be?

14. You are using a small tower packed with Berl saddles to effect absorption of 99% of the H_2S in an effluent gas. The equilibrium line for the absorbing liquid you are using is

$$y_1^* = 8x_1$$

The HTU in this tower is expected to be around 1.2 m; the gas enters with 630 ppm. The absorbing liquid enters pure. The gas flux per liquid flux G/L is 0.025. (a) How many transfer units are needed? (b) What height of tower is required?

15. You have successfully developed a small column which removes 99% of the H_2S in an effluent stream. The HTU in this tower is around 1.2 m; the gas enters with 630 ppm. The absorbing liquid enters pure. The gas flux per liquid flux G/L is 0.025. While this tower is running well, you now need to remove 99.9% of the H_2S to meet governmental requirements. To do so, you plan to use a new absorbing liquid which reacts instantaneously with the H_2S, reducing the liquid-side mass transfer resistance by 90%; this liquid-side resistance had been 80% of the total. The new equilibrium line with this reactive liquid is $y^* = x$. (a) What is the new HTU? b) Assume the tower is 8 m high. What flow of gas per reactive liquid flux G/L will be required now?

Further Reading

Kister, H. Z., Scherffins, J., Afshar, K., and Abkar, E. (2007). Realistically predict capacity and pressure drop for packed columns. *Chemical Engineering Progress*, **103**(7), 28.

McCabe, W., Smith, J. C., and Harriott, P. 2005. *Unit Operations of Chemical Engineering*, 7th ed. New York: McGraw-Hill.

Perry, R. H., and Green, D. W. (1997). *Chemical Engineers Handbook*, 7th ed. New York: McGraw-Hill.

Seader, J. D., and Henley, E. J. 2006. *Separation Process Principles*, 2nd ed. New York: Wiley.

CHAPTER 11

Mass Transfer in Biology and Medicine

The ideas of mass transfer covered in the previous three chapters provide a strong framework for describing mass transfer in a wide variety of situations. One obvious example is gas absorption, which is one route to controlling the carbon dioxide emissions that contribute to global warming. The ideas of mass transfer are also basic to many physiologic functions. For example, they are key to respiration, to digestion, and to drug metabolism.

The descriptions of mass transfer developed in this book are detailed and accurate because they were developed for well-defined problems in the chemical industry. In that highly competitive industry, small improvements in chemical processing can mean large increases in profits. This promise of higher profits led engineers to examine the details of mass transfer and to get highly quantitative results.

The descriptions of mass transfer developed in biology and medicine took place largely independently of those in chemistry and engineering. These biologically and medically based descriptions have often provided good insight into rate processes in living systems. However, these descriptions rarely take advantage of insights provided by the deeper engineering analyses. There are two reasons for this. First, the accuracy of data in living systems is often uncertain because living systems vary more widely. After all, the weight of a person varies more than the weight of a nitrogen molecule. As a result, details of mass transfer known from engineering may not be that useful. For example, if biological data are only accurate to 30 percent, then an analysis offering improvements of 10 percent has little value.

The second reason that those in medicine and biology take little advantage of the engineering results is that those results are buried within dense, highly mathematical engineering curricula. These curricula assume that any student entering the introductory class will normally take all the subsequent classes offered. As a result, outsiders trained in other fields – like biology – can find it hard suddenly to dip into the engineering curriculum to study, for example, mass transfer or reaction kinetics. They conclude – correctly, I think – that the only way to learn the stuff is to take the entire program, and that it is not worth it.

This chapter tries to bridge this gap. It gives an overview of mass transfer for those with little mathematics beyond elementary differential equations. This chapter does not completely stand alone; it probably will require the complete novice to consult other sections to understand completely the basic ideas involved. Still, it can give the biologist an easier introduction, and the engineer a good review.

We begin the review as follows. Mass transfer describes the amount of solute moving from one region to another. For example, it describes how a solute like glucose moves from the lumen of the small intestine into the blood. The solute flux N_1 is the amount of solute per area per time. It is given by

$$N_1 = K\Delta c_1 \qquad\qquad (11.0\text{-}1)$$

where K is an overall mass transfer coefficient and Δc_1 is a concentration difference. Each of these quantities is complex and merits review.

Any flux includes transport both by convective flow and by diffusion. However, in general, one of these routes is much more important than the other. For example, glucose transport along the length of the small intestine is almost completely by flow, but glucose transfer across the intestinal wall is almost completely by diffusion.

The overall mass transfer coefficient K is a rate constant, a measure of how fast the process occurs. It is a close parallel to the rate constant of a first-order chemical reaction, or to the half life of radioactive decay. It is different from these chemical rate constants in two important ways. First, K is defined per unit area, and chemical rate constants are normally defined per unit volume. One consequence is that we will sometimes work with Ka, where a is the interfacial area per system volume. The product Ka has the same units of reciprocal time as a first-order chemical rate constant.

The second important difference between K and a chemical rate constant is that the former varies with physics and the latter with chemistry. Thus K is a function of stirring but changes little with temperature. A chemical rate constant is independent of stirring and a strong function of temperature. If these ideas are unfamiliar, you may wish to review Section 8.1.

The overall mass transfer coefficient K often involves diffusion across interfaces. For example, in the lung, oxygen is transported from the air in the alveolus to the alveolus wall, across that wall, and into the blood. These different steps are often described as different mass transfer resistances. Thus, $(1/K)$ is the overall resistance to mass transfer, and equals the sum of the other mass transfer resistances. If this idea is new or not completely clear, then you may wish to review Section 8.5.

The concentration difference Δc_1 given in Equation 11.0-1 is the hardest idea in mass transfer. It is not normally the concentration on one side of an interface minus that on the other side of the interface. To see why this is not true, imagine you are studying transport of a lipophilic drug from water into fat. The drug is initially present in the water but not in the fat. As time proceeds, the drug will diffuse from the water into the fat. After a while, the concentration in the water drops, and that in the fat rises. But transport does not stop when the concentrations are equal; instead, the concentration in the water continues to drop, and that in the fat continues to rise. At large times, the small concentration in the water may reach equilibrium with the large concentration in the fat. The difference Δc_1 may be defined as the concentration in the water c_1 minus that which would be in the water if it were in equilibrium with the fat c_1^*. This difficult point, also covered in Section 8.5, will be discussed throughout the chapter.

The chapter itself is organized as follows. In Section 11.1, we discuss experiments in which data can be organized using mass transfer coefficients. Section 11.2 describes blood oxygenators and artificial kidneys, mass transfer devices whose geometry is exactly known. Section 11.3 discusses the role of mass transfer in pharmacokinetics, where the system's geometry is unknown, lumped into other parameters which may include the mass transfer coefficient. Thus the chapter provides an introduction for life scientists to important engineering ideas.

11.1 Mass Transfer Coefficients

This book is based on two complementary models which describe mass transfer. One of these is diffusion, described by the diffusion coefficient. This model, widely used in science, describes how the concentration of a solute varies with position and time. For example, it can describe how a drug penetrates human tissue.

The second model for describing mass transfer is a mass transfer coefficient, an engineering idea which is infrequently used in science. This is unfortunate because, for fluid systems, it is probably of greater value than models using a diffusion coefficient. The mass transfer coefficient describes solute movement from one region into another region. The solute concentration in each region is assumed to have an average value, and the flux between the two regions is related to the difference between these average concentrations. Because of this use of average concentrations, mass transfer coefficients are used to describe how the concentration changes with position or how concentration changes with time. For example, mass transfer can describe how fast a drug is taken from the blood into tissue. Thus mass transfer coefficients are a form of "lumped parameter model," while diffusion coefficients are an example of a "distributed parameter model."

In this section, we use four examples to illustrate the use of mass transfer coefficients. These examples, which supplement those given in Chapter 8, have a basis in biology and medicine. They serve to introduce the ideas involved. They are an overview, and for the complete novice, may require referring to earlier sections of the book.

Example 11.1-1: Intestinal uptake   Starch in the human diet is digested in the stomach and small intestine into monosaccharides, which are the only form which can be absorbed. For lactose, enzymatic activity limits the rate of absorption. For other sugars, diffusion is rate limiting. For one set of experiments, the rate of uptake of glucose is 1.6×10^{-10} mol/cm^2 sec from a solution containing 2.7×10^{-4} M. What is the mass transfer coefficient?

Solution   The mass transfer coefficient K is defined by

$$N_1 = K(c_{1'} - c_1^*)$$

If the amount of glucose in the blood is relatively small, the concentration c_1^* is near zero and

$$1.6 \cdot 10^{-10} \frac{\text{mol}}{\text{cm}^2 \text{ sec}} = K\left(\frac{2.7 \cdot 10^{-4} \text{ mol}}{1000 \text{ cm}^3} - 0\right)$$

$$K = 6 \cdot 10^{-4} \text{ cm/sec}$$

This is a typical value.

Three points about this definition merit discussion. First, the concentration c_1^* may be near zero even when the solute concentration in the blood is more than zero. This is especially true for solutes whose properties, like solubility, are pH dependent. For the case of glucose, however, the solubility in chyme and blood is about the same, and so c_1^* will be close to the actual glucose concentration in blood.

The second point about this example is that the flux is given per a projected area of intestine. Because of the brush border, we could argue that a more appropriate flux would be defined per villi area. Either definition will work; the only danger is that we do not specify which we are using.

The third point about this analysis is that the results are rarely interpreted in this fashion. Instead, they are reported as an "unstirred layer thickness" l, defined as

$$K = \frac{D}{l}$$

where D is the solute's diffusion coefficient. This is an identical concept to that used successfully for the film theory of mass transfer described in Section 9.1. It's a good concept, useful because it gives a simple picture of mass transport. The only danger is that we will forget that the film thickness does not actually exist. It is simply an alternative way of expressing the mass transfer coefficient.

To illustrate this, imagine that we are comparing the absorption of glucose with a small peptide. If we measure mass transfer at the same infusion rate (i.e., at equal intestinal flow), we will usually find different unstirred layer thicknesses. There are several reasons why this could occur. First, while glucose is controlled by uptake in the intestinal lumen, peptide uptake may be controlled by transport across the intestinal wall. Even if mass transfer of both solutes is controlled by rates in the lumen, the mass transfer coefficient rarely depends linearly on the diffusion coefficient D. A more typical dependence is on the square root of D. If this is so, then the unstirred layer thickness l depends on D, which makes little sense physically.

I feel that data are better correlated as mass transfer coefficients K, where our approximations are acknowledged. I feel that using the idea of an unstirred layer pretends more knowledge than we have.

Example 11.1-2: Lung capacity One method to evaluate lung capacity is to ask the patient to take a deep breath of air spiked with traces of gases like neon and carbon monoxide. The patient holds his breath for a standard time and then quickly exhales into a gas chromatograph. The concentrations of Ne and CO are then measured.

Explain how these data can be used to find the mass transfer coefficient in the lung.

    Solution Answering this question requires mass balances on each of the trace gases. Because neon (species 2) is not absorbed, its concentrations allow finding the residual volume in the lung:

$$(\text{neon in}) = (\text{neon out})$$
$$c_{20} V = c_2(V + V_0)$$

where c_{20} and c_2 are the neon concentrations in the inhaled and exhaled gases, respectively; V is the volume of the breath; and V_0 is the volume of the lungs before taking the breath. We can measure the breath volume V and the concentrations c_2 and c_{20}; thus we can calculate the lung volume V_0.

Note that this mass balance implies that taking the breath produces complete mixing in the lungs. Parenthetically, we could check this assumption by measuring the concentration after a second breath, which is given from the mass balance

$$c_2(\text{one breath}) V_0 = c_2 (\text{two breaths}) (V + V_0)$$

If the values of V_0 found after one and two breaths are the same, we have justified the assumption of perfect mixing.

With the lung volume V_0, we can now find the mass transfer coefficient from a mass balance on carbon monoxide while the patient is holding his breath:

$$\begin{bmatrix} \text{carbon monoxide} \\ \text{lost from lung} \end{bmatrix} = \begin{bmatrix} \text{carbon monoxide} \\ \text{absorption into the blood} \end{bmatrix}$$
$$(V + V_0) \frac{dc_1}{dt} = -KA \left(c_1 - c_1^* \right)$$

where K is the overall transfer coefficient in the lung, A is the total lung area, c_1 is the concentration in the lung, and c_1^* is that lung concentration which would exist at equilibrium. Because the trace of CO present reacts irreversibly with the relatively large amount of hemoglobin in the blood, c_1^* is zero. This mass balance is subject to the initial condition

$$t = 0, \quad c_1 = c_{10}$$

Integrating the mass balance subject to this condition, we have the familiar result

$$\frac{c_1}{c_{10}} = e^{-kAt}$$

Rearranging

$$KA = \frac{1}{t} \ln\left[\frac{c_{10}}{c_1}\right]$$

Because all quantities on the right-hand side of this equation are known, the quantity KA is easily found.

Again, some points about this result deserve attention. First, the product KA is normally called the "diffusing capacity," which is descriptive but conceals its relation to mass transfer. Second, K includes mass transfer resistances in the lung gas, across the alveoli walls, and into the blood. The resistance in the gas is probably small because diffusion in gases is fast. The resistance in the blood may be small because of the fast reaction between CO and hemoglobin which accelerates mass transfer; this acceleration is discussed in detail in Chapter 17. The resistance across the membrane may be rate controlling.

Finally, we note that KA appears only as a product. Thus we will not normally know whether a relatively slow uptake indicates a patient with a compromised lung area A or a slow mass transfer coefficient K. Clinically this doesn't matter. If the value of the product KA is relatively small, then the patient's breathing will be compromised.

Example 11.1-3: Mass transfer to leaves   Leaves are the chief route by which plants absorb carbon dioxide and reject oxygen during photosynthesis. They provide the plants with the large surface area per volume required to make this transfer efficient. Not surprisingly, plants grow leaves until the mass transfer around the leaves is not a major resistance, at least in well-ventilated canopies. For plants growing in more protected environments, like greenhouses, this resistance may be more significant.

What do engineering correlations of mass transfer suggest about the effects of diffusion, velocity, and leaf size for plants growing in sheltered environments?

Solution   To explore this issue, we must first choose a characteristic geometry for the leaves. The common choice is to assume a flat plate, for which the mass transfer coefficient k is given by

$$\left(\frac{kl}{D}\right) = 0.646 \left(\frac{lv}{\nu}\right)^{1/2} \left(\frac{\nu}{D}\right)^{1/2}$$

in which l is the size of the plate in the direction of the flow v, and ν and D are the kinematic viscosity of the air and the diffusion coefficient in the air, respectively. This classical result, derived in Section 9.4, has an unusually strong theoretical basis. The terms in parentheses, which are dimensionless, occur frequently and so are called the Sherwood, Reynolds, and Schmidt numbers, respectively.

We seek how k varies with D, v, and l. The correlation says that k is proportional to the two-thirds power of D. While this expectation is almost always quoted, it has a weak experimental basis, although it does have a strong theoretical basis. However, because D varies little, this uncertainty is not important.

The variation of k and v is more interesting and more complicated. It is valid only at Reynolds numbers below about 20,000. In this case, the flow is laminar. Because the Reynolds number itself depends on l, this means that even at the same flow the correlation above may work for a small leaf, but not for a big one. For large leaves with Reynolds numbers above 20,000, a better correlation is:

$$\left(\frac{kl}{D}\right) = 0.036 \left(\frac{lv}{\nu}\right)^{0.8} \left(\frac{\nu}{D}\right)^{1/3}$$

The velocity has a larger effect at high flow than at low flow. In addition, at very low flow, the wind velocity v is less important than that caused by small differences in density due to adsorption or desorption. Now, the correlation is

$$\left(\frac{kl}{D}\right) = 0.5 \left(\frac{l^2 g(\Delta\rho/\rho)}{\nu D}\right)^{1/4}$$

where $(\Delta\rho/\rho)$ is the fractional density difference caused by the mass transfer itself. Thus a plot of the logarithm of k vs. the logarithm of v will have a slope of zero at small v, of 0.5 at intermediate v, and of 0.8 at large v. Delightfully, this is true over an enormous range of experimental conditions.

Finally, we consider how l varies with leaf shape. After all, leaves aren't really flat plates, and it may seem startling that we have had the success so far with such a simple geometry. To see what the effect of leaf shape could be, consider flow across the leaf shown in Fig. 11.1-1. Moderate flow at the position A will effect a smaller mass transfer per area than flow at position B. The reason is the k is proportional to $l^{-1/2}$, and l for A is larger than l for B. At the same time, flow at A will contact the leaf for a longer time because l is larger at the position. This suggests an average value of l given by

$$l = \left[\frac{\int x(y)dy}{\int \sqrt{x(y)}dy}\right]^2$$

This seems to describe the available data for moderate flow. For high flow or very low flow, the results will differ.

Example 11.1-4: Skin transport Mass transfer across skin is another good example, illustrating both the relation between the mass transfer coefficient and the diffusion

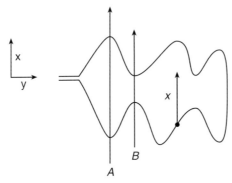

Fig. 11.1-1. Mass transfer from a leaf. Mass transfer rates can be accurately estimated from boundary layer theory.

coefficient and the complexity possible in an apparently simple case. We will start with the simplest case, and then add complexity.

Imagine that we expose a sample of human skin to a dilute antiseptic gas in air. We want to estimate the uptake of the gas into tissue which originally contains no antiseptic. The flux of the gas N_1 is given by

$$N_1 = K(c_1 - c_1^*)$$

where K is an overall mass transfer coefficient; c_1 is the antiseptic concentration in the gas; and c_1^* is proportional to the concentration in the tissue. Since the concentration in the tissue is zero, c_1^* is zero.

The mass transfer coefficient in this case is often dominated by transport across the skin itself. In this limit, the overall coefficient K is given by (cf. Section 9.1)

$$K = K_{skin} = \frac{DH}{l}$$

where D is the diffusion coefficient in the skin; H is the partition coefficient, the quotient of the concentration in the skin divided by the concentration in the adjacent air; and l is the thickness of the skin.

Mass transfer coefficients like that directly above occur commonly. As a result, they are often described by a separate vocabulary codified for membrane transport. The product DH is called the permeability; the variable (DH/l) is called the permeance. Note that the permeability is a physical property of the skin, but the permeance refers to a specific skin sample.

We can make additional estimates of transport for different gases by considering the properties of the skin itself. Skin is a layered structure. The top layer, the stratum corneum, is commonly believed to be the greatest resistance to transdermal transfer. This layer includes relatively impermeable protein flakes immersed in a lipid continuum. Because the flakes are aligned parallel to the surface of the skin, the diffusion across this layer is (cf. Section 6.4):

$$\frac{D_0}{D} = 1 + \frac{\alpha^2 \phi^2}{1 - \phi}$$

where D_0 is the diffusion coefficient of solute in the pure lipids, α is the aspect ratio, i.e. the shape of the flakes, and ϕ is the volume fraction of flakes. Note that the right-hand side of this equation is independent of D_0. Thus the diffusion coefficient in the skin changes by the same relative amount for any solute.

Diffusion coefficients in dense, noncrystalline solids don't change that much (cf. Sections 5.2 and 5.3). Thus any difference in mass transfer must come more from the partition coefficient H than from the diffusion coefficient D. These differences are often approximated as the relative solubility in octanol, chosen to approximate the skin lipids. If a second test gas has a lipid solubility ten times higher than the first, then we expect the mass transfer coefficient K, equivalent to the permeance DH/l, to be 10 times larger for the second more soluble gas than for the first.

11.2 Artificial Lungs and Artificial Kidneys

In this section, we discuss two types of artificial organs. Artificial lungs, or "blood oxygenators," are used during surgery to add oxygen and remove carbon dioxide from blood. Artificial kidneys, "blood dialyzers," are used to remove toxins, including urea, which accumulate in blood as the result of renal failure. In their current mode, artificial lungs and kidneys are not intended permanently to replace the patient's organs. Artificial lungs are used only during and directly after surgery, for perhaps twelve hours. Artificial kidneys are used longer, until a kidney suitable for transplantation can be located. This time will normally be several months and is rarely longer than three years.

Both artificial lungs and artificial kidneys are designed using the ideas of mass transfer taught in this book. Improvements in design will also derive from the same ideas. Moreover, because these artificial organs have exact physical dimensions and known chemical behavior, their design can be unusually quantitative. This does not mean that biology and medicine are not important: for example, the damage caused to blood proteins and any resulting clotting is extremely important. However, the known properties and dimensions of the artificial organs make these designs unusually exact.

In this section, we first consider the overall mass transfer coefficient in these devices. This coefficient includes three resistances, each of which can be quickly estimated. We then develop our design based on which resistances are most important. For blood oxygenators the only important resistance is in the blood, and so the design is straightforward. For blood dialyzers, there are several important resistances, so the design includes more subtle compromise. Details of these ideas are given below.

Fluxes in These Devices

Both blood oxygenators and blood dialyzers seek a large mass transfer flux in a small volume. The large flux is desired to easily add oxygen to or remove toxins from the patient's blood. In more quantitative terms, the flux per volume $N_1 a$ is given by

$$N_1 a = K a (c_1 - c_1^*) \tag{11.2-1}$$

where K is the overall mass transfer coefficient, a is the area per volume, and $(c_1 - c_1^*)$ is a concentration difference. For blood oxygenators, a is the interfacial area per volume between the air and the blood. For blood dialyzers, a is the interfacial area between the blood and the dialyzing solution, which is normally saline. We want to build devices with a large a.

The concentration difference $(c_1 - c_1^*)$ can be a more difficult concept. For dialyzers, it is the concentration of a toxin like urea in the blood minus that which would exist in blood which was in equilibrium with the dialysate. Because both blood and dialysate are largely water, this is often just the toxin concentration in blood minus that in dialysate. That doesn't seem like a difficult idea.

The difficulty is more obvious for the blood oxygenator. There, $(c_1 - c_1^*)$ is the concentration of oxygen in air minus that in air which is in equilibrium with the blood. The concentration of oxygen in air is of course 21% at 37 °C, or

$$\frac{0.21 \text{ mol}}{22.4 \text{ l}} \left(\frac{273 \text{ K}}{273 + 37 \text{ K}} \right) = 8.3 \cdot 10^{-3} \text{ mol/l} \tag{11.2-2}$$

The mole fraction of oxygen in water in equilibrium with pure oxygen is $1.9 \cdot 10^{-5}$; thus the concentration of oxygen in equilibrium with air at this temperature is

$$0.21 \, x_{O_2} c = 0.21 \left(1.9 \cdot 10^{-5} \right) \frac{\text{mol}}{0.0181}$$

$$= 2.2 \cdot 10^{-4} \frac{\text{mol}}{1} \tag{11.2-3}$$

At equilibrium, oxygen is $(8.3 \cdot 10^{-3}/2.2 \cdot 10^{-4} = 37)$ times more concentrated in air than in water. Thus in this case,

$$c_1 - c_1^* = \left[\frac{\text{conc in}}{\text{air}} \right] - 37 \left[\frac{\text{conc in}}{\text{water}} \right] \tag{11.2-4}$$

Similar results for blood will be functions of the hematocrit.

We now turn to the specific geometries used for blood oxygenators and dialyzers. These devices are now commonly based on hollow fiber membranes assembled in modules as shown schematically in Fig. 11.2-1. Hollow fibers are like small pipes with permeable walls. One fluid flows up the bore of the fibers – the "lumen" – as shown in the figure. The other fluid washes around the outside of the fibers, called "the shell side" by analogy with heat exchangers of similar geometry.

Two characteristics of hollow fiber membranes are important in the design of these devices. First, we seek a large area per volume a, because Eq. 11.2-1 says that this will give a large flux. For the cylindrical hollow fibers, this area is given by

$$a = \frac{\text{membrane area}}{\text{fiber volume}} = \frac{\text{fiber volume}}{\text{module volume}} = \frac{\pi d l}{\frac{\pi}{4} d^2 l} \cdot \phi = \frac{4\phi}{d} \tag{11.2-5}$$

where d and l are the fiber diameter and length, respectively; and ϕ is the volume fraction of hollow fibers. Typically, d is about 400 μm and ϕ is around 0.3, so a is perhaps 30 cm^2/cm^3.

The second characteristic of the hollow fiber membranes is the overall mass transfer coefficient K. For blood oxygenators, this coefficient describes diffusion of oxygen from the air to the membrane, from one side of the membrane to the other side, and from the membrane–blood interface into the bulk blood. For blood dialyzers, the coefficient K

One oxygenator design

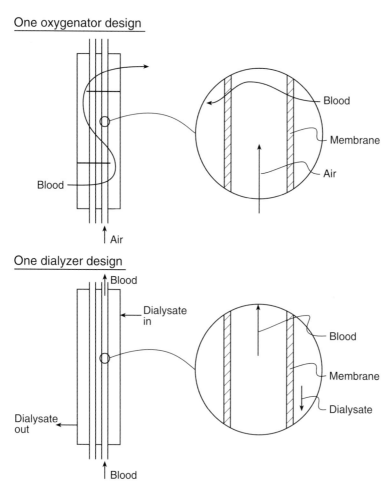

One dialyzer design

Fig. 11.2-1. Blood oxygenators and kidney dialyzers. Most designs use hollow fibers, small tubes with permeable walls giving a large area per volume for mass transfer.

describes diffusion of toxins from the blood to the membrane, across the membrane, and into the dialysate. How we get large values of K is described next.

Mass Transfer Coefficients in Blood Oxygenators

As mentioned above, the heart and lungs often need to be bypassed during open-heart surgery. The heart is replaced by a pump, and the lungs are replaced by a blood oxygenator. The oxygenator must add oxygen, remove carbon dioxide, and minimally damage the blood. Adding oxygen is harder than removing carbon dioxide and so will be the focus here.

Building a blood oxygenator requires getting a large oxygen flux, which in turn requires a large area per volume a and a large overall mass transfer coefficient K, just as suggested by Eq. 11.2-1. While the obvious way to get a large a is to form small air bubbles in the blood, this also causes considerable damage to the blood. This damage results in blood clots, which can cause strokes. As a result, current oxygenators are based on membranes, which separate the air and the blood. These membranes dramatically reduce clotting at the blood–air interface. Most current designs are based on hollow fibers, like that shown schematically in Fig. 11.2-1.

We want to maximize the overall mass transfer coefficient in this type of device. The overall mass transfer coefficient involves three sequential steps. Each step retards the mass transfer. These steps parallel those of getting dressed in the morning: getting out of bed, putting on your pants, lacing up your shoes. If getting out of bed takes much longer than all the other steps, then it is the most important, "rate-limiting," step. The mass transfer resistances contribute similarly to the overall rate. In quantitative terms, this can be expressed by

$$\frac{1}{K} = \frac{1}{k(\text{air})} + \frac{1}{k(\text{membrane})} + \frac{1}{k(\text{blood})} \tag{11.2-6}$$

where, for example, $[1/k(\text{air})]$ is the resistance to oxygen transfer in air, and $k(\text{air})$ is the mass transfer coefficient in the air next to the membrane.

We can use the large engineering literature to make estimates of these three coefficients. To begin, we recognize that

$$k(\text{air}) = \frac{D(\text{air})}{l(\text{air})} \tag{11.2-7}$$

where $l(\text{air})$ is an unstirred layer or boundary layer thickness, which is typically about 0.01 cm. Because the diffusion coefficient in gases is typically about $0.1 \text{ cm}^2/\text{sec}$,

$$k(\text{air}) = \frac{0.1 \text{ cm}^2/\text{sec}}{0.01 \text{ cm}} = 10 \text{ cm}/\text{sec} \tag{11.2-8}$$

Because the hydrophobic membrane typically contains a void fraction ε of 30% pores which have an effective length l of 0.01 cm, the mass transfer coefficient for the membrane is

$$k(\text{membrane}) = \frac{D(\text{air})\,\varepsilon}{l'} = \frac{0.1 \text{ cm}^2/\text{sec}\,(0.3)}{0.01 \text{ cm}}$$
$$= 3 \text{ cm}/\text{sec} \tag{11.2-9}$$

Thus the membrane's resistance is about three times larger than that in the feed air.

We now need only calculate the resistance for oxygen being transferred from the membrane into the blood. We will do this in two steps: as a quick approximation, and then as a more quantitative effort. Quickly, as before,

$$k(\text{blood}) = \frac{D(\text{blood})}{l''H} \tag{11.2-10}$$

where l'' is again a boundary layer, and H is the concentration of oxygen in air divided by that in blood (cf. Section 2.1, 8.5, or 9.1). The diffusion in liquids is about $10^{-5} \text{ cm}^2/\text{sec}$, l'' is still around 0.01 cm, and H was estimated above to be 37. Thus

$$k(\text{blood}) = \frac{10^{-5} \text{ cm}^2/\text{sec}}{0.01 \text{ cm}\,(37)} = 3 \cdot 10^{-5} \text{ cm}/\text{sec} \tag{11.2-11}$$

The resistance in blood is 100,000 times bigger than the other resistances. It is the only resistance that matters.

This result raises other questions, of which we will discuss three. First, hemoglobin will react with oxygen and accelerate oxygen mass transfer. This acceleration can be a factor of ten, which is certainly significant but not enough to change the blood resistance so it is not the key factor. Second, carbon dioxide is about 30 times more soluble in water than oxygen, so H is of order one. This means that $k(CO_2$ in blood) is thirty times larger than that for oxygen, but it is still the dominant resistance. Finally, note that new membranes which were more permeable would not change the mass transfer of oxygen significantly. Making k(membrane) ten times bigger wouldn't change K much, and so would not alter N_1a.

Mass Transfer Coefficients in Blood Dialyzers

Like a blood oxygenator, a blood dialyzer is usually based on hollow fiber membranes. Blood flows down one side of each membrane, usually in the hollow fiber bore, or lumen. An aqueous solution, or "dialysate," normally containing salts at about the same concentration as in the blood, flows on the other side of the membrane, normally the outside. The membrane itself is somewhat hydrophilic, swollen with water, so that urea and other toxins can diffuse out of the blood across the membrane and into the dialysate. Any dialysate loaded with toxins is discarded.

The mass transfer in blood dialyzers also is characterized by an overall mass transfer coefficient K, which is in turn the result of three resistances:

$$\frac{1}{K} = \frac{1}{k(\text{blood})} + \frac{1}{k(\text{membrane})} + \frac{1}{k(\text{dialysate})} \tag{11.2-12}$$

where k(blood), k(membrane), and k(dialysate) are the mass transfer coefficients out of the blood, across the membrane, and into the dialysate. Parenthetically, k(membrane) is sometimes called the membrane's "permeance," or archaically, its "permeability." Equation 11.2-12 is a close parallel to Eq. 11.2-6 for blood oxygenators. However, unlike blood oxygenators, the three mass transfer coefficients for blood dialyzers are all about the same size. No single coefficient dominates mass transfer. Thus if by some invention, we made mass transfer in the blood infinitely fast, we would make k(blood) much larger, but we would increase K only around 30 percent. If we made the membrane extremely thin and thus made k(membrane) very large, we still would increase K only about 30 percent.

Oxygenator and Dialyzer Performance

We now have estimates of mass transfer coefficients of these devices. We can use these estimates to see how either oxygenators or dialyzers will behave. As an example, we consider the dialyzer of length l shown schematically in Figure 11.2-2. Blood enters at the bottom of this dialyzer with toxins at concentration c_{1l} and leaves with toxins at the reduced concentration c_{10}. Dialysate flows countercurrently, i.e., in the opposite direction, entering without toxins ($C_{10} = 0$) and leaving with toxins at the concentration C_{1l}.

We want to find how the toxin concentrations change with the module length l and the mass transfer coefficient K. To do so, we write three equations: an overall mass balance, an expression of thermodynamic equilibrium, and a rate equation. We have already

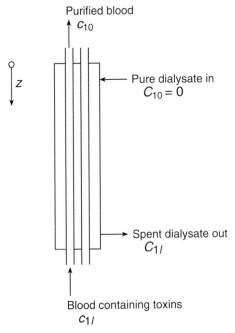

Purified blood
c_{10}

Pure dialysate in
$C_{10} = 0$

Spent dialysate out
C_{1l}

Blood containing toxins
c_{1l}

Fig. 11.2-2. Blood dialyzer performance. Mass balances and equilibrium constraints allow estimation of dialyzer behavior.

successfully used this strategy to describe gas absorption in the previous chapter. Here, the mass balance is written on both blood and dialysate between the top of the dialyzer and some arbitrary position:

$$B(c_1 - c_{10}) = D(C_1 - 0)$$

$$c_1 = c_{10} + \frac{D}{B} C_1 \tag{11.2-13}$$

where B and D are the superficial velocities of blood and of dialysate. The equilibrium relation is simple:

$$c_1^* = C_1 \tag{11.2-14}$$

At equilibrium, the toxin concentrations in blood and dialysate are equal because each is a dilute aqueous solution. Finally, the rate of toxin removal is found from a steady-state mass balance on the blood alone:

$$\begin{bmatrix} \text{accumulation} \\ \text{of toxin} \end{bmatrix} = \begin{bmatrix} \text{toxin flow in} \\ \text{minus that out} \end{bmatrix} - \begin{bmatrix} \text{toxin transferred} \\ \text{from blood to dialysate} \end{bmatrix}$$

$$0 = B\frac{dc_1}{dz} - Ka(c_1 - c_1^*) \tag{11.2-15}$$

where z is the distance from the top of the dialyzer downwards (cf. Fig. 11.2-2).

We can integrate this rate equation by substituting C_1 for c_1^* using the equilibrium relation in Eq. 11.2-13, and then substituting $C_1(c_1)$ using the overall balance given in Eq. 11.2-14. This integration is subject to the boundary conditions

$$z = 0, \quad c_1 = c_{10}, \quad C_1 = 0 \tag{11.2-16}$$

$$z = l, \quad c_1 = c_{1l}, \quad C_1 = C_{1l} \tag{11.2-17}$$

The result is

$$\frac{c_{10}}{c_{1l} - C_{1l}} = e^{-Kal\left(\frac{1}{B} - \frac{1}{D}\right)} \tag{11.2-18}$$

This is equivalent to the gas absorption result but with $m = 0$ and $x_0 = 0$. It is a key for the design of blood dialyzers.

Three characteristics of this result may clarify its physical significance. First, when the dialysate flow D is very high, C_1 is always near zero and Eq. 11.2-18 becomes

$$\frac{c_{10}}{c_{1l}} = e^{-\frac{Kal}{B}} \tag{11.2-19}$$

This is equivalent to what happens in a blood oxygenator. There, air is free, and so excess air is used to oxygenate the blood as fast as possible. Here, dialysate costs money, so we probably will not use a huge dialysate flow. The second characteristic of Eq. 11.2-18 occurs when the blood and the dialysate flows are equal. In this limit, $(c_1 - C_1)$ is a constant, equal to c_{10}, and Equation 11.2-15 can be directly integrated to give

$$\frac{c_{10}}{c_{1l}} = \frac{1}{1 + \frac{Kal}{B}} \tag{11.2-20}$$

Larger flows will remove less toxin because the time for mass transfer is reduced.

The third characteristic of Eq. 11.2-18 is a geometric representation, rather than an algebraic one. Equation 11.2-14 makes a parallel with gas absorption, and so is most useful for those trying to understand both processes. This characteristic is an "equilibrium line," the 45° diagonal on a plot of c_1 vs. C_1. Equation 11.2-13 is an "operating line," which on the same c_1 vs. C_1 coordinates has a positive intercept and a slope of (B/D). The vertical distance between these lines $(c_1 - c_1^*)$ is the driving force responsible for mass transfer. As you become skilled in this analysis, you may find a sketch of these lines gives you physical insight into the analysis not only of blood dialyzers but of any mass transfer device.

Example 11.2-1: The best location for blood flow in an oxygenator Knowing that the mass transfer in a blood oxygenator is dominated by the resistance in the blood immediately raises another question: should the blood flow be inside or outside of the hollow fibers? Make estimates to answer this question.

    Solution We recognize that the module will be run with excess air flow, so that the oxygen concentration in the air will always be around 21 percent. Thus the oxygen concentration of the air is the same everywhere. In more quantitative terms, the value of c_1^* is a constant.

With this in mind, we answer this question using correlations developed for mass transfer, some of which were summarized in Section 8.3. In particular, for blood flow inside the hollow fibers

$$\frac{kd}{D} = 1.62 \left(\frac{d^2 v}{Dl} \right)^{1/3}$$

where k is the mass transfer coefficient in the blood, D is the diffusion coefficient, and v is the average velocity through a fiber of diameter d and length l. This result, which has an unusually strong theoretical and experimental basis, is derived in Section 9.4. To make it more specific, imagine that the liquid velocity is 1 cm/sec through fibers 400 μm in diameter and 30 cm long. Then we can estimate k as

$$\frac{k \left(400 \cdot 10^{-4} \text{ cm} \right)}{10^{-5} \text{ cm}^2/\text{sec}} = 1.62 \left(\frac{\left(400 \cdot 10^{-4} \text{ cm} \right)^2 1 \text{ cm/sec}}{10^{-5} \text{ cm}^2/\text{sec} \, (30 \text{ cm})} \right)^{1/3}$$

$$k = 0.71 \cdot 10^{-3} \text{ cm/sec}$$

This is a typical value for mass transfer in the liquid.

Alternatively, we can design the blood oxygenator so that the blood flows outside of the fibers. Blood flow outside and parallel to the fibers is usually compromised by uneven hollow fiber spacing. Blood flow outside and perpendicular to the fibers tends to be more reliable. In this geometry, the mass transfer coefficient k is correlated by

$$\frac{kd}{D} = 0.8 \left(\frac{dv}{\nu} \right)^{0.47} \left(\frac{\nu}{D} \right)^{1/3}$$

where ν is the kinematic viscosity. Using the typical values suggested above, we find

$$\frac{k \left(400 \cdot 10^{-4} \text{ cm} \right)}{10^{-5} \text{ cm}^2/\text{sec}} = 0.8 \left(\frac{\left(400 \cdot 10^{-4} \text{ cm} \right)^2 1 \text{ cm/sec}}{0.01 \text{ cm}^2/\text{sec}} \right)^{0.47} \left(\frac{0.01 \text{ cm}^2/\text{sec}}{10^{-5} \text{ cm}^2/\text{sec}} \right)^{1/3}$$

$$k = 3.8 \cdot 10^{-3} \text{ cm/sec}$$

Thus having the blood flow across the outside of the hollow fibers gives about five times faster mass transfer. This is the strategy used by most but not all blood oxygenators.

Example 11.2-2: Toxin removal vs. dialysate flow   A hollow fiber dialyzer is 30 cm long, 3.8 cm in diameter, and contains a volume fraction of hollow fibers ϕ of 0.2 which are 200 μm in diameter. The overall mass transfer coefficient κ in these fibers is $3.6 \cdot 10^{-4}$ cm/ sec, and can be assumed independent of blood and dialysate flows.

If the blood flow is 4.1 cm/sec, what percentage of toxins is removed for a dialysate flow equal to the blood flow, twice the blood flow, and much greater than the blood flow?

Solution From the data given, the area per volume a can be calculated from Equation 11.2-5

$$a = \frac{4\phi}{d} = \frac{4(0.2)}{200 \cdot 10^{-4} \text{ cm}} = 40 \text{ cm}^2/\text{cm}^3$$

The blood flux B is given by the volumetric flow divided by the module's cross-sectional area

$$B = \frac{4.1 \text{ cm}^3/\text{sec}}{\frac{\pi}{4}(3.8 \text{ cm})^2} = 0.36 \text{ cm}/\text{sec}$$

Thus

$$\frac{kal}{B} = \frac{\left(3.6 \cdot 10^{-4} \frac{\text{cm}}{\text{sec}}\right) \frac{40}{\text{cm}} (30 \text{ cm})}{0.36 \text{ cm}/\text{sec}} = 1.2$$

We now can find the fraction of toxins removed. When D equals B, Equation 11.2-20 gives

$$\frac{c_{10}}{c_{1l}} = \frac{1}{1 + 1.2} = 0.45$$

so 55% of the toxins are removed. When D equals $2B$, Equations 11.2-13 and 11.2-14 show that C_{1l} equals $(c_{1l} - c_{10}/2)$. Equation 11.2-18 then says that

$$\frac{2c_{10}}{c_{1l} + c_{10}} = e^{-kal/2B} = e^{-0.6}$$

$$\frac{c_{10}}{c_{1l}} = 0.38$$

so 62% of the toxins are removed. When L is large, Equation 11.2-19 gives

$$\frac{c_{10}}{c_{1l}} = e^{-1.2} = 0.30$$

Seventy percent of the toxins are removed. The price of greater removal is of course more dialysate used.

11.3 Pharmacokinetics

Pharmacokinetics is the study of the distribution and metabolism of drugs in living organisms. It tries to estimate the drug concentrations in blood and tissue as a function of dosage and time. The basic tools, parallels of those in chemical reaction engineering, are mass balances written on regions or "compartments" in the system being studied. In some cases, the compartments may refer to particular organs, like the stomach or the brain; more frequently, they just refer to approximate regions, like the blood or the tissue.

The mass balances used in pharmacokinetics often contain rate constants which may include overall mass transfer coefficients and interfacial areas per volume. In cases like this, the relation between drug mass transfer and the more exact engineering coefficients is not known because the geometry implied in the pharmacokinetic models may not be known. In this case, we cannot use engineering correlations to predict pharmacokinetic behavior. We can only see whether drug concentration vs. time varies in roughly the manner expected for mass transfer.

In this section we review some of the simplest pharmacokinetic models which contain mass transfer coefficients. We do so only to illustrate the basic ideas involved; more complete analyses for specific drugs are beyond the scope for this book. To begin, we consider how we can easily measure blood flow in one artery. One way is to inject a known mass M of a nonadsorbing solute into the artery and measure the downstream concentration c_1 spectrophotometrically as a function of time. From a mass balance

$$M = \int_0^\infty Q c_1 dt = Q \int_0^\infty c_1 dt \qquad (11.3\text{-}1)$$

where Q is the volumetric flow in the artery. If we divide the area under a plot of concentration vs. time into the total solute mass M, we find the flow Q. This can be the start of our analysis.

We now want to approximate the distribution of a drug or other marker in the body. As the simplest case, we consider an inert, nonabsorbing marker injected into one organ of volume V. We will also make the conventional assumption that the contents of the organ are well mixed, just to give us a simple starting point. If blood flows in and out of the organ at a volumetric flow rate Q, then a mass balance on the marker in the organ gives an equation for the drug concentration c_1:

$$\begin{bmatrix} \text{drug} \\ \text{accumulation} \end{bmatrix} = \begin{bmatrix} \text{drug} \\ \text{flow in} \end{bmatrix} - \begin{bmatrix} \text{drug} \\ \text{flow out} \end{bmatrix}$$

$$V\frac{dc_1}{dt} = 0 - Q c_1 \qquad (11.3\text{-}2)$$

This is subject to the initial condition

$$t = 0, \quad c_1 = c_{10} \qquad (11.3\text{-}3)$$

Integrating

$$\frac{c_1}{c_{10}} = e^{-\frac{Q}{V}t} \qquad (11.3\text{-}4)$$

The drug concentration decays exponentially. The rate constant for this decay (Q/V), which is the reciprocal of the residence time in the organ, has nothing to do with mass transfer. Thus it should be the same for any marker, independent of, for example, the marker's size or hydrophilicity.

The second example involves removing a drug from an organ which has two compartments. One compartment of volume V is perfused with a steady flow Q in and out. The second compartment of volume V' quickly absorbs and desorbs the drug.

However, the drug is significantly metabolized in neither compartment. The mass balance is now

$$\left[\begin{array}{c} \text{drug} \\ \text{accumulation} \end{array}\right] = \left[\begin{array}{c} \text{drug} \\ \text{flow in} \end{array}\right] - \left[\begin{array}{c} \text{drug} \\ \text{flow out} \end{array}\right]$$

$$\frac{d}{dt}\left[c_1 V + c'_1 V'\right] = 0 - Q c_1 \tag{11.3-5}$$

The drug concentration c'_1 in the unperfused compartment is in rapid equilibrium with the concentration c_1 in the perfused compartment

$$c'_1 = H c_1 \tag{11.3-6}$$

where H is a partition coefficient, which may be much greater than one. Again, the mass balance is subject to an initial condition

$$t = 0, \quad c_1 = c_{10} \tag{11.3-7}$$

Integration gives

$$\frac{c_1}{c_{10}} = e^{-\left(\frac{Q}{V + HV'}\right)t} \tag{11.3-8}$$

where $(V + HV')$ is a virtual volume of the two compartments. Obviously, Eqs. 11.3-4 and 11.3-8 have the same mathematical form. Thus to tell if we in fact have one or two compartments, we must make experiments with two solutes which have different values of H.

Finally, we consider the same two compartments but with mass transfer which is not instantaneous. We still consider drug metabolism to be relatively slow. The mass balances on the two compartments are now

$$V\frac{dc_1}{dt} = -Q c_1 + KA\left(\frac{c'_1}{H} - c_1\right) \tag{11.3-9}$$

$$V'\frac{dc'_1}{dt} = KA\left(c_1 - \frac{c'_1}{H}\right) \tag{11.3-10}$$

where A is the area for mass transfer between the two compartments, the overall mass transfer coefficient K is

$$\frac{1}{K} = \frac{1}{k(\text{compartment } V)} + \frac{1}{Hk(\text{compartment } V')} \tag{11.3-11}$$

While we don't know much about particular values for the individual mass transfer coefficients k, we do know that they do not vary much. Thus most of the change in the mass transfer resistance is likely to come from the value of H. Parenthetically, this is the same as in liquid–liquid extraction, where the overall mass transfer coefficient is most influenced by the partition coefficient (cf. Section 14.3).

The integration of Eqs. 11.3-9 to 11.3-11 is straightforward but complicated. It leads to a plot of concentration vs. time which is a weighted sum of two decaying exponentials. To understand what this means, we are better off to consider a couple of special cases.

First, if $KA/Q \gg 1$, then mass transfer is fast enough to let c_1 and c_1' approach equilibrium. Then this example reduces to the second one, described by Eq. 11.3-8.

The more interesting case is the antithesis, when $KA/Q \ll 1$. Now, the concentration c_1 in the perfused compartment is small, and the concentration in the compartment regulated by mass transfer is

$$\frac{c_1'}{c_{10}'} = e^{-KAt/V'} \tag{11.3-12}$$

As a result, the small exiting concentration c_1 is approximately

$$\frac{c_1}{c_{10}'} = \frac{KA}{Q} e^{-KAt/V'} \tag{11.3-13}$$

Now, the apparent residence time (V'/KA) is a function of the mass transfer, and not of the flow.

This glimpse of pharmacokinetics shows both the advantage and the shortcoming of the approach. The advantage is that we can easily generate a mathematical model which can explain drug distribution. This model includes parameters which are physically plausible, describing topics like the flow, the mass transfer, and the partition between two regions. Moreover, the predictions of this mathematical model often give a good fit to the experimental data, which are often subject to considerable error. This is not a criticism because data on living systems are often hard to obtain.

The shortcoming of this approach is equally major. While the agreement between the model and the data may be good, the parameters found are often impossible to check independently. For example, we may not be able to distinguish between the residence times of $[(V + HV)/Q]$ and $[V'/KA]$ (Eqs. 11.3-8 and 11.3-13, respectively). We often will not really know what these residence times mean. Thus we run the risk of curve fitting our data without any real understanding. In this case, the analysis of data for living systems is not aided by the knowledge of diffusion and mass transfer developed for nonliving systems, the knowledge of which is the subject of the other chapters in this book.

11.4 Conclusions

The ideas of mass transfer developed in Chapters 8–10 can be effectively used in biology and medicine. They can be used to organize rates of digestion or breathing. They can serve as a guide for the design of blood oxygenators or the effectiveness of blood dialyzers. In cases where the geometry of the mass transfer device is well known, the new device can often be effectively designed by borrowing results for mass transfer in other, nonliving systems.

Mass transfer ideas can also be used in other *in vivo* situations. Their use often provides successful strategies for correlating data. However, because the geometry and chemistry of living systems is not often well defined, the physical meaning of the parameters used is much less clear. Knowledge of *in vitro* mass transfer, the subject of earlier and later chapters of this book, cannot be easily and effectively used in these other cases. We must make judgments to be effective.

Questions for Discussion

1. Will the mass transfer coefficient be larger for big leaves or for small ones?
2. Would the mass transfer correlation suggested for leaves work for tall grasses?
3. Would inhaling at different speeds change measurements of the apparent lung volume and of the lung mass transfer coefficient?
4. Why are better membranes a poor goal for research on blood oxygenators?
5. To maximize mass transfer, where should blood flow in a blood oxygenator?
6. Where should it flow in a hollow fiber dialyzer?
7. How can mass transfer be affected by the intestinal villi, i.e., the uneven surfaces of the small intestine?
8. The dissolution of many drugs is said to be diffusion controlled. How could you test this for a new drug?
9. Describe experiments which let you test if a drug's dissolution is partly controlled by diffusion and partly by a surface reaction.
10. How can mass transfer in the blood and in the tissue ever be separated in pharmacokinetic studies?

Problems

1. Find the dissolution rate of a cholesterol gallstone 1 cm in diameter immersed in a solution of bile salts. The solubility of cholesterol in this solution is about $3.5 \cdot 10^{-3}$ g/cm^3. The density difference between the bile saturated with cholesterol and that containing no cholesterol is about $3 \cdot 10^{-3}$ g/cm^3; the kinematic viscosity of this solution is about 0.06 cm^2/sec; the diffusion coefficient of cholesterol is $1.8 \cdot 10^{-6}$ cm^2/sec. *Answer:* 0.2 g per month.

2. You need to estimate an overall mass transfer coefficient for solute adsorption from an aqueous solution of density 1.3 g/cm^3 into hydrogel beads 0.03 cm in diameter. The coefficient sought K_y is defined by

$$N_1 = K_y(y - y^*)$$

where N_1 has the units of g/cm^2 sec, and the y's have units of solute mass fraction in the water. The mass transfer coefficient k_S in the solution is 10^{-3} cm/sec; that within the beads is given by

$$k_B = \frac{6D}{d}$$

where d is the particle diameter and D is the diffusion coefficient, equal here to $3 \cdot 10^{-6}$ cm^2/sec. Because the beads are of hydrogel, the partition coefficient is one. Estimate K_y in the units given.

3. Giardia, a microbial disease, infects most large animals in the western United States. When you canoe in northern Minnesota, you risk this "beaver fever" if you drink lake water infected with giardia cysts. A commercial device to remove these cysts consists of a small bed of ion exchange beads loaded with I_3^-. When you suck water through the bed into your mouth, you quickly kill the cysts. (a) Write a differential equation giving the percent of cysts killed versus flow. (b) Imagine the bed is of

constant volume and operates at a given flow. Explain why a tall skinny bed is better than a short fat bed. (c) In fact, the short fat bed is used. Explain why.

4. In the lab, we have been testing artificial gills for breathing underwater. In one of these gills, exhaled air containing 10% oxygen flows at 2 cm/sec past 20-cm long hollow fibers containing excess air saturated water. Oxygen in the water is transferred across the fibers into the air. The fibers have a surface area per gill volume of 16 cm^2/ cm^3. The mass transfer coefficient across the fibers is 0.01 cm/sec. The air saturated water contains $y^* = 10^{-7}$ mol air per mol total. (a) Derive an equation giving the oxygen concentration coming out. (b) Estimate this exit concentration.

Further Reading

Datta, A.K. (2002). *Biological and Bioenvironmental Heat and Mass Transfer*. Boca Raton, FL: CRC Press.

Maina, J.N. (2002). *Fundamental and Structural Aspects and Features in the Bioengineering of Gas Exchangers*. Berlin: Springer.

Minuth, W.M., Strebl, R., and Schumacher, K. (2005). *Tissue Engineering: From Cell Biology to Artificial Organs*. New York: Wiley.

Vogel, S. (1988). *Life's Devices*. Princeton, NJ: Princeton University Press.

CHAPTER 12

Differential Distillation

Distillation is the process of heating a liquid solution to drive off a vapor and then collecting and condensing this vapor. It is the most common method of chemical separation, the workhorse of the chemical process industries. Distillation columns are ubiquitous; they are the brightly lighted towers that rise from chemical plants, and they are the stills used by moonshiners.

Distillation is carried out in two ways: differential distillation and staged distillation. The difference is what is inside of the distillation column. In differential distillation, the column contains packing like that used for gas absorption. In small laboratory columns, this packing is usually random, of Raschig rings or even glass beads. Such distillation aims to provide small amounts of very pure chemicals. In larger differential distillation columns, the column internals are usually structured packing. These distillations aim to produce large amounts of commodity chemicals at the lowest possible cost.

The second way to effect these separations is staged distillation. In staged distillation, the column internals are completely different than those normally used for gas absorption. Now these internals consist of a series of compartments or "trays," where liquid and vapor are contacted intimately, in the hope that they will approach equilibrium. Now, the liquid and vapor concentrations in the column do not vary continuously, but discretely, jumping to new values on each tray. Staged distillation was an innovation for commodity chemicals a century ago, and was the standard during the rapid growth of the chemical industry. While it is still the standard in universities, it has been eclipsed by differential distillation in many areas of industrial practice.

I will describe differential distillation in this chapter, and I will develop staged distillation in the next chapter. I have tried to make the chapters separate, so that either may be read without the other. At the same time, I have also tried to keep them parallel, so that ideas for one of the methods may be reinforced by comparison with the other method.

In this chapter on differential distillation, we cover in Section 12.1 what is distilled and what equipment is used. In Section 12.2, we consider only distillation of very pure products, which is a close parallel to dilute absorption. Because we begin with dilute distillation, we can focus on the physics and chemistry without complex mathematics. In Section 12.3 we discuss the changes caused by a feed in the center of the column. In Section 12.4, we analyze concentrated distillation, which normally requires numerical computation.

12.1 Overview of Distillation

12.1.1 What is Distilled

In distillation, we heat a liquid to make a vapor enriched in the more volatile component. By doing this repeatedly, we can separate the more volatile and less volatile species.

Distillation is a ubiquitous process, used for a huge variety of chemical separations. In the popular imagination, it is associated with the production of ethanol. Wine is distilled to make brandy; beer is distilled to make whiskey; and corn mash is distilled to make gin, vodka and other white spirits. Distillation originated in efforts to increase the alcohol concentration of wine, thus making the wine more stable to heat. Many cultures seem to have been involved: the stills shown on the labels of Irish whiskey bottles are "aliquitaras," a form of Spanish still based on designs by the Moors, who are believed to have gotten their ideas from the Chinese.

Today, distillation is a key to production of most commodity chemicals. It is used to make gasoline from crude oil. It is basic to the four most important separations of organics: aliphatic from aromatic hydrocarbons, linear from branched hydrocarbons, olefins from alkanes, and alcohols from water. One essay on choosing separation processes starts with the question "Why not distillation?" After then discussing over fifty other processes, it repeats, "Are you sure you do not want to use distillation?"

In this environment, we should discuss which species are not successfully separated by distillation. The normal rule is any species which is nonvolatile or harmed by heat. This will include most drugs, dyes, and detergents. It will include many high value-added materials with molecular weights above perhaps 700 daltons. For such solutes, separation by adsorption or crystallization will be preferred. For other mixtures, we should always consider distillation.

12.1.2 The Distillation Process

Distillation normally takes place in one of three different ways, as shown in Fig. 12.1-1. In the simplest, in Fig. 12.1-1(a) a saturated vapor is fed to the bottom of a column. The vapor passes up the column and is collected and condensed at the top. Part of this condensed distillate is passed back into the column where it flows countercurrently downwards, past the rising vapor. The separation occurs because the more volatile species evaporates out of the falling liquid into the vapor, and the less volatile species condenses from the vapor into the liquid.

The distillation shown in Fig. 12.1-1(a) is a close parallel to the process of gas absorption. The two other forms of distillation shown in Fig. 12.1-1 are similar. In Fig. 12.1-1(b), a saturated liquid feed enters the top of the column and flows downwards into a heated kettle, called a reboiler. The reboiler produces a vapor, which passes up the column, countercurrently to the falling liquid. The liquid "bottoms" product drawn out of the reboiler is normally in thermodynamic equilibrium with the vapor leaving the reboiler to go up the column. This type of distillation would be effective in removing a trace of impurity, like a "volatile organic compound (VOC)" from a less volatile solvent like water.

The final form of distillation, in Fig. 12.1-1(c), is just the other two set on top of each other. This is the most important form of distillation because it can produce highly purified distillate and bottoms out of a concentrated feed. This process is the workhorse of most of the commodity chemical industry. Because its analysis is somewhat more complex, we will analyze the other two cases in the next section, and defer development of this harder case until Section 12.3.

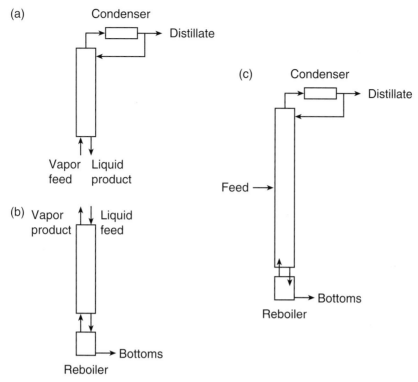

Fig. 12.1-1. Three forms of distillation. The process in (c) is most common and most important. The processes in (a) and (b), for highly pure products, are easier to analyze and so are the starting point here.

12.1.3 Distillation Equipment

As described above, the basic equipment has three parts: a reboiler, a condenser, and the column itself. The reboiler is nothing more than a kettle, most commonly heated with steam. The condenser, which is equally standard, most commonly liquifies all of the vapor.

The column internals for differential distillation are usually selected from the same group as those used for gas absorption. Again, the key question is the estimation of the maximum flows which can be tolerated without flooding. Pressure drop is important. These problems are resolved by adjusting the column's cross-sectional area, and hence changing the convective fluxes. This is the same way in which the fluid mechanics of gas absorption was resolved.

In spite of these parallels, there has been a major recent change in the internals of distillation columns. Many of those who are responsible for column performance have replaced the stages in large columns with structured packing. This has typically produced 10 to 30 percent increased capacity and allowed the same separations with less reboiler heat (i.e., with a reduced reflux ratio.) Some have hesitated in making these retrofits because of the high cost of structured packing. Because some of the patents on these packings have expired, the price of the packing is expected to fall.

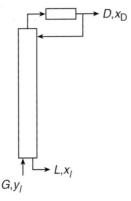

Fig. 12.2-1. Distillation producing a very pure condensate. This column is fed at the bottom with a saturated vapor.

12.2 Very Pure Products

We now switch from questions of the column's cross-section to the column's height. In other words, we have chosen how fat the tower is; we want to estimate how tall it should be. To do so, we first consider the details of a distillation column like that in Fig. 12.2-1. In this column, a saturated vapor enters the tower producing a distillate and a bottoms. As in absorption, we want to analyze the concentration changes occurring in the column. Now, however, we will follow only the concentration of the more volatile species. We will describe this concentration as a mole fraction in the gas y and as a mole fraction in the liquid x. We will use subscripts on these concentrations to indicate where in the column we are, and not the component we are considering. Thus feed contains a high concentration of the more volatile species y_I: this mole fraction may be close to one. The distillate concentration x_D is higher still, but the bottoms concentration x_I must, of course, be less pure.

We want to calculate the height of column needed to make this separation. To do so, we need three key equations: a mole balance on the more volatile species, an energy balance, and a rate equation. The overall mole balance is easy

$$G = L + D \tag{12.2-1}$$

where G and L are the vapor and liquid fluxes, respectively; and D is the distillate per column cross-section. Expressing each flow per column cross-section simplifies the form of the equations which follow. The balance on the more volatile species, written as a column height Δz, is not much harder:

$$\frac{d}{dz}(Gy) = \frac{d}{dz}(Lx) \tag{12.2-2}$$

If we assume that the fluxes G and L are constants, we have

$$\frac{dy}{dx} = \frac{L}{G} \tag{12.2-3}$$

This is subject to the condition that at the top of the column

$$z = 0 \quad y_0 = x_0 = x_D \tag{12.2-4}$$

These concentrations are equal because the condenser liquefies all the vapor, returning some condensate to the top of the column. Integrating, we find

$$y = \left(y_0 - \frac{L}{G}x_0\right) + \frac{L}{G}x \tag{12.2-5}$$

This is commonly rewritten as

$$y = \frac{x_D}{1 + R_D} + \frac{R_D}{1 + R_D}x \tag{12.2-6}$$

where $R_D\,(= L/D)$ is the reflux ratio of the column. This ratio can be adjusted to change the way the column is operating.

Eq. 12.2-6 is an important result called the "operating line" of the column. Like the operating line for absorption, it is based on mole balances. It has two important features. First, the reflux ratio R_D which appears in this equation is a major factor in the control of this type of column. Obviously, we can vary R_D between zero and infinity by changing the amount of condensate we send back into the column. If we send none back, R_D is zero, and we get little separation. If we send all of it back except one tiny drop of distillate, then R_D is infinity and the tiny drop will be the purest product possible with this column.

The second important feature of Eq. 12.2-6 is the assumption that the vapor and liquid fluxes G and L are constants. I never understand why this should be a good assumption, but it turns out in practice to be remarkably good. For the time being, we will not worry about why it works, but be content that it does.

In addition to the operating line, based on mole balances, we need a second key equation based on energy balances. This says that the vapor concentration y^* in equilibrium with the local liquid concentration x is given by

$$y^* = a + mx \tag{12.2-7}$$

where a and m are constants. Note that m is closely related to a Henry's law constant, except for a solution concentrated in the more volatile component. Note that when $x = 1$, $y^* = 1$; hence a and m add up to one. Obviously, the linear relation in Eq. 12.2-7 is an approximation, which will need to be replaced when the components being distilled are present in more equal amounts. Finally, remember that this line expresses a vapor–liquid equilibrium at constant pressure and so is implicitly at a varying temperature.

We need our third key equation, a mole balance on the vapor phase alone. Again, we consider the moles of the more volatile species entering and leaving the vapor in a column section $A\Delta z$:

$$[\text{accumulation}] = \left[\begin{array}{c}\text{moles}\\\text{flowing in}\end{array}\right] - \left[\begin{array}{c}\text{moles flowing}\\\text{out}\end{array}\right] + \left[\begin{array}{c}\text{moles}\\\text{evaporating}\end{array}\right]$$

$$0 = (GAy)_{z+\Delta z} - (GAy)_z + K_y a\,(A\Delta z)\,(y^* - y) \tag{12.2-8}$$

As expected, we divide by the volume $(A\Delta z)$ and take the limit as this volume goes to zero:

$$0 = G\frac{dy}{dz} + K_ya(y^* - y)$$

(12.2-9)

This is subject to the boundary conditions

$$z = 0, \quad y = y_0 = x_D$$

(12.2-10)

$$z = l, \quad y = y_l$$

(12.2-11)

We can now integrate this equation. In this integration, we find y^* as a function of x from Eq. 12.2-7, and x as a function of y from Equation 12.2-6. The result is

$$
l = \int_0^l dz = \frac{G}{K_ya} \int_{y_l}^{y_0} \frac{dy}{y^* - y}
$$

$$
= \left[\frac{G}{K_ya}\right] \left(\frac{1}{1 - \dfrac{mG}{L}} \ln \left(\frac{y_l^* - y_l}{y_0^* - y_0}\right)\right)
$$

$$
= \left[\frac{G}{K_ya}\right] \left(\frac{1}{1 - \dfrac{m(R_D + 1)}{R_D}} \ln \left(\frac{(a + mx_l) - y_l}{(a + mx_D) - x_D}\right)\right)
$$

(12.2-12)

Just as in gas absorption, this result is often rewritten in different terms:

$$l = [\text{HTU}] \cdot (\text{NTU})$$

(12.2-13)

The height of a transfer unit HTU, given in square brackets in these equations, measures the column efficiency

$$\text{HTU} = \frac{G}{K_ya}$$

(12.2-14)

A small HTU is evidence of efficient column internals. The number of transfer units NTU, given in parentheses, describes the difficulty of the separation

$$\text{NTU} = \frac{1}{1 - \dfrac{m(R_D + 1)}{R_D}} \ln \frac{y_l^* - y_l}{y_0^* - y_0}$$

(12.2-15)

A small NTU signals an easy distillation.

As with gas absorption, the characteristics of this distillation can be clarified graphically. To do so, we make a plot of y vs. x as given by the operating line in Eq. 12.2-5 or Eq. 12.2-6. This plot passes through the 45° diagonal at the point (x_D, x_D), as shown in Fig. 12.2-2; its slope is (L/G), or $(R_D/(R_D + 1))$. On the same coordinates, we plot y^* vs. x, as given by the equilibrium line in Eq. 12.2-7. This plot passes through the point $(1,1)$ and has slope m.

The difference between the two lines $(y^* - y)$ in Fig. 12.2-2 is the driving force for distillation. When this difference is large, the vapor concentration will change rapidly

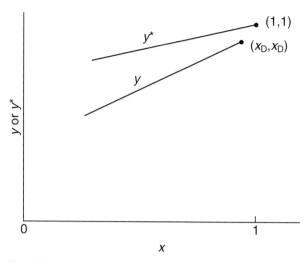

Fig. 12.2-2. Performance of a column producing a very pure condensate. All mole fractions involved are close to one. Note the equilibrium line is above the operating line.

over the column length. When the difference is small, the concentration will change more slowly because the mass transfer will be slower. If the operating and equilibrium lines cross, the separation is not possible. This type of plot often gives important insight into what is happening in the distillation.

This plot can also provide important information about the operation of the column. Normally, we will know the feed concentration, and we will want to specify the distillate concentration. We are given an equilibrium line by the chemistry of the system. If we have an infinitely tall column, the liquid exiting the column would be in equilibrium with the feed. In graphical terms, this means that the operating line passes through the points (x_D, x_D) and $(((y_I - a)/m), y_I)$. The slope of this operating line corresponds to the minimum reflux at which the separation is possible. We will often design the column to operate at a reflux ratio which is 1.1 to 1.8 times larger than this minimum reflux. These ideas are illustrated in the examples that follow.

Example 12.2-1: Mass transfer in a small still We want to remove acetic acid from acetone to be used to rinse electronic devices. Our still is 1.22 m high and 0.088 m in diameter, fed with saturated vapor at about 0.026 mol/sec. With a feed of 1.1% acetic acid, the best that we can get from a distillate is 0.04% acetic acid. We believe that the equilibrium line for this system is

$$y^* = 0.93 + 0.07x$$

where y^* and x are acetone mole fractions. (a) What is the mass transfer coefficient per volume $K_y a$ in this still? (b) How can we get a purer distillate?

    Solution (a) We will solve this problem in terms of mole fractions of acetone, which is the more volatile species. To get the best separation, we must make the concentration difference $(y^* - y)$ as large as possible. We do so by operating at infinite reflux, with G equal to L. Thus we are collecting a vanishingly small amount of distillate,

and the operating line is the 45° diagonal. In this case, the vapor and liquid concentrations x and y must be equal everywhere in the column. Equation 12.2-12 now reduces to

$$l = \left[\frac{G}{K_y a}\right] \left(\frac{1}{1 - \dfrac{mG}{L}} \ln \left(\frac{(0.93 + 0.07 y_l) - y_l}{(0.93 + 0.07 y_0) - y_0}\right)\right)$$

$$= \left[\frac{G}{K_y a}\right] \left(\frac{1}{1 - m} \ln \left(\frac{1 - y_l}{1 - y_0}\right)\right)$$

Inserting the values given

$$1.22 \text{ m} = \left[\frac{0.026 \text{ mol/sec}}{\dfrac{\pi}{4}(0.088 \text{ m})^2}}{K_y a}\right] \left(\frac{1}{1 - 0.07} \ln \left(\frac{0.011}{0.0004}\right)\right)$$

Thus

$$K_y a = 12.5 \text{ mol/m}^3 \text{ sec}$$

Note that this is mass transfer per volume, not the more normal mass transfer per area.
   (b) To improve this separation, we must make the still more efficient. More quantitatively, we see from Eq. 12.2-13 that because l is fixed, we must reduce the HTU to
increase the NTU. The easy way to do this is to decrease the feed flux G. We should
remember that such a decrease will also reduce K_y and potentially a, so we may not get as
much benefit as we hope.

Example 12.2-2: Benzene purification   You are distilling modest quantities of a 95 percent benzene feed to get a purer feedstock for catalytic experiments. The equilibrium line
for this concentration and higher is approximately

$$y^* = 0.58 + 0.42x$$

(a) You want to make a 99% pure product using a reflux which is 1.5 times the minimum.
How many transfer units will you need? (b) You have a 2 m column with an HTU of 0.34 m.
What is the best separation of this feed that is possible with this column?
   Solution   (a) When the reflux ratio is the minimum possible, the operating line
must run from the point (x_D, x_D) to the point (x_l^*, y_l). In quantitative terms, it runs from
the point (0.99, 0.99) to the point $\left(\dfrac{0.95 - 0.58}{0.42}, 0.95\right)$. The slope of this line is

$$\frac{\Delta y}{\Delta x} = \frac{L}{G} = \frac{R_D(\text{min})}{1 + R_D(\text{min})} = \frac{0.99 - 0.95}{0.99 - 0.881}$$

Thus

$$R_D(\text{min}) = 0.579$$

and under our operating conditions,

$$R_D = 1.5\, R_D(\min) = 0.869$$

From an overall balance, we can find the concentration of the exiting liquid:

$$Gy_l = Lx_l + Dx_D$$

$$(R_D + 1)y_l = R_D x_l + x_D$$

$$1.869(0.95) = 0.869 x_l + 0.99$$

As a result

$$x_l = 0.904$$

Finally, from Eq. 12.2-15

$$NTU = \dfrac{1}{1 - \dfrac{m(R_D + 1)}{R_D}} \ln \left(\dfrac{y_l^* - y_l}{y_0^* - y_0} \right)$$

$$= \dfrac{1}{1 - \dfrac{0.42(1.869)}{0.869}} \ln \left(\dfrac{(0.58 + 0.42\,(0.904)) - 0.95}{(0.58 + 0.42\,(0.99)) - 0.99} \right)$$

$$= 5.3$$

This is a typical number.

(b) To find the best possible separation, we first recognize from Equation 12.2-13 that

$$NTU = l/HTU = 2.0/0.34 = 5.88$$

Under these conditions, we run a total reflux, so R_D is infinite. Equation 12.2-15 then gives

$$5.88 = \dfrac{1}{1 - 0.42} \ln \left(\dfrac{(0.58 + 0.42\,(0.95)) - 0.95}{(0.58 + 0.42\,(x_D)) - x_D} \right)$$

Thus

$$x_D = 0.998$$

At total reflux, we can get a better separation than that in part (a).

Example 12.2-3: Differential stripping   Extend the analysis for the vapor-fed rectifying column in Figs. 12.1-1(a) and 12.2-1 to a liquid-fed stripping column shown in Fig. 12.1-1(b). In particular, develop equations for the operating line, the equilibrium line, and the rate equation. Combine these to find the HTU, the NTU, and the column height.

Solution We begin with an overall balance, a parallel to Equation 12.2-1

$$L = B + G$$

where B is the bottoms stream, again normalized by dividing by the cross-section of the column. We then make a species balance from the bottom part of the column to some arbitrary position to find

$$G(y - y_l) = L(x - x_l)$$

$$y = y_l + \frac{L}{G}(x - x_l)$$

In this analogue of Eq. 12.2-5, we note that L/G is equal to or greater than one. In the earlier case, it was equal to or less than one. The equilibrium line is now

$$y^* = mx$$

a contrast with Eq. 12.2-7. Note that y_l^* equals mx_B because the reboiler usually functions at very near equilibrium.

We now turn to the rate equation, which is the same as Eq. 12.2-9

$$0 = G\frac{dy}{dz} + K_y a(y^* - y)$$

We integrate this between the ends of the column to find

$$l = \left[\frac{G}{K_y a}\right]\left(\frac{1}{1 - \dfrac{mG}{L}} \ln\left(\frac{y_l^* - y_l}{y_0^* - y_0}\right)\right)$$

$$= [\text{HTU}] \cdot (\text{NTU})$$

The concentrations at the top of the column y_0 and x_0 will normally be known or can be calculated from an overall mass balance. Once x_0 is known, y_0^* is found from the equilibrium line. The concentration y_l is in equilibrium with the bottoms concentration x_B. The concentration x_l can often be found by a mass balance on the column alone, and y_l^* is then found from x_l using the equilibrium line. Thus the NTU and the column height l can be calculated.

12.3 The Column's Feed and its Location

In most cases, we will want to separate a feed of two components into a distillate which is largely the more volatile component and a bottoms which is mostly the less volatile component. To do so, we will normally use a distillation column fitted with both a condenser and a reboiler. Thus we commonly will be interested in more complex cases than in the previous section. There, a column was fed at the bottom with a vapor, or at the top with a liquid.

In the more common case, we will feed the column somewhere in the middle. Such a column, which was shown schematically in Fig. 12.1-1(c), requires a more complex analysis than that for a column fed at the top or the bottom. All three key equations basic

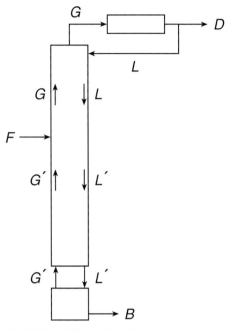

Fig. 12.3-1. Differential distillation separating a concentrated feed. The central feed changes the flows of vapor and liquid in the column.

to the analysis are usually altered. The operating line, based on mole balances, changes because the feed alters the flows within the column. The equilibrium line becomes non-linear, a result of the more concentrated solutions that are involved. The rate equation is different because the fluxes like G can change because of the more central feed. Changes in fluxes can also change the individual mass transfer coefficients, and hence the overall mass transfer coefficient K_y.

In this section, we focus on changes to the operating line. We defer discussing the other changes to the description of concentrated differential distillation in Section 12.4. To understand the changes in operating line, we first refer to the schematic of a column shown in Fig. 12.3-1. By convention, that part of the column above the feed is called the "rectifying section"; that below is termed the "stripping section." The vapor and liquid fluxes above the feed location are called G and L; those below the feed are called G' and L'.

We begin with the case of a saturated liquid feed. Such a feed does not change the vapor flow up the column; it only alters the liquid flow down the column. In other words, G equals G' and L' equals $(L+F)$ where F is the feed per time per column cross-section. This means that the slope of the operating line must change at the feed point, from (L/G) to $((L+F)/G)$.

However, other types of feed are obviously possible, as illustrated schematically in Fig. 12.3-2. For example, if the feed is a cold liquid that is below the saturation temper-ature at the composition on that stage, then the feed will necessarily condense some of the vapor rising up through the column. In this case, G' will be greater than G. In the same sense, if the feed is a superheated vapor, then it will vaporize some of the liquid moving down the column, and L will be greater than L'.

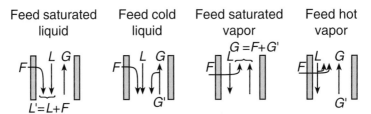

Fig. 12.3-2. The effects of different feeds. The conditions at which the feed enters affect the flows in the column as shown.

To see how these new feeds alter the distillation, we define a new quantity q as equal to the moles of liquid produced per mole of feed:

$$q = \frac{L'-L}{F} \qquad (12.3\text{-}1)$$

By comparison with Fig. 12.3-2, we see that $q = 1$ if the feed is a saturated liquid. Similarly, q equals zero if the feed is a saturated vapor. However, for feeds that are subcooled liquids or superheated vapors, we need a more complete energy balance. For a feed of subcooled liquid, we recognize that

$$\begin{bmatrix} \text{energy supplied} \\ \text{by condensation} \end{bmatrix} = \begin{bmatrix} \text{energy needed to heat the liquid feed to} \\ \text{its saturation temperature} \end{bmatrix} \qquad (12.3\text{-}2)$$

More specifically,

$$\begin{bmatrix} \text{moles vapor} \\ \text{condensed} \end{bmatrix} \begin{pmatrix} \text{heat of} \\ \text{vaporization } \lambda \end{pmatrix} = \begin{bmatrix} \text{heat capacity per} \\ \text{mole liquid } \tilde{C}_p \end{bmatrix}$$

$$\times \begin{pmatrix} \text{bubble} \\ \text{temperature } T_B \end{pmatrix} - \begin{pmatrix} \text{feed} \\ \text{temperature } T_F \end{pmatrix} \begin{pmatrix} \text{moles} \\ \text{feed} \end{pmatrix} \qquad (12.3\text{-}3)$$

Rearranging, we see

$$\frac{\text{moles vapor condensed}}{\text{moles feed}} = \frac{\tilde{C}_p(T_B - T_F)}{\lambda} \qquad (12.3\text{-}4)$$

From the definition of q, we know

$$q = \frac{\text{moles feed} + \text{moles vapor condensed}}{\text{moles feed}} \qquad (12.3\text{-}5)$$

Combining, we obtain the desired result for a subcooled liquid feed:

$$q = 1 + \frac{\tilde{C}_p(T_B - T_F)}{\lambda} \qquad (12.3\text{-}6)$$

Remember that the heat capacity $\tilde{C}_p$ is that for the liquid. Note also that when the feed temperature equals the temperature at the bubble point, $q = 1$, consistent with our expectation above.

We can obtain similar results for superheated vapor feed by arguments that are completely parallel. The result is

$$q = \frac{\tilde{C}_p(T_D - T_F)}{\lambda} \qquad (12.3\text{-}7)$$

in which $\tilde{C}_p$ is now the heat capacity of the vapor, T_D is the temperature at the dew point, and T_F is the temperature of the feed. Because for a superheated vapor the feed temperature is greater than the temperature at the dew point, q is negative for this case. However, when the feed is a vapor fed at saturation, $T_F = T_D$ and q is zero, again consistent with our qualitative argument above.

We now want to describe how different feeds affect the operating line. We begin by writing mass balances from the feed stage up to and including the ends of the column. For the upper "rectifying" section, we have an operating line

$$Gy = Lx + Dx_D \qquad (12.3\text{-}8)$$

For the lower "stripping" section,

$$L'x = G'y + Bx_B \qquad (12.3\text{-}9)$$

Our interest is in the intersection of these two operating lines, where x has the same value in the rectifying and stripping sections, and y does also. Thus our interest is in the sum of the last two equations:

$$\begin{aligned}(G - G')y &= [Dx_D + Bx_B] + (L - L')x \\ &= [Fx_F] + (L - L')x \qquad (12.3\text{-}10)\end{aligned}$$

where x_F is the concentration in the feed of the more volatile component.

In this expression, we have used the overall solute balance to rewrite the quantity in square brackets in terms of the feed. From Eq. 12.3-1, we expect that $(L - L') = (-qF)$; from an overall balance of the feed location we see that

$$\begin{aligned}G - G' &= (L - L') + F \\ &= -qF + F \qquad (12.3\text{-}11) \\ &= F(1 - q)\end{aligned}$$

By combining Eqs. 12.3-10 and 12.3-11 and rearranging, we get a new "feed line" which gives the intersection of the operating lines

$$y = \frac{x_F}{1 - q} + \left[-\frac{q}{1 - q}\right]x \qquad (12.3\text{-}12)$$

Note that this equation is a straight line passing through the point (x_F, x_F).

The new feed line has simple behavior for a number of special cases. For a saturated liquid feed, $q = 1$ and the feed line is vertical. For the case of a saturated vapor feed, $q = 0$ and the feed line is horizontal. For the case of a subcooled liquid, $q > 1$ and the slope of the feed line is positive. For the case of a superheated vapor, $q < 0$ and the feed line again has a positive slope. Finally, for the case of the feed which is a mixed vapor and liquid in equilibrium with each other, the feed line has a negative slope.

The feed line marks the change from the operating line for the upper, rectifying section and that for the lower, stripping section. In other words, it passes both through the point (x_F, x_F), as suggested by Eq. 12.3-12, and through the intersection of the feed lines. We will use this feature in calculating the column height in the next section.

12.4 Concentrated Differential Distillation

We now turn to the full complexities of differential distillation of a concentrated feed into a distillate which is highly enriched in the more volatile component and a bottoms stream which contains very little of this volatile component. We do so by effectively placing a rectifying column like that in Fig. 12.1-1(a) on top of a stripping column sketched in Fig. 12.1-1(b). The feed enters at the junction between these two columns.

Like the analysis of dilute differential distillation, the analysis of concentrated differential distillation depends on three equations: an operating line, an equilibrium line, and a rate equation. Each is changed when the feed is concentrated. The changes to the operating lines, which were detailed in the previous section, are greatest. Above the feed, the operating line is

$$
\begin{aligned}
y &= \frac{Dx_D}{G} + \frac{L}{G} x \\
&= \frac{x_D}{1 + R_D} + \frac{R_D}{1 + R_D} x
\end{aligned}
$$
(12.4-1)

where $R_D (= L/D)$ is the reflux ratio. Below the feed, the operating line is

$$
y = -\frac{Bx_B}{G'} + \frac{L'}{G'} x
$$
(12.4-2)

where L' and G' are the liquid and vapor fluxes in the lower, stripping portion of the column. Remember that in general, these fluxes will differ from those above the feed. For example, for the common case of a saturated liquid feed, L' is $(L + F)$, even though G' equals G.

The changes in the equilibrium line and the rate equation are equally major, but they may be easier to understand. The equilibrium line is now nonlinear

$$
y^* = y^* (x)
$$
(12.4-3)

At small x, we expect

$$
y^* = mx
$$
(12.4-4)

and at large x, we can assume

$$
y^* = a + m'x
$$
(12.4-5)

However, m and m' will be far from equal. Often, we will expect m to be much greater than m'. This complicates our analysis but still involves the same thinking.

The rate equation for concentrated cases still has the familiar form of a steady-state mass balance on the volatile component alone:

$$
[\text{accumulation}] = [\text{moles flowing in}] - [\text{moles flowing out}] + [\text{moles transferred}]
$$

$$
0 = G\frac{dy}{dz} + K_y a(y^* - y)
$$

(12.4-6)

For the upper section of the column, this is subject to

$$z = 0, \qquad y = x_D \tag{12.4-7}$$

$$z = l - l', \quad y = y' \tag{12.4-8}$$

where $(l - l')$ is the height of just the rectifying section and y' is the vapor concentration at the feed point. As we showed in the previous section of this chapter, y' will be given by the intersection of the feed line with the operating line in the upper section. This equation can be integrated to give

$$l - l' = \int_0^{l-l'} dz = \left[\frac{G}{K_y a}\right] \left(\int_{x_D}^{y'} \frac{dy}{y^* - y}\right) \tag{12.4-9}$$

As before, the quantity in square brackets corresponds to the HTU, and that in parentheses is the NTU for the rectifying section of the column. As before, the HTU measures the efficiency of the distillation equipment and the NTU indicates the difficulty of the separation.

The results for the lower, stripping section of the column are analogous. A steady-state mass balance gives

$$0 = G' \frac{dy}{dz} + K_y a (y^* - y) \tag{12.4-10}$$

where the vapor flux G' may differ from that in the top of the column. This is subject to the constraints

$$z = l - l', \quad y = y' \tag{12.4-11}$$

$$z = l, \qquad y_l = y^* (x_B) \tag{12.4-12}$$

The first of these constraints asserts that the vapor concentration varies continuously from the rectifying to the stripping sections of the column. Note that this implies no major perturbation caused by the feed: it enters and mixes very quickly, causing a sudden change in flow but not in concentration. The second of these conditions repeats our expectation that the reboiler acts like a single equilibrium stage. Thus the vapor concentration y_l leaving the reboiler is in equilibrium with the liquid concentration x_B leaving the reboiler.

The rate equation for the stripping section can also be integrated

$$l' = \int_{l-l'}^{l} dz = \left[\frac{G'}{K_y a}\right] \left(\int_{y'}^{y_l(x_B)} \frac{dy}{y^* - y}\right) \tag{12.4-13}$$

Again, the quantity in parentheses is the NTU, now for the stripping section of the column. The quantity in square brackets is the HTU. We would normally expect that the HTU in the top of the column will be somewhat different than in the bottom. This occurs for three reasons. First, because the flows of vapor and liquid will change, the individual mass transfer coefficients k_y and k_x can change. Second, because the slope of the equilibrium line m is different, the weighting of mass transfer resistances in gas and liquid will be different. Third, unless the feed is a saturated liquid, the vapor flows G' and G will be different. These factors are often ignored, and the HTU is assumed constant over the height of the column.

When the HTU is constant, we can combine Eq. 12.4-9 and 12.4-13 to find the total height of the tower:

$$l = \int_0^l dz = \text{HTU} \cdot \text{NTU} \tag{12.4-14}$$

where

$$\text{NTU} = \int_{x_D}^{y_l(x_B)} \frac{dy}{(y^* - y)} \tag{12.4-15}$$

Once we know the equilibrium and operating lines, we now just evaluate $(y^* - y)$ vs. y and perform this integration numerically. Remember that a large NTU signals a difficult separation, and a small NTU is an easy one. Note also that by integrating numerically between $y_0 (= x_D)$ and an intermediate value of y, we can find how the vapor composition varies with tower position z.

In the analysis of distillation, we will normally specify the feed, distillate and bottoms concentrations. We will specify the reflux ratio, usually as between 1.2 and 1.8 times the minimum. With this information, we can evaluate the integral in Eq. 12.4-15, and hence the total height of the column. These ideas are illustrated in the following example.

Example 12.4-1: Benzene–toluene distillation in a packed tower You want to separate a saturated liquid feed of 3500 mol/hr containing 40% benzene into a distillate containing 98% benzene and a bottoms with 2% benzene. The reflux ratio should be 1.5 times the minimum; the packing has an HTU of 0.2 m.

How tall a tower is needed?

Solution We begin our analysis with overall balances for a basis of one hour and a column of cross-sectional area A

$$3500 \, \text{mol/hr} = DA + BA$$

$$0.4(3500 \, \text{mol/hr}) = 0.98 \, DA + 0.02 \, BA$$

Thus

$$DA = 1400 \, \text{mol/hr}$$

$$BA = 2100 \, \text{mol/hr}$$

Next, we calculate the feed line, which for a saturated liquid feed with $q = 1$ can be found from Eq. 12.3-12:

$$x = 0.40$$

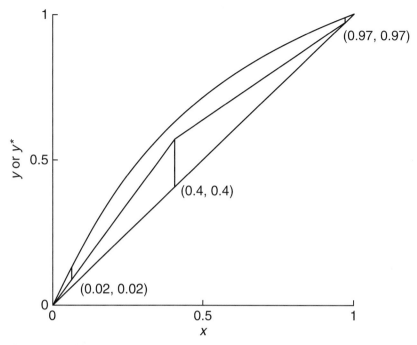

Fig. 12.4-1. Differential distillation of benzene and toluene. The objective is finding the number of transfer units from value of $(y^* - y)$. A parallel problem using stages is given in Example 13.3-1.

The equilibrium line giving the mole fraction of benzene in the vapor y^* vs. that in the liquid x is the curve shown in Fig. 12.4-1. The feed line contacts this equilibrium line at the point (0.40, 0.625). When the reflux ratio R_D has its minimum value, the operating line above the feed must pass through the point (0.40, 0.625) and through the point (x_D, x_D), or in this case (0.97, 0.97). From Eq. 12.4-1, we see

$$\frac{R_D(\text{min})}{1 + R_D(\text{min})} = \frac{\Delta y}{\Delta x} = \frac{0.97 - 0.625}{0.97 - 0.40}$$

Hence

$$R_D(\text{min}) = 1.59$$

This is the smallest reflux ratio at which a separation is possible, and requires an infinitely tall column. We want to operate at 1.5 times this minimum:

$$R_D = 1.5(1.59) = 2.39$$

Thus, in the upper rectifying part of the column, LA and GA are 3330 and 4730 mol/hr, respectively. The operating line for this region is found from Eq. 12.4-1:

$$y = 0.287 + 0.704x$$

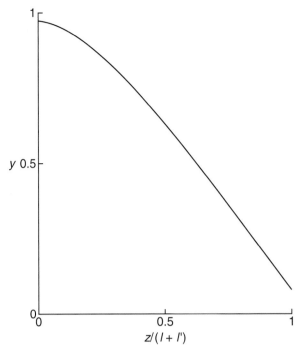

Fig. 12.4-2. Benzene vapor composition vs. tower position. The concentration changes more slowly near the top of the column ($z = 0$) because the driving force ($y^* - y$) is smaller.

Below the feed, in the rectifying part of the column, the flows $L'A$ and $G'A$ are 6830 and 4730 mol/hr, respectively; G' equals G because the feed is a saturated liquid. The operating line for this stripping part of the column is given by Eq. 12.4-2:

$$y = -0.009 + 1.444x$$

Together, these operating lines show how the vapor and liquid concentrations vary along the column.

We can now use the operating and equilibrium lines to perform the integration in Eq. 12.4-15. Because x_B is 0.02, we find the concentration of vapor leaving the reboiler $y^*(x_B)$ is 0.085. This is the concentration of the vapor entering the bottom of the column, and sets the upper limit of the integral. Making the integration, we find

$$NTU = 13.9$$

Thus the tower height for this separation

$$l = 0.2\,m\,(13.9) = 2.8\,m$$

This is what we seek.

We can gain more insight into this separation by calculating how the separation varies with column height, as shown in Fig. 12.4-2. As expected, the vapor concentration is 0.97 at the top of the column, when $z/z(l + l')$ is zero. It is 0.02 at the bottom. It varies more slowly with position at the top, near the feed, and at the bottom, because these are the points where the driving force $(y^* - y)$ is smaller. In this case, these differences are not major; in other separations, they may be. Parenthetically, this same separation is analyzed for staged distillation in Example 13.3-1.

12.5 Conclusions

Distillation separates a feed solution by means of volatility differences. When the feed is at the top or the bottom of the column, the analysis parallels that for gas absorption. In the most common cases, the feed is in the middle of the column, and the feed is separated into a distillate enriched in the volatile component and a bottoms depleted in this component. In all cases, the analysis begins with mole balances, called operating lines; and with free energy balances, called equilibrium lines.

How the analysis of distillation proceeds further depends on the internals of the distillation column equipment. In almost all laboratory columns and in many modern commercial columns, these internals are random or structured packing. The packing in small columns is often the same as the random packing used for absorption. The packing in commercial columns is more commonly structured because it permits higher flows at modest pressure drops. In both cases, the analysis uses a mass balance on one phase. This mass balance depends on rates of mass transfer summarized as an overall mass transfer coefficient. The analysis then is a close parallel to that used for absorption. When the distillation involves a dilute solution, the column size can often be estimated analytically. When the distillation involves a concentrated solution, the estimation of column size is similar but requires numerical integration.

This form of distillation is an example of "differential contacting." Sometimes, the column contains discrete, separate compartments called "stages." Such "staged contacting" is discussed in the next chapter.

Questions for Discussion

1. What are some specific chemicals that are separated by distillation?
2. What effects allow distillation to be effective?
3. Why would you not choose distillation to separate a mixture?
4. When will distillation of two volatile species fail?
5. How could you tell if a differential distillation column will flood?
6. How are the vapor and liquid concentrations at the top of a distillation column related?
7. How are the concentrations in and out of the reboiler related?
8. What is the reflux ratio?
9. What is an advantage and a disadvantage of a large reflux ratio?
10. What is an advantage and a disadvantage of increasing the amount reboiled?
11. Why is there one operating line in some distillation columns and two operating lines in others?
12. Are the operating lines in a distillation column dependent on column internals?
13. Discuss the limits of the equilibrium line when the mole fraction of the volatile species x is near zero and near one.

14. Would it ever make sense to run a distillation with a partial condenser?
15. Could the height of a transfer unit change over the length of a column?
16. Compare differential distillation with gas absorption.

Problems

1. Distillation is being carried out in a packed column to produce 99.9% benzene from a feed of 99% benzene and 1% toluene fed directly to the reboiler. The equilibrium line over this concentration range is

$$y^* = 0.58 + 0.42\,x$$

The feed is 100 mol/hr, and the bottoms is 16 mol/hr. The condenser is not total, but produces equal amounts of a product and a liquid returned to the top of the tower. These streams are approximately in equilibrium. (a) How much distillate is produced? (b) What is the reflux ratio? (c) What is the vapor concentration coming out of the top of the tower? (d) What is the vapor concentration going into the bottom of the tower? (e) What number of transfer units (NTU) is involved?

2. You want to remove traces of a volatile solvent and make ultrapure water by distillation in a column filled with structured packing and fitted with a condenser and a reboiler. Note the reboiler operates as an equilibrium stage. Your bottoms should contain less than 10^{-5} mole fraction solvent; your distillate should be 0.02 mole fraction solvent; and the equilibrium curve is

$$y^* = 6.1x$$

You plan to operate at high reflux. How many transfer units are needed?

3. You have a saturated vapor feed containing 5% water and 95% methanol. You want to feed this to the bottom of a column to make 99.99% product methanol and a waste of 90% methanol. The equilibrium line in this case is

$$y^* = 0.60 + 0.40x$$

(a) What is the reflux ratio in the column? (b) You want to use differential distillation with packing, which gives an HTU of 0.3 m. How tall will the distillation tower be?

4. You are making a pigment for ink-jet printing in a stirred tank reactor containing one nonvolatile reagent dissolved in a halogenated aromatic. You are slowly adding a second nonvolatile reagent dissolved in acetone and keeping the total volume small by boiling off extra solvent. The acetone concentration in this mixture is 0.12; the equilibrium line is

$$y^* = 0.82 + 0.18x$$

Unfortunately, you are losing too much aromatic. To solve this, you plan to add a 1.2 m packed column and a condenser to the top of this reactor. You expect the HTU for the column to be 0.26 m. If you run with a high reflux, what will the mole fraction of acetone be in the distillate?

5. The wash liquid from your paint shop produces 3000 mol/day of a 40% solution of methanol in water. You have been dumping this in the sewer as a 10% solution, just by

diluting this. Still, methanol's expensive so you'd like to recover some of it, at least as a 90% solution. You can buy an old 6-m packed column including a reboiler, with an HTU of 0.33 m. It has the feed directly to the reboiler and a reflux ratio of six. Equilibrium data for this system are as follows:

x	y
0.10	0.41
0.20	0.57
0.30	0.67
0.40	0.73
0.50	0.77
0.60	0.84
0.80	0.93
1.00	1.00

(a) What is the composition of the bottoms stream? (b) Assuming a saturated liquid feed, what would it be if you switched the feed to the optimum location?

6. Your company produces *ca.* 7000 kg mol/day of a 10% solution of methanol–water, for which the vapor–liquid data are in the previous problem. This methanol stream is to be distilled to produce concentrates for two purposes: a pigment precipitation using 200 kg mol/day of 90% methanol, and a recrystallization using the rest as 40% methanol. As a result, your boss suggests a column like that shown below.

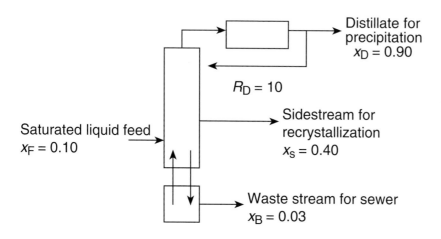

Saturated liquid feed
$x_F = 0.10$

$R_D = 10$

Distillate for precipitation
$x_D = 0.90$

Sidestream for recrystallization
$x_S = 0.40$

Waste stream for sewer
$x_B = 0.03$

Your job is to size the column. (a) How much new methanol do you need per day? (b) What are the values of the liquid and vapor flows? (c) What are the slopes of the operating lines? Plot the operating lines on the y-x diagram. (d) How many transfer units are needed? (e) Where should the column be fed?

Further Reading

Kister, H. Z. (1992). *Distillation Design*. New York: McGraw Hill.

Kister, H. Z. (2001). *Distillation Troubleshooting*. New York: Wiley.

McCabe, W., Smith, J. and Harriott, P. (2004). *Unit Operations of Chemical Engineering*, 7th ed. New York: Mc22Graw-Hill.

Petlyuk, F. (2004). *Distillation Theory*. Cambridge: Cambridge University Press.

Wankat, P. C. (2006). *Separation Process Engineering*, 2nd ed. Englewood Cliffs: Prentice Hall.

CHAPTER 13

Staged Distillation

      As outlined in the previous chapter, distillation is a separation based on volatility differences. It is by far the most important separation. In North America alone, distillation consumes over one million barrels of oil per day, or about four percent of the continent's energy consumption. It is estimated to be only eleven percent efficient, and so offers an opportunity for increased energy efficiency.

      Distillation normally involves three pieces of equipment: a column, a condensor on top of the column, and a reboiler at the bottom of the column. The reboiler, often a steam-jacketed kettle, is heated so that much of its contents evaporate and flow upwards through the column. The vapors passing out of the column are liquefied in the condensor and much of this condensate is sent back downwards through the column. This countercurrent flow of vapor and condensate is common to all forms of distillation.

      The internals of the column itself can differ dramatically. In many columns, including almost all found in the laboratory, the columns' internals are random or structured packing. These packed columns were described in detail in Chapter 12. In many other columns, especially older, large-scale equipment, the column's internals are "stages," volumes providing close contact between vapor and condensate. In most cases, the stages are designed so that liquid and vapor leave to each stage nearly in equilibrium with each other.

      This chapter discusses staged distillation. In such a process, the concentrations of vapor and liquid do not change continuously from one end of the column to the other as they do in adsorption or differential distillation. Instead, the vapor and liquid concentrations have only discrete values, with new values on each stage. As shown below, such "staged contacting" is analyzed very differently than "differential contacting" discussed in the previous chapters.

      We normally begin each chapter on separation processes with a discussion of what species needs to be separated. However, we already had such a discussion in the previous chapter, on differential distillation. Instead, we begin in Section 13.1 with a discussion of the diameter of a column needed for a given distillation. Estimating this tower diameter is based on fluid mechanics, and so parallels the estimation of the diameter of an absorption tower given in Section 10.2.

      We next turn to estimating the tower's height. In practice, the height of each stage is normally fixed, in North America at 0.6 m. As a result, estimating the height is equivalent to estimating the number of stages required. In addition, the number of stages is often approximated first as being that at equilibrium, and then later corrected with a mass transfer-dependent efficiency. It is like a calculation in physics, which is first made neglecting friction, and then is later corrected for the presence of friction.

      We calculate the number of stages for a dilute feed in Section 13.2. In Section 13.3, we describe the effect of a concentrated feed, commonly located somewhere in the middle of the column; and we use this description to calculate the number of stages for concentrated, staged distillation. We explore how mass transfer compromises this "equilibrium stage" model in Section 13.4. In all these sections, we are implicitly calculating the height of the distillation tower, because this height is just the number of stages times 0.6 m.

13.1 Staged Distillation Equipment

The design of distillation columns is commonly phrased in terms of vapor and liquid flows. For example, we will speak of vapor flow of 600 mol/sec of 99.7% propylene coming out of the top of a propylene–propane distillation tower. We will consider the effect of changing the product purity by changing the reflux ratio, and hence the ratio of liquid to vapor flows in the column. We will evaluate the effect of partially vaporizing some of our feed, and hence reducing the liquid flow in the lower, stripping section of our column.

These concerns are important, but they do not explore how large the equipment should be. We do not worry about the effect of 6000 mol/sec instead of 600 mol/sec. We don't focus on the effect of cutting the feed in half. In other words, our common distillation design aims at the number of stages or how tall the column is. Our common design does not focus on the column's diameter, on how fat the column is.

In this section, we want to emphasize column diameter. Doing so is not a question of mass transfer or of thermodynamics, but of fluid mechanics. It is not a question of mass and energy balances, but of physics. To see why, we consider the schematic drawing of liquid and vapor in Fig. 13.1-1. The liquid, which is shown shaded, is flowing down the column. It flows across each tray to the downcomer, where it gushes down to the tray below. Vapor flows upwards, passing through holes in each tray and forming bubbles which froth upwards. The whole thing has the tumultuous turbulence of a waterfall in the forests of northern Minnesota.

The performance of a staged distillation column is thus a precarious balance between the rising vapor and the falling liquid. The liquid flow should be kept on the trays, and the vapor flow must be high enough to keep the liquid from leaking through the holes in the trays, but not so high that liquid drops get entrained in the rising vapor.

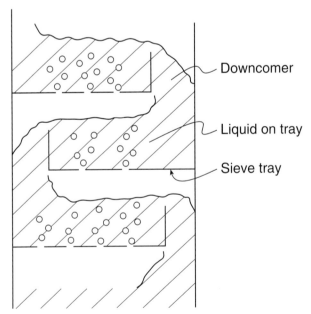

Fig. 13.1-1. Fluid mechanics on sieve trays. Liquid condensate flows across each stage, while vapor bubbles upwards, through the sieves and the flowing liquid.

Not surprisingly, there is a vivid vocabulary to explain this precarious balance. This includes

(i) *entrainment flooding*, when liquid drops are swept upwards;
(ii) *downcomer flooding*, when the liquid can't flow downwards;
(iii) *weeping*, when liquid drops fall through the holes in the sieve;
(iv) *dumping*, when the drops become streams; and
(v) *turndown ratio*, which is the maximum vapor flow divided by the minimum vapor flow.

These descriptors always seem as appropriate for a dating service as for a distillation column.

The two more important column characteristics are entrainment flooding and turndown ratio. To see when entrainment of droplets will be a problem, we make a force balance on a drop which is suspended, without movement:

$$[\text{gravity force down}] = [\text{drag force up}]$$

$$\frac{\pi}{6} d^3 (\rho_L - \rho_G) g = f \left(\frac{1}{2} \rho_G v_G^2 \right) \left(\frac{\pi}{4} d^2 \right) \tag{13.1-1}$$

were d is the drop diameter; ρ_L and ρ_G are the densities of liquid and vapor, respectively; g is the acceleration due to gravity; f is the friction factor; and v_G is the vapor velocity. Rearranging this

$$v_G = [C] \left(\frac{\rho_L - \rho_G}{\rho_G} \right)^{1/2} = \left[\left(\frac{4}{3} \frac{dg}{f} \right)^{1/2} \right] \left(\frac{\rho_L - \rho_G}{\rho_G} \right)^{1/2} \tag{13.1-2}$$

where C is the capacity factor, the parameter in square brackets. Note that in this parameter, both the droplet size d and the friction factor f are normally not known.

Values of C can be measured experimentally. Results are normally correlated vs. the flow parameter, defined as

$$\frac{L''}{G''} \sqrt{\frac{\rho_G}{\rho_L}} = \left[\frac{\frac{1}{2} \rho_L v_L^2}{\frac{1}{2} \rho_G v_G^2} \right]^{\frac{1}{2}} \tag{13.1-3}$$

where L'' and G'' are the *mass* fluxes of liquid and vapor. The use of mass rather than mole fluxes is a consequence of the basis of C in fluid mechanics, not in vapor–liquid equilibrium used to find the height of the tower.

Typical experimental values are shown in Fig. 13.1-2 for a variety of tray spacings. Note that the values of C have the dimensions of velocity; note also that the curve for 0.6 m (i.e., 24 in) is the standard for North America and hence the most important line on the graph. Not surprisingly, there are many efforts to improve this figure, like corrections for surface tension, for foaming, and for tower cross-section.

The physical significance of the variations shown in Fig. 13.1-2 is complicated, but certainly related to the friction factor f. From Eq. 13.1-2, we note that C is inversely proportional to f. At high gas flow, we expect that f will be constant because turbulence

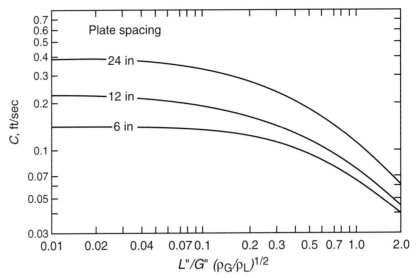

Fig. 13.1-2. Capacity factor vs. flow parameter for sieve trays. The common tray spacing is 24 in, or 0.6 m.

and form drag are most important. At high gas flow, we then expect that C will be constant, first as is reported in Fig. 13.1-2 and as suggested by Eq. 13.1-2. At low gas flow, we expect that f will vary inversely with velocity. Because C in turn varies inversely with f, we expect that C should be proportional with gas velocity. Figure 13.1-2 plots C vs. the reciprocal of gas velocity, and shows that at low gas flow, C is proportional to velocity, just as expected on the basis of friction factor. However, our self-satisfaction at this variation of C with f must be tempered by our neglect of the changes in bubble size d and liquid flux L''.

In addition to entrainment flooding, we need to be concerned with the turndown ratio, the maximum vapor flow divided by the minimum vapor flow. This ratio is important because it determines the flexibility of our distillation column. In an ideal world, we would want to operate our separation at full capacity all the time. In a real world, we will not always have a high, steady demand for our product, or we may have a fluctuating feedstock. In this real world, we may want to run at less than full capacity without the possibility of problems like weeping.

Our flexibility, expressed as this turndown ratio, will depend on the type of trays which we chose. Not surprisingly trays which are expensive have a higher turndown ratio. From experiments, bubble-cap trays have a turndown ratio of about eight, valve trays of about five, and sieve trays of two. Frustratingly, sieve trays are the cheapest, the most efficient and the least flexible.

Example 13.1-1: Pentane–heptane distillation   You want to separate 10 mol/sec of 50% pentane in heptane into a distillate of 90% pentane and a bottoms of 15% pentane. Following laboratory studies, you expect to use a column with six stages. You plan to use a saturated liquid as a feed, with a reflux ratio of 3.5. This means that the liquid flow in the top of the column will be 3.5 times the distillate.

What diameter of column should be used?

Solution We first must make mole balances to find the flows in the column. We then convert these into mass flows and calculate the flow parameter. Using Fig. 13.1-2, we find the capacity factor and hence the gas velocity. This gives us the column diameter.

Beginning with mole balances,

$$10 = D + B$$
$$0.50(10) = 0.90D + 0.15B$$

Thus

$$D = 4.67 \, \text{mol/sec}$$
$$B = 5.33 \, \text{mol/sec}$$

The total molar flows are thus

$$L = 3.5D = 16.3 \, \text{mol/sec}$$
$$G = L + D = 21.0 \, \text{mol/sec}$$

Below the feed, the liquid flow is increased by the feed, but the vapor flow is the same

$$L' = 26.3 \, \text{mol/sec}$$
$$G' = 21.0 \, \text{mol/sec}$$

For these larger flows, the flow parameter is

$$\frac{L'\tilde{M}}{G'\tilde{M}}\sqrt{\frac{\rho_G}{\rho_L}} = \frac{26.3 \, \text{mol/sec}}{21.0 \, \text{mol/sec}}\sqrt{\frac{3.1 \cdot 10^{-3} \, \text{g/cm}^3}{0.65 \, \text{g/cm}^3}} = 0.086$$

From Fig. 13.1-2 at a tray spacing of 0.6 m (i.e., 24 in), C is 0.35 ft/sec or 0.11 m/sec. Thus from Eq. 13.1-2

$$v_G = \frac{0.11 \, \text{m}}{\text{sec}}\left[\frac{0.65}{3.1 \cdot 10^{-3}}\right]^{\frac{1}{2}} = 1.6 \, \text{m/sec}$$

Finally, we find the column diameter

$$1.6 \, \text{m/sec} \left(\frac{\pi}{4}d^2\right) = 21 \, \text{mol/sec} \left(\frac{22.4 \cdot 10^{-3}\text{m}^3}{\text{mol}}\right)\frac{340 \, \text{K}}{273 \, \text{K}}$$
$$d = 0.7 \, \text{m}$$

Note that this result is based on the lower section of the column, where the flows are larger.

13.2 Staged Distillation of Nearly Pure Products

The simplest case of staged distillation is that of a nearly pure feed from which we want to make an even purer product. Often that product is a distillate, as shown in

Fig. 13.2-1(a); sometimes it is a bottoms stream, as shown in Fig. 13.2-1(b). In either case, the other stream contains more of the impurities than the feed. We will emphasize the case of the highly pure distillate; the case of a highly pure bottoms is an easy analogue.

As before, we base our calculation with a mole balance, called an operating line, and an energy balance, called an equilibrium line. For the moment, we will assume that the stages reach equilibrium. This implies that the mass transfer is very fast. In this limit, we have no need for a rate equation.

We begin by defining the concentrations on a single stage as shown in Fig. 13.2-2. The concentration of the more volatile species in the liquid leaving a particular stage n is x_n; that in the vapor leaving the same stage is call y_n. Note that for staged separations, the subscripts indicate the location in the column. This is a different meaning than that used earlier in this book, where the subscripts refer to a specific species, like a solute or a solvent. Note also that some of the concentrations may not exist in the column itself. For example, in Fig. 13.2-1(a), the concentration y_{N+1} does not exist in a column which

(a) Vapor feed at bottom of column

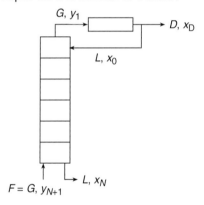

(b) Liquid feed at top of column

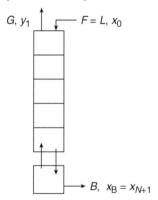

Fig. 13.2-1. Staged distillation producing very pure products. These processes are not common, but are easier to analyze and hence a good starting point.

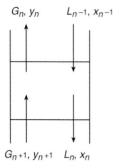

Fig. 13.2-2. Concentrations in a single stage. The concentrations y_n and x_n are defined as those of vapor and liquid leaving stage n.

has a total number of stages N. The concentration y_{N+1} is that in the vapor feed entering the column.

With these definitions, we can write operating lines around stages "1" through "n" in Fig. 13.2-1(a). An overall mole balance gives

$$G_{n+1} + L_0 = G_1 + L_n \tag{13.2-1}$$

where G_n and L_n represent the total vapor and liquid fluxes leaving stage n. For now we can define these either as fluxes, in moles per area per time, or as flows, in moles per time. Either definition works well. Later, when we will worry about nonequilibrium stages, we will need to make this distinction. In addition, because we are dealing with flows of nearly pure materials, we will assume these fluxes are constant, that for example,

$$G_{n+1} = G_n = \ldots = G_1 = G \tag{13.2-2}$$

This will greatly simplify our analysis.

We next consider a mass balance on the more volatile species in stages "1" to "n" at the top of our column:

$$G y_{n+1} + L x_0 = G y_1 + L x_n \tag{13.2-3}$$

where x_n and y_n refer to the mole fractions of the more volatile component on stage n. Rearranging,

$$y_{n+1} = \left(y_1 - \frac{L}{G} x_0 \right) + \frac{L}{G} x_n \tag{13.2-4}$$

However, because y_1 equals x_0 and x_D, this can be rewritten as

$$y_{n+1} = \frac{D}{G} x_D + \frac{L}{G} x_n = \left(\frac{1}{1 + R_D} \right) x_D + \frac{R_D}{1 + R_D} x_n \tag{13.2-5}$$

This operating line has the same form as Eq. 12.2-6 for differential distillation. Remember that the subscripts ($n+1$) and n now refer to particular stages in the tower.

We now need an equilibrium line, which for nearly pure streams of the more volatile species has an especially simple form

$$y_n = y_n^* = a + m x_n \tag{13.2-6}$$

where a and m are constants. Note that because y_n^* is the equilibrium value, this idealizes the stages as being unaffected by mass transfer. The quantity m is closely related to the partition coefficients and Henry's law coefficients used elsewhere to describe transport across interfaces. Note that for a nearly pure material, a and m sum to one.

We can now combine the operating and equilibrium lines in Eqs. 13.2-5 and 13.2-6 to give

$$y_{n+1} = \left(y_1 - \frac{L}{G} \left[\frac{y_0 - a}{m} \right] \right) + \frac{L}{G} \left[\frac{y_n - a}{m} \right] \tag{13.2-7}$$

or

$$y_{n+1} = y_1 - A y_0 + A y_n \tag{13.2-8}$$

where the reciprocal of A ($= mG/L = m(1 + R_D)/R_D$) is often called the absorption factor. Note that y_0 is the vapor concentration that would be in equilibrium with the liquid concentration x_0. This vapor concentration y_0 is hypothetical: it does not physically exist.

We now write out Eq. 13.2-8 for each of the stages in the column. For the first stage ($n = 1$),

$$y_2 = y_1 - A y_0 + A y_1 = (1 + A) y_1 - A y_0 \tag{13.2-9}$$

For the second ($n = 2$),

$$y_3 = y_1 - A y_0 + A y_2 = \left(1 + A + A^2 \right) y_1 - \left(A + A^2 \right) y_0 \tag{13.2-10}$$

For the entire cascade ($n = N$),

$$y_{N+1} = y_1 - A y_0 + A y_N$$
$$= \left(1 + A + A^2 + \cdots + A^N \right) y_1 - \left(A + A^2 + \cdots + A^N \right) y_0$$
$$= \left[\frac{1 - A^{N+1}}{1 - A} \right] y_1 - \left[\frac{A(1 - A^N)}{1 - A} \right] y_0 \tag{13.2-11}$$

This result is often easier to use in rearranged form. From an overall balance (i.e., $n = N$ in Eq. 13.2-8):

$$\frac{y_{N+1} - y_1}{y_N - y_0} = \frac{L}{mG} = A \tag{13.2-12}$$

Combining the last two equations leads to

$$N = \frac{\ln\left[\dfrac{y_{N+1} - y_N}{y_1 - y_0}\right]}{\ln A} \tag{13.2-13}$$

or

$$N = \frac{\ln\left[\dfrac{y_{N+1} - y_N}{y_1 - y_0}\right]}{\ln\left[\dfrac{y_{N+1} - y_1}{y_N - y_0}\right]} \tag{13.2-14}$$

This result lets us estimate the number of ideal stages for separating a less volatile impurity from a more volatile solute.

Equations 13.2-11, 13.2-13, and 13.2-14 are called the Kremser equations, relations for estimating the number of equilibrium stages for a linear equilibrium line. The results are most useful for highly dilute solutions, though they can be applied over any small concentration range over which the equilibrium line can be approximated as linear.

This development has been phased completely in terms of concentrations in the vapor. Obviously, we could have phased the development in terms of concentrations in the liquid. Had we done so, we would derive a completely parallel set of equations containing a stripping factor, which is no more than the reciprocal of the absorption factor. When I try to use this second set of equations, I often confuse myself. As a result, I find it easier simply to recognize that labeling the phases is arbitrary, and just plug into the Kremser equation in the form given.

These ideas are best illustrated in examples.

Example 13.2-1: Benzene purification   We wish to produce 99.99% benzene from a feed containing 99% benzene and 1% toluene. We plan to use a staged column operated with a total condenser. Over this concentration range, the equilibrium line is approximately

$$y_n^* = 0.58 + 0.42x_n$$

where the mole fractions are those of the more volatile benzene.

If the reflux ratio is three, what is the number of ideal stages required for the separation?
    Solution   We can find the number of stages from the Kremser equation. To do so, we first find the slope of the operating line

$$\frac{L}{G} = \frac{R_D}{1 + R_D} = \frac{3}{1 + 3} = 0.75$$

Thus A equals $(0.75/0.42)$. We then find the various vapor concentrations. We are given $y_1 = 0.9999$. For a total condenser, x_0 equals y_1; thus from the equilibrium line,

$$y_0 = 0.58 + 0.42(0.9999)$$
$$= 0.999958$$

We are also given $y_{N+1} = 0.99$. To find x_N and hence y_N, we make an overall mass balance parallel to Eq. 13.2-5:

$$0.99 = \left(\frac{1}{1+3}\right)0.9999 + \left(\frac{3}{1+3}\right)x_n$$

$$x_n = 0.9867$$

From the equilibrium line

$$y_N = 0.58 + 0.42(0.9867)$$
$$= 0.99441$$

Inserting these values into Eq. 13.2-13,

$$N = \frac{\ln\left[\dfrac{0.99 - 0.99441}{0.9999 - 0.999958}\right]}{\ln\left[\dfrac{0.75}{0.42}\right]} = 7.5$$

We will require eight stages for this purification.

Example 13.2-2: Hydrocarbon removal from process water   We have a process water containing a mole fraction of a volatile hydrocarbon equal to $x_0 = 0.0082$ which we must reduce to a bottoms concentration $x_B = 10^{-4}$. To do so, we plan to feed the process water as a saturated liquid to the top of a distillation column mounted above a reboiler. The equilibrium line for this system is $y^* = 36x$. If we adjust the heat added to the reboiler to give a vapor flux three times the minimum, how many stages will we need to achieve this separation?

   Solution   At the minimum reboiler heating, the exiting vapor will be in equilibrium with the entering liquid:

$$y_0 = 36x_0 = 0.2952$$

As a basis, we assume the liquid flow L equals one mole. Thus overall and hydrocarbon balances give

$$1 = G + B$$
$$0.0082 = 0.2952G + 10^{-4}B$$

or

$$G = 0.0274\,\text{mol}$$

We want to operate at three times this flow:

$$G = 0.0823\,\text{mol}$$

From an overall balance, the actual bottoms flow is

$$B = 1 - G = 0.9177$$

We now find the concentration of the vapor effluent leaving the column

$$1 \, mol \, (0.0082) = 0.0823 \, mol \, (y_1) + 0.9177 \, mol \, (10^{-4})$$
$$y_1 = 0.0985$$

We know the vapor entering the column is in equilibrium with the liquid in the bottoms

$$y_{N+1} = 36x_B = 0.0036$$

A species balance on the column without the reboiler gives the unknown concentration x_N

$$0.0823 \, mol \, (0.0985 - 0.0036) = 1 \, mol \, (0.0082 - x_N)$$
$$x_N = 0.00039$$

Thus

$$y_N = 36x_N = 0.0140$$

Finally, we can apply one of the Kremser equations, Eq. 13.2-14, to find

$$N = \frac{\ln\left[\dfrac{0.0036 - 0.0140}{0.0985 - 0.2952}\right]}{\ln\left[\dfrac{0.0036 - 0.00985}{0.0140 - 0.2952}\right]} = 0.8$$

We will need only one stage, plus the reboiler.

13.3 Concentrated Staged Distillation

Most staged distillations separate a concentrated feed into a distillate enriched in the more volatile species and a bottoms enriched in the less volatile species. The equipment consists of a staged column, with a total condenser mounted on the top, and a heated reboiler below. The condenser usually liquefies all the vapor coming out of the top of the column and returns most of the condensate back down the column. The reboiler boils much of the liquid coming out of the bottom of the column and sends the vapor back into the column. The feed normally enters somewhere in the middle of the column.

We want to analyze the performance of such a common column. We may want to improve the separation achieved, or reduce the energy required, or design a larger column for the same separation. Like other separation processes, our analysis will be grounded in operating lines (i.e., mole balances) and in equilibrium lines (i.e., free energy balances). For a separation using stages which reach equilibrium, which is the ideal assumed here, we don't even need the rate equation required for gas absorption or for differential distillation.

The equilibrium line for such a distillation is not linear. Normally, it starts at the point $(x = 0, y = 0)$ with a slope greater than one; and it ends at the point $(x = 1, y = 1)$ with a slope less than one. In most cases, the equilibrium line will be the result of careful experiments. In some cases, the equilibrium line can be calculated with software developed for thermodynamic estimates. These calculations are especially reliable for simple hydrocarbons. Here, we will assume that this equilibrium line is known.

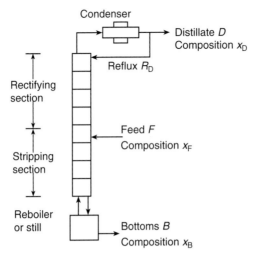

Fig. 13.3-1. Staged distillation separating a concentrated feed. Because the feed changes the flows within the column, the operating lines above and below the feed stage differ.

The operating line, again based on mole balances, is complicated by the feed on a stage somewhere in the middle of this column. This feed means that the flows of vapor and liquid are different above and below the feed stage. But this is not hard; it just is a complication which we must consider carefully.

We begin with overall and species mole balances on the entire column using the notation in Fig. 13.3-1:

$$F = D + B \qquad\qquad\qquad\qquad (13.3\text{-}1)$$
$$Fx_F = Dx_D + Bx_B \qquad\qquad\qquad (13.3\text{-}2)$$

Arbitrarily, we assume that these mole fractions refer to the more volatile component. In addition, the feed F, distillate D, and bottoms B can be treated either as flows, in moles per time, or as fluxes, in moles per tower cross-section per time. For equilibrium-staged separations this distinction does not matter. These overall balances must hold, no matter what happens inside the column.

We next develop mole balances over parts of the column itself using the notation suggested by Fig. 13.3-2. We initially consider only what is happening between the top of the tower and nth stage, which is above the feed stage. In this "rectifying" section of the column, we assume that the vapor flux G and the liquid flux L are constant between the top of the tower and this nth stage. We define x_n and y_n as the liquid and vapor mole fractions of the more volatile component leaving the nth stage. Remember that these definitions are the same as those used earlier in this chapter but are different than those used in other chapters in this book, where subscripts like these refer to component n or to position z. Because the more volatile component is moving up the column, x_n is greater than x_{n+1}, and y_n is greater than y_{n+1}. Consistent with this convention, we call the composition of the vapor leaving the first stage y_1 and the composition of the liquid entering the first stage x_0.

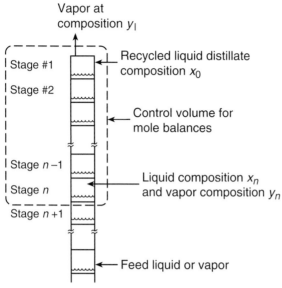

Fig. 13.3-2. The distillation column itself. The analysis depends on mole balances between the top of the column and the *n*th stage (i.e., over that part of the system surrounded by the dotted line). The analysis given is simplified by the assumption that the vapor flux G and the liquid flux L are constants.

We now write a mole balance on the more volatile component between the *n*th stage and the top of the column, that is, on the control volume in Fig. 13.3-2. Because the column is operating in steady state, this is

$$Gy_{n+1} + Lx_0 = Gy_1 + Lx \tag{13.3-3}$$

or

$$y_{n+1} = \left[y_1 - \frac{L}{G} x_0 \right] + \left(\frac{L}{G} \right) x_n \tag{13.3-4}$$

This important result relates the vapor composition on the $(n+1)$th stage y_{n+1} to the liquid composition on the *n*th stage x_n. It says that y_{n+1} varies linearly with x_n; the slope of this line is L/G, and the intercept is the quantity in square brackets.

Equation 13.3-4 can be simplified by considering the condenser in more detail. In many cases, all the vapor leaving the top of the column is condensed. Part of the condensate is the product distillate; the remainder, the reflux L, is returned to the top of the tower. In this normal case, the vapor composition y_1 leaving the top plate will equal the liquid composition entering the top plate x_0.

Occasionally, the condenser at the top of the tower may condense just part of the vapor and reflux this more easily condensed fraction. A condenser operating in this way has a delightful name: a "dephlegmator." The remaining vapor becomes the distillate. In most elementary texts, the liquid and vapor leaving the dephlegmator are assumed in equilibrium. This is equivalent to assuming that the partial condenser is an additional stage, and this is a significant approximation. We refer those interested in the effects of

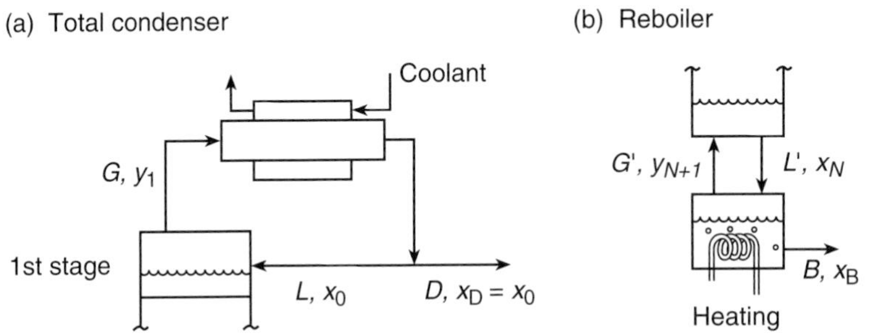

(a) Total condenser   (b) Reboiler

Fig. 13.3-3. The ends of the distillation column. The vapor leaving the top of the column is condensed, as suggested by (a). Part of this condensate is recycled to the column as liquid. The liquid leaving the bottom of the column is partially vaporized in the reboiler, shown in (b). This reboiler effectively serves as an extra stage in the column.

partial condensers to the more specialized books and restrict our discussion here to total condensers.

For the case of a total condenser, we can rewrite Eq. 13.3-4 in a form that will be useful later. From Fig. 13.3-3, we see that $x_D = x_0 = y_1$ and that

$$G = D + L \tag{13.3-5}$$

When we combine this with Eq. 13.3-4, we obtain

$$y_{n+1} = \left(\frac{1}{1 + R_D}\right) x_D + \frac{R_D}{1 + R_D} x_n \tag{13.3-6}$$

where R_D $(= L/D)$ is the reflux ratio. This important result is called "the operating line of the rectifying section." Essentially a mole balance, it is more useful than Eq. 13.3-4 because it is written in terms of experimentally controlled quantities. The concentration x_D is usually specified, and the reflux ratio R_D can be changed with the twist of a value. In addition, because R_D is positive, Eq. 13.3-6 says that a plot of y_{n+1} versus x_n has a positive intercept and a slope less than unity. We shall come back to this equation later.

These mole balances on the rectifying section have parallels in the lower stripping section. Differences exist because the feed results in different flow rates of vapor G' and liquid L'. Parallel to the foregoing, we assume that G' and L' are constant through the stripping section. The mole balance that results is

$$y_{n+1} = \left[y_{N+1} - \frac{L'}{G'} x_N \right] + \left(\frac{L'}{G'}\right) x_n \tag{13.3-7}$$

where N is the total number of stages in the column, y_{N+1} is the vapor composition entering the bottom of the column, and x_N is the liquid composition leaving this stage.

We next consider what happens in the reboiler at the bottom of the column. There, the liquid leaving the column is boiled and partially evaporated. The vapor is returned to the column, and the liquid is removed as product. To a reasonable approximation, vapor and liquid are in equilibrium, so the reboiler simply acts as an additional stage. The notation used in this region is shown in Fig. 13.3-3(b).

We find it useful to rewrite Eq. 13.3-5 in terms of the flows around the reboiler. Mole balances give

$$L' = G' + B \tag{13.3-8}$$
$$L'x_N = G'y_{N+1} + Bx_B \tag{13.3-9}$$

Combining with Eq. 13.3-7, we obtain

$$y_{n+1} = -\left(\frac{B}{G'}\right)x_B + \left(1 + \frac{B}{G'}\right)x_n \tag{13.3-10}$$

This result, called the "operating line of the stripping section," is also important and will be used later. Note that in contrast to the rectifying section, a plot of y_{n+1} versus x_n in the stripping section has a negative intercept and a slope greater than unity.

Finally, we consider the feed plate in more detail. The nature of the feed strongly influences what happens in the column. If the feed is entirely saturated liquid, it simply joins the liquid flowing down the column; if it is entirely saturated vapor, it will join the vapor stream and flow upward. If the feed is a liquid cooled below saturation, then it will both join the liquid stream and condense some vapor to produce additional liquid.

These changes in column flows are best described in terms of the variable q, defined as

$$q = \frac{\text{amount liquid produced}}{\text{amount feed}} = \frac{L' - L}{F} \tag{13.3-11}$$

This quantity was also used to analyze concentrated differential distillation in Section 12.4. Here we can use an analysis identical to that given earlier to find the locus of intersection between the operating and equilibrium lines above and below the feed:

$$y_n = \frac{x_F}{1 - q} - \left[\frac{1}{1 - q}\right]x_n \tag{13.3-12}$$

This "q-line" changes with changes in the flows within the column, including those caused by the feed. Note that this result passes through the point (x_F, x_F). When the feed is a saturated liquid, q is one and the q-line is vertical. When the feed is a saturated vapor, q is zero and the q-line is horizontal. Other cases were discussed in Section 12.4.

We can now calculate the number of ideal stages needed for a particular separation. This calculation often assumes that the reflux ratio R_D is between 1.2 and 1.8 times the minimum required. We will assume a typical value of 1.5 times the minimum. Then the number of stages needed can often be found from the following template:

1. Plot the equilibrium line $y^*(x)$. Locate the points (x_D, x_D), (x_F, x_F), and (x_B, x_B).
2. Plot the q-line in given Eq. 13.3-12. Draw this from the point (x_F, x_F) until it intersects the equilibrium line. For example, for a saturated liquid feed, this intersection occurs at $(x_F, y^*(x_F))$.
3. Connect this intersection with the point (x_D, x_D) to find the operating line for minimum reflux. Calculate this minimum reflux ratio using Eq. 13.3-6. Then find the reflux ratio which will actually be used.

4. Draw the actual operating line for the upper rectifying section, running through (x_D, x_D) with a slope of $R_D/(1 + R_D)$. This operating line stops when it intersects the q-line.

5. Draw the operating line for the lower, stripping section of the column between this q-line intersection and the point (x_B, x_B).

6. From these operating and equilibrium lines, we can now find the number of stages and the concentrations on each stage as detailed below.

I do not maintain that this template is always the best way to find the number of stages. However, I have found that it is often the best way to learn this standard analysis, often called the McCabe–Thiele method.

Stepping off the stages in the last step of the template is illustrated by the detail in Fig. 13.3-4. In this figure, the top curve is the equilibrium line, the next line is the operating line, and the lower line is the 45° diagonal, included only as a reference. We begin with the point on the operating line $(x_0 = x_D, y_1 = x_D)$. We read horizontally to the point on the equilibrium line (x_1, y_1), which is the concentration on the first stage. We then read vertically to the point (x_1, y_2) on the operating line and then horizontally to find (x_2, y_2), the composition on the second stage. We continue along the column until we reach a stage concentration lower than x_B. This is the total number of stages needed, including the reboiler. The number of stages needed in the column is one less than this total.

Before we turn to examples, I want to review the results summarized by the template and by Fig. 13.3-4 in more detail. I do so partly because they can be superficially clear, and still obscure some of the information implicit in the analysis. To disperse some of the obscurity, I'll discuss four specific points: the optimum feed stage location, the effect of a different feed location, the minimum number of plates required, and the minimum reflux ratio. Each of these four points is instructive.

To examine the location of the feed stage, we consider one of the steps in Fig. 13.3-4 in the greater detail in Fig. 13.3-5. The actual concentration differences in this detailed figure represent the concentration changes from stage to stage in the column. We want these changes to be as large as possible, so we want to switch from one operating line to the other as soon as possible. Thus we want to locate the feed on the stage where the two operating lines meet.

The second point worth discussing is the effect of feeding at some other location than the optimum. If the feed were below the optimum location, then the upper, rectifying section of the column would still have the same vapor and liquid flows and hence the same operating line as if the feed were optimal. As a result, the concentrations at the stages in the upper part of the column would still be given by the same type of graphical analysis, exemplified by the dashed lines in Fig. 13.3-6. When the feed stage is finally reached, the concentrations tumble onto those given by the operating line for the lower, stripping section in the column. This results in requiring more stages than those dictated by the optimal feed location.

The third implication of the McCabe–Thiele analysis is the estimation of the minimum number of stages. The minimum number of stages will occur when the concentration changes between stages are largest. This implies that the distance between the equilibrium and operating lines is as large as possible, which in turn suggests that the reflux ratio R_D $(= L/D)$ be infinite. Under these conditions, the operating lines for both the upper rectifying section and the lower stripping section collapse to the forty-five degree diagonal. Such a reflux ratio means that the vapor and liquid flows are equal; such total reflux means that

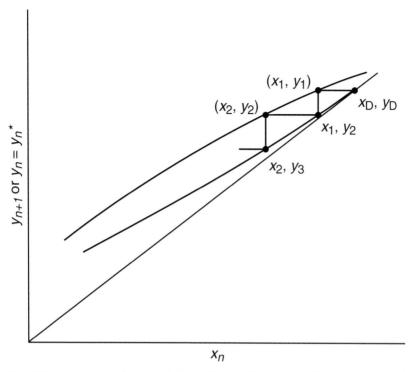

Fig. 13.3-4. Determining the concentrations of the equilibrium stages. The concentrations (x_D, $x_D = y_1$, x_0) are given by the operating line; the concentrations (x_1, y_1, $= y_1^*$) are given by the equilibrium line; etc.

the amount of feed, distillate, and bottoms are near zero. All the vapor going up the column is condensed and sent down again; all the liquid running into the reboiler is evaporated.

At first glance, it may seem silly to operate a distillation column under these conditions. After all, we are paying a lot to run the column, and we are making next to nothing as product. However, if our purpose in a research laboratory is to make a few drops of especially pure product, we may wish to run our distillation column under these conditions just to get the maximum purification possible with the number of stages that we actually have. Note that under these conditions, we can add our few drops of feed

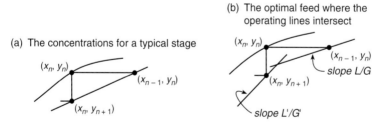

Fig. 13.3-5. Details of a single stage. Each point in the diagram either relates equilibrium concentrations or reflects mole balances.

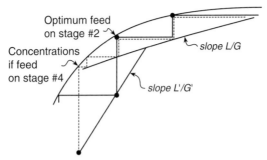

Fig. 13.3-6. The effect of different feed locations. When the feed is in the optimal location, the concentration differences are larger and the number of stages required is reduced.

anywhere because the operating lines are both the same diagonal. These conditions define the minimum number of stages required for the separation.

The antithesis of this case of the minimum number of stages is the case of a minimum reflux ratio. To explore this case, imagine that we decrease the reflux ratio R_D coming out of the condenser. This decrease will reduce the slope of the operating line in the rectifying section and increase that in the stripping section. Eventually, the intersection of these lines will collide at the equilibrium line. The slopes of the operating lines at this point give the minimum reflux at which a separation is possible. This reflux is never used, for it requires an infinite number of stages. However, columns are often specified at reflux ratios which are about 1.2 to 1.8 times the minimum, depending on the relative costs of capital and energy. If capital is relatively cheap, we want to use a low reflux, which implies using a larger column with more stages but a lower energy requirement in the reboiler. If energy is relatively cheap, we will prefer a smaller column with fewer stages, a large reflux, and hence a larger energy requirement in the reboiler.

Example 13.3-1: Distillation of benzene and toluene We wish to distill 3500 mol/hr containing 40 mol% benzene into streams containing 97 mol% benzene and 98 mol% toluene. The column uses a total condenser and a reflux ratio of 3.5, and the feed is a saturated liquid. How many stages will be required?

Solution We begin by making mole balances on the entire column. From Eqs. 13.3-1 and 13.3-2,

$$3500 \, \text{mol/hr} = D + B$$
$$0.4(3500 \, \text{mol/hr}) = 0.97D + 0.02B$$

Thus D is 1400 mol/hr, and B is 2100 mol/hr. Within the column, L is 3.5(1400) = 4900 mol, L' is (3500 + 4900) = 8400 mol, and G and G' are (1400 + 4900) = 6300 mol. From these values the operating lines, found from Eqs. 13.3-6 and 13.3-10, are plotted as shown in Fig. 13.3-7. The equilibrium line is added, and the difference calculation is made as shown. The result is that the column should have fourteen stages, including the reboiler. The feed should be added to the seventh stage.

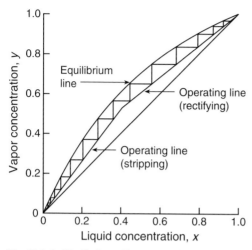

Fig. 13.3-7. Distillation of benzene and toluene. The column is calculated to require 13 stages plus the reboiler. The calculation made here assumes that vapor and liquid are in equilibrium on each stage, an assumption relaxed in Section 13.4.

13.4 Stage Efficiencies

The design strategies given above for stage-wise contacting depend on one huge assumption: that the streams leave each stage in equilibrium. In particular, the basic analysis of staged distillation developed in Sections 13.2 and 13.3 assumes that the vapor composition y_n and the liquid composition x_n leave the nth stage in equilibrium with each other. This assumption is the basis of the entire strategy, including the graphical methods exemplified by Fig. 13.3-4.

In fact, the concentration of the more volatile species leaving a particular stage may fall well short of the equilibrium value. To see why, imagine we have a liquid and vapor in a closed container, as shown in Fig. 13.4-1(a). The initial concentration in the vapor y_{n+1} rises with time towards an equilibrium limit y_n^*. How close we approach this limit depends on the time and how quickly the rise takes place. This in turn depends on the mass transfer coefficient and the interfacial area between liquid and vapor.

A similar situation for this closed container will exist on a typical stage, as shown in Fig. 13.4-1(b). Again, the concentration rises from the value flowing in y_{n+1} towards an equilibrium value of y_n^*. Now, the plot is not vs. time, but rather vs. some residence time on the stage. The rate of rise is again a function of the mass transfer coefficient and the interfacial area between liquid and vapor.

We can put these ideas on a more quantitative basis by defining stage efficiency. If the efficiency is equal to one, the stage will be at equilibrium; if it is less than one, it will operate without achieving equilibrium. Three definitions of stage efficiency are common. First, we can define an overall efficiency

$$\eta_0 = \frac{\text{number of equilibrium stages}}{\text{number of actual stages}} \qquad (13.4\text{-}1)$$

For example, if the McCabe–Thiele analysis says that we need six stages, and we discover that we actually need twelve, then our overall stage efficiency is 0.5, or 50%.

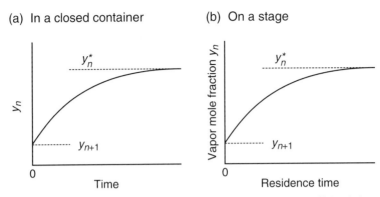

(a) In a closed container

(b) On a stage

Fig. 13.4-1. Concentrations approaching equilibrium. If the time is sufficiently long, the vapor concentration reaches equilibrium, and the analysis in the previous sections is valid.

This definition of overall efficiency is more empirical than I want to use in this text. However, I am impressed with a dimensional correlation suggested by Lockett for η_0:

$$\eta_0 = 0.49\,(\alpha\mu_L)^{-0.25} \tag{13.4-2}$$

where α is the relative volatility of the species being separated, and μ_L is the liquid viscosity, in centipoises. For example, for benzene–toluene, α is about 2.5, and μ_L is around 0.3 cp, so η_0 is about 0.5. This illustrates how far from ideal equilibrium stages we will normally operate.

A second useful definition is the Murphree efficiency, named after a director of research at Standard Oil of New Jersey. This definition assumes that each fluid is well mixed on every stage, and so can be described by a single average concentration. We can then define

$$\eta = \frac{y_n - y_{n+1}}{y_n^* - y_{n+1}} \tag{13.4-3}$$

Clearly, this efficiency is the fraction of the concentration difference between feed and equilibrium which the stage manages to obtain. Remember that the concentrations are averages over the entire volume of the stage. The Murphree efficiency is a valuable concept, but the use of average concentrations is a serious approximation, which can in unusual cases lead to measured efficiencies over 100% when vapor and liquid are contacted countercurrently.

We can avoid this embarrassment by defining a third, local efficiency based on local concentrations. However, these local concentrations are almost always unknown, so these efficiencies can rarely be used. Rather, we can only recognize that the average of the local efficiency over the stage volume is the Murphree efficiency, and the average of the Murphree efficiency over the column height is the overall efficiency.

The Murphree efficiencies will be the focus of the development in this section. They have two characteristics that we want to discuss. First, they are obviously related to the speed of diffusion, expressed as mass transfer coefficients. Second, they affect the design of distillation columns outlined in the previous section. We discuss these characteristics sequentially.

13.4.1 Stage Efficiencies and Mass Transfer

We can begin to see how the Murphree efficiency is related to diffusion and mass transfer by returning to the concentration variations shown in Fig. 13.4-1(b). This figure plots the vapor phase concentration y_n as a function of the residence time on the stage. The vapor enters from the $(n + 1)$th stage with a mole fraction y_{n+1}. This mole fraction rises, first quickly and then more slowly.

We will get the most effective separation if the concentration y_n reaches the equilibrium limit y_n^*. This implies building a column whose stages permit long residence times. However, such a column will be big and hence expensive. We may prefer to build a smaller, cheaper column. We may be willing to accept a smaller concentration change per stage if it means a much smaller equipment cost.

With this potential savings in mind, we turn to the detailed shape of the curve in Fig. 13.4-1(b). We want that curve to jump upward fast, to rise from y_{n+1} to near y_n^* with a small residence time on the stage. In other words, we want to have high Murphree efficiency for a short residence time. Intuitively, we expect that the Murphree efficiencies should be a function of mass transfer and of flow. If mass transfer between liquid and vapor is fast, these phases should almost be in equilibrium, and the efficiency should approach unity. If the liquid and vapor flow slowly past each other, then these phases again should almost be in equilibrium, and again the efficiency should be about 100 percent.

Converting this intuition into a more quantitative form requires a more detailed physical model. Two such models are frequently used. In the first, simpler model, both the liquid and vapor are assumed well mixed. In the second model, the liquid is again assumed to be well mixed, but the gas is assumed to move in plug flow. Each is detailed below.

When both liquid and vapor are well mixed, the relation between the Murphree efficiency and the mass transfer coefficient is especially straightforward. To find this relation, we make a mass balance on the vapor:

$$(\text{accumulation}) = \begin{pmatrix} \text{solute in minus} \\ \text{that out in vapor} \end{pmatrix} + \begin{pmatrix} \text{solute gained by} \\ \text{mass transfer from liquid} \end{pmatrix}$$

$$0 = GA(y_{n+1} - y_n) + K_y aAl(y_n^* - y_n) \tag{13.4-4}$$

where G is the constant vapor flux up the column, in moles per area per time; A and l are the cross-sectional area and the depth of the liquid–vapor froth, respectively; a is the interfacial area per volume between the froth of gas and liquid on the stage; and K_y is an overall mass transfer coefficient. Like the mass transfer coefficients used in gas absorption, K_y is based on a gas-phase concentration difference expressed in terms of mole fractions.

We rewrite this equation as

$$G(y_n - y_{n+1}) = K_y al((y_n^* - y_{n+1}) - (y_n - y_{n+1})) \tag{13.4-5}$$

Rearranging

$$(G + K_y al)(y_n - y_{n+1}) = K_y al(y_n^* - y_{n+1}) \tag{13.4-6}$$

By definition

$$\eta = \frac{y_n - y_{n+1}}{y_n^* - y_{n+1}} = \frac{1}{1 + \dfrac{G}{K_y al}} \tag{13.4-7}$$

This is the desired result. Note that the Murphree efficiency is high at low molar flow G (and hence at high residence time). It is high for rapid mass transfer and for a high interfacial area per stage volume between the vapor and liquid. This prediction of Murphree efficiencies is consistent with our intuitive expectations about the effects of residence time and mass transfer.

Our alternative approach to the well-mixed result given in Eq. 13.4-7 is the partially mixed model. In this model, the liquid on the stage is well mixed; but the gas bubbles change their compositions as they rise through the liquid. To examine the effect of this change, we make a mass balance on the volatile species in a differential volume located at z and Δz thick:

$$(\text{accumulation}) = \left(\begin{array}{c} \text{volatile species in minus} \\ \text{that out by convection} \end{array} \right) + \left(\begin{array}{c} \text{species gained by mass} \\ \text{transfer from liquid} \end{array} \right) \tag{13.4-8}$$

or

$$0 = GA(y|_z - y|)_{z+\Delta z} + K_y a(A\Delta z)(y_n^* - y) \tag{13.4-9}$$

where y is the local mole fraction in the vapor, A is the cross-sectional area of the stage, and the other variables are parallel to those defined in Eq. 13.4-4. Dividing by $A\Delta z$ and taking the limit as this volume goes to zero, we find

$$0 = G\frac{dy}{dz} + K_y a(y_n^* - y) \tag{13.4-10}$$

This is subject to the condition that

$$z = 0, \ y = y_{n+1} \tag{13.4-11}$$
$$z = l, \ y = y_n \tag{13.4-12}$$

where l is again the depth of the liquid–vapor froth on the stage. Integrating, we easily find

$$\frac{y_n^* - y_n}{y_n^* - y_{n+1}} = e^{-K_y al/G} \tag{13.4-13}$$

When we combine this with the definition of the Murphree efficiency given in Eq. 13.4-3, we see that

$$\eta = 1 - e^{-K_y al/G} \tag{13.4-14}$$

Like Eq. 13.4-7, this predicts high efficiencies for low vapor flux G, for large mass transfer coefficient K_y, and for large surface area per volume a. We can also derive corresponding relations for other assumed flows for cases when neither liquid nor vapor is well mixed.

All of these analyses show that the approach to equilibrium in distillation is a strong function of the overall mass transfer coefficient. The analyses also use an overall mass transfer coefficient which varies with the Henry's law coefficient m, which in turn changes with the changing concentrations along the length of the column. However, all of the analyses depend on inexact assumptions about the nature of the flow and the mixing on each stage. Often, these flows aren't known, so the analyses are hard to apply.

As a result, we should focus on how to use experimentally measured Murphree efficiencies. We should appreciate how these efficiencies change with K_y and other relevant variables. This qualitative appreciation can be based on simple models. We should not expect accurate a-priori predictions of these efficiencies.

13.4.2 Stage Efficiencies and Column Design

We now want to show how Murphree efficiencies are used to improve our calculations of the number of stages needed for a given separation. To begin, we graph the equilibrium and operating lines as usual. We remember that for a given value of x_n, the vertical distance between the equilibrium and operating lines is $(y_n^* - y_{n+1})$, as shown in the inset of Fig. 13.4-2. But from Eq. 13.4-4, this distance times the Murphree efficiency is $y_n - y_{n+1}$. Thus we can plot a new, nonequilibrium line, shown as the dashed curve in Fig. 13.4-2. When we design a distillation column, we can use this dashed curve and the operating lines to calculate the number of stages required. This calculation is illustrated in the second of the examples that follow.

Calculations based on Murphree efficiencies are about as far as mass transfer models can be pushed. These calculations may not always be reliable, even though they are based on a huge number of experimental results. The reason is that a single overall mass transfer coefficient may be inadequate to describe all aspects of the flow and diffusion occurring in a single stage. Still, the value of any scientific effort is the product of the importance of the problem and the quality of the solution. Distillation is very important; although concepts of efficiency are certainly imperfect, they seem to me to remain valuable.

Example 13.4-1: Finding mass transfer coefficients from stage efficiencies   On one tray of an acetone–water distillation we find that y_n equals 0.84, x_n equals 0.70, and y_{n+1} equals 0.76. The stage is at 59 °C and 1 atm; it has a vapor flow of 0.14 kg mol/sec and a froth volume of 0.04 m³. Assuming that both vapor and liquid are well mixed, estimate the Murphree efficiency and the mass transfer coefficient on this stage.

Solution   From vapor–liquid equilibrium data for acetone–water, we find that when x_n equals 0.70, y_n^* equals 0.874. The Murphree efficiency is then found from Eq. 13.4-3:

$$\eta = \frac{y_n - y_{n+1}}{y_n^* - y_{n+1}}$$

$$= \frac{0.84 - 0.76}{0.874 - 0.76}$$

$$= 0.70$$

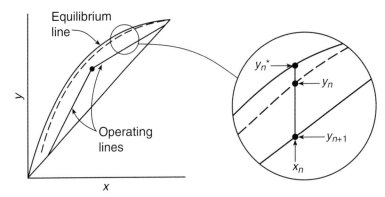

Fig. 13.4-2. Using Murphree efficiencies. Murphree efficiencies effectively lower the solid equilibrium line to the dashed line. They do so because the vapor composition y_n leaving a particular stage is less than the equilibrium value y_n^*. Remember that these efficiencies may not have the same value in each stage.

If the stage is well mixed, then the relation between the efficiency and the mass transfer coefficient is that in Eq. 13.4-7, which is easily rearranged to give:

$$K_y a = \frac{GA}{Al\left(\frac{1}{\eta} - 1\right)}$$

$$= \frac{0.14 \, \text{kg mol/sec}}{0.04 \, \text{m}^3 \left(\frac{1}{0.70} - 1\right)}$$

$$= 8.2 \, \text{kg mol/m}^3 \text{sec}.$$

We cannot calculate the mass transfer coefficient itself without knowing the surface area per volume a.

Example 13.4-2: Distillation design using Murphree efficiencies   A solution containing 47 mol% carbon disulfide in carbon tetrachloride is to be separated in a distillation tower operated at very high reflux. The distillate and bottoms should contain 97 mol% CS_2 and five mol% CS_2, respectively. The average molar volume in the liquid is about 80 cm³/mol. The trays in the tower, which are of a proprietary design, were installed to separate another system. Experiments on this other system suggest operating with a gas flow of 59 mol/sec gives a froth volume of 10 l/stage. These experiments also show that the mass transfer is not fast enough to reach equilibrium, but is characterized by a $k_G a$ of 440 per second and a $k_L a$ of 1.7 per second. These coefficients are inferred from efficiencies measured for the other system by assuming the liquid is well mixed but the vapor is not.

(a) Find the number of stages that would be required in this column if mass transfer was fast enough to reach equilibrium. (b) Estimate the number of stages required if the stages are not ideal but are described by Murphree efficiencies estimated from the available mass transfer data. In solving the second part of this problem, you may assume that the total molar concentrations are constant, equal to the value at 60 °C and at the feed concentration.

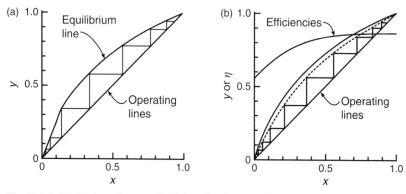

Fig. 13.4-3. Distillation of carbon disulfide and carbon tetrachloride. If liquid and vapor reached equilibrium, this separation would require seven stages, including the reboiler. When the stage efficiencies are considered, the separation requires nine stages. Note how the stage efficiency drops at the lower end of the column as the result of higher values of the Henry's law parameter H.

Solution (a) Because the system is operating at a very high reflux, both operating lines become equal to the diagonal. We then can plot these lines and the equilibrium line as shown in Fig. 13.4-3. By making the usual graphical calculation, we find that the system should contain six stages plus the reboiler.

(b) When the mass transfer is considered, we must correct the equilibrium line for the inefficiencies of the distillation. To do this, we must first find the overall mass transfer coefficients. From an analysis like that leading to Eq. 8.5-7, we have

$$\frac{1}{K_y a} = \frac{RT}{k_G a p} + \frac{m}{k_L a C_L}$$

$$= \frac{\dfrac{8.31 \text{ kg m}^2}{\text{sec}^2 \text{mol K}} (333 \text{ K})}{\dfrac{440}{\text{sec}} \left(101 \cdot 10^2 \dfrac{\text{kg}}{\text{m sec}^2}\right)} \frac{10^6 \text{cm}^3}{\text{m}^3} + \frac{m}{\left(\dfrac{1.7}{\text{sec}}\right)\left(\dfrac{\text{mol}}{80 \text{ cm}^3}\right)}$$

$$= 62 \frac{\text{cm}^3 \text{sec}}{\text{mol}} + 47 \frac{\text{cm}^3 \text{sec}}{\text{mol}} m$$

We can find the values of m for different values of x using the equilibrium curves in Fig. 13.4-3. For example, for $x = 0.2$, $m = 1.41$, and

$$K_y a = 7.8 \text{ mol/l sec}.$$

This and other values are then used to find the Murphree efficiency η, which requires assuming the nature of the flow on the stages. If we assume plug flow of the liquid, we can calculate the efficiencies from Eq. 13.4-14. For example, for $x = 0.2$,

$$\eta = 1 - \exp^{\left(\frac{-K_y a V_s}{G A}\right)}$$

$$= 1 - \exp\left[-\frac{\left[7.8 \frac{\text{mol}}{l \text{ sec}} 10\, l\right]}{59 \frac{\text{mol}}{\text{sec}}}\right]$$

$$= 0.73$$

where we have recognized that the stage volume V_s equals the cross-sectional area of the stages A times the depth l. Other values are shown in Fig. 13.4-3(b). Using these values, we calculate the new y–x curve in this same figure and we find that the number of stages required is now eight plus the reboiler.

13.5 Conclusions

Distillation, the most important separation process for commodity chemicals, can be accomplished using two methods. In the first, the vapor and liquid move countercurrently through a packed bed, and the analysis closely parallels that for gas absorption. This analysis was the focus of the previous chapter, and rests directly on the rates dictated by mass transfer coefficients.

In the second method, the vapor and liquid again move countercurrently, but through a series of well-mixed stages. The goal is to have vapor and liquid approach equilibrium in each stage. When this equilibrium is closely approximated, the analysis results in expressions like the Kremser equations and the McCabe–Thiele method. In many cases, however, the stages do not approach equilibrium. In these cases, the analysis usually makes a correction using an efficiency, which is a function of mass transfer coefficients. The corrections suggested by these efficiencies are large, often around a factor of two. Unfortunately, they are rarely well enough known to be predictive, but are instead a way to rationalize and organize experimental results.

However, while ideal equilibrium stages may be a poor approximation, they are usually the language in which distillation problems are discussed. Thus anyone who studies distillation should analyze any target separation in terms of a McCabe–Thiele analysis even though he knows that this analysis may not be that close to what actually happens. The graphical picture of what is happening is just too good to abandon.

Questions for Discussion

1. Why is distillation such a widely used separation process?
2. What is a Kremser equation?
3. Could you run a distillation column without a reboiler?
4. How in distillation do you get the maximum separation?
5. Any regions where the operating and equilibrium lines are close together are called "pinch points." Discuss why.
6. Why does sending some liquid back down the column improve the separation by distillation?
7. Most distillation is carried out adiabatically. Why?
8. When would structured packing be a better choice for distillation?
9. When would stages be a better choice?
10. Will increases in reflux ratio always improve the separation?
11. How could you define the efficiency of a distillation?
12. Define a Murphree efficiency.
13. Plot the Murphree efficiency vs. the overall mass transfer coefficient.
14. Will diffusion-induced convection be important in concentrated distillation?

Problems

1. You want to distill a feed of fifty percent saturated vapor containing 30% n-pentane and 70% n-heptane, for which vapor–liquid equilibrium data are

x (pentane)	y (pentane)
0.0	0.0
0.059	0.271
0.145	0.521
0.254	0.701
0.398	0.836
0.594	0.925
0.867	0.984
1.000	1.000

The overhead should contain 95 mol% pentane, and the bottoms 10 mol% pentane; the reflux ratio should be 1.3 times the minimum. (a) How many ideal stages are required? (b) On which ideal stage should you feed the vapor? (c) Sketch the effect on the equilibrium line of a Murphree efficiency of 67%, but don't use this for parts (a) and (b).

2. You want to separate methanol and water into a distillate containing 90 mol% methanol and a bottoms containing 5 mol% methanol. Vapor–liquid equilibrium data for this system are:

For this purpose, you want a column designed for a recycle 1.5 times the minimum. All

x (methanol)	y (methanol)
0.0	0.0
0.10	0.417
0.20	0.579
0.30	0.669
0.40	0.727
0.50	0.780
0.60	0.825
0.70	0.871
0.80	0.915
0.90	0.959
1.00	1.000

stages in this column are near-ideal. The trouble is that you have two feeds to this column. One is 100 mol/sec, a saturated liquid containing 60 mol% methanol. The second is 70 mol/sec, a saturated vapor containing 30 mol% methanol. (a) How much distillate is produced? (b) Write an operating line above both feed plates. (c) Find an operating line below both feed plates. (d) How many stages, including the reboiler, will be needed? (e) Where should each feed enter?

3.   As part of applying paint, you produce a large amount of methanol vapor. You recover the methanol by absorbing the vapor in water, and then distilling the water–methanol mixture in an old six-stage column plus reboiler. (Vapor–liquid equilibria are given in the previous problem.) You have an old description of the column which says that the Murphree efficiency of the top three stages is 50%, but that of the bottom three and the reboiler is 100%. The man who wrote the report died last year, only eleven months after retiring. Company policy dictates a reflux 1.5 times the minimum. You need to process 1200 kg/hr of a 32 mol% methanol feed. You want a distillate that is 90 mol% methanol. What is the concentration of the bottoms?

4.   You want to distill acetone and ethanol, whose vapor–liquid equilibrium data are

x (acetone)	y (acetone)
0.0	0.0
0.05	0.15
0.10	0.26
0.20	0.41
0.30	0.54
0.40	0.60
0.60	0.74
0.70	0.80
0.80	0.87
0.90	0.93
1.00	1.00

You plan to produce a bottoms product containing 90 mol% ethanol using another old six-plate column (plus reboiler and a total condenser) which has a Murphree efficiency of 75%; you plan to use a very high reflux. The feed, containing 50 mol% acetone, is 36 mol% vapor. (a) What will the distillate concentration be? (b) How much bottoms product will be produced per mole of feed?

5.   A saturated liquid solution of 40 mol% acetone and 60 mol% acetic acid is fed at a rate of 100 lb mol/hr to a distillation column. The desired separation of acetone is a 96 mol% distillate concentration and a 5 mol% bottoms concentration. The column is operated at 1.6 times the minimum reflux ratio, so $R_D = 0.44$. The average column temperature is 95 °C and the average liquid density is 0.95 g/cm^3. The liquid depth on a tray is 1.5 cm, and the tower diameter is 30 cm. The mass transfer coefficients are $k_G a = 505$ sec^{-1} and $k_L a = 1.8$ sec^{-1}. Equilibrium data for acetone and acetic acid are

x (acetone)	y (acetone)
0.0	0.0
0.05	0.162
0.10	0.306
0.20	0.557
0.30	0.725
0.40	0.840
0.50	0.912
0.60	0.947
0.70	0.969
0.80	0.984
0.90	0.993
1.00	1.000

Determine the Murphree efficiency from Eq. 13.4-7, and then find the number of stages and the feed tray location for this column. (D. McCullum).

6. To make a photochemically active pigment, you are feeding reagent plus 1 mol/min methanol into a batch reactor containing dilute solutes, 10 mol% methanol, and 90 mol% chlorobenzene. This reaction mixture is boiling, producing a vapor containing 82 mol% methanol and 18 mol% chlorobenzene. Above this methanol concentration, the y–x diagram is near linear. Unfortunately, this chlorobenzene loss means that you presently must continuously feed extra chlorobenzene, an inconvenience and a risk (because chlorobenzene is a carcinogen). To reduce this inconvenience, you plan to put some distillation stages on top of the reactor. The stages are to be nearly ideal and use a reflux that is large. (a) How much chlorobenzene must you feed to keep constant the solvent concentration in the reboiler without any stages? (b) How much must be fed with one stage? (c) How much must be fed with three stages?

Further Reading

Geankoplis, G. J. (2003). *Transport Processes and Separation Process Principles*, 4th ed. Upper Saddle River, NJ: Prentice Hall.

Humphrey, J. L. and Keller, G. E. (1997). *Separation Process Technology*. New York: McGraw-Hill.

McCabe, W. L. Smith, J. C., and Harriott, P. (2004). *Unit Operations of Chemical Engineering*, 7th ed., New York: McGraw-Hill.

Seader, J. D. and Henley, E. J. (2006). *Separation Process Principles*, 2nd ed. New York: Wiley.

Wankat, P. C. (2006). *Separation Process Engineering*. Upper Saddle River, NJ: Prentice Hall.

CHAPTER 14

Extraction

Extraction treats a feed with a liquid solvent to remove and concentrate a valuable solute. When the feed is a liquid, the process is called "liquid–liquid extraction," or more commonly just "extraction." When the feed is a solid, the process is called "solid–liquid extraction," or more commonly "leaching." In either case, the original solution is commonly called the feed; after the extraction, this stream is called the raffinate. Similarly, the second solvent is called the extract once it contains solute.

Extraction is almost never the first choice as a separation process. If the solute of interest is a gas, then we will first try gas absorption or stripping. If the solute of interest is volatile under convenient conditions, then we will attempt distillation. We will normally try extraction only after we fail at absorption and distillation. Still, we have included a separate chapter on extraction for two reasons. First, it is an important process, central to some petrochemical, pharmaceutical, and metallurgical processes. We discuss these in Section 14.1. Second and more importantly, extraction gives an extended example of the generalization of the analyses of absorption and distillation. When extraction is carried out in differential contactors like packed towers, its analysis is similar to gas absorption. When extraction is carried out in staged contactors, its analysis parallels staged distillation. Thus we can test our understanding of absorption and distillation by discussing extraction.

At the same time, we want to focus on the role of diffusion in extraction, for that is the subject of this book. As a result, we emphasize the case of a dilute solute being extracted between two immiscible liquids. This defers complicated issues of ternary phase equilibria to more specialized texts and lets us focus on the issues of mass transfer, which can be obscured in those texts. Specifically, we discuss extraction equipment in Section 14.2, we analyze differential extractors as a parallel to gas absorption in Section 14.3, and we describe staged extraction in Section 14.4. Leaching, which can be either staged or differential, is treated in Section 14.5. The result is a brief summary that emphasizes the role of mass transfer.

14.1 The Basic Problem

Extraction is a common separation process used where distillation and gas absorption fail. Most obviously, extraction can be used for nonvolatile components like metal ions. It is effective for valuable solutes like flavors, which can be unstable at distillation temperatures. Less obviously, extraction is useful for volatile solutes that have nearly equal boiling points or that show azeotropes.

Some common extractions are listed in Table 14.1-1. When I look at this table, I think of three specific applications as a way to organize my thinking. The first specific application is the dewaxing of lubricants. Lubricants are made from particular fractions collected during distillation of crude oil. These fractions are the most valuable part of

Table 14.1-1 *Important applications of extraction*

Industry	Objective	Typical feed	Typical solvents	Remarks
Petroleum and petrochemicals	Dewaxing lubricating oils	Crude lube stocks	Glycols, furfural, cresol, liquid SO_2	Wide variety of solvents have been used
	Higher octane aromatic fuels	Aliphatic–aromatic mixtures	Glycols, sulfolane	Demand affected by legislation
	Desulfurization for reduced emissions	Sour distillates	Dilute aqueous base	Product is elemental sulfur after further reaction
	Butadiene–butene separation	Incompletely dehydrogenated feed	Aqueous copper complexes	Unreacted butene is recycled
Pharmaceuticals and foods	Concentrating impure antibiotics	Filtered fermentation beer	Amyl acetate, methylene chloride	Penicillin is a good example
	Refining fats and oils	Soybeans	Propane, hexane	Supercritical CO_2 is often suggested
	Sugar	Beets	Water	Major domestic sugar source
Metals	Concentrating copper for electrowinning	Acidic leach liquors	Hydroxyoximes in kerosene	pH changes are key
	Gold	Low-grade ore	Sodium cyanide solutions	Can be environmental problem
	Uranium and rare earth separations	Acidic leach liquors	Tertiary amines in kerosene	Future depends on nuclear power

the barrel. Crude oil produced in Pennsylvania historically had a larger fraction of lubricants than crude oil produced in other locations, which is why the names of several motor oil companies refer to Pennsylvania.

Unfortunately, these lubricating fractions contain a significant amount of linear hydrocarbons. When such a fraction becomes cold, as in a Minnesota winter, these linear hydrocarbons precipitate as wax crystals. When the wax crystals are present in motor oil, the oil is a poor lubricant, so removing these waxes is important. While a huge number of solvents – including supercritical solvents – have been used for this purpose, this remains a competitive area, for it represents the largest single extraction process.

The second specific application is purification of penicillin. Penicillin is produced by fermentation. The result is a cloudy broth, which looks much like a beer made at home or in a small microbrewery. The microbes are first removed by filtration. The clarified broth is then extracted with an alkyl acetate, commonly amyl acetate. If the aqueous broth is acetic, the penicillin partitions into the acetate. If the organic solution is then contacted with dilute aqueous base, the penicillin will partition back into the base. After about four such transfers, the purified penicillin is crystallized to produce the final product.

The third specific application of extraction that merits emphasis is the concentration of copper. Originally, copper was produced by roasting copper sulfide to produce cupric oxide. The cupric oxide was dissolved in acid and further purified by electrowinning. This process is no longer practiced because current ores have too small a concentration of copper sulfide to be economically roasted. Instead, low-grade ores are leached with sulfuric acid to produce dissolved copper. This dissolved copper is present at such a low concentration that it cannot be economically recovered without further concentration. It is concentrated by extraction from the dilute acid solution into a kerosene solution of liquid ion exchangers. The copper is recovered from this kerosene solution by a second extraction with still more concentrated acid. The result is a copper solution concentrated enough for economical electrowinning. Similar processes are used for uranium and other rare earths.

These three examples – dewaxing lubricants, isolating penicillin, and concentrating copper – suggest the criteria for how an extraction solvent should be chosen. First, the solvent should have a favorable equilibrium: the solute concentration in the extract should be greater than the solute concentration in the raffinate. Second, the extract should be easy to separate. For example, in the extraction of oils from soybeans, the hexane in the extract is easily removed by distillation. The third criterion for choosing an extraction solvent must be the degree of environmental insult. Solvents that are suspected carcinogens, such as methylene chloride and toluene, should be used sparingly.

These generalizations will always have exceptions. For example, acetic acid is extracted from water using methyl ethyl ketone, even though the equilibrium is unfavorable. In other words, the concentration of the acetic acid in water is higher at equilibrium than the concentration in the ketone. In this case, the extract of acetic acid and the ketone is much more easily separated by distillation than are extracts with alternative solvents. Methyl ethyl ketone is chosen in spite of its poor equilibrium.

Similar concerns have also inhibited the adoption of unconventional solvents. Two good examples are supercritical carbon dioxide and two-phase aqueous systems. Supercritical carbon dioxide can be effectively used to decaffeinate coffee. However, its broad adoption has been inhibited by the major capital cost of the equipment. Two-phase aqueous extraction most often uses an aqueous solution of a polymer like dextran and a second aqueous solution of phosphates. These solutions are immiscible even though each contains more than 90 percent water. When several proteins are dissolved in these solutions, they partition unequally, so that single proteins can be separated. The separation is gentle, and often does not denature the proteins. However, this gentleness is compromised by the major difficulty of later separating the protein from the dextran.

The chemistry of extraction frequently involves nonlinear equilibria between the raffinate and extract. This reflects the extraction chemistry, which is often more complicated than that for absorption or dilute distillation. The nonlinear equilibria frequently result from specific chemical reactions. For example, benzoic acid can be extracted from water into benzene. The benzoic acid in water can ionize to form a mixture of benzoic acid, benzoate anions, and protons. Benzoic acid in benzene can dimerize. The ionization and dimerization may result in an equilibrium that is nonlinear and strongly dependent on concentration, pH, and temperature.

In a similar way, much metal purification involves ion exchange. For example, in the case of copper extraction mentioned above, the equilibrium is:

$$Cu^{2+}(\text{aqueous}) + 2HX \text{ (in kerosene)} \rightleftharpoons CuX_2(\text{in kerosene})$$
$$+ 2H^+(\text{aqueous}) \tag{14.1-1}$$

where HX represents the liquid ion exchanger dissolved in kerosene and X represents the anion of that exchanger. Thus any equilibrium between copper in water and in kerosene should include variables like the concentration of the ion exchanger.

In spite of these complexities, most analyses of extraction assume linear equilibria. The detailed chemistry appears as a concentration-dependent partition coefficient. While such concentration-dependent partition coefficients are beyond the scope of this book, they are well understood and discussed in detail in more specialized references.

14.2 Extraction Equipment

Liquid–liquid extraction and leaching use different equipment, even though the analysis of this equipment can be similar. Liquid–liquid extraction can be accomplished either in differential contactors or in staged extractors (Godfrey and Slater, 1995). The differential contactors are analyzed in ways that parallel the analysis of gas absorption; the staged extractors depend heavily on ideas developed for distillation. In both cases, an enormous variety of equipment is used, with specific apparatus often being optimized for particular separations.

Rather than survey the many types, we will discuss the four characteristic types shown in Fig. 14.2-1. The first three types are differential contactors. The simplest is the spray column shown in Fig.14.2-1(a). In the configuration shown, the light extraction solvent is pumped through a sparger – an inverted shower nozzle – into a column filled with the aqueous feed. Drops of the solvent rise slowly through heavy feed. At the same time, the feed moves downward through the column. The result is countercurrent contacting between the lighter extraction solvent and the heavier feed.

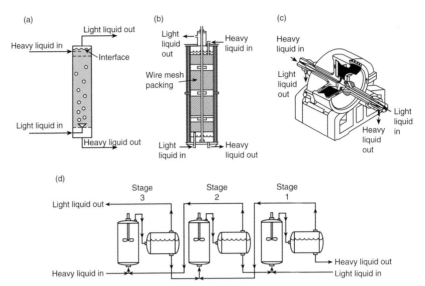

Fig. 14.2-1. Four important types of extraction equipment. Types (a) – (c) are differential contactors, described in the same manner as gas absorption. Type (d), a three-stage mixer-settler, depends on stages, as does distillation.

Similar contacting can use a packed column like that shown in Fig. 14.2-1(b). In general, a packed column has better mass transfer but lower capacity than a spray column. As before, the heavy feed enters the top of the column and the light solvent enters the bottom. Now, a larger interfacial area is generated between the feed and solvent by the packing. The packing should be wet by the phase that we want to be continuous, and not by the phase that we want to be discontinuous. Choosing which phase we want to be continuous depends on which configuration will give us the faster mass transfer.

Both spray columns and packed columns are seriously compromised by flooding. In flooding, the feed and solvent streams do not flow evenly and countercurrently past each other, but both simply gush out one end of the column. Flooding is a more serious risk in extraction than in absorption because of the smaller density difference between the two fluids. This density difference is typically less than 0.1 g/cm^3, about 10 times less than that common in gas absorption. As a result, countercurrent flows that are routine in gas absorption will be difficult to realize in liquid–liquid extraction.

Both spray columns and packed columns routinely perform much less well than expected. There are two main reasons for this. First, the two liquid phases may not flow evenly past each other but mix more randomly together. In that case, they will behave more like a stirred tank than a packed bed, and have very few numbers of transfer units (NTU) or equivalent stages. Indeed, the fact that spray columns can show less mass transfer than even one equilibrium stage signals first how poor the contact between the two liquids can be.

The second reason that spray and packed columns perform poorly is their risk of bypassing caused by using large drops of the discontinuous fluid. While these drops

Table 14.2-1 *Characteristics for common extraction equipment*

| Type | Features | Applications | Apparatus suitable for conditions given | | |
			Number of stages	Flow rate, m^3/hr	Density difference, g/cm^3
Spray columns	Low capital and operating cost; handles corrosive material	Petrochemical, chemical	< 1	< 100	≥0.05
Packed columns	Less capacity but more mass transfer than spray columns; restricted by flooding	Petrochemical, pharmaceutical	< 10	< 50	≥0.05
Centrifugal extractors	High capital cost; short contact times	Pharmaceutical, nuclear	< 5	< 10	≥0.01
Mixer-settlers	High capacity and flexibility; handles high viscosity	Petrochemical, metallurgical	Any value	> 250	≥0.10

Note: Abstracted from Lo and Baird, 1993.

flow faster, they have a smaller surface area per volume and hence reduced mass transfer. Small drops, which have a higher area per volume and hence faster mass transfer, flow more slowly and risk flooding. To reduce the risk of flooding, we can use centrifugal extractors, like that shown in Fig. 14.2-1(c). In this case, the two liquids are forced past each other by centrifugal force. Because centrifugal extraction has modest capacity and high equipment cost, it tends to be used only for high value added products.

When constraints like flooding are severe, we are driven to look at mixer-settlers like those shown in Fig. 14.2-1(d). These are nothing more than tanks whose contents are stirred to reach equilibrium. The contents are then pumped into unstirred tanks where the two fluid phases separate slowly. Such tanks always work and are not expensive. The advantages and disadvantages of this and the other three types are summarized in Table 14.2-1. We will refer to this table in the analysis of extraction problems given in the next two sections.

14.3 Differential Extraction

Differential liquid–liquid extraction, like differential distillation a form of rate-dependent, differential contacting, is analyzed in ways that parallel the analysis of countercurrent gas absorption. This analysis is complicated by two factors. First, the solute concentrations at equilibrium in the raffinate and the extract are not linearly proportional. This nonlinear behavior most often reflects a chemical change, like ionization or chemical reaction. The second factor complicating the analysis of extraction is that the feed and the solvent used for extraction are partly miscible. For example, in the extraction of aqueous acetic acid, the water is significantly soluble in the methyl ethyl ketone commonly used as the extraction solvent. This mutual solubility means that the analysis depends on ternary-phase equilibria, often expressed on ternary coordinates.

We minimize these complications here by considering only the extraction of dilute solutes between immiscible solvents. Though this limit misses some problems of practical interest, it focuses our attention on dilute solutions of expensive materials. In the dilute limit, we again have three key equations:

(1)   a mass balance on both liquids, called the operating line;
(2)   a free energy balance for the solute, again called the equilibrium line; and
(3)   a rate equation, derived from a mass balance on only one liquid phase.

These three equations are the basis for our analysis of dilute extraction.

The mass balance on both liquid phases is written on a control volume, located from some arbitrary position z to the end of the extractor where the solvent enters. For convenience, we will call the mass flux of feed and raffinate streams G, and the solvent and extract streams L; this notation stresses the parallel with gas absorption. The mass balance is

$$Gy + Lx_0 = Gy_0 + Lx \qquad (14.3\text{-}1)$$

where y and x are solute concentrations of the feed and solvent phases, respectively; and y_0 and x_0 are the corresponding values where the solvent enters at the end of the extractor.

In absorption and distillation, y and x were usually mole fractions. This may also be true for extraction. Often, however, y and x will be expressed as mass fractions or mass ratios. For example, x may be defined as mass of solute per mass of solvent. This is especially true when the solute is a chemical mixture, like a wax in glycol. The mass balance in Eq. 14.3-1 is easily rearranged to give

$$y = \left(y_0 - \frac{L}{G} x_0 \right) + \frac{L}{G} x \qquad (14.3\text{-}2)$$

This mass balance or "operating line" is the mathematical equivalent of Eq. 10.3-5, developed for gas absorption. As in that case, this operating line says that y varies linearly with x, with a slope of (L / G).

In addition to this operating line, we need a free-energy balance or equilibrium line. For the dilute case, this is still a simple linear relation:

$$y^* = mx \qquad (14.3\text{-}3)$$

where m is a partition coefficient. As before, this partition coefficient is dimensionless.

The third key relation for dilute extraction is a rate equation, a mass balance written on only one liquid phase. For example, for the feed, this steady-state mass balance is written on a differential volume ($A\Delta z$) located at an arbitrary position z measured from the end of the extractor:

$$\begin{bmatrix} \text{solute} \\ \text{accumulation} \end{bmatrix} = \begin{bmatrix} \text{solute flow} \\ \text{in} - \text{that out} \end{bmatrix} - \begin{bmatrix} \text{amount} \\ \text{extracted} \end{bmatrix}$$

$$0 = GAy|_{z+\Delta z} - GAy|_{z} - K_y a(A\Delta z)(y - y^*) \tag{14.3-4}$$

in which K_y is an overall mass transfer coefficient and a is a liquid–liquid interfacial area per extractor volume. Dividing by the differential volume ($A\Delta z$) and taking the limit as this volume goes to zero, we find

$$0 = G\frac{dy}{dz} - K_y a(y - y^*) \tag{14.3-5}$$

subject to the conditions

$$z = 0, \quad y = y_0, \quad x = x_0 \tag{14.3-6}$$

$$z = l, \quad y = y_l, \quad x = x_l \tag{14.3-7}$$

Equations 14.3-2, 14.3-3, and 14.3-5 are the basis for analyzing dilute absorption.

We can now find the size l of the differential extractor by integrating Eq. 14.3-5. To do so, we first rearrange this equation to get

$$\int_0^l dz = \frac{G}{K_y a} \int_{y_0}^{y_l} \frac{dy}{y - y^*} \tag{14.3-8}$$

or

$$l = \frac{G}{K_y a} \int_{y_0}^{y_l} \frac{dy}{y - mx} = \frac{G}{K_y a} \left[\frac{1}{1 - \dfrac{mG}{L}} \ln\left(\frac{y_l - mx_l}{y_0 - mx_0}\right) \right] \tag{14.3-9}$$

The quantity L / mG, often of order one, is called the extraction factor E. For liquid extraction, this parallels Eq. 10.3-12 for gas absorption.

In this discussion, we have repeatedly stressed the mathematical parallels between extraction and absorption because these help understanding the developments. However, these mathematical parallels should not obscure physical differences. Some are straightforward: The concentrations for extraction are commonly expressed as mass fractions, but those in absorption are more frequently mole fractions. Other differences are more subtle. The equilibrium constant m is a good example. Obviously, m is a linear approximation that can be inaccurate if the chemistry includes ionization or complex formation. Less obviously, the dimensionless m requires careful definition. For example, imagine we are given

$$\begin{bmatrix} \text{mole fraction} \\ \text{in aqueous phase} \end{bmatrix} = m' \begin{bmatrix} \text{mole fraction} \\ \text{in solvent} \end{bmatrix} \tag{14.3-10}$$

We want to find the m defined by

$$\begin{bmatrix} \text{mass fraction} \\ \text{in aqueous phase} \end{bmatrix} = m \begin{bmatrix} \text{mass fraction} \\ \text{in solvent} \end{bmatrix} \tag{14.3-11}$$

For dilute solutions, we can show that

$$m = m' \begin{bmatrix} \text{molecular weight of } L \\ \text{molecular weight of } G \end{bmatrix} \tag{14.3-12}$$

The relation is more complex for concentrated solutions.

Other subtle differences occur in the mass transfer coefficient. By arguments that parallel those in Section 8.5, we may show that

$$\begin{aligned}
\frac{1}{K_y} &= \frac{1}{k_y} + \frac{m}{k_x} \\
&= \frac{1}{k_G \rho_G} + \frac{m}{k_L \rho_L}
\end{aligned} \tag{14.3-13}$$

where k_y and k_x are defined in terms of differences of mass fractions, k_G and k_L are defined in terms of differences of mass concentrations, and ρ_G and ρ_L are the densities of the raffinate and extract phases, respectively. In practice, k_G and k_L are of similar magnitude, around 10^{-3} cm/sec; and ρ_G and ρ_L are obviously also about equal, around 1 g/cm^3.

Thus the limiting resistance to mass transfer depends most critically on m. If m is much less than one, which is normally what we seek in an extraction solvent, then the rate of extraction will be controlled by the mass transfer in the raffinate phase. If m is much more than one, then the extraction rate will be dominated by mass transfer in the extract phase. This situation is different than that in absorption, where the mass transfer in the gas is usually so fast that it doesn't affect the overall mass transfer coefficient. We explore these ideas further in the following example.

Example 14.3-1: Steroid extraction   We are trying to isolate a steroid like sitosterol from an aqueous vegetable feed. To do so, we pump 6.5 kg/hr of the feed upward through a packed column 0.61 m high and 0.1 m in diameter. We spray 3.0 kg/hr of pure methylene chloride to trickle downward through the bed. We find by experiment that m is 0.14; even so, we get only a 53% recovery. (a) What is the value of $K_y a$? (b) How long should we make the tower for a 90% recovery?

Solution   (a) We begin by calculating the exit concentration in the methylene chloride from Eq. 14.3-1

$$x_l = \frac{6.5}{3.0} (y_l - 0.47 y_l)$$

$$x_l = 1.15 y_l$$

Thus from Eq. 14.3-9

$$0.61 \text{ m} = \left[\frac{(6.5 \text{ kg})/\left(3600 \text{ sec } \pi/4 (0.1 \text{ m})^2\right)}{K_y a} \right] \cdot \left\{ \frac{1}{1 - \frac{0.14(6.5)}{3.0}} \ln\left[\frac{y_l - 0.14(1.15 y_l)}{0.47 y_l - 0} \right] \right\}$$

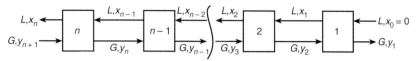

Fig. 14.4-1. An idealized staged countercurrent extraction. The feed H and extractant L flow at constant rates in the limit of the dilute solutions considered here. As in distillation, the concentrations in these streams are identified by the stage where each originates.

Thus

$$K_y a = 0.31 \text{ kg/m}^3 \text{sec}$$

(b) As in (a), we first calculate the exit concentration x_0, but for 90% recovery:

$$3.0x_l = 6.5(y_l - 0.1yl)$$
$$x_l = 1.95y_l$$

We then use Eq. 14.3-9 to find

$$\frac{l(90\% \text{ recovery})}{l(53\% \text{ recovery})} = \frac{\ln[(y_l - 0.14(1.95y_l))/0.1y_l]}{\ln[(y_l - 0.14(1.15y_l))/0.47y_l]} = 3.4$$

Thus

$$l(90\% \text{ recovery}) = 2.1 \text{ m}$$

We achieve a better extraction by making the column longer.

14.4 Staged Extraction

The alternative to differential extraction is staged extraction. This frequently uses the mixer-settlers shown in Fig. 14.2-1(d), but it can also use column extractors containing the equivalent of sieve trays. In these cases, extraction is analyzed using the same concepts of equilibrium stages used in the analysis of distillation. Such an equilibrium stage analysis depends on mass transfer only tangentially. As a result, we will quickly review the analysis and then discuss how mass transfer is involved.

To begin, we consider the staged cascade shown schematically in Fig. 14.4-1. As before, G and L are the fluxes of the immiscible feed and extract, respectively; y and x are the concentrations of the dilute solute being transferred from G to L. We again label the solute concentrations with subscripts giving the stage where they originate. Thus for stage n, the entering mass fractions are x_{n-1} and y_{n+1}; the exiting mass fractions are x_n and y_n.

We are interested first in the limit where the exiting concentrations are in equilibrium. Because the solute is dilute, we expect this equilibrium is linear:

$$y_n = y_n^* = mx_n \qquad (14.4\text{-}1)$$

We also recognize that each stage is subject to a mass balance. For example, for the first stage,

$$Gy_2 + L(0) = Gy_1 + Lx_1 \tag{14.4-2}$$

Combining this with the equilibrium, we find

$$y_2 = \left(1 + \frac{L}{mG}\right) y_1$$
$$= (1 + A) y_1 \tag{14.4-3}$$

where $A\,(= L/mG)$ is the reciprocal of an absorption factor like that used in Section 13.2. Other analogous factors can be used as well. For example, A is often called an extraction factor E or a stripping factor S. Here, we will use A so that all our equations for staged operations remain closely parallel.

As an analogue to Eq. 14.4-3 for the first stage, we can write for the second stage,

$$Gy_3 + Lx_1 = Gy_2 + Lx_2 \tag{14.4-4}$$

This leads to

$$y_3 = (1 + A) y_2 - Ay_1$$
$$= \left(1 + A + A^2\right) y_1 \tag{14.4-5}$$

For N stages, the result is

$$y_{N+1} = \left(1 + A + A^2 + \cdots A^N\right) y_1$$
$$= \left[\frac{1 - A^{N+1}}{1 - A}\right] y_1 \tag{14.4-6}$$

This is the desired result, a special case of Eq. 13.2-11 for distillation. This result allows estimation of the number of equilibrium stages required to achieve a desired separation.

In many cases, Eqs. 14.4-5 and 14.4-6 are reasonably accurate, for each stage comes close to equilibrium. This is because the expense of the solute makes long times on each stage attractive. In these cases, we will find it advantageous to use an absorption factor that is close to one. In other cases, the time spent in individual stages is not long enough to reach equilibrium. In these cases, we can parallel our analysis for distillation to again define a Murphree efficiency η

$$\eta = \frac{y_n - y_{n+1}}{y_n^* - y_{n+1}} \tag{14.4-7}$$

where y_n is the actual exiting concentration and y_n^* is the concentration that would exit at equilibrium. If mass transfer is rapid, we will be near equilibrium and η will equal one; if the mass transfer is slow, y_n and y_{n+1} will be almost the same and η will be near zero.

The relation between the Murphree efficiency and the mass transfer coefficient is simplest when the stages are well mixed. In this case, we can parallel the development leading to Eq. 13.4-7 to find

$$\eta = \frac{1}{1 + \dfrac{GA}{K_y aV}} \tag{14.4-8}$$

where GA is the total raffinate flow, in mass per time; a is the liquid–liquid surface area per mixer volume; V is the mixer volume; and K_y is the overall mass transfer coefficient. The value of the overall mass transfer coefficient K_y, and hence the Murphree efficiency, is most strongly influenced by the partition coefficient m. To see why, we return to the definition of K_y, given by

$$\frac{1}{K_y} = \frac{1}{k_y} + \frac{m}{k_x}$$
$$= \frac{1}{k_G \rho_G} + \frac{m}{k_L \rho_L} \tag{14.4-9}$$

For extraction, the values of k_G and k_L are usually around 10^{-3} cm/sec, and the densities of feed ρ_G and extract ρ_L are often around 1 g/cm^3. Thus if we know m, we can estimate K_y, and hence make guesses on the stage volume required to give high Murphree efficiencies.

Sometimes, however, the extractions in staged cascades are much less effective than expected even when the mass transfer is fast. In other words, the Murphree efficiencies are much less than one even when the quantity $(K_y aV/GA)$ is expected to be much greater than one. Such poor performance is often due to backflow of either the raffinate or the extract. Such backflow may occur because of entrained liquid. While the cause varies, the effect is the same: a lower apparent stage efficiency. We explore these ideas in the example that follows.

Example 14.4-1: Actinomycin extraction A clarified fermentation broth (G) containing 260 mg/l of actinomycin is to be extracted using butyl acetate (L). Because the beer's pH is 3.5, the equilibrium constant m is 0.018. You plan to set G at 450 l/hr and L at 37 l/hr. You want to recover 99% of the antibiotic in the feed. (a) How many equilibrium stages will you need to accomplish this separation? (b) How many will you need if the Murphree efficiency is 60%?

Solution (a) The factor A can be calculated from the values given:

$$A = \frac{L}{mG} = \frac{37 \text{ l/hr}}{0.018(450 \text{ l/hr})} = 4.57$$

The number of stages is easily found from Eq. 14.4-6:

$$\frac{y_{N+1}}{y_1} = \left[\frac{1 - A^{N+1}}{1 - A}\right]$$

$$100 = \left[\frac{1 - 4.57^{N+1}}{1 - 4.57}\right]$$

Thus

$$N = 2.9$$

We need about three ideal stages.

(b) To find the result in terms of inefficient stages, we parallel Eqs. 14.4-1, 14.4-2, and 14.4-7 as follows:

$$G(y_2 - y_1) = Lx_1$$

$$= \frac{Ly_1^*}{m}$$

$$= \frac{L}{m}[y_2 + \frac{1}{\eta}(y_1 - y_2)]$$

Rearranging, we find

$$y_2 = \left[1 + \frac{1}{\left(\frac{mG}{L} + \frac{1}{\eta} - 1\right)}\right]y_1$$

This is analogous to Eq. 14.4-3 but with the new factor shown in parentheses. Other mass balances give similar results. Because

$$\left[\frac{mG}{L} + \frac{1}{\eta} - 1\right]^{-1} = \left[\frac{0.018(450)}{37} + \frac{1}{0.60} - 1\right]^{-1} = 1.13$$

we find from Eq. 14.4-6 that

$$N = 21 \text{ stages}$$

The separation which looked easy has become harder.

14.5 Leaching

We now turn from liquid–liquid extraction to solid–liquid extraction. In more casual terms, we turn from extraction to leaching. As expected, we can use the same ideas of differential and staged separations to handle this case as well. We begin with target separations, we list separation equipment, and we then describe differential and staged leaching. The summary leads to the case of unsteady-state leaching, a transitional case best handled by parallels with the analysis of adsorption in the next chapter.

The target separations achieved by leaching usually involve feedstocks from nature. Many of these feedstocks are agricultural products. Soybean oil, the main cooking oil in North America, is obtained by leaching soybeans with hexane. Sugar is leached out of beets or sugar cane with water. Caffeine is removed from coffee beans with methylene chloride, ethyl acetate, or supercritical carbon dioxide. Less benignly, cocaine is recovered from cocoa leaves with acetone.

The other major group of solutes recovered by leaching are minerals. Gold is recovered from low-grade ores with aqueous sodium cyanide. Copper is recovered by leaching the ore with acid and then extracting the dissolved copper with kerosene solutions of oximes. Nickel can be leached with combinations of sulfuric acid and ammonia.

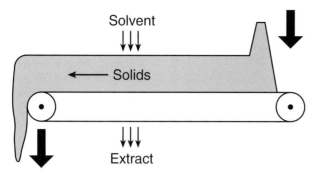

Fig. 14.5-1. Soybean leaching. The beans, moving on a screen, are washed with fresh solvent flowing in cross flow.

Characteristic equipment for leaching is shown in Fig. 14.5-1. The soybean oil extractor shown places the flaked soybeans in a shallow layer on a conveyer belt, and washes them with a solvent in cross flow. Sugar-beet extraction is similar; sliced beets move countercurrently to water as the sugar is extracted.

Staged leaching is used in the washing of solids. The equipment used, called a "thickener," first suspends the solid in the solvent, and then allows the solid to settle out under gravity. The resulting solid paste entrains a significant amount of liquid, which will normally have the same concentration of the desired solute as the clear solution on top of it. Thus in these stages, the entrained solution is what is really being washed; the solid just goes along for the ride.

Differential Leaching

We can put these ideas on a more quantitative level by considering a differential process shown in Figure 14.5-2(a). In this idealization, the dry solid feed enters with a flux F but is contacted with solvent to produce a flux G and a solute concentration y_l. This wet feed is extracted with pure solvent entering at flux L, but leaving at a smaller flux E.

We can use the description in Fig. 14.5-2(a) to estimate the size of the apparatus. The simplest case occurs for a dilute solute, so that solid and solvent flows are constant within the equipment. In this case, the size is estimated just as it would be for dilute adsorption:

$$l = [\mathrm{HTU}] \cdot \{\mathrm{NTU}\}$$

$$= \left[\frac{G}{K_y a}\right] \cdot \left\{\frac{1}{1 - \dfrac{mG}{L}} \ln \frac{y_l - y_l^*}{y_0 - y_0^*}\right\} \tag{14.5-1}$$

where HTU and NTU are again the height and number of transfer units; G and L are the solid and solvent fluxes; m is the slope of the equilibrium line; and the y's are the concentrations. Remember that this analysis presumes constant L and G within the apparatus. Normally, the solid will retain some solvent, so G will be greater than the dry feed F, and L will be greater than the extract removed E.

E, x_I L L, x_0

F, y_I G G, y_0

(a) Differential leaching

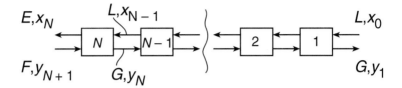

E, x_N L, x_{N-1} L, x_0

F, y_{N+1} G, y_N G, y_1

(b) Staged leaching

Fig. 14.5-2. Two idealizations of leaching. The flows of solids G and solvent L change because some liquid is entrained by the solids.

While the analysis of differential leaching is much like that of differential gas absorption or differential distillation, the definitions of the concentrations vary. The concentration x in the solvent stream is easy enough: it is just the amount of solute per amount of solvent. It may be expressed as a mole fraction, a mass fraction, or a mass ratio. The concentration y in the feed is harder. When the solute is being extracted from the solid itself, y will be expressed as amount of solute per amount of solid. When the solute completely dissolves in the solvent to leave none in the solid, then the concentration y will be expressed as solute per retained solvent, and the slope of the equilibrium line will be one. We will return to this point below, in the discussion of staged leaching.

Staged Leaching

The second case of leaching uses stages, that is, mixer-settlers or thickeners. These are analyzed in a way similar to staged distillation, but without the complexities introduced by a central feed, a reboiler, and a condenser. The basic scheme, shown in Fig. 14.5-2(b), can use either countercurrent flow or cross flow. As expected, countercurrent flow gives a more efficient separation, but cross flow is simpler and is sometimes used for expensive products. We will stress the countercurrent case here.

As before, our normal goal is to find the number of stages required for a given separation. The most important special case of countercurrent leaching occurs when all the solute quickly dissolves in the solvent. The spent solid itself then has no solute left and is just moving through the process as an inert. In this case, the concentration in the extracting solution leaving a particular stage equals the concentration in the solution retained by the solid leaving that stage. The reason is that the concentrations are really in the same solution; the only difference is that part of this solution has been trapped within voids between solid particles. As a result,

$$y_n = x_n$$

i.e., the equilibrium line is the 45° diagonal. This part of the analysis of staged leaching is easy.

The part of the analysis that is not easy centers on the Nth stage. The reason is that the feed normally enters dry but usually leaves containing solvent. As a result, the mass balances on this nth stage must normally be made separately. Once these are complete, the remaining number of stages may be calculated either from a McCabe–Thiele analysis or from one of the Kremser equations.

Unsteady Leaching

The third type of leaching is by far the simplest. In this type, we simply pour a solvent on top of a packed bed of solids, and collect the solution, which trickles through. The leaching of low-grade ores and coffee are two good examples. If we plot the concentration of eluted solutes vs. time, we usually get a breakthrough followed by an exhaustion, just as occurs in adsorption, detailed in the next chapter. This curve is a function of the mass transfer, whose rate-controlling step may change as the solute in the solid particles become depleted. This type of solid–liquid extraction is best analyzed by extending the methods used for adsorption in the next chapter.

Example 14.5-1: Gold recovery from waste electronics   We are trying to recover 99% of the gold in a chipped solid, electronic waste which contains 80 ppm gold. We plan to use a cyanide-containing acid solution as a solvent. This solvent, which initially contains no gold, enters at 5 kg per kg of dry waste. While the waste enters dry, it leaves with 2 kg of solvent for each kilogram of solid.

How many equilibrium stages would be required? Should we expect the stages to approach equilibrium?

Solution   Assuming a basis of 1 kg waste, we first make an overall balance on the solvent

$$5 \text{ kg fed} = \left[\frac{2 \text{ kg entrained}}{\text{kg waste}} \right] 1 \text{ kg waste} + E$$

$$E = 3 \text{ kg extract}$$

From the problem statement, we find the concentration in the raffinate y_1:

$$0.01(80 \text{ ppm}) 1 \text{ kg waste} = (y_1) 2 \text{ kg solvent}$$

$$y_1 = 0.04 \text{ ppm}$$

We can now find the solvent concentration x_N leaving the feed stage

$$(80 \text{ ppm}) 1 \text{ kg waste} = x_N (3 \text{ kg solvent}) + 0.4 \text{ ppm} (2 \text{ kg solvent})$$

$$x_N = 25.4 \text{ppm}$$

A balance on the gold in the feed stage gives x_{N-1}

$$(80 \text{ ppm}) 1 \text{kg waste} = x_{N-1} (5 \text{ kg solvent})$$

$$= 26.4 \text{ ppm} (3 \text{ kg extract}) + 26.4 \text{ ppm} (2 \text{ kg entrained solvent})$$

$$x_{N-1} = 10.4$$

Finally, we find the number of stages from Equation 13.2-13:

$$N - 1 = \ln\left[\frac{y_N - y_{N-1}}{y_1 - y_0}\right] \Big/ \ln\left[\frac{L}{mG}\right]$$

$$= \ln\left[\frac{x_N - x_{N-1}}{y_1 - 0}\right] \Big/ \ln\left[\frac{L}{G}\right]$$

$$= \ln\left[\frac{26.4 - 10.4}{0.4 - 0}\right] \Big/ \ln\left[\frac{5}{2}\right]$$

$$= 4.0$$

We need a little more than five stages, including the feed stage. We can probably get the stages to approach equilibrium if all the gold is on exposed surfaces within the solid waste. If so, it will dissolve quickly. If not, this analysis may be seriously in error.

14.6 Conclusions

The synopsis of extraction presented in this chapter is important for two reasons. First, extraction is an important separation process in its own right. It is used for petrochemicals, pharmaceuticals, and metals. Because it uses solvents which are often environmentally threatening, it is under close scrutiny, especially for unconfined applications existing in, for example, some mining operations. However, it will probably remain central to many fine-chemical purifications, including those of antibiotics.

The second reason that extraction is important is intellectual, not practical. We can analyze extraction in two different ways. First, we can treat extraction with an analysis like that used for gas absorption in Chapter 10. This form of analysis, developed for extraction in Section 14.3, is called *differential contacting*. Alternatively, we can treat extraction with an analysis like that basic to staged distillation, as described in Chapter 13. This form of analysis, given for extraction in Section 14.4, is called *staged contacting*.

The analysis of extraction in these two different ways exemplifies the analysis of almost any separation process. To be sure, one form of analysis may be strongly implied by the physical situation. For example, if we had enthusiastically stirred mixer-settlers, we would be sensibly inclined to analyze these as staged contacting. Frequently, the choice is less obvious. For example, if we had a large number of inefficient mixer-settlers, we might get a simpler correlation of data via the analysis of differential contacting. Thus extraction is important not only for itself but as an illustration of these alternative strategies.

Questions for Discussion

1. What is the difference between extraction and leaching?
2. What concentration variables are best for extraction?

3. Differential extractors are much less common than staged extractors. Discuss why.
4. Is a mixer-settler a differential or a staged contactor?
5. Is a Podbielniak extractor a differential or a staged contactor?
6. Is a coffee maker a differential or a staged contactor?
7. How will the analysis of extraction change if the operating and equilibrium lines are parallel?
8. How will extraction vary with temperature?
9. Extraction is often accused of being environmentally abusive. Explain why.
10. What concentration variables are best for leaching?
11. The leaching of soybeans with hexane depends critically on finely flaking the soybeans. Discuss why.
12. Leaching is uses acid as a leachant; less commonly, it uses base. Discuss why.
13. How would you analyze leaching from a large pile of low-grade ore sprayed with acid?

Problems

1. You are planning to extract a mushroom flavor from the hexane solution made by leaching beans. Your extraction solvent is a modified cyclodextrin solution in an aqueous base. This has an extremely small partition coefficient m ($= y/x$), so that at this pH, the adsorption is nearly irreversible.

    Because of this, you plan to use a differential extractor 1.85 m tall. When the hexane solution flows through at 0.16 m/s, you know from the model experiments that the overall mass transfer coefficient $K_G a$ (based on the hexane feed) is 0.14 s^{-1}. What percent recovery of the flavor do you expect?

2. You want to extract 99% of a new anti-inflammatory drug from a dilute reaction mixture. You will do this with mixer-settlers operated countercurrently, using pure solvent feed. The equilibrium line is

    $$y_n^* = 0.2 x_n$$

    (a) What is the minimum ratio for solvent flow to raffinate flow?
    (b) If the solvent flow is twice the minimum, how many stages will you need?

3. You have a three-stage extractor with which you are isolating a therapeutic protein with physical properties similar to myoglobin from 9400 l of clarified beer. The extraction solvent is an organic solution of inverted micelles of Aerosol OT in dodecanol, which at pH 6.2 has a partition coefficient given by

    $$m = \frac{y^*}{x} = 0.0056$$

    How much solvent will you need to recover 95% of the protein?

4. You are extracting a highly dilute rare earth from aqueous solution with mixed organic phosphates that enter free of rare earths. You can get a 63% recovery in three stages. What percent recovery can you get in twice as many stages?

5. You have 30,000 l of a fermentation broth containing 66 mg/l of an enzyme used to control biofilms in the paper industry. You plan to extract this enzyme with an iso-osmotic polyethylene glycol solution which does not dissolve significantly in the broth. The equilibrium for this extraction at the operating pH of 6.6 is

$$y = 103x$$

where y and x are the concentrations in polyethylene glycol and beer, respectively. You plan to extract 99% of this enzyme with staged *cross flow* (not countercurrent) extraction. In such an extraction, you use pure, enzyme-free extraction in each stage. How much extractant should you use to achieve your goal in four stages?

6. Sugar in beets containing 0.26 kg sugar/kg vegetable matter is leached using pressurized vessels called "diffusers" operating about 75 °C. The extracting liquid is pure water, fed at a rate of 2 kg water/kg vegetable matter. In one test, these diffusers are operated countercurrently, yielding a waste stream containing 0.5 kg entrained water/kg vegetable matter. The beets are fed without entrained water. How many stages are needed to recover 99% of the sugar?

7. You are extracting two rare earths with the four-stage cascade shown in part (a) of the figure below. For the first rare earth A,

$$y = y^* = 1.5x$$

For the second B,

$$y = y^* = 0.5x$$

The aqueous feed enters with equal concentrations of the rare earths, and the ratio G/L is one. Each stage has a Murphree efficiency of 0.80. (a) What is the fraction recovered of each rare earth? (b) What is the purity $y_{4A}/(y_{4A} + y_{4B})$? (c) To improve this purity, you attach a second four-stage cascade as shown in part (b) of the figure below. These stages have the same properties as the first four. What is the fraction recovered now? (d) What is the purity now?

(a)

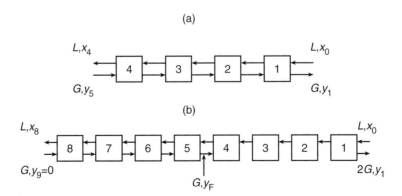

(b)

Further Reading

Belter, P. A., Cussler, E. L., and Hu, W. S. (1988). *Bioseparations*. New York: Wiley.

Godfrey, J. C., and Slater, M. J., eds. (1994). *Liquid–Liquid Extraction Equioment*. Chichester: Wiley.

Rydberg, J., Cox, M., Musikas, C., and Choppin, G.R. (2004). *Solvent Extraction Principles and Practice*, 2nd ed. Boca Raton FL: CRC.

Taylor, L. T. (2001). *Supercritical Fluid Extraction*. New York: Wiley.

Treybal, R. E. (1980). *Mass Transfer Operations*, 3rd ed. New York: McGraw Hill.

CHAPTER 15

Adsorption

Adsorption is very different from absorption, distillation, and extraction. These three processes, detailed in the five previous chapters, typically involve two fluids flowing steadily in opposite directions. In absorption, a gas mixture flows upward through a packed column while an absorbing liquid trickles down. In distillation, a liquid mixture is split into a more volatile liquid distillate and a less volatile bottoms stream. In extraction, two liquid streams move countercurrently to yield an extract and a raffinate. To be sure, in some cases, the contacting may involve near-equilibrium states, and in other cases it may be described with nonequilibrium ideas like mass transfer coefficients. Still, all three units operations involve two fluids at steady state.

In contrast, adsorption is almost always an unsteady process involving a fluid and a solid. The use of a solid is a major difference because solids are hard to move. They abrade pipes and pumps; they break into fine particles which are hard to retain. As a result, we usually pump the feed fluid through a stationary bed of solid particles to effect a separation by adsorption.

Thus adsorption asks a different kind of question than the questions asked in absorption, distillation, or extraction. In absorption or differential distillation, the basic question is how tall a tower is needed. This question is answered with a mass transfer analysis, including an operating line and an equilibrium line. The mass transfer analysis includes overall and individual coefficients summarized by dimensionless correlations. In staged distillation or extraction, the basic question is how many stages are needed. This question is resolved largely with operating and equilibrium lines, with the mass transfer aspects conveniently compressed into an efficiency. In absorption, distillation, and extraction, the analysis is sufficiently reliable to answer the questions without experiment.

In adsorption, the analysis is less reliable and is rarely made without experiment. The initial experiment, made on a small scale, leads to the basic question of adsorption: how will a large bed behave? Answering this question commonly presumes a knowledge of adsorbents and isotherms like that given in this chapter. In Section 15.1, we summarize the systems where adsorption is often chosen for separation. In Section 15.2, we describe the solid adsorbents themselves and how the amount they capture depends on concentration. This relation between the amount adsorbed on the solid and the concentration in the fluid is called an isotherm. In Section 15.3, we turn to the basic behavior of adsorption, which is usually summarized as a plot of solute concentration eluted from the bed as a function of time. This plot is called a breakthrough curve, implying that solute has forced its way past an adsorption zone. Section 15.4 describes how breakthrough curves are affected by mass transfer, and Section 15.5 concludes the chapter with more complex cases. The entire chapter is thus an easy introduction to this different separation process.

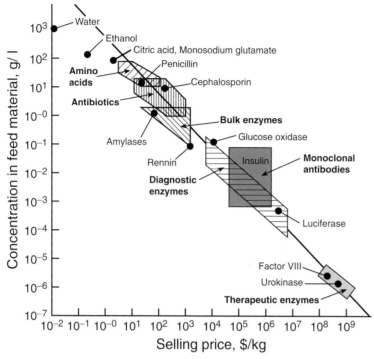

Fig. 15.1-1. Feed concentration vs. price of final product. Concentration and price are inversely related over almost twelve orders of magnitude.

15.1 Where Adsorption is Important

Adsorption is important because it can be effective in dilute solution. Many other separations are not. The ability to treat dilute solutions easily is uncommon and is what makes adsorption especially valuable. To illustrate this, consider the costs of the enormous variety of solutes shown in Fig. 15.1-1. The prices of these solutes shown on the abcissca vary by over 10 orders of magnitude, from about \$0.01/kg to \$1,000,000,000/kg. These prices are strikingly well correlated with the feed concentrations of the various products, shown on the ordinate. The correlation between feed concentration and price is almost perfectly inverse, i.e., the feed concentration varies with the product price to the (−1.0) power. Therefore, separation processes which can concentrate valuable products from dilute solutions will have special value.

Adsorption is often the best choice for separation from a dilute solution. Not surprisingly, it is expensive because the free-energy changes from dilute to concentrated solution are larger than those involved in most distillations or absorptions. This higher price is often felt to be justified in the relatively simple process. Adsorption can be used to separate more concentrated mixtures as well, as suggested by Table 15.1-1. These processes have considerable value. However, many highly valuable adsorption-based separations are used for the expensive compounds in Fig. 15.1-1, and these are harder to list in a table. These are where new interest in adsorption will lie. It is for these expensive compounds that we will need the analysis given in the remainder of this chapter.

Table 15.1-1 *Typical large scale adsorption processes*

Separation	Typical adsorbent
1. Liquid purifications	
Color (e.g., in sugars)	Carbons
Antibiotics	Ion exchangers
Water softening	Ion exchangers
2. Gas Purification	
Water removal (drying)	Silica, alumina
Solvent stripping	Activated carbons
Odor removal	Activated carbons
SO_2 capture	Zeolites
3. Gas bulk separations	
Air	Zeolites
Hydrocarbon isomers	Zeolites
Water–ethanol	Zeolites

Note: By "purification," we imply that only small solute concentrations are removed. By "bulk separations," we imply high solute concentrations.

Adsorption uses a bed of porous solid particles. The particles' pores are small, giving a surface area of several hundred square meters per gram. When a solution of liquid or gas flows through the bed, solutes are adsorbed from the solution onto the surface of the particles.

Many learning about this topic want to know more details about why solutes bond to adsorbents. To be sure, solutes bond because the free energy of the adsorbed species is lower than that in solution, but this truth supplies little physical insight. To get such insight, we can imagine five types of bonds in adsorption. First, adsorption can occur because of van der Waals bonds in a spectrum of hydrophobic–hydrophillic interactions. In simpler words, greasy solutes stick to greasy surfaces, and hydrophilic solutes adsorb on more polar, hydrated surfaces. Second, ions will often adsorb on fixed ion-exchange sites of opposite charge. Examples are Ca^{2+} adsorbing onto sulfonated polystyrene, and Cl^- adsorbing onto resins containing cross-linked tetraalkylammonium groups. Third, adsorption can involve the formation of strong covalent bonds between the solute and sites on the adsorbents' surfaces. These bonds will require large changes of temperature or pH to desorb the solute and hence regenerate the ion exchanger.

In addition to these three simple forms of adsorption, two other types of bonds are important but harder to explain. In some cases, adsorption depends on solute shape. One good example is adsorption of antibiotics on custom-synthesized ion-exchange resins. There, successful separation can involve solute adsorption on several adjacent, non-ionic sites. Such multiple site adsorption tends to reflect the solute's molecular shape and, hence, is of special value for expensive solutes.

Finally, some adsorption depends on rates of uptake, rather than only on the equilibrium amount adsorbed. One dramatic example is the adsorption of oxygen and nitrogen on zeolites. If the adsorption is run slowly using zeolite NaLiX, then nitrogen is adsorbed more than oxygen, and the separation is controlled by equilibrium. In contrast, if the adsorption is run quickly using zeolite A, then oxygen is adsorbed more rapidly than nitrogen, and the separation is controlled by kinetics. This enormous variety in what is adsorbed and under what conditions is why we will treat adsorption phenomenologically, starting with experimental data taken at small scales and then trying to design separations for much larger scales.

15.2 Adsorbents and Adsorption Isotherms

Molecules adsorb on virtually all surfaces. The amount they adsorb is roughly proportional to the amount of the surface. As a result, commercial adsorbents are highly porous, with surface areas typically of several hundred square meters per gram. Some specialized adsorbents have surface areas as high as 3000 m^2/g.

In discussing these highly porous materials, we have two goals. First, we must describe what the materials actually are. Though there are a very wide variety of adsorbents, this description must be brief to be within the scope of the book. Second, we are interested in the isotherm, in how the amount adsorbed varies with the concentration in solution. Both adsorbents and isotherms are discussed in the following paragraphs.

15.2.1 Adsorbents

Adsorbents are conveniently divided into three classes: carbons, inorganic materials, and synthetic polymers. The carbons have nonpolar surfaces that are used to adsorb nonpolar molecules, especially hydrocarbons. They are manufactured from both organic and inorganic sources, including coal, petroleum coke, wood, and coconut shells. Decolorizing carbons tend to be based on a mixture of sawdust and pumice. Carbons used for gas adsorption can be made from vegetable sources like coconut shells and fruit pits. Activated carbons, which use manufacturing conditions to control pore size more exactly, can be used to recover solvent vapors, to filter gases, and to purify water. Overall, carbons are a broad and important class of adsorbents.

Inorganic materials vary widely. Activated alumina, which has a polar surface, is used largely as a dessicant. It is also used for laboratory-scale chromatography. Silica gel, consisting of amorphous silicon dioxide, is also used as a dessicant. Clays are used as inexpensive adsorbents; for some petroleum-based applications, they have in the past been used once and discarded. Fuller's earth is used to purify oils, an echo of its original purpose to adsorb lanolin from fleece.

The most important class of inorganic adsorbents is the zeolites, a subclass of molecular sieves. These are crystalline aluminosilicates with specific pore sizes located within small crystals. Two common classes have simple cubic crystals (type A) or body-centered cubic crystals (type X). Sometimes, the type is assigned a number equal to a nominal pore size in the crystals. For example, zeolite 5A with a nominal 5 Å pore size is used to separate normal from branched paraffins.

Adsorbents based on synthetic polymers also vary widely. Ion-exchange polymers with a fixed negative charge are most commonly made by treating styrene–divinylbenzene copolymers with sulfuric acid. These polymers, as well as acrylic ester polymers, are used for water treatment. Polymers with a fixed positive charge are frequently based on alkylammonium groups. In either case, the adsorbing polymers tend to capture polyvalent ions in preference to monovalent ones. They are also surprisingly useful for

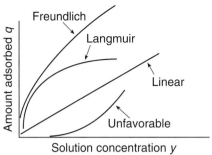

Fig. 15.2-1. Typical schematic isotherms. While the Freundlich isotherm often gives the best fit of data, the linear isotherm is assumed in most simple theories.

adsorbing hydrophobic materials, including highly valued solutes like drugs and pigments.

15.2.2 Isotherms

The isotherms for these various materials are shown schematically in Fig. 15.2-1. Any isotherm with a downward curvature is referred to as favorable, and any isotherm with an upward curvature is referred to as unfavorable. These terms imply that adsorption will frequently be used to capture small amounts of solutes from dilute solutions. A highly favorable isotherm will be especially effective in dilute solutions, whereas a highly unfavorable isotherm will be particularly ineffective under those conditions. While these terms are useful, one must remember that an isotherm that is strongly favorable for adsorption will be strongly unfavorable when it is time to elute the adsorbed species. Such elution is necessary if the adsorbent is to be reused.

Three commonly cited isotherms are the linear, Langmuir, and Freundlich types. Each merits discussion. The simple linear isotherm assumes:

$$q = Ky \qquad\qquad\qquad (15.2\text{-}1)$$

where q is the concentration in the adsorbent and y is the concentration in the solution. As in other separation processes, the units of these concentrations are tricky. The solution concentration y is most commonly given as moles of solute per volume of solution. The concentration on the adsorbent q is commonly in units of moles of solute per dry mass of adsorbent. Thus the equilibrium constant K will often have units of a reciprocal density.

There is one important case of the linear isotherm which merits more discussion. This is ion exchange, when one ion of a given charge is exchanged for a second ion, usually of a different charge. To discuss this case, we will focus on the exchange of Ca^{2+} for Na^{+} bound to a cation exchanger made of cross-linked, sulfonated polystyrene. This particular exchange is the key to water softening.

To begin our discussion, we consider the ionic concentrations in aqueous solution near the surface of the exchanger. These are shown schematically in Fig. 15.2-2. The sodium ion concentration is elevated near the exchanger surface; and the chloride concentration is reduced. Far from the surface, the concentrations of Na^{+} and Cl^{-} are equal.

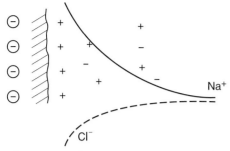

Fig. 15.2-2. Ionic concentrations near the surface of an ion-exchange resin. For the negatively charged surface shown, the cation concentration is elevated, and the anion concentration is depressed.

The sodium concentrations near to and far from the surface area are in equilibrium, so if sodium is species "1,"

$$\mu_1(\text{bulk}) = \mu_1(\text{surface})$$
$$\mu_1^0 + k_B T \ln c_1(\text{bulk}) = \mu_1^0 + k_B T \ln c_1(\text{surface}) + z_1 \phi e \tag{15.2-2}$$

where k_B is Boltzmann's constant, c_1 is the Na^+ ion concentration, z_1 is the ionic charge, ϕ is the surface potential, and e is the electronic charge. Rearranging this,

$$\frac{c_1(\text{surface})}{c_1(\text{bulk})} = e^{-z_1 \phi e / k_B T} \tag{15.2-3}$$

The electronic charge is $1.6 \cdot 10^{-19}$ J/V; φ is typically around –0.1 V for an anionic exchanger; and the temperature is normally about 298 K. Thus

$$\frac{c_1(\text{surface})}{c_1(\text{bulk})} = e^{4z_1} \tag{15.2-4}$$

This implies the values shown in Table 15.2-1. This is why an anionic exchanger loaded with Na^+ will adsorb Ca^{2+} and release Na^+, even when the calcium is dilute.

We now return from this special case of the linear isotherm to the common, favorable Langmuir isotherm. The Langmuir isotherm has a clear theoretical basis. This isotherm assumes that the limited number of sites on the adsorbent are subject to a mass balance

$$\begin{bmatrix} \text{total} \\ \text{sites} \end{bmatrix} = \begin{bmatrix} \text{filled} \\ \text{sites} \end{bmatrix} + \begin{bmatrix} \text{empty} \\ \text{sites} \end{bmatrix} \tag{15.2-5}$$

Moreover, these sites are assumed to be subject to a chemical equilibrium

$$\begin{bmatrix} \text{bulk} \\ \text{solute} \end{bmatrix} + \begin{bmatrix} \text{empty} \\ \text{site} \end{bmatrix} \rightleftharpoons \begin{bmatrix} \text{filled} \\ \text{site} \end{bmatrix} \tag{15.2-6}$$

or in more quantitative terms

$$\begin{bmatrix} \text{filled} \\ \text{site} \end{bmatrix} = K' \begin{bmatrix} \text{empty} \\ \text{site} \end{bmatrix} \begin{bmatrix} \text{bulk} \\ \text{solute} \end{bmatrix} \tag{15.2-7}$$

Table 15.2-1 *Characteristic concentrations for water softening*

Salt	Bulk concentration	Surface concentration
10^{-3} M NaCl	10^{-3} M Na$^+$ 10^{-3} M Cl$^-$	$55 \cdot 10^{-3}$ M Na$^+$ $0.018 \cdot 10^{-3}$ M Cl$^-$
10^{-3} M CaCl$_2$	10^{-3} M Ca^{2+} $2 \cdot 10^{-3}$ M Cl$^-$	$2980 \cdot 10^{-3}$ M Ca^{2+} $0.034 \cdot 1^{-3}$ M Cl$^-$

Note: These imply a negatively charged sulfonated polystyrene–divinylbenzene copolymer, the most common case.

where K' is the equilibrium constant for the chemical equilibrium. Combining these relations, we find

$$
\begin{bmatrix} \text{filled} \\ \text{site} \end{bmatrix} = \frac{K' \begin{bmatrix} \text{total} \\ \text{sites} \end{bmatrix} \begin{bmatrix} \text{bulk} \\ \text{solute} \end{bmatrix}}{1 + K' \begin{bmatrix} \text{bulk} \\ \text{solute} \end{bmatrix}}
\tag{15.2-8}
$$

This is more easily written in terms of the concentrations defined earlier

$$
q = \frac{q_0 y}{K + y}
\tag{15.2-9}
$$

In this equation q_0 is the total concentration of sites and K is the reciprocal of the equilibrium constant K'. Note that Eq. 15.2-9 implies that the reciprocal of q should vary linearly with the reciprocal of y. The intercept on this plot is q_0^{-1}; the slope on the plot is (K/q_0). We can use such a plot to test whether an adsorbent exhibits a Langmuir isotherm.

The third of the common isotherms, called the Freundlich isotherm, is given by

$$
q = K y^n
\tag{15.2-10}
$$

where both K and n are empirical constants. In cases where the isotherm is favorable, n is less than one. This isotherm suggests that a plot of q versus y should be linear on log–log coordinates. This is the most successful common isotherm, even though its theoretical base is vague. How this isotherm compares with others is shown in the example that follows.

Example 15.2-1: Selective copper adsorption   Cross-linked copolymers of allylacetyla-cetone and hydroxyethylmethylacrylate (HEMA) selectively bind Cu(II) preferentially over other cations such as Mg(II) and Ni(II). This mechanism differs from normal ion exchange, which is almost completely based on charge. Equilibrium data are shown below.

Which isotherm best fits these data?

[Cu(II)] in solution (M)	[Cu(II)] in polymer (M)
$9.03 \cdot 10^{-6}$	$6.13 \cdot 10^{-3}$
$1.90 \cdot 10^{-5}$	$6.76 \cdot 10^{-3}$
$3.91 \cdot 10^{-5}$	$8.72 \cdot 10^{-3}$
$5.00 \cdot 10^{-4}$	$1.92 \cdot 10^{-2}$
$1.00 \cdot 10^{-3}$	$2.11 \cdot 10^{-2}$

Solution Plots testing the linear, Langmuir, and Freundlich isotherms are given in Fig. 15.2-3. The linear isotherm is a flop, fitting only data in very dilute solution. The Langmuir isotherm seems somewhat better, at least for concentrated solutions. The Freundlich isotherm works best.

15.3 Breakthrough Curves

Adsorption is normally carried out in a packed bed. A fluid solution is forced into one end of the bed; solute is adsorbed out of the solution within the bed, and an eluent with very little solute flows out of the other end of the bed. Adsorption in a packed bed normally gives a much better separation than adsorption in a stirred tank. Adsorption in a packed bed is like differential distillation, and adsorption in

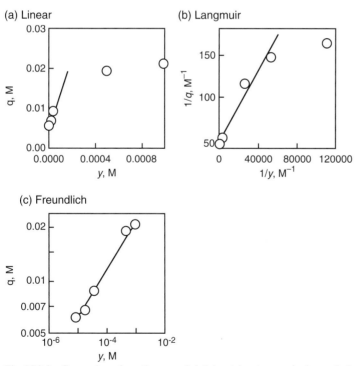

Fig. 15.2-3. Copper ion adsorption on poly(allylacetylacetone-co-hydroxyethylmethylacrylate). The Freundlich isotherm gives the best description.

a stirred tank is like flash distillation. Differential distillation normally gives better separation than flash distillation.

We can illustrate the major gain from adsorption in a packed bed from the following idealized example. Imagine that we have 30 l of a solution containing 10^{-4} g/l gentamicin, an antibiotic used to treat eye infections. We want to recover 99% of this material using a custom ion-exchange resin for which the isotherm is

$$q = \left(\frac{20\,l}{g}\right)y \tag{15.3-1}$$

where q and y are the concentrations in the exchanger and solution, respectively. We want to know how much of the expensive exchanger we will need if we run the adsorption in a stirred tank or in a packed bed.

For stirred-tank adsorption, we know that for 99% recovery, only 1% of the antibiotic is left in solution, i.e., y is 10^{-6} g/l. From the isotherm, the concentration in the adsorbent q is $20 \cdot 10^{-4}$ g/g. We can find the weight of adsorbent W from a mass balance:

(Total antibiotic) = (antibiotic in solution) + (antibiotic in adsorbent)

$$30\,l\left(\frac{10^{-4}\,g}{1}\right) = 30\,l\left(10^{-6}\,\frac{g}{1}\right) + W\left(20 \cdot 10^{-6}\,\frac{g}{g}\right) \tag{15.3-2}$$

$$W = 150\,g$$

We need about 0.15 kg adsorbent.

For packed-bed adsorption, we need to make an additional assumption. We expect that we will feed solution to the packed bed and collect an effluent from it. We assume that there is no solute in this effluent until the bed is completely saturated. Thus, the concentration in solution within the bed is the feed concentration 10^{-4} g/l, and that in the adsorbent is $20 \cdot 10^{-4}$ g/g. We expect that the amount of solution in the packed bed will be small, close to zero. A mass balance now gives

(Total antibiotic) = (antibiotic in effluent) + (antibiotic in adsorbent)

$$30\,l\left(\frac{10^{-4}\,g}{1}\right) = 30\,l\left(0\,\frac{g}{1}\right) + W\left(20 \cdot 10^{-4}\,\frac{g}{1}\right) \tag{15.3-3}$$

$$W = 1.5\,g$$

Using a packed bed may mean using one hundred times less of the expensive adsorbent. This is normally a major saving.

However, the advantages of a packed bed rest squarely on our additional assumption, that no solute comes out until the bed is saturated. We need to examine this assumption carefully in two steps. In the first step in the next few paragraphs, we look at the accuracy of this assumption in terms of the concentration of the effluent vs. time, called a "breakthrough curve." Later, in the second step of our examination, we quantify the approximation in terms of a new idea, an "unused bed length." These two steps are basic to our analysis of adsorption.

15.3.1 Ideal vs. Actual Breakthrough Curves

We start our analysis of packed bed adsorption by imagining a plot of effluent concentration vs. time. If adsorption were completely efficient, the effluent concentration

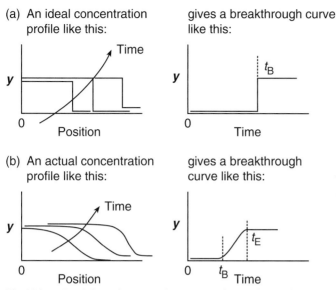

(a) An ideal concentration profile like this:

gives a breakthrough curve like this:

(b) An actual concentration profile like this:

gives a breakthrough curve like this:

Fig. 15.3-1. Breakthrough curves. The concentration profile within the bed, suggested at the left of these figures, leads to the concentrations eluted that are shown on the right of these figures. The sudden step in (a) is less common than the gentle rise in (b).

would be zero for a long time. Then at some breakthrough time t_B, the concentration would suddenly jump to the value in the feed. At this point, the bed would be completely saturated, and adsorbent throughout the bed would be in equilibrium with solution which has the feed's concentration.

In this ideal, completely efficient limit, the breakthrough curve would behave like the step function shown schematically in Fig. 15.3-1(a). This step function at the end of the column implies the concentration profiles shown within the column. Each of these is also a step function: at small times, the concentration drops to zero near the column's entry; at larger times, the step function pushes its way through the column. Note that the rate at which the bed gets saturated is usually much slower than the rate that the feed solution is flowing through the bed.

Of course, real breakthrough curves are not completely efficient and don't have the shape of step functions. The effluent concentration still stays zero for a while, then becomes significant at some time t_B, and finally reaches the feed concentration at some longer "exhaustion time" t_E. An example is shown in Fig. 15.3-1(b). In a few cases, the real breakthrough curve will be close to a step; but in many more cases, it will be much more gradual, so t_E may be 30 to 100% bigger than t_B. Often, the breakthrough will initially be fast and then increase more slowly.

A real breakthrough curve implies a history of concentration profiles like those shown in the figure. As expected, when solution is first fed, it is like a step function, but this rapidly becomes a dispersed ramp. However, stunningly, the amount the ramp disperses may not continue to get worse as it proceeds down the bed. To be sure, the ramp is not perfectly sharp, but it doesn't get worse either. It is self-sharpening. We will discuss why this happens later in this section.

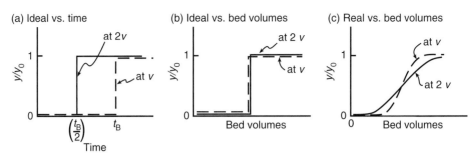

Fig. 15.3-2. Simplifying breakthrough curves. If bed behavior is nearly ideal, curves can be
simplified by plotting vs. bed volumes, rather than vs. time. This is less effective for non ideal
breakthrough curves.

Before doing this, we should point out that plotting breakthrough curves as eluent
concentration vs. time is the norm only in some branches of engineering. In many areas
of science, breakthrough curves are plotted as eluent concentration vs. number of bed
volumes. This number is simply the volumetric flow rate times the time divided by the
volume of the adsorbent bed itself. This method is especially useful in analytical chromatography, where the adsorbent particle size and volumetric flow are restricted to
a narrow range. It is less useful for scale-up of adsorption.

Example 15.3-1: Alternative plots of breakthrough curves   You are planning a series of test
experiments using a new adsorbent. You hope to use this adsorbent at a commercial scale,
and so plan to let many process variables vary widely. In particular, you plan to determine
breakthrough curves at the manufacturer's recommended velocity and at twice this velocity. (a) If the breakthrough curves at different velocities are ideal, without dispersion, what
will they look like when plotted vs. time? (b) What will they look like plotted vs. number of
bed volumes? (c) How will they change if the breakthrough curves do show dispersion?
     Solution   The results are shown in Fig. 15.3-2. If there is no dispersion, dou-
bling the velocity means that the breakthrough occurs at half the time. Plotting vs. number
of bed volumes will superimpose these curves. However, if dispersion is significant, the
breakthrough will often be velocity dependent, and superposition will not occur.

15.3.2 Why Breakthrough Curves Disperse

     In practice, the sharp breakthrough curves shown in Fig. 15.3-2 are not a goal
but a dream. They will sometimes be approached under laboratory conditions, but they
will be rare in commercial operations. In most cases, breakthrough will be much more
gradual and will often tail off, only slowly creeping up to the feed concentration.
     There are four important phenomena which are responsible for the extent of disper-
sion. Each merits discussion:
     (a) Axial diffusion. Most obviously, solute flowing into an adsorbent bed will diffuse
in the direction of flow. After all, it is being fed at high concentration into a bed which is
originally solute free. It will diffuse down its concentration gradient. However, while this
is true, the dispersion from axial diffusion which results is rarely significant, especially at
an industrial scale.

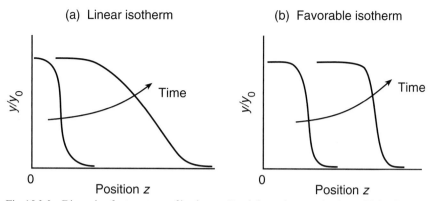

Fig. 15.3-3. Dispersion for two types of isotherms. Breakthrough curves for favorable isotherms are self-sharpening, and so don't disperse more in longer beds (i.e., larger z).

(b) Dispersion. This can occur for two reasons. Clearly, it can result from a carelessly packed bed, one with large voids or an uneven distribution of solids. While these packing problems were major in the past, modern adsorbents are of nearly equal particle size, and effective packing procedures are well known.

Dispersion can also result from the Taylor–Aris mechanism discussed in detail in Section 4.4. This important mechanism involves a coupling between axial flow, which broadens the front, and radial diffusion, which tends to sharpen it. This effect is a major reason why breakthrough curves blur.

(c) Adsorption kinetics. If the rate of adsorption is slow, the breakthrough curve will spread. This is occasionally due to slow chemical kinetics between the adsorbing solute and the adsorbent's sites. It is more often caused by slow mass transfer, i.e., by the long time that it takes for the solute to diffuse through the pores to the sites. This effect is sometimes important, especially for large adsorbent particles.

(d) The isotherms themselves. This important factor is surprising. Linear and unfavorable isotherms blur the breakthrough curves; but favorable isotherms tend to make breakthrough sharper. Because most isotherms are favorable, this is a very important and positive effect. Indeed, most breakthroughs are a balance beween the sharpening caused by the favorable isotherm and the blurring caused by dispersion.

We need to consider why a favorable isotherm tends to correct all of the problems caused by axial diffusion, Taylor–Aris dispersion, and slow mass transfer into the adsorbent particles. To see why this occurs, imagine a concentrated solution flowing through a bed of solute-free adsorbent. The solution flowing very near to the adsorbent particles' surface is adsorbed, but that flowing far from this surface is swept ahead. This blurs the front. In other words, flow near the wall is slow; flow far from the wall is faster; thus the front spreads.

If the isotherm is linear or unfavorable, this front spreads more and more as the flow continues through the bed. However, if the isotherm is favorable, the solute swept ahead to fresh adsorbent is especially avidly adsorbed. The solute left behind is in contact with nearly saturated adsorbent and so is not retarded that much. For a favorable isotherm, the front doesn't spread. This difference, illustrated schematically in Fig. 15.3-3, is a key to designing adsorption columns, the topic of the next paragraphs.

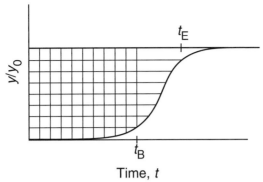

Fig. 15.3-4. Analyzing breakthrough curves. The horizontally shaded region is proportional to the bed's capacity. The vertically shaded region measures the amount adsorbed at breakthrough.

15.3.3 Designing Adsorption Columns

The design of adsorption columns uses a breakthrough curve measured at small scale to estimate performance at a large scale. The experiment is essential because the effects of the isotherm and the dispersion cannot normally be predicted. In this regard, adsorption is less well understood than staged distillation or dilute absorption. For staged distillation, the key information is thermodynamic, summarized as a y–x diagram. For simple absorption, the information is both thermodynamic (a Henry's law coefficient) and kinetic (a mass transfer correlation), but the dilute analysis leads to simple, reliable answers. Adsorption is more complex and requires that first experiment.

The result of the first experiment is a breakthrough curve, a plot of concentration vs. time like that shown in Fig. 15.3-4. We can learn a lot about the adsorption from the shape of this curve. The total amount fed to the bed from time zero to time t is of course

$$\text{total fed} = \int_0^t Q\,y_0\,dt = Q\,y_0 t \tag{15.3-1}$$

where Q is the volumetric flow rate. The total adsorbed when the bed is completely saturated is the amount fed minus the amount that escapes:

$$\begin{bmatrix} \text{total} \\ \text{adsorbed} \\ \text{at saturation} \end{bmatrix} = \int_0^\infty Q\,(y_0 - y)\,dt \tag{15.3-2}$$

This integral corresponds to the horizontally shaded region in the figure. The total adsorbed until the breakthrough time is

$$\begin{bmatrix} \text{total adsorbed} \\ \text{at breakthrough} \end{bmatrix} = \int_0^{t_B} Q\,(y_0 - y)\,dt \tag{15.3-3}$$

This corresponds to the vertically shaded part of the figure. Thus the fraction θ of the bed saturated at t_B is

$$\theta = \frac{\text{vertically shaded area}}{\text{horizontally shaded area}} \tag{15.3-4}$$

These equations are approximate because they do not consider in detail the concentration of the solution between the adsorbent particles. Because the adsorbent normally has much more solute than the solution – that is why it is a good adsorbent – this approximation is minor.

The fraction saturated θ is easily calculated from numerical integration of the breakthrough curve. In practice, some adsorbent manufacturers suggest an easier short cut. They assume that the breakthrough curve is a ramp, starting at a breakthrough time t_B and ending at an exhaustion time t_E. Then the amount adsorbed at breakthrough in Eq. 15.3-3 is

$$\begin{bmatrix} \text{total adsorbed} \\ \text{at breakthrough} \end{bmatrix} = Q y_0 \, t_B \tag{15.3-5}$$

This approximates the vertically shaded part of Fig. 15.3-4 as a rectangle. The amount adsorbed at exhaustion is

$$\begin{bmatrix} \text{total adsorbed} \\ \text{at exhaustion} \end{bmatrix} = Q y_0 \, t_B + \frac{1}{2} Q y_0 \, (t_E - t_B) \tag{15.3-6}$$

This approximates the horizontally shaded region as the rectangle plus a triangle. Thus the fraction of the bed saturated is

$$\theta = \frac{t_B}{t_B + \frac{1}{2}(t_E - t_B)} = \frac{2 t_B}{t_B + t_E} \tag{15.3-7}$$

When the breakthrough is reasonably sharp, this manufacturer-endorsed relation is surprisingly accurate.

At this point, the analysis of adsorption takes a wonderful, intuitive leap. We assume that we can approximate the concentration in the column at two zones, a large saturated zone, and a smaller "unused" zone. We then assert that when the column gets longer, the size of the saturated zone will change, but the size of the unused zone will not. In the terms commonly used, the " length of the unused bed" is constant.

The basis of this intuitive assertion is the shape of the concentration profiles in Fig. 15.3-3. On the left-hand side of this figure, the results for a linear isotherm show that dispersion grows with time as the solute moves down the bed. On the right-hand side, the results for a favorable isotherm show that the dispersion stays the same as the adsorption continues. This favorable isotherm is self-sharpening, with the effects of dispersion and adsorption in balance. Thus the length of unused bed l' is a constant, found easily from the fraction of bed saturated, θ

$$l' = l(1 - \theta) \tag{15.3-8}$$

where l is the actual length of the bed. Indeed, l' is the main information we get from our basic experiment. If we made a second experiment with a longer bed, we would expect a new θ but the same l'.

We use this idea as the basis for the design of adsorption colums, as illustrated in the following examples. The examples suggest that long beds are better than short beds. After all, in a short bed we do not use that part of the bed summarized as the length of unused bed. In a long bed, we waste about the same unused bed length, but this is a smaller percentage of the total bed length. Thus, in a longer bed the percentage of the bed used will be larger. Moreover, for a given amount of adsorption, a long bed will have a smaller diameter and hence be easier to feed.

In practice, long beds are not always better because greater length means greater pressure drop. When adsorbent particles are large, this greater pressure drop is easily tolerated, but adsorbent particles are often made small ($<$ 300 μm) to minimize dispersion. In addition, adsorbent particles are sometimes not rigid but deform at higher pressures. In particular, doubling the pressure may not double the flow but still sharply increase the dispersion. Still, the basic conclusion suggested above is often correct: long adsorbent beds are better than short ones.

Example 15.3-2: Selective ferric adsorption   A packed bed of ion exchanger is selective for ferric ion over nickel ion. In the separation of these two species, the nickel breaks through at 1.5 min and the ferric ion at 23 min. The ferric ion exhausts the bed at 33 min. If the bed is 120 cm long, what length is unused?

Solution   From Eq. 15.3-7,

$$\theta = \frac{2\,t_B}{t_B + t_E} = \frac{2(23)}{23 + 33} = 0.8$$

Thus from Eq. 15.3-8

$$l' = l(1 - \theta) = 120 \text{ cm } (1 - 0.8) = 24 \text{ cm}$$

The length of unused bed should be constant, even with a longer bed. Such simple characterization is common.

Example 15.3-3: Cesium adsorption   Tests of a new selective ion exchanger for cesium adsorption use a bed 12 cm long with a 1 cm^2 cross-section. These tests show a breakthrough time of 10 min and an exhaustion time of 14 min. We hope to scale up this process 10,000 times. (a) How much adsorbent is needed if the bed is kept 12 cm deep? (b) How much is needed if the bed is 10 m long?

Solution   We begin by calculating the length of the unused bed from our tests. From Eq. 15.3-7, we see that the fraction of the bed saturated θ is

$$\theta = \frac{2\,(10 \text{ min})}{10 \text{ min} + 14 \text{ min}} = \frac{5}{6}$$

The length of the unused bed is found from Eq. 15.3-8

$$l' = l(1 - \theta) = 12 \text{ cm} \left(1 - \frac{5}{6}\right) = 2 \text{ cm}$$

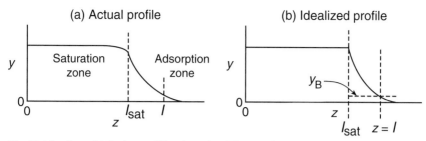

Fig. 15.4-1. A model for irreversible adsorption. The actual concentration profile is approximated by the two zones shown. The third zone, related to the time to reach saturation, is
relatively short, and so is not shown. The concentration y_B is that at breakthrough.

We can now estimate the amount of adsorbent for the two cases. (a) For the 12 cm
"pancake" bed, the amount of the bed which is used is still 10 cm. The volume of adsorbent used is thus 10,000 times that used in one test. Thus the total volume needed is

$$V = 10,000 \, (1 \, \text{cm}^2) \, (10 + 2) \, \text{cm} = 0.12 \, \text{m}^3$$

This implies a bed about 357 cm in diameter. The pressure drop in the bed is unchanged.
(b) For a bed 10 m long, the length of unused bed is still 2 cm, so the fraction of bed which
is used is 0.998. The volume of adsorbent used is still 10,000 times the 10 cm^3 in our test,
or 0.1 m^3. Since the length of bed used is 9.98 m, we can find the bed diameter d

$$9.98 \, \text{m} \left(\frac{\pi}{4} d^2 \right) = 0.1 \, \text{m}^3 \quad d = 11.3 \, \text{cm}$$

The total amount of adsorbent needed will be

$$V = 0.1 \, \text{m}^3 \left(\frac{10 \, \text{m}}{9.98 \, \text{m}} \right) = 0.1002 \, \text{m}^3$$

By building the longer bed, we need about 20% less adsorbent. We also have a narrower
bed which will be much easier to feed. However, to sustain the same bed velocity, we will
need a pressure drop which is about 100 times larger.

15.4 Mass Transfer Effects

We now turn to estimating the effects of mass transfer on the breakthrough
curves. These effects are complicated because of the unsteady-state process and the
nonlinear isotherms. This means that the analysis is more elaborate than for the separation processes discussed in earlier chapters. Because of this complexity, we consider
only one approximate analysis: we assume that the adsorption is irreversible. While the
analysis is not especially valuable quantitatively, it does let us estimate how the quality of
the separation will change with variables like solution flow and adsorbent particle size.

To begin this analysis, we consider not the breakthrough curve but the concentration
profiles inferred within the adsorbent bed. These profiles, sketched in Fig. 15.4-1, can be
idealized as three zones. First, the bed takes some time to reach saturation at all. Second, at
larger times, the bed contains a growing region where the adsorbent is saturated. The

saturation zone normally grows much more slowly than solution is flowing through the bed. Finally, at the leading edge of this saturation zone, the adsorption has not yet approached saturation. It is in this adsorption zone where mass transfer and dispersion are important.

We can estimate each of these three zones. First, the time t_{sat} to saturate even the entrance of the bed can be found from a mass balance:

$$\left[\frac{\text{amount adsorbed}}{\text{mass adsorbent}}\right]\left[\frac{\text{mass adsorbent}}{\text{volume bed}}\right][\text{volume bed}]$$

$$= [\text{flux}]\left[\frac{\text{area}}{\text{volume}}\right][\text{volume bed}][t_{sat}]$$

$$q_0[\rho(1-\varepsilon)] = k(y-0)a\,t_{sat}$$

$$t_{sat} = \frac{q_0\,\rho\,(1-\varepsilon)}{k\,a\,y_0} \tag{15.4-1}$$

where q_0 is the solute concentration in the saturated adsorbent; ρ is the adsorbent density; ε is the bed's volume fraction; k is the solute's mass transfer coefficient; a is the adsorbent's area per volume; and y_0 is the feed concentation. This equation makes sense: for example, if the saturation concentration q_0 is small or if the mass transfer coefficient k is large, then t_{sat} should be small.

Estimating the velocity of the saturation zone v_{sat} is also based on a mass balance

$$(\text{amount solute fed}) = (\text{amount adsorbed})$$

$$y_0 v\,(\text{cross-sectional area})\,(t - t_{sat}) = q_0\,\rho(1-\varepsilon)\,v_{sat}(\text{cross-sectional area})\,(t - t_{sat})$$

$$v_{sat} = v\left(\frac{y_0}{q_0\,\rho\,(1-\varepsilon)}\right) \tag{15.4-2}$$

As mentioned above, the saturation velocity v_{sat} is often much less than the feed velocity v because y_0 is much less than q_0. Equation 15.4-2 also lets us estimate how far the saturation zone extends:

$$l_{sat} = v_{sat}\,(t - t_{sat}) \tag{15.4-3}$$

where l_{sat} is the length of the saturation zone. When l_{sat} reaches the end of the column, the column is exhausted.

Finally, we turn to the trickiest part of this estimate, the calculation of the adsorption zone. The calculation is tricky because of the implications of the assumptions made in the calculation, not because of the mathematics. We begin with a mass balance on a small volume within the adsorption zone:

$$(\text{accumulation}) = (\text{flow in}) - (\text{flow out}) - (\text{amount adsorbed})$$

$$\varepsilon\frac{\partial y}{\partial t} = -v\frac{\partial y}{\partial z} - ka(y - y^*) \tag{15.4-4}$$

We now assume that the accumulation in solution is small relative to the amount of adsorption so that the left-hand side of this equation is zero. This assumption will be good for most favorable adsorbents, which do try dramatically to concentrate the solute (i.e., $q_0\rho \gg y_0$). We also assume that adsorption is irreversible, so the concentration at equilibrium y^* is zero. This assumption means that our predicted breakthrough curve will fit the data only near the breakthrough time. Finally, we assume that Eq. 15.4-4 is subject to the constraint that

$$z = l_{sat} \qquad y = y_0 \qquad\qquad (15.4\text{-}5)$$

This implies that the adsorption zone is moving much more slowly than the fluid is flowing.

With these assumptions, the concentration in the adsorption zone is easily calculated

$$\frac{y}{y_0} = e^{-\frac{ka}{v}(z - l_{sat})} \qquad\qquad (15.4\text{-}6)$$

If desired, we can now combine this relation with Eq. 15.4-1 to 15.4-2 to find our estimate of the concentration profile in the bed. This estimate says that the profile will rise exponentially with time until it reaches y_0. This is a significant approximation, as Fig. 15.4-1 shows.

However, the real point of Eq. 15.4-6 is not its accuracy but what it says about the adsorption zone. It says that the key to this zone is a quantity (v/ka), which is a close parallel to the length of unused bed l and to the height of transfer unit (HTU) used for differential adsorption and distillation:

$$\frac{v}{ka} \doteq l' \doteq \text{HTU} \qquad\qquad (15.4\text{-}7)$$

Like the HTU, the quantity (v/ka) is a measure of the efficiency of the adsorption. A small value of (v/ka) indicates an efficient adsorption, and a large value signals an inefficient one.

We then must consider how (v/ka) or l' will vary with process conditions. This depends most directly on the mechanism responsible for k. In deriving Eq. 15.4-6, we implied that k was a mass transfer coefficient, but we now should realize that it depends on all factors which change the shape of the breakthrough curve. As we discussed in Section 15.3, these factors include mass transfer to and within the adsorbent particles and dispersion in the bed itself. As the particle size increases, the mass transfer coefficient decreases and the dispersion increases, so the adsorption zone gets longer. As the velocity increases, the mass tranfer coefficient may go up a little, but the time for adsorption goes down, so the adsorption zone gets longer. As the velocity increases, the dispersion increases, so the adsorption zone gets longer. These factors, explored in the examples which follow, are discussed in more detail in the next section.

Example 15.4-1: Odor removal   We have been removing odor from a Minnesota resort's water supply using a bed of activated carbon. The bed's breakthrough is in 38 days, and its exhaustion is in 46 days. Because of seasonal demand, we want to double the water flow through the bed. We suspect that the mass transfer coefficient will be constant. If it is, when will breakthrough occur now?

   Solution   To estimate the breakthrough, we will combine the concept of l' and (v/ka) summarized by Eq. 15.4-7. From Eq. 15.3-7,

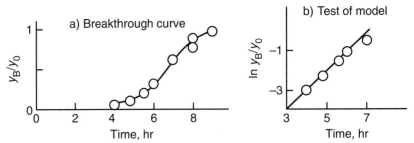

Fig. 15.4-2. Steroid adsorption vs. time. This breakthrough curve is the basis of Example 15.4-2.

$$\theta = \frac{2(38)}{38 + 46} = 0.90$$

From Equation 15.3-8,

$$\frac{l'}{l} = 0.10$$

When the velocity is doubled, we expect from Equation 15.4-7,

$$\frac{l'}{l} = 0.20 = \frac{2\,t_B}{t_B + t_E}$$

The capacity of the bed is unchanged by the higher flow:

$$y_0\,Q\left[38 + \frac{1}{2}(46 - 38)\right] = y_0\,(2Q)\left[t_B + \frac{1}{2}(t_E - t_B)\right]$$

Solving these two equations, we find

$$t_B = 4.2\ \text{days}$$

Note that this is less than the 21 days expected from the increased flow without the increased l' caused by altered dispersion.

Example 15.4-2: Steroid recovery   We have made a significant amount of chemically modified steroid for use as a topical anti-inflammatory like hydrocortizone, which is used for reducing allergic reactions such as that to poison ivy. When we measure the adsorption of this hormone, we obtain the data shown in Fig. 15.4-2(a). These data are for a superficial velocity of 1 cm/min through a packed bed of 310 μm spheres with a void fraction of 0.4. The saturation concentration per volume $q_0\rho$ is 300 times the feed concentration y_0. (a) Are these data consistent with the model for irreversible adsorption? (b) What is the rate constant ka? (c) Is this rate constant consistent with literature estimates?

   Solution   (a) The irreversible model, given by Eqs. 15.4-1 to 15.4-3 and 15.4-6, predicts that the logarithm of concentration should vary linearly with time. As Fig. 15.4-2(b) shows, it does, supporting the irreversible model. Note that it does so only at small times, and that only half the data in Fig. 15.4-2(a) fit this irreversible limit. Still, for a valuable product like this, we will probably stop the adsorption at a short time when the irreversible limit still works.

(b) The data in Fig. 15.4-2(b) follow the relation

$$\ln\frac{y_B}{y_0} = -6.72 + \left(\frac{0.93}{\text{hr}}\right)t$$

From the time dependence of the breakthrough curve

$$\frac{0.93}{\text{hr}} = \frac{ka}{v}(l_{\text{sat}}) = \frac{ka\,y_0}{q_0\rho(1-\varepsilon)}$$

$$= \frac{ka}{300\,(1-0.4)}$$

$$ka = 0.047\ \text{sec}^{-1}$$

(c) We compare this with values from an earlier correlation in Table 8.3-2:

$$\frac{kd}{D} = 1.17\left(\frac{dv}{\nu}\right)^{0.58}\left(\frac{\nu}{D}\right)^{0.33}$$

We estimate the diffusion coefficient D to be $5 \cdot 10^{-6}$ cm^2/sec; we assume the kinematic viscosity ν is 0.01cm^2/sec. Using these values,

$$\frac{k(310.10^{-4}\,\text{cm})}{5.10^{-6}\,\text{cm}^2/\text{sec}} = 1.17\left(\frac{310.10^{-4}\,\text{cm}\frac{1\,\text{cm}}{60\,\text{sec}}}{0.01\,\text{cm}^2/\text{sec}}\right)^{0.58}\left(\frac{0.01\,\text{cm}^2/\text{sec}}{5.10^{-6}\,\text{cm}^2/\text{sec}}\right)^{0.33}$$

$$k = 4.1 \cdot 10^{-4}\text{cm}/\text{sec}$$

The particle area per bed volume is

$$a = \left[\frac{\pi d^2}{\frac{4}{3}\pi\left(\frac{d}{2}\right)^3}\right](1-\varepsilon) = \frac{6}{d}(1-\varepsilon) = \frac{6(1-0.4)}{310.10^{-4}\text{cm}} = \frac{116\,\text{cm}^2}{\text{cm}^3}$$

Thus

$$ka = 0.048\ \text{sec}^{-1}$$

This value is almost identical to that found for that data in Fig. 15.4-2, analysed in part (b), suggesting that the adsorption is controlled by mass transfer from the bulk solution to the surface of the particles.

15.5 Other Characteristics of Adsorption

The earlier material in this chapter reviews the standard topics that are covered in adsorption: isotherms, the length of unused bed, and the effects of mass transfer. In this section, we want to cover three additional topics. First, we will discuss the more general behavior of an adsorption column including not only mass transfer but also dispersion and nonlinear isotherms. Each of these can contribute to the spread of the breakthrough curves. Second, we will discuss elution chromatography, a technique common in chemistry but uncommon in engineering. Elution chromatography is like adsorption but with a different initial condition. Finally, we will qualitatively discuss pressure swing adsorption, a technique for gas separation which was slow to develop but has become common; and simulated moving beds, a method for drug purification that has developed because of the ease of electronic control.

15.5.1 General Behavior

We first explore how we might obtain better approximations to the breakthrough curve than those possible with the analysis in earlier sections. Such a general analysis depends upon the equation

$$\varepsilon \frac{\partial y}{\partial t} = -v \frac{\partial y}{\partial z} - ka(y - y*) + (D + E) \frac{\partial^2 y}{\partial z^2} \tag{15.5-1}$$

As in Eq. 15.4-4, the left-hand side of this relation represents the accumulation of the solute within the solution in a differential column volume. The first term on the right-hand side represents solute convection in minus out and is also common with the earlier analysis. The second term on this side represents material lost from the solution into the adsorbent. The overall mass transfer coefficient k may include not only diffusion from the bulk solution to the adsorbent but both diffusion and reaction within the adsorbent. The surface area per volume a usually is taken to represent external adsorbent area per volume of bed; however, if the mass transfer is dominated by reaction within the particle, this area can be more conveniently defined as the actual surface area of the microporous adsorbent per volume of bed. The concentration $y*$ is as usual that concentration that would exist in the liquid if it were at equilibrium with the solid. It is this concentration that includes the nonlinear isotherms.

The third term on the right-hand side is new. It describes axial diffusion and dispersion in the column during the adsorption. Radial dispersion also occurs, but this radial dispersion is presumed to be so great that the concentration across the bed's radius is about constant. The dispersion coefficient E that appears in this term has the same dimensions as the diffusion coefficient D, that is, length squared per time. It is a strong function of the physical processes in the bed, with only a weak dependence on any solute chemistry. It can be estimated using the techniques summarized in Chapter 4. We will return to this estimation later.

The solution to Eq. 15.5-1 is, in general, complicated. It can be approximated by a breakthrough curve of the form

$$\frac{y - \bar{y}}{y_0 - \bar{y}} = \mathrm{erf} \frac{t - \bar{t}}{\sqrt{2}\,\sigma} \tag{15.5-2}$$

where t is the time, $\bar{t}$ is the time required to elute the average concentration $\bar{y}$, and σ is the standard deviation, the square root of the variance σ^2. Both $\bar{t}$ and σ merit discussion.

The time $\bar{t}$ necessary to elute the average concentration is most easily found from a mass balance by assuming the breakthrough curve is a step function

$$\begin{bmatrix} \text{total amount} \\ \text{fed into bed} \end{bmatrix} = \begin{bmatrix} \text{amount left in} \\ \text{saturated bed} \end{bmatrix}$$

$$y_0 A v \bar{t} = [\varepsilon y_0 + (1 - \varepsilon)q_0] Al \qquad (15.5\text{-}3)$$

where l is the length and A is the cross-sectional area of the bed. Thus the average time $\bar{t}$ is given by

$$\bar{t} = l\left[\frac{\varepsilon}{v} + \frac{q_0\rho(1 - \varepsilon)}{y_0 v}\right] \qquad (15.5\text{-}4)$$

The first term within brackets is the reciprocal of the actual velocity in the bed, not the superficial velocity. The second term in the brackets is the reciprocal of the velocity of the saturation zone given in Eq. 15.4-2.

The standard deviation σ gives the spread of the breakthrough curve. If it equals zero, a step function into the column will be eluted as a step function out of the column. If σ is large, then the breakthrough out of the bed will be gradual and the adsorbent in the bed will not be efficiently used. The physical significance of σ is most easily seen if we rewrite this quantity as a dispersion coefficient, like those developed in Chapter 4. In particular, we can rewrite the concentration profile in terms of position, rather than time

$$\frac{y - \bar{y}}{y_0 - \bar{y}} = \text{erf}\,\frac{z - \bar{z}}{\sqrt{4E\bar{t}}} \qquad (15.5\text{-}5)$$

By comparing this with Equation 15.5-2, we see that

$$\sqrt{2}\,v\sigma = \sqrt{4E\bar{t}} \qquad (15.5\text{-}6)$$

as $v = z/\bar{t}$.

We will find it easiest to frame our remaining discussion in terms of the dispersion coefficient E. Like the standard deviation, the dispersion coefficient describes the spread of the breakthrough curve. If it is zero, the breakthrough is a step function; if it is large, the breakthrough is gradual.

The dispersion coefficient has three physical causes in adsorption: axial diffusion, adsorption kinetics, and Taylor–Aris dispersion:

$$E = E_{\text{diffusion}} + E_{\text{kinetics}} + E_{\text{Taylor Aris}} \qquad (15.5\text{-}7)$$

The contribution from axial diffusion is most straightforward

$$E_{\text{diffusion}} = \varepsilon D\left[1 + k'\right] \qquad (15.5\text{-}8)$$

where

$$k' = \frac{q_0\rho(1 - \varepsilon)}{y_0\varepsilon} \qquad (15.5\text{-}9)$$

where k' is called the "capacity factor," that is, the capacity for the solute in the solid per bed volume divided by the capacity for the solute in the solution, also per bed volume. In physical terms, the quantity εD is just the effective axial diffusion coefficient in the adsorbent bed. This contribution to dispersion is normally minor.

The contribution to dispersion from kinetics can be shown to be

$$E_{kinetics} = \frac{v^2}{\varepsilon^2 ka} \left(\frac{k'}{1 + k'} \right) \tag{15.5-10}$$

Remember that the mass transfer coefficient k will include both diffusion to and into the particle, and any chemical reaction within the particle. Often, k will depend inversely on the particle diameter d; the adsorbent area per bed volume a also depends inversely on d; so the dispersion caused by kinetics often depends on the square of the particle size. Big particles mean big dispersion and, hence, gradual breakthrough.

The final source of dispersion, that associated with Taylor and Aris, is often the largest. In the early days of adsorption, dispersion often occurred because of unevenly packed adsorbent beds, but this cause has largely been minimized by careful packing. Now, the dispersion comes from the coupling of axial flow and radial diffusion detailed in Section 4.3. The result is

$$E_{Taylor-Aris} = \frac{d^2 v^2}{192 \varepsilon^2 D} \frac{\left(1 + 6k' + 11(k')^2 \right)}{1 + k'} \tag{15.5-11}$$

Note how dispersion varies inversely with the diffusion coefficient D, which is exactly the opposite of the axial diffusion mechanism given in Eq. 15.5-8. Big particles and big flows cause more dispersion. All these mechanisms in Eq. 15.5-7 tend to spread out the concentration front; only the favorable isotherm works against this trend. This balance between dispersion and self-sharpening is why the length of unused bed is constant as long as the flow and particle size are fixed.

15.5.2 Chromatography

We now turn to chromatography, where we must first define the process involved. The definitions for this process differ in engineering and in chemistry. In engineering, the common process is the adsorption just described. In this process, the feed concentration is a step function and the concentrated feed is continued until breakthrough. In engineering, the uncommon process, called chromatography, uses a pulse of feed with several solutes. This pulse is then washed through the column with additional solvent to effect a separation.

In chemistry, both these processes are called chromatography. The process that uses a step function in the feed is referred to as frontal chromatography. In chemistry, frontal chromatography is relatively uncommon. The more common chemical process, which uses a pulse of feed solution, is called elution chromatography. Unfortunately, the word chromatography is sometimes used without distinguishing between these two processes so that one must read the details of the experiments to discover which case is involved.

The analysis of elution chromatography is a close parallel to that for adsorption. The best theoretical development, due to Golay, assumes the process takes place not in a packed bed but in a thin cylindrical tube with adsorbent-coated walls. Solvent moves steadily in laminar flow through the tube. At time zero a pulse of solutes is injected at one end of the tube. Each solute elutes with a concentration profile given by Eq. 4.4-31

$$c_1 = \frac{M/A}{\sqrt{4\pi E \bar{t}}} e^{-\frac{(z - v\bar{t})^2}{4 E \bar{t}}} \qquad (15.5\text{-}12)$$

where M is the amount of solute injected, A is the cross-section of the tube, and $\bar{t}$ is the time where the maximum concentration exits, given by Eq. 4.4-32.

Because the solutes are dilute, the concentration profile for a second solute would have the same mathematical form as the first, although values like k' and D would of course be different. Now, for two solutes, we seek peaks which are not broadly spread but eluted separately. This is the goal of this process.

Sharp peaks in chromatography, like sharp breakthrough curves in adsorption, depend on the dispersion coefficient E, given in Eq. 4.4-34.

$$E = D(1 + k') + \frac{\delta^2 v^2}{3D'}\left(\frac{k'}{1 + k'}\right) + \frac{d^2 v^2}{192 D}\left(\frac{1 + 6k' + 11(k')^2}{1 + k'}\right) \qquad (15.5\text{-}13)$$

where D' is the diffusion coefficient in the adsorbent. This equation is a complete analogue to Eqs. 15.5-8, 15.5-10, and 15.5-11: the three terms on the right-hand side correspond to dispersion from axial diffusion, from mass transfer, and from Taylor–Aris dispersion.

The results of elution chromatography are frequently described in terms of ideas similar to those used for absorption and distillation. The numbers of theoretical plates N, a rough parallel to the number of transfer units (NTU), can be defined by

$$N = \text{NTU} = \frac{l^2}{4 E \bar{t}} = \frac{lv}{4 E (1 + k')} \qquad (15.5\text{-}14)$$

Thus a very sharp peak, that is, a small amount of dispersion, implies a large number of theoretical plates. Values of several million plates are not uncommon in analytical systems. This plate number is used to define the height of an equivalent theoretical plate (HETP), a parallel to the height of a transfer unit:

$$\text{HETP} = \text{HTU} = \frac{l}{\text{NTU}} \qquad (15.5\text{-}15)$$

This term is used to define a reduced plate height h defined as

$$h = \text{HETP}/d \qquad (15.5\text{-}16)$$

Note that the characteristic length for the reduced plate height is the tube diameter d. Finally, this reduced plate height is frequently written in terms of a reduced velocity,

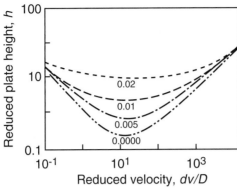

Fig. 15.5-1. Plate height versus velocity in cylindrical capillaries. The increased height at low flow is due to axial diffusion; that at high flow results from Taylor–Aris dispersion. The lowest line is for capillaries of exactly the same diameter, and the upper lines are for polydisperse diameters given as the ratio of the standard deviation of the diameter divided by the diameter itself. (Schisla et al., 1993).

which is another name for a Péclet number. To calculate this height, we combine Eqs. 15.5-13 and 15.5-16 to obtain

$$h = O\left(\frac{dv}{D}\right) + O'\left(\frac{D}{dv}\right) \tag{15.5-17}$$

The result of this type of analysis is commonly given by a plot like that in Fig. 15.5-1. The minimum in this plot represents the smallest reduced plate height and hence the largest number of theoretical plates. Alternatively, one could say it represents the smallest amount of dispersion and the sharpest peaks. This minimum is the goal of elution chromatography, though not necessarily of adsorption.

15.5.3 Other Adsorption Processes

Two other separations that are based on adsorption merit mention even though their details are beyond the scope of this book. In each case, the processes' efficiencies are compromised by mass transfer effects like those that compromise conventional adsorption and chromatography.

These two processes are the simulated moving bed and pressure swing adsorption. The simulated moving bed uses the same packed bed as conventional adsorption, but changes the feed point. This altered feed point means that the stationary adsorbent has a concentration profile more like that which would exist if the feed point were fixed and the adsorbent were moving. The results approximate those that would be possible with countercurrently moving adsorbent but without the curse of particle attrition. This process, developed especially by the American chemical company UOP, is used for the separation of liquid *n*-paraffins from branched paraffins. It is also used to separate xylene, cymene, and cresol isomers, and aqueous solutions of racemates.

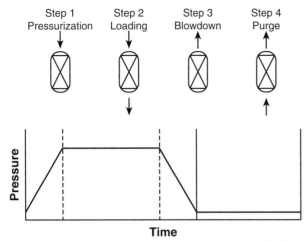

Fig. 15.5-2. Pressure swing adsorption. This process, an effective way to separate gases without cryogenic distillation, involves the four steps shown and described in the text.

The second process, pressure swing adsorption or PSA, is used as an alternative to cryogenic distillation for separating gases. Its original form, which is still in use, is the basis of several similar processes. It offers exceptionally high purity because of a large number of equivalent stages.

One typical process, shown in Fig. 15.5-2, has four steps. In the first step, the bed is pressurized with the feed gas mixture. At the beginning of the second step, a valve at the exit end of the bed is opened to allow a continuous flow of feed and product at the elevated pressure. This product is enriched in the weakly adsorbing species. The third and fourth steps constitute the regeneration of the adsorbent bed. During the third "blowdown" step, the bed is countercurrently depressurized to release the adsorbed species. During the fourth step, a low-pressure gas dilute in the adsorbed component is fed countercurrently to the feed direction in order to purge the adsorbed species from the bed. The blowdown and purge steps, run countercurrently to the feed, move the concentration front of the adsorbed species away from the product end of the bed. After the fourth step, the column is repressurized and the cycle begins again. Although this particular example contains a single bed, the original Skarstrom process consists of two beds operated at one-half cycle apart. This two-bed process enables a portion of the product gas from one bed to serve as the purge gas for the second bed.

In the original form of the process, pressure swing adsorption was run slowly, near equilibrium. The concentrations obtained were then less a function of mass transfer and hence largely independent of diffusion. More recently, pressure swing adsorption beds have operated much more rapidly. For example, in Skarstrom's original work, the cycle time for the beds was around 20 minutes; in more recents efforts, the cycle time can be 10 seconds. With these rapidly cycled beds, the separation depends not only on differences in adsorption but also on differences in mass transfer. The extensions of this rapid process, beautifully detailed by Yang, are beyond the scope of this chapter. I did want to mention the process because I believe it is still emerging.

15.6 Conclusions

Adsorption is a different separation process from absorption, distillation, and extraction discussed in earlier chapters. Adsorption takes place at unsteady state with nonlinear isotherms. Because of these complexities, adsorption depends more on experiments than other separations. For the common case of a favorable isotherm, these experiments are used to determine the "length of unused bed," a measure of the separation's efficiency. This length is independent of the total bed length and so serves as the basis of adsorption scale-up.

The length of unused bed results from the combined effects of the isotherm and of dispersion. A favorable isotherm tends to sharpen the concentration profile in the bed, and dispersion tends to blur it. The dispersion comes from axial diffusion, mass transfer into the adsorbent, and coupled axial convection and radial diffusion (Taylor–Aris dispersion). Understanding these effects is the key to improving the efficiency of adsorption.

Questions for Discussion

1. What is the difference between an isotherm and an equilibrium line?
2. What is a good adsorbent for removing water vapor? What is a good one for removing color?
3. What are typical units for q and y?
4. How can you tell if your equilibrium data fit a Langmuir isotherm?
5. How can you tell if they fit a Freundlich isotherm?
6. Why is packed-bed adsorption usually superior to stirred-tank adsorption?
7. When would packed-bed adsorption be less effective than stirred-tank adsorption?
8. What is the length of unused bed? Why is it often constant with increasing bed length?
9. When will the length of unused bed increase with increasing bed length?
10. Would you ever deliberately use an unfavorable isotherm?
11. If you double the flow rate, what will happen to the breakthrough curve?
12. If you double the adsorbent particle size, what will happen to the breakthrough curve?
13. Why are moving beds of adsorbents used rarely?
14. Explain PSA to someone not trained in engineering or science.
15. Compare separations by means of adsorption and extraction.

Problems

1. A modified dextran will adsorb up to $8 \cdot 10^{-8}$ mol of immunoglobulin G per cm^3 dextran. The adsorption follows a Langmuir isotherm with a constant K equal to 2×10^{-8} mol/l. How much dextran do you need to adsorb 90% of the protein in 1.2 l of solution initially containing $4 \cdot 10^{-6}$ mol/l?

2. You need to adsorb phenol in 5300 kg of an unexpected waste stream, reducing the concentration to meet the environmental standard of 10 ppm (i.e., $10 \cdot 10^{-6}$ kg/kg). You plan to use an activated carbon for which

$$q = 0.53 \, y^{0.23}$$

where y is in kg phenol/kg solution and q is in kg phenol/kg carbon. (a) If the feed concentration is 730 ppm, how much carbon do you need if you carry out this contacting in a well-stirred tank? (b) How much carbon will you need if you use a packed bed but get no dispersion? (In other words, the breakthrough curve is a step function.)

3. Data for drying nitrogen with molecular sieve type 4A are given below:

t, hr			15	15.4	15.6	16	
c, ppm			<1	6	30	140	
t, hr	16.4	16.6	16.8	17	17.2	18	24
c, ppm	430	610	800	990	1100	1400	1500

The nitrogen feed is 12.0 kg mol/hr m^2; the bed density is 0.66 g/cm^3; its length is 4.2 cm. Calculate the saturation capacity from the breakthrough curve, and determine the fraction of unused bed based on a breakthrough concentration y/y_0 of 0.05.

4. You want to adsorb a brown color from a toner solution by letting the solution flow through a bed of activated carbon. In tests on a 32 cm bed of 1 cm^2 cross-section, you find that color breakthrough occurs after a volume of 140 cm^3, and that the bed is exhausted after 280 cm^3. The breakthrough curve is about linear. You can keep the same flow in the bed up to a bed length of 175 cm, but you want 1000 times greater capacity. (a) What is the percent saturation of the test bed? (b) What is its unused length? (c) What should be the diameter of the new bigger bed? (d) What volume of adsorbent should you buy?

5. You want to scale up an adsorption which, in a lab column 1 cm in diameter and 20 cm long, showed a breakthrough at 40 min and exhaustion at 52 min. Other experiments suggest that the unused bed length is about equal to an HTU, and you expect mass transfer coefficients in this bed to vary with the square root of the flow. In your new column, you must have 10,000 times greater capacity and a bed 130 cm long (to minimize pressure drop). You plan to increase the flow 17 times. What should the bed diameter be?

6. You are carrying out the separation of a pharmacologically active polypeptide by adsorption from dilute aqueous solution on beads of a cross-linked dextran gel. The gel beads, which have a diameter around 620 μm, work well in the lab. Using a pressure drop of 80 kPa, you find with a cylindrical column of 1.0 cm diameter and 20 cm length that you have an unused bed length of 8 cm and can process 600 cm^3 in 2 hr. You now want to filter 2000 l of the same concentration of peptide solution in 7 hr. The obvious solution is to increase the bed length and run at larger pressure drop, expecting that the unused bed length will stay the same. This is especially attractive because the cross-linked dextran beads are expensive, so you don't want to use more than you must. However, when you call the dextran gel supplier, she says that while the unused bed length should remain constant, the gel is compressible, so that the specific cake resistance is given by

$$\alpha = \alpha_0 \, \Delta p^{0.43}$$

where Δp is in kPa. She suggests that the pressure drop should always be below 1000 kPa/m bed. What length and diameter should you use for this bed?

7. Your company has recently developed a microorganism which produces large amounts of an enzyme for oxidizing phenolics. Such an enzyme can be used in a fixed-bed reactor to improve the taste of drinking water. For such a use, it must be unusually pure. To isolate the enzyme, you plan to adsorb it on a custom-synthesized ion-exchange resin. The isotherm of the resin is such that, at a liquid velocity of 10 cm/hr, you get no solute out of the bed for 3 hr and the feed concentration at 4 hr. The outlet concentration between these limits is nearly linear. Everything's great. Then because of unexpected demand, you are asked to double the velocity yet lose only 10% of the feed. Your experiments show that the slope of the linear region of a plot of exit concentration versus time varies inversely with velocity. How long should you run the bed? *Answer:* 2.5 hr.

8. You have isolated a protein to be used as a vaccine by absorption from a buffer on a packed bed of a custom-synthesized ion-exchange resin. The resin consists of 0.011 cm spheres, with a void fraction of 0.37. It is packed in a 100 cm column, 83 cm in diameter, fed at a velocity of 0.052 cm/sec. Under these conditions, the protein is adsorbed with a mass transfer coefficient of 6×10^{-6} cm/sec and an adsorption equilibrium constant of 27. After this bed is completely loaded, you plan to elute it with a more acidic buffer. This new buffer has an adsorption constant of 0.22, but all other conditions are unchanged. How long must you elute until the concentration is 10% of the maximum? *Answer:* 40 min.

Further Reading

Gembicki, S. A., Oroskar, A. R., and Johnson, J. A. (1991). Adsorption: Liquid Separation. In *Encyclopedia of Chemical Technology*, ed., J. I. Kroschwitz. New York: Wiley.

Golay, M. J. E. (1958). Theory of Gas Chromatography. In *Gas Chromatography-1958*, ed., D. H. Desty. London: Butterworth.

Masel, R. I. (1996). *Principles of Adsorption and Reaction on Solid Surfaces*. New York: Wiley.

Ruthven, D. M. (1984). *Principles of Adsorption and Adsorption Processes*. New York: Wiley.

Ruthven, D. M. (2001). Adsorption, Fundamentals. In *Kirk-Othmer Encyclopedia of Chemical Technology*. New York: Wiley.

Schisla, D. K., Ding, H., Carr, P. W., and Cussler, E. L. (1992). Polydisperse tube diameters comprise multiple open tubular chromatography. *AICHE J.*, **39**: 6, 946.

Yang, R. T. (1987). *Gas Separation by Adsorption Processes*. Boston: Butterworth.

Yang, R. T. (2003). *Adsorbents: Fundamentals and Applications*. New York: Wiley.

PART IV

Diffusion Coupled With Other Processes

CHAPTER 16

General Questions and Heterogeneous Chemical Reactions

In the previous chapters, we have discussed how diffusion involves physical factors. We calculated the gas diffusion through a polymer film, or sized a packed absorption column, or found how diffusion coefficients were related to mass transfer coefficients. In every case, we were concerned with physical factors like the film's thickness, the area per volume of the column's packing, or the fluid flow in the mass transfer. We were rarely concerned with chemical change, except when this change reached equilibrium, as in solvation.

In this chapter, we begin to focus on chemical changes and their interaction with diffusion. We are particularly interested in cases in which diffusion and chemical reaction occur at roughly the same speed. When diffusion is much faster than chemical reaction, then only chemical factors influence the reaction rate; these cases are detailed in books on chemical kinetics. When diffusion is not much faster than reaction, then diffusion and kinetics interact to produce very different effects.

The interaction between diffusion and reaction can be a large, dramatic effect. It is the reason for stratified charge in automobile engines, where imperfect mixing in the combustion chamber can reduce pollution. It is the reason for the size of a human sperm. It can reduce the size needed for an absorption tower by 100 times. The interaction between diffusion and reaction can even produce diffusion across membranes from a region of low concentration into a region of high concentration.

In this and the following chapters, we explore interactions between diffusion and chemical reaction. For heterogeneous reactions, we shall find that diffusion and reaction occur by steps in series, steps that can produce results much like mass transfer across an interface. For homogeneous reactions, we shall find that diffusion and reaction occur by steps partially in parallel, steps that are different than processes considered before. In both cases, we shall find that non-first-order stoichiometries lead to unusual results.

We begin with two surprisingly subtle questions. First, in Section 16.1, we discuss whether a chemical reaction is heterogeneous or homogeneous. The answer turns out to be a question of judgement; we must decide which aspects of the chemistry to ignore and choose the type of description leading to the simplest result. This choice leads in Section 16.2 to the concept of a diffusion-controlled reaction, an idea with as many manifestations as Vishnu. After these general questions, we turn to a more explicit discussion of heterogeneous reactions. In Section 16.3, we calculate simplest results for first-order heterogeneous reactions. We extend these results to reactions producing ash in Section 16.4, and to different stoichiometries in Section 16.5. The results supply a synopsis of the heterogeneous results.

16.1 Is the Reaction Heterogeneous or Homogeneous?

We first want to discuss the difference between heterogeneous and homogeneous reactions. On first inspection, this difference seems obvious. Heterogeneous reactions must involve two different phases, with the chemical reactions occurring at the interface. Homogeneous reactions take place in a single phase, and the reaction occurs throughout.

In practice, this distinction is less obvious. As an example, imagine a spherical coal particle burning in a fluidized bed. All the reaction initially takes place on the sphere's surface. This initial reaction is best modeled as heterogeneous. If it were first-order in, for example, oxygen, then the rate equation might be

$$
\begin{pmatrix} \text{combustion rate} \\ \text{per particle} \\ \text{area} \end{pmatrix} = \kappa_1 \begin{pmatrix} \text{oxygen concentration} \\ \text{at the surface} \end{pmatrix}
\tag{16.1-1}
$$

in which κ_1 is the heterogeneous rate constant. If the oxygen concentration is in moles per volume, the constant κ_1 has dimensions of length per time, the same dimensions as the mass transfer coefficient.

However, as the combustion proceeds, the particle may become porous, and the chemical reaction may occur not only at the surface but on all pore walls throughout the particle. In some cases, the pore area may far exceed the particle's superficial surface area. The combustion is now occurring throughout the particle as if the reaction were homogeneous; its rate is best modeled as

$$
\begin{pmatrix} \text{combustion rate} \\ \text{per particle} \\ \text{volume} \end{pmatrix} = \kappa_1 \begin{pmatrix} \text{oxygen concentration} \\ \text{per volume} \end{pmatrix}
\tag{16.1-2}
$$

The oxygen concentration can be defined either per pore volume or per particle volume. In either case, the homogeneous rate constant has units of reciprocal time. Such constants are fixtures of elementary chemistry textbooks.

Thus combustion of the coal particle can be modeled as heterogeneous or homogeneous, depending on how the coal is burning. The choice of a model for the reaction is usually subjective, but rarely explicitly stated in the research literature. Instead, it must be inferred. The key for inference is the continuity equation. If this equation has the form

$$
\frac{\partial c_1}{\partial t} = D\nabla^2 c_1 - \nabla \cdot c_1 \mathbf{v}^0
\tag{16.1-3}
$$

then any reactions appear in the boundary conditions. Such reactions are being modeled as heterogeneous. If the continuity equation is

$$
\frac{\partial c_1}{\partial t} = D\nabla^2 c_1 - \nabla \cdot c_1 \mathbf{v}^0 + r_1
\tag{16.1-4}
$$

the reactions r_1 can occur in every differential volume in this system. Such a reaction is being modeled as homogeneous.

In your own work, you must use your judgement about which model is most appropriate. This judgement can be tested with the following spectrum of examples:

(1)   Particles of low-grade lead sulfide are roasted to produce a porous lead oxide ash. At the center of the ore particles is a core of unreacted sulfide. The oxygen permeability in this core is much less than in the ash. Thus this reaction is best modeled as heterogeneous, occurring at the interface between ore and ash.

(2)   Traces of ammonia are scrubbed out of air with water. The reaction to produce ammonium hydroxide will take place within the liquid phase, and the reaction is best modeled as homogeneous.

(3)   Aspirin dissolves in chyme. The aspirin hydrates at the solid surfaces and then diffuses away into the solution. This case is best treated as a heterogeneous hydration followed by diffusion.

(4)   Ethane is dehydrogenated on a single platinum crystal. The reaction, which takes place on the crystal's surface, is best modeled as heterogeneous.

(5)   Ethane is dehydrogenated on a porous platinum catalyst. Here, the ethane can diffuse through the catalyst pores at a rate similar to diffusion in the surrounding gas. The reaction takes place on pore walls within the catalyst pellet. This reaction is usually treated as homogeneous, even though the chemistry is similar to the previous example.

(6)   Sulfur oxides are scrubbed out of stack gas with an aqueous lime slurry. Here, sulfur oxides and dissolved lime both will have about the same diffusion coefficients in the water. Lime is not very soluble, so its concentration is low. You need to know more chemistry before you can guess where the reaction takes place, and whether it is best modeled as heterogeneous or homogeneous.

We discuss models of heterogeneous reactions in this chapter and models of homogeneous reactions in the next chapter. This artificial division finesses the embarrassing question of which model we should use. This is usually not a major problem. We shall find that the choice between heterogeneous and homogeneous models is often obvious, one that we shall make almost automatically.

16.2 What is a Diffusion-Controlled Reaction?

Throughout science and engineering we find references to diffusion-controlled mass transfer and diffusion-controlled chemical reaction. Those using these terms often have very specific cases in mind or are not aware of how broadly and loosely these terms are used. In this short section, we want to describe the cases to which these terms most commonly refer. Each of these cases will be analyzed in detail later.

A diffusion-controlled process always involves various sequential steps. For example, the dehydrogenation of ethane on a single platinum crystal involves the diffusion of the ethane to the solid followed by the reaction on the solid surface. The reaction between protons and hydroxyl ions in water first has these species diffusing together, and then reacting.

Reactions like these are said to be diffusion-controlled when the diffusion steps take much longer than the reaction steps. Four cases in which this is true are shown in Fig. 16.2-1. For the heterogeneous reaction in Fig. 16.2-1(a) reagent diffuses to the surface; the

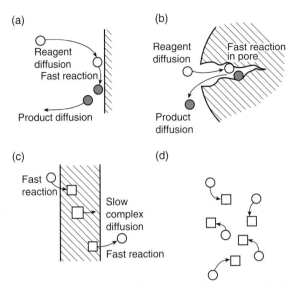

Fig. 16.2-1. Four types of diffusion-controlled reactions. In (a), a reagent slowly diffuses to a solid surface and quickly reacts there; this case occurs frequently in electrochemistry. In (b), a reagent slowly diffuses into the pores of a catalyst pellet, quickly reacting all along the way; this reaction is modeled as if it were homogeneous. In (c), a circular solute quickly reacts with a mobile carrier, thus facilitating the solute's diffusion across the membrane. In (d), two solutes rapidly react by diffusing together after a perturbation caused by fast reaction methods.

reagent quickly reacts to form product, and the product diffuses away (Aris, 1975). The overall rate is determined by the diffusion steps weighted by the equilibrium constant of the surface reaction. Cases like these, exemplified by ethane dehydrogenation on a single catalyst crystal, are the subject of the later sections of this chapter.

A second, very different type of diffusion-controlled reaction occurs in the case of a porous catalyst, shown schematically in Fig. 16.2-1(b). Here, the reagent must diffuse into the pores to reach catalytically active sites, where it reacts quickly. The overall rate of reaction depends on this diffusion. This case, central not only to catalysis but also to many scrubbing and extraction systems, is a subject of the next chapter.

Both these processes can be diffusion controlled, but the ways in which the control is exerted are very different. One way to see this difference is to examine electrical analogues of the two processes. For the heterogeneous reaction, this analogue is just the three resistors in series, shown at the top of Fig. 16.2-2. For the porous catalyst, the analogue is the much more elaborate arrangement shown at the bottom of Fig. 16.2-2. Unlike the single-crystal case, this combination is not simply a sequence of diffusion and reaction steps in series. It has some characteristics of diffusion and reaction steps in series. It has some characteristics of diffusion and reaction in parallel. In my own lectures, I have sometimes urged students to think in these helpful but inexact terms.

In each of these cases, the reactions are said to be diffusion-controlled if the resistance to diffusion is much greater than the resistance to reaction. However, these cases will clearly lead to very different combinations of these resistances.

Heterogeneous reaction on a flat catalyst surface

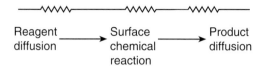

Reagent _____→ Surface _____→ Product
diffusion   chemical   diffusion
                       reaction

"Homogeneous" reaction within a porous catalyst

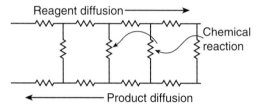

Fig. 16.2-2. Electrical analogues of two reactions affected by diffusion. These two cases correspond to those shown in Fig. 16.2-1(a) and 16.2-1(b), respectively. If the reactions are diffusion-controlled, then the resistances to chemical reaction will be relatively small.

In addition to these two cases, other very different situations can also be diffusion-controlled. Two more examples are the case of facilitated diffusion across membranes and that of reactions controlled by Brownian motion. In the facilitated-diffusion case, shown schematically in Fig. 16.2-1(c), one solute quickly reacts with a second carrier solute to form a complex; this complex then diffuses across the membrane. The overall transport rate is governed by complex diffusion weighted by the equilibrium constant for complex formation. This case is discussed in Section 18.5.

Still another diffusion-controlled process is called in chemistry a diffusion-controlled reaction, although it is very different from the other cases. In this case, shown schematically in Fig. 16.2-1(d), the system is an initially homogeneous mixture of two types of molecules. These species react instantaneously whenever they collide, so that their reaction rate is controlled by their molecular motion, that is, by their diffusion. This process is described in detail in Section 17.4; a similar dispersion-controlled process is described in Section 17.5.

By this point, we should try to find the common thread through this tweed of diffusion control. The key feature in all these cases is the coupling between chemical kinetics and diffusion. In every case, the overall rate is a function of the diffusion coefficient. Sometimes this rate depends on little else; more frequently, it also includes aspects of chemical dynamics. In any case, the idea of diffusion control is obviously indefinite without reference to a more specific situation. Make sure you know which definition is being implied before trying to understand what is happening.

16.3 Diffusion and First-Order Heterogeneous Reactions

After these general concerns, we turn to the analysis of diffusion and heterogeneous chemical reaction. The simplest case is the first-order mechanism shown schematically in Fig. 16.3-1. The reaction mechanism, the standard against which other ideas are measured, depends on three sequential steps. First, reagent diffuses to the surface; second, it reacts reversibly at the surface; finally, product diffuses away from the surface. The first and third steps depend on physical factors like reagent flow and fluid viscosity. The second step depends largely on chemistry, including adsorption and electron transfer.

This particular case occurs with surprising frequency. The most common practical example is an electrochemical reaction. For example, anions diffuse to the anode; these anions react there, and any products diffuse away. Because this kind of reaction often takes place in aqueous solution and at moderately high voltage, its rate is often governed by the diffusion steps.

We want to calculate the overall rate of the reaction shown in Fig. 16.3-1. To do so, we must calculate the rate of each of the three steps shown. More specifically, at steady state, the overall reaction rate per area r_2 equals the diffusion fluxes:

$$r_2 = n_1 = k_1(c_1 - c_{1i})$$

$$= -n_2 = k_3(c_{2i} - c_2) \tag{16.3-1}$$

in which c_1 and c_2 are the bulk concentrations, c_{1i} and c_{2i} are the concentrations at the surface, and k_1 and k_3 are the mass transfer coefficients of steps 1 and 3 in Fig. 16.3-1. In passing, note that the concentrations c_{1i} and c_{2i} have the same units as c_1 and c_2. For example, they might be moles per cubic decimeter.

The surface reaction, step 2 in Fig 16.3-1, is first order:

$$\text{species 1} \underset{\kappa_{-2}}{\overset{\kappa_2}{\rightleftharpoons}} \text{species 2} \tag{16.3-2}$$

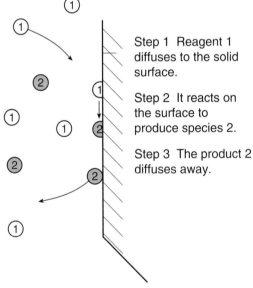

Step 1 Reagent 1 diffuses to the solid surface.

Step 2 It reacts on the surface to produce species 2.

Step 3 The product 2 diffuses away.

Fig. 16.3-1. Diffusion and heterogeneous chemical reaction. The reaction involved is first-order and reversible. The overall reaction rate depends on a sum of resistances, not unlike those involved in interfacial mass transfer.

The rate constants κ_2 and κ_{-2} refer to the forward and reverse reactions, respectively. In the rest of the chapter, we will use κ to signal chemical rate, and k to indicate the effect of diffusion, a physical rate process. Such a reaction is described by the rate equation

$$r_2 = \kappa_2 c_{1i} - \kappa_{-2} c_{2i} \qquad (16.3\text{-}3)$$

Because this reaction rate has units of moles per area per time and the concentrations have units of moles per volume, the rate constants both have units of length per time.

Two interesting points in these equations merit emphasis. First, the units used in these equations bother some readers, who feel that a surface reaction should be written in terms of surface concentrations. Such concentrations would have units of moles per area. Some make a fuss over this; they rewrite everything in terms of these surface concentrations, or claim that the surface concentrations are in equilibrium with c_{1i} and c_{2i} in the bulk. Because none of this affects the form of the final result, the argument is tangential.

The second interesting idea comes from comparing Eq. 16.3-3 with Eq. 16.3-1. In Eq. 16.3-3, the forward and reverse reaction rate constants are different. These differences lead to an equilibrium constant for the chemical reaction:

$$K_2 = \frac{\kappa_2}{\kappa_{-2}} \qquad (16.3\text{-}4)$$

In contrast, in Eq. 16.3-1, the rate constants of the forward and reverse steps are the same. As a result, the equilibrium constant for the diffusion step 1 is

$$K_1 = \frac{k_1}{k_1} = 1 \qquad (16.3\text{-}5)$$

In other words, diffusion is like a heterogeneous reversible first-order reaction with an equilibrium constant of unity.

We now return to our objective, finding the overall reaction rate. We can measure the bulk concentrations c_1 and c_2, but not the surface concentrations c_{1i} and c_{2i}. Accordingly, we combine Eqs. 16.3-1 and 16.3-3 to eliminate these unknowns. This combination is a complete parallel to that in Section 8.5, simple but algebraically elaborate. The result is

$$r_2 = n_1 = [K]\left(c_1 - \frac{c_2}{K_2}\right) \qquad (16.3\text{-}6)$$

in which the overall rate constant K is given by

$$K = \frac{1}{1/k_1 + 1/\kappa_2 + 1/k_3 K_2} \qquad (16.3\text{-}7)$$

This is resistances in series again. It looks just like mass transfer across an interface; the heterogeneous reaction is just another step in the series.

Some parallels between heterogeneous reaction and interfacial mass transfer are obvious. Steps 1, 2, and 3 do occur in series, and resistances like $1/k_1$ and $1/\kappa_2$ do add to the total resistance $1/K$. Moreover, the quantity c_2/K_2 is chemically equivalent to the concentration of species 1 that would be in equilibrium with the existing bulk concentration of species 2. I find this chemical equivalence easier to grasp than the physical one of c_1^* and p_1^*, those elusive pseudo concentrations that characterize interfacial transport.

Another parallel between heterogeneous reaction and interfacial transport is more subtle. The reaction equilibrium constant K_2 is roughly parallel to the Henry's law coefficient H that characterizes phase changes. Both K_2 and H vary widely, easily covering a range of 10^6 or more for different systems. Thus it is the equilibrium constant K_2 that determines the relative impact of k_1 and k_3. The implications of this are best seen from the examples that follow.

Example 16.3-1: Limits of a first-order heterogeneous reaction   What is the overall rate for a first-order heterogeneous reaction under each of the conditions: (a) fast stirring, (b) high temperature, and (c) an irreversible reaction? Express this rate as r_2.

Solution   Each of these cases is a limit of Eqs. 16.3-6 and 16.3-7. For rapid stirring (a), k_1 and k_3 become very large. Thus

$$r_2 = [\kappa_2]\left(c_1 - \frac{c_2}{K_2}\right)$$

In this case, physics is unimportant and chemistry is omnipotent. For the case of high temperature (b), κ_2 and κ_{-2} become much larger than k_1 and k_3. Thus

$$r_2 = \left[\frac{1}{1/k_1 + 1/k_3K_2}\right]\left(c_1 - \frac{c_2}{K_2}\right)$$

The effect of the reaction is still very much there, but as the equilibrium constant. Finally, for an irreversible reaction (c), K_2 becomes infinite, and

$$r_2 = \left[\frac{1}{1/k_1 + 1/\kappa_2}\right]c_1$$

Only in this case are the resistances so simply additive. In other cases, these resistances are weighted with equilibrium constants.

Example 16.3-2: The rate of ferrocyanide oxidation   We are studying electrochemical kinetics using a flat platinum electrode 0.3 cm long immersed in a flowing aqueous solution. In one series of experiments, the solution is 1-M KCl containing traces of potassium ferrocyanide. The ferrocyanide is reduced by means of the reaction

$$Fe(CN)_6^{4-} \rightarrow Fe(CN)_6^{3-} + e^-$$

When the solution is flowing at 70 cm/sec and the potential is at some fixed value, the overall mass transfer coefficient is 0.0087 cm/sec. Estimate the rate of constant of this reaction, assuming that the solution has the properties of water.

Solution   The overall mass transfer coefficient K found experimentally is

$$K = \frac{1}{1/k_1 + 1/\kappa_2}$$

We want to find κ_2, so must calculate k_1. We can do this from Eq. 9.4-53 and the diffusion coefficient in Table 6.1-1:

$$k_1 = 0.646 \left(\frac{D}{L}\right) \left(\frac{Lv^0}{\nu}\right)^{1/2} \left(\frac{\nu}{D}\right)^{1/3}$$

$$= 0.646 \left(\frac{0.98 \cdot 10^{-5} \text{cm}^2/\text{sec}}{0.3 \text{ cm}}\right) \left(\frac{(0.3 \text{ cm})(70 \text{ cm/sec})}{10^{-2} \text{cm}^2/\text{sec}}\right)^{1/2}$$

$$\left(\frac{10^{-2} \text{cm}^2/\text{sec}}{0.98 \cdot 10^{-5} \text{cm}^2/\text{sec}}\right)^{1/3} = 0.97 \cdot 10^{-2} \text{cm/sec}$$

Inserting this value and that for K into the preceding equation, we find

$$0.0087 \text{ cm/sec} = \frac{1}{\dfrac{\text{sec}}{0.0097 \text{ cm}} + \dfrac{1}{\kappa_2}}$$

$$\kappa_2 = 0.08 \text{ cm/sec}$$

Obviously, other values will be found at other potentials.

This type of experiment can be used to give reliable values of the mass transfer coefficient, but not the rate constant. The reason is that the electrode surface is usually contaminated in some fashion. Instead, electrochemists commonly use the more reliable method of cyclic voltametry, in which the potential is not held constant, but cycled sinusoidally. The results are complex, but they give considerable qualitative information about the chemistry involved.

Example 16.3-3: Cholesterol solubilization in bile Bile is the body's detergent, responsible for solubilization of water-insoluble materials. It is the key to fat digestion and the principal route of cholesterol excretion. Indeed, the failure of bile to effect excretion of available cholesterol is implicated in the formation of cholesterol gallstones.

Pharmacological experiments have shown that gallstones can be dissolved without surgery by feeding patients specific components of bile. These experiments have sparked the study of the dissolution rates of these gallstones. These rates are conveniently studied with a spinning disc of radioactively tagged cholesterol, like that described in Example 3.4-3. In one experiment with such a disc, the cholesterol dissolution rate was found to be $5.37 \cdot 10^{-9}$ g/cm^2 sec in a solution containing 5 wt% sodium taurodeoxycholate, a 4 : 1 molar ratio of this bile salt to lecithin, and 0.15-M NaCl. The solubility of cholesterol in this solution is $1.48 \cdot 10^{-3}$ g/cm^3, and the diffusion coefficient is about $2 \cdot 10^{-6}$ cm^2/sec. The disc was 1.59 cm in diameter, spinning rapidly with a Reynolds number of 11,200. The kinematic viscosity of this model bile is about 0.036 cm^2/sec, and the density of a cholesterol-saturated solution is $1.0 \cdot 10^{-5}$ g/cm^3 greater than bile containing no cholesterol.

Find the rate of the surface reaction, assuming that this reaction is irreversible. Then find the dissolution rate for a 1-cm gallstone in unstirred bile.

Solution As in the previous example, we first need to find the mass transfer coefficient in this solution. From Table 8.3-3,

$$k_1 = 0.62 \frac{D}{d} \left(\frac{d^2 \omega}{\nu} \right)^{1/2} \left(\frac{\nu}{D} \right)^{1/3}$$

$$= 0.62 \frac{2 \cdot 10^{-6} \text{cm}^2/\text{sec}}{1.59 \text{ cm}} (11,200)^{1/2} \left(\frac{0.036 \text{ cm}^2/\text{sec}}{2 \cdot 10^{-6} \text{ cm}^2/\text{sec}} \right)^{1/3}$$

$$= 2.16 \cdot 10^{-3} \text{cm}/\text{sec}$$

The overall rate constant is the sum of this resistance and that of the surface reaction:

$$K = \frac{1}{1/k_1 + 1/\kappa_2}$$

$$\frac{5.37 \cdot 10^{-9} \text{g}/\text{cm}^2 \text{ sec}}{1.48 \cdot 10^{-3} \text{g}/\text{cm}^3} = \frac{1}{\dfrac{\text{sec}}{2.16 \cdot 10^{-3} \text{cm}} + \dfrac{1}{\kappa_2}}$$

Thus,

$$\kappa_2 = 3.6 \cdot 10^{-6} \text{cm}/\text{sec}$$

Note that the experimentally measured rate is dominated by the surface reaction.

We now want to use this surface rate constant to find the dissolution rate if a spherical stone is immersed in this unstirred model bile. This bile is probably affected by free convection. Thus, from Table 8.3-3,

$$\frac{k_1 d}{D} = 2 + 0.6 \left(\frac{d^3 \Delta \rho g}{\rho \nu^2} \right)^{1/4} \left(\frac{\nu}{D} \right)^{1/3}$$

$$\frac{k_1(1 \text{ cm})}{2 \cdot 10^{-6} \text{cm}^2 \text{ sec}} = 2 + 0.6 \left(\frac{(1 \text{ cm})^3 (1.0 \cdot 10^{-5} \text{g}/\text{cm}^3)(980 \text{ cm}/\text{sec})}{(1 \text{ g}/\text{cm}^3)(0.036 \text{ cm}^2/\text{sec})^2} \right)^{1/4}$$

$$\left(\frac{0.036 \text{ cm}^2/\text{sec}}{2 \cdot 10^{-6} \text{ cm}^2/\text{sec}} \right)^{1/3}$$

$$k_1 = 5.6 \cdot 10^{-5} \text{cm}/\text{sec}$$

We then find the overall rate as before:

$$K = \frac{1}{\dfrac{\text{sec}}{5.6 \cdot 10^{-5} \text{cm}} + \dfrac{\text{sec}}{3.6 \cdot 10^{-6} \text{cm}}}$$

$$= 3.4 \cdot 10^{-6} \text{ cm}/\text{sec}$$

In unstirred bile, the rate is also controlled by reaction.

16.4 Finding the Mechanism of Irreversible Heterogeneous Reactions

The simple ideas of the previous section are the benchmark of this entire chapter, the standard against which more complex concepts are compared. Our basic strategy has been to assume a simple mechanism and calculate the overall reaction rate for this situation. The result is a close parallel to the results for interfacial mass transfer.

However, in some ways, the strategy of the previous section is misleading, for it implies that the mechanism is known. This is often not the case. In many situations, we already have experimental results, and we want to find which mechanisms are consistent with these results. In other words, the arguments of the previous section are backwards.

In this section, we want to explore how the mechanism of an irreversible reaction can be inferred from the overall reaction rate, instead of the other way around. This exploration can be complicated, hampered by elaborate algebra. As a result, we consider only special cases of the two types of heterogeneous reactions shown in Fig. 16.4-1. These types differ in the products produced by the reaction. In some cases, these products are fluid, and hence diffuse away. More commonly, the products form a layer of ash around an unreacted core; this second case is sometimes called a topochemical model.

To reduce algebraic complexity, we consider only limits in which the overall rate is controlled by a single diffusion or reaction step. The five limits we consider are tabulated in Table 16.4-1. In cases A and C, we assume that the surface reaction

$$\begin{pmatrix} \text{gaseous} \\ \text{species 1} \end{pmatrix} + \begin{pmatrix} \text{solid} \\ \text{species 2} \end{pmatrix} \rightarrow \begin{pmatrix} \text{various} \\ \text{products} \end{pmatrix} \tag{16.4-1}$$

is described by the rate equation

$$r_2 = -\kappa_2 c_2 c_1 \tag{16.4-2}$$

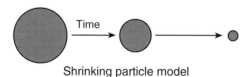

Shrinking particle model

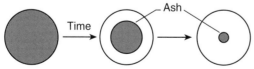

Shrinking core or "topochemical" model

Fig. 16.4-1. Two basic models for heterogeneous reaction. Many solid–gas non-catalytic reactions follow one of these two limiting models. Note that "ash" can be any solid product. Some characteristics of these models are given in Table 16.4-1.

Table 16.4-1 Common types of heterogeneous reactions

	Physical situation	Rate-controlling step	Size $R = f$ (time, reagent)	Size $= f$ (temperature)	Size $= f$ (flow)	Remarks
A	Shrinking particle	Reaction	$R \propto c_1 t$	Strong temperature variation	Independent of flow	Other reaction stoichiometries can be found easily
B	Shrinking particle	External diffusion	$R^2 \propto (c_1 t)$ (small particles) $r^{3/2} \propto (c_1 t)$ (larger particles)	Weak temperature variation	Independent for small particles only	The exact variation with flow depends on the mass transfer coefficient
C	Shrinking core[a]	Reaction	$R \propto c_1 t$	Strong	Independent of flow	This is the same as case A, except for ash formation
D	Shrinking core[a]	External diffusion	$R \propto c_1 t$	Weak	Usually about square root of flow	This case is uncommon
E	Shrinking core[a]	Ash diffusion	$R \propto (c_1 t)^{1/2}$	Weak	Independent of flow	This case is common, an interesting contrast with the previous one

Notes: [a]This is often called the topochemical model. The size R refers to the cone.

Note the concentration of the solid, c_2, is essentially constant, so that the reaction behaves as if it were first order. In such an equation, the rate constant κ_2 has dimensions of $L^4/\text{mol } t$. We now can write a mass balance on one particle of radius R:

$$\frac{d}{dt}\left(\frac{4}{3}\pi R^3 c_2\right) = 4\pi R^2 r_2 \qquad (16.4\text{-}3)$$

Combining the previous two equations,

$$\frac{d}{dt}R = -\kappa_2 c_1 \qquad (16.4\text{-}4)$$

This equation, subject to the conditions that c_1 is constant and r is initially R_0, is easily integrated:

$$R = R_0 - \kappa_2 c_1 t \qquad (16.4\text{-}5)$$

The particle size is proportional to time.

The other three cases are more interesting. When diffusion outside of a shrinking particle is rate controlling (case B), the key variable becomes the mass transfer coefficient. This coefficient is a function of particle size; for example, for a single particle, it is often assumed to be (see Table 8.3-3)

$$\frac{kd}{D} = 2.0 + 0.6\left(\frac{dv}{\nu}\right)^{1/2}\left(\frac{\nu}{D}\right)^{1/3} \qquad (16.4\text{-}6)$$

where d is the particle diameter, D is the diffusion coefficient of the reacting gas, v is the fluid's velocity, and ν is the fluid's kinematic viscosity. For very small particles, this implies

$$k = \frac{D}{R} \qquad (16.4\text{-}7)$$

a relation derived in Section 2.4. The mass balance on a single particle is now

$$\frac{d}{dt}\left(\frac{4}{3}\pi R^3 c_2\right) = -(4\pi R^2)\frac{Dc_1}{R} \qquad (16.4\text{-}8)$$

Again, this can be integrated, with the condition that r is initially R_0, to give

$$R^2 = R_0^2 - \left(\frac{2Dc_1}{c_2}\right)t \qquad (16.4\text{-}9)$$

For larger particles, the mass transfer coefficient is

$$k = \left(\frac{0.42 v^{1/2} D^{2/3}}{\nu^{1/6}}\right)\left(\frac{1}{R^{1/2}}\right) \qquad (16.4\text{-}10)$$

and the resulting variation of R is

$$R^{3/2} = R_0^{3/2} - \left(\frac{0.64v^{1/2}D^{2/3}}{v^{1/6}}\right)\left(\frac{c_1}{c_2}\right)t \qquad (16.4\text{-}11)$$

Equations 16.4-9 and 16.4-11 are both for a diffusion-controlled reaction on the surface of a shrinking particle.

The results are different for a particle of constant size with a shrinking core of unreacted material. In such a topochemical model, there are two diffusional resistances in series. First, material must diffuse from the bulk to the particle's surface; second, it must diffuse from the surface through ash to the unreacted core.

When diffusion in the bulk controls (case D), the reaction rate is again determined by the mass transfer coefficients around the particle. Because the particle size is constant, this mass transfer coefficient is also constant. The mass balance is still

$$\frac{d}{dt}\left(\frac{4}{3}\pi R^3 c_2\right) = -(4\pi R^2)kc_1 \qquad (16.4\text{-}12)$$

This can be easily integrated using the same initial size of particle R_0:

$$R = R_0 - \left(k\frac{c_1}{c_2}\right)t \qquad (16.4\text{-}13)$$

Although this result has the same variation with time as do the cases where surface reaction controls, it shows a square-root dependence on flow. It also shows a smaller variation with temperature, for mass transfer coefficients vary much less with temperature than reaction-rate constants.

When ash diffusion controls (case E), the mass transfer coefficient depends on the thickness of the ash layer. The usual assumption is that this coefficient is

$$k = \frac{D}{R_0 - R} \qquad (16.4\text{-}14)$$

in which D is now an effective value for diffusion through the ash. This assumption is tricky, for it implies that diffusion across the ash layer is a steady-state process. At the same time, we are assuming that the particle size varies with time. These assumptions imply that the diffusion through the ash is much faster than the combustion of the entire particle.

The mass balance on the particle now becomes

$$\frac{d}{dt}\left(\frac{4}{3}\pi R^3 c_2\right) = -(4\pi R^2)\frac{Dc_1}{R_0 - R} \qquad (16.4\text{-}15)$$

Integrating this result yields, after some rearrangement,

$$R = R_0 - \left(\frac{2Dc_1}{c_2}t\right)^{1/2} \qquad (16.4\text{-}16)$$

The particle size now varies with the square root of time. The differences between these cases and others in this section are considered in the example that follows.

Example 16.4-1: Mechanisms of coal gasification with steam   The gasification of coal particles using steam is being studied in a batch fluidized-bed reactor. In this reactor, the steam concentration is held constant, and the average particle size is monitored versus time. A plot of the logarithm of this size versus time has a slope of 0.6; the slope does show some variation with the gas flow used to maintain fluidization. What is the rate-controlling step for this gasification?

   Solution   The variation of reaction rate with flow indicates that the process is controlled by diffusion outside of particle. By inference, the resistance of any ash formed is apparently negligible. At the same time, the variation of particle radius with time is in the range suggested by case B in Table 16.4-1; checking this point further requires using a mass transfer correlation in fluid beds parallel to Eq. 16.4-6. Although no dependence on temperature is mentioned, we would expect this dependence to be small.

16.5 Heterogeneous Reactions of Unusual Stoichiometries

   In the previous sections we discussed how the overall rate of reaction was affected by the rates of diffusion and reaction. The rate of diffusion could be altered by changes in factors like fluid flow or diffusion coefficient or ash thickness. The overall rate of reaction was always assumed to be first order, always doubling when the reagent concentration was doubled.

   In this section we want to consider two examples of other chemistries that can alter the simple combinations of diffusion and reaction developed earlier. The first example is an irreversible second-order reaction. The second involves fast reactions of concentrated reagents and products.

16.5.1 A Second-Order Heterogeneous Reaction

   The case considered here, shown schematically in Fig. 16.5-1, involves two sequential steps. The first of these steps is simple mass transfer, but the second step is a second-order irreversible chemical reaction. As before, we want to calculate the overall rate r_1 of this heterogeneous reaction. In mathematical terms, this rate is

$$r_1 = n_1 = k_1(c_1 - c_{1i}) \tag{16.5-1}$$

where c_1 and c_{1i} are again the bulk and interfacial concentrations, respectively. This rate is also

$$r_1 = \kappa_2 c_{1i}^2 \tag{16.5-2}$$

We can combine these two equations to find the unknown interfacial concentration:

$$c_{1i} = \frac{k_1}{2\kappa_2}\left(\sqrt{1 + 4\kappa_2 c_1/k_1} - 1\right) \tag{16.5-3}$$

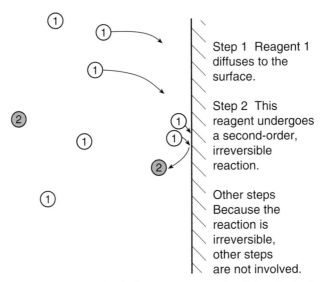

Fig. 16.5-1. A second-order heterogeneous reaction. When the simple stoichiometry used earlier in this chapter is not followed, the overall rate no longer depends on a simple sum of the resistances of the various steps.

This can, in turn, be used to find the overall reaction rate:

$$r_1 = k_1 c_1 \left[1 - \frac{k_1}{2\kappa_2 c_1} \left(\sqrt{1 + 4\kappa_2 c_1 / k_1} - 1 \right) \right] \qquad (16.5\text{-}4)$$

This expression is obviously very different from the corresponding result for a first-order reaction, given by Eq. 16.3-6.

The conclusion drawn from this result is that resistances in series are no longer additive. This is true whenever any of the resistances is not first order. In electrical or thermal systems, resistances are almost always first order. However, in chemical systems, resistances will not be first order when there are non-first-order chemical reactions. This occurs frequently.

16.5.2 Heterogeneous Reactions in Concentrated Solutions

In the simple cases in Sections 16.3 and 16.4, we assumed that the mass transfer coefficient k_1 was independent of the reaction rate. This is actually an implicit approximation, valid only in dilute solution or for reagents producing one mole of product for every mole of reagent. To see where this approximation might be inaccurate, consider the two solid–gas reactions shown in Fig. 16.5.-2. In the first, a reagent is split into many smaller parts:

$$\text{species 1} \xrightarrow{\kappa_2} \nu\,(\text{species 2}) \qquad (16.5\text{-}5)$$

where ν is a stoichiometric coefficient. In the second, the converse occurs:

$$\nu\,(\text{species 1}) \xrightarrow{\kappa_2} \text{species 2} \qquad (16.5\text{-}6)$$

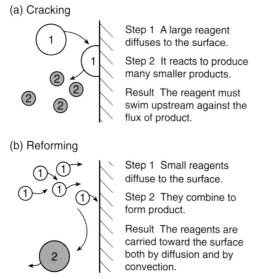

(a) Cracking

Step 1 A large reagent diffuses to the surface.

Step 2 It reacts to produce many smaller products.

Result The reagent must swim upstream against the flux of product.

(b) Reforming

Step 1 Small reagents diffuse to the surface.

Step 2 They combine to form product.

Result The reagents are carried toward the surface both by diffusion and by convection.

Fig. 16.5-2. Diffusion-induced convection can alter heterogeneous reaction rates. Because cracking increases the number of moles in the system, reagents must diffuse against a convective flow out of the surface, as shown in (a). Because reforming decreases the number of moles, reagents are swept towards the surface, as shown in (b).

The first of these is an idealization of a cracking reaction, and the second of a reforming reaction.

The difficulty in both cases is that the mass transfer step must include both diffusion and convection. In the case of the cracking reaction, this convection is away from the surface, so that the reacting species must diffuse against the current, swimming upstream to reach the reactive surface. For the reforming reaction, the opposite is true; the reacting species are buoyed along, swept toward the surface by the reaction.

To estimate the size of the effect, we idealize the region near the reactive surface as a thin stagnant gas film of thickness l, as in Section 9.5. At the outside of this film, located at $z = 0$, the reagent concentration is the bulk value, and the product concentration is zero. At the solid surface, at $z = l$, the reaction occurs. The overall rate of reaction r_1 across this film is

$$r_1 = n_1 = -\frac{n_2}{\nu} = \kappa_2 c_{1i} \tag{16.5-7}$$

If the stoichiometric coefficient ν is greater than unity, we have the cracking reaction; if ν is less than unity, we have the reforming reaction; if ν equals unity, we have the simple case discussed in Section 16.3.

We now must calculate the flux n_1. In dilute solutions, this hinged on Eq. 16.3-1; but in these concentrated solutions, we must return to the continuity equation

$$0 = -\frac{dn_1}{dz} \tag{16.5-8}$$

Integrating and combining with Fick's law,

$$n_1 = -D\frac{dc_1}{dz} + c_1 v^0 \tag{16.5-9}$$

where v^0 is the volume average velocity, which in this case is the convection caused by the reaction. For a gas,

$$cv^0 = n_1 + n_2 = (1 - v)n_1 \tag{16.5-10}$$

where c is the total concentration. Inserting this result into the previous equation and integrating to find n_1, we have

$$r_1 = n_1 = -\frac{k_1 c}{(v-1)} \ln\left(\frac{1 + (v-1)n_1/\kappa_2 c}{1 + (v-1)c_{10}/c}\right) \tag{16.5-11}$$

in which $k_1 \, (= D/l)$ is the mass transfer coefficient and c_{10} is the concentration of reagent in the bulk. The rate relative to that in dilute solution is shown for a variety of stoichiometries in Fig. 16.5-3 for the limit of diffusion control (i.e., $\kappa_2 c_{10}/n_1$ is large).

Example 16.5-1: Limiting behavior of a second-order heterogeneous reaction   Describe what happens to Eq. 16.5-4 if the dimensionless group $\kappa_2 c_1/k_1$ is either very large or very small.
   Solution   When $\kappa_2 c_1/k_1$ is very large, the reaction will become diffusion-controlled. Under these conditions,

$$\lim_{\kappa_2 c_1/k_1 \to \infty} r_1 = k_1 c_1 \left\{ 1 - \frac{k_1}{2\kappa_2 c_1} \sqrt{4\kappa_2 c_1/k_1} \right\} = k_1 c_1$$

On the other hand, when $\kappa_2 c_1/k_1$ is small, the reaction will be unaffected by diffusion:

$$\lim_{\kappa_2 c_1/k_1 \to 0} r_1 = k_1 c_1 \left\{ 1 - \frac{k_1}{2\kappa_2 c_1} \left[1 + \frac{1}{2}\left(\frac{4\kappa_2 c_1}{k_1}\right) - \frac{1}{8}\left(\frac{4\kappa_2 c_1}{k_1}\right)^2 \right. \right.$$

$$\left. \left. + \frac{1}{16}\left(\frac{4\kappa_2 c_1}{k_1}\right)^3 - \cdots - 1 \right] \right\}$$

$$= \kappa_2 c_1^2 \left(1 - \frac{2\kappa_2 c_1}{k_1} + \cdots \right)$$

In this case, the chemical reaction controls the overall rate.

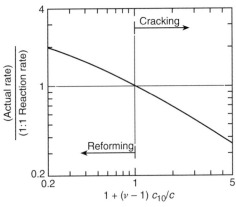

Fig. 16.5-3. Effect of diffusion-induced convection. This graph gives the change in reaction rate for a diffusion-controlled reaction producing v moles of product per mole of reagent. The product concentration in the bulk is zero; that of the reagent is c_{10}.

Example 16.5-2: Thermal cracking of gas oil Thermal cracking of a gas oil is being studied on a hot plate immersed in a rapidly flowing gas stream. The plate is so hot that the reaction is essentially diffusion-controlled. The molecular weight of the product is only 23% of that of the reagent. By how much will convection introduced by cracking change the reaction rate?

Solution In this case, the change in molecular weight implies that

1 molecule of gas oil → 1/0.23 molecules of smaller size

Moreover, because the gas oil is undiluted, c_{10}/c is unity. Thus, from Fig. 16.5-3, the actual rate will be about 45% of that expected from reactions or mass transfer coefficients measured in dilute solution.

16.6 Conclusions

This chapter has two principal parts. In the first part (Sections 16.1 and 16.2), the focus is on the modeling of systems containing reaction and diffusion. One question concerns whether a reaction is heterogeneous or homogeneous; the answer depends more on the physical geometry involved and less on the chemistry at a molecular level. A second question concerns what diffusion-controlled reactions are; the answer is that they are reactions in which the time for diffusion is more than that for chemical change. However, how these times are combined depends on the specific situation involved.

The second part of the chapter (Sections 16.3 through 16.5) is concerned with heterogeneous reactions. The key point is that the overall rate frequently varies with a sum of resistances in series. The results are similar to those involved in interfacial mass transport, but with chemical equilibrium constants replacing the Henry's law constants. Although this simplicity can be compromised by unusual stoichiometry or by concentrated solutions, the analogy with interfacial mass transfer is useful and worth remembering.

Questions for Discussion

1. What is a heterogeneous reaction?
2. What is a homogeneous reaction?
3. If the mass balance used in a research paper is,

$$0 = \frac{D}{r} \frac{\partial}{\partial r} \left(r \frac{\partial c_1}{\partial r} \right) - r_1$$

    is the reaction being treated as heterogeneous or homogeneous?
4. Acid–base reactions are said to be "diffusion controlled." What does this mean?
5. The oxidation of ammonia in excess air over a silver wire catalyst makes NO_x, the precursor to nitric acid. This reaction is described as "irreversible" and "diffusion controlled." What does this mean?
6. Suggest electrical circuits which are analogous to each of the four cases in Fig. 16.2-1.

7. For a heterogeneous reaction influenced by mass transfer, how will the rate change as the temperature is raised?

8. For a first-order heterogeneous chemical reaction, will the rate-controlling step change as the reagent concentration increases?

9. For a second-order heterogeneous chemical reaction, will the rate-controlling step change as the reagent concentration increases?

10. Heterogeneous reactions within the surface of porous catalysts are modeled as if they are homogeneous. How is this reflected in the analysis?

11. Compare the effect of the reaction equilibrium constant for a heterogeneous reaction with the partition coefficient involved in interfacial transport.

12. Digestion in humans is enhanced by villi, small protrusions on the intestinal wall which increase the intestine's area 30 times. Discuss why villi have evolved.

Problems

1. The solubilization rates of ^{14}C-tagged linoleic acid can be measured in 1% sodium taurodeoxycholate using a spinning liquid disc, for which

$$\frac{kd}{D} = 0.62 \left(\frac{d^2\omega}{\nu}\right)^{1/2} \left(\frac{\nu}{D}\right)^{1/3}$$

where k is the mass transfer coefficient, D is the diffusion coefficient, d and ω are the disc diameter and rotation speed, and ν is the kinematic viscosity. These data can be explained as a heterogeneous reaction followed by mass transfer. The solubility of linoleic acid in this solution is $2.23 \cdot 10^{-3}$ g/cm^3. Find the rate constant of the heterogeneous chemical reaction. *Answer:* $7 \cdot 10^{-4}$ cm/sec.

2. The oxidation

$$Ce^{3+} \rightarrow Ce^{4+} + e^-$$

has a rate constant of $4 \cdot 10^{-4}$ cm/sec when effected on platinum in 1-M H_2SO_4. You carry out this reaction by suddenly applying a potential across a large stagnant volume of this solution. Estimate how long you can reliably measure the kinetics before diffusion becomes important. *Answer:* 20 sec.

3. As part of a study of electrochemical kinetics, you insert a gold electrode into a solution at 25 °C containing 0.5-M H_2SO_4 and small amounts of ferrous ion. You then apply a potential between this electrode and a second, reversible electrode, so that at the gold the iron is oxidized:

$$Fe^{2+} \rightarrow Fe^{3+} + e^-$$

You measure the current density i (amp/cm^2) under a fixed potential and find that, to a first approximation,

$$\frac{c_1}{i} = 2 + 50t^{1/2}$$

where c_1 is the ferrous concentration in moles per liter and t is the time in minutes. Find the rate constant of the surface reaction and the diffusion coefficient of the ferrous ion. *Answer:* $D = 0.81 \cdot 10^{-5}$ cm^2/sec.

4. A single potassium chloride crystal about 0.063 cm in diameter, which is immersed in a 5.2% supersaturated solution containing about 25 wt% potassium chloride, is growing at a rate of 0.0013 cm/min. If the system is well mixed, this growth is second order, presumably because both potassium and chloride ions are involved. In our case, the solution may not be well mixed; it flows past the crystal at 6 cm/sec. The solution's viscosity is about 1.05 cp; its density is 1.2 g/cm^3, and the crystal's density is 1.984 g/cm^3. Does diffusion influence the rate of crystal growth? *Answer:* Diffusion is about 25% of the total resistance.

5. When copper and silicon are placed together, a layer of Cu_3Si grows at the interface. W. J. Ward and K. M. Carroll [*J. Electrochem. Soc.*, **129**, 227 (1982)] reported that the thickness of this layer l at 350 °C is

$$l = (1.4 \cdot 10^{-4} cm/ sec^{1/2})t^{1/2}$$

They argued that the layer forms by reaction of Cu at the Cu_3Si–Si interface. They also maintained that this reaction is controlled by diffusion of Cu through Cu_3Si. (a) Show that the variation of thickness versus time is consistent with a diffusion mechanism. (b) Discuss what reaction stoichiometry would be required to produce this same variation. (c) Calculate the diffusion coefficient and compare it with other values for diffusion in solids. In this experiment, the driving force of Cu_3Si is believed to be 1.10 mol% Cu. *Answer:* $2 \cdot 10^{-6}$ cm^2/sec.

6. Diffusion out of the intestinal lumen may be governed by two resistances in series [K. W. Smithson, D. B. Millar, L. R. Jacobs, and G. M. Gray, *Science*, **214**, 1241 (1981)]. The first is mass transfer in the lumen itself, which is described by $j_1 = k(c_1 - c_{1i})$, where k is the mass transfer coefficient. The second is mass transfer across the intestinal wall, which in this case is governed by a rate equation:

$$j_1 = \frac{\nu_{max} c_{1i}}{K_m + c_{1i}}$$

where ν_{max} and K_m are parameters measured in well-mixed experiments unaffected by mass transfer. (a) To avoid complexities, these authors assumed that

$$j_1 = \frac{\nu_{max} c_1}{K_a + c_1}$$

where K_a is found from the slope of a plot of $1/j_1$ versus $1/c_1$. Discuss the approximations in this plot. (b) These authors then found the mass transfer coefficient k from "Winne's equation":

$$k = \frac{0.5\nu_{max}}{K_a - K_m}$$

Justify this equation and comment on its accuracy.

7. Petrochemical processing involves two major types of chemical reactions: cracking and reforming. In cracking, large hydrocarbon molecules of perhaps 20 carbons are broken ("cracked") into much smaller ones, with about 4 carbons. In reforming, these are put together again to make C_8's, the standard for gasoline. Imagine that you are studying a very active solid catalyst for reforming. This catalyst is so active that the reaction is both fast and irreversible. Thus C_4's diffuse from an almost pure stream of C_4's of total concentration c to the solid surface, where they react instantaneously.

The products – C_8's – diffuse away. This diffusion takes place across a film of thickness l between the bulk and the surface, which always has the same value. (a) What is the total flux of C_4's towards the surface? (b) What would it be if the system were diluted a lot with nitrogen?

8.   Part of the manufacture of a long-lived light bulb consists of the high-temperature system shown below. Barium oxide is steadily produced by a zero-order reaction in the ceramic. It then diffuses both out into the sink and through the interface to the support. At the support, it reacts rapidly and irreversibly at the support's surface. There is no reaction in the interface, and the diffusion coefficient there is different than that in the ceramic. What is the flux of barium oxide at the support?

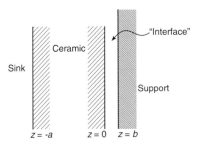

9.   While he was practicing for his oral qualifying exam, Ranil Wickramasinghe was asked by a classmate about interfacial resistances to mass transfer. Such resistances are rare, usually due to a molecularly compact interfacial film. Transfer across such an interface is like a reversible heterogeneous first-order reaction. For example, for gas absorption, it is

$$\begin{bmatrix} \text{solute in gas} \\ \text{at interface} \end{bmatrix} \underset{\kappa_{-2}}{\overset{\kappa_{2}}{\rightleftharpoons}} \begin{bmatrix} \text{solute in liquid} \\ \text{at interface} \end{bmatrix}$$

In addition, there is resistance to mass transfer in the gas $(1/k_1)$ and in the liquid $(1/k_3)$. What is the mass transfer for such a process?

10.   Obsidian is a volcanic glass used by primitive peoples for arrowheads, knife blades, and the like. The depth of water penetration into artifacts made of obsidian can be measured by cleaving the object and examining its surface under a microscope. Because the hydrate has a greater specific volume, there is a stress crack at the hydrated–unhydrated interface. This depth of the hydrate is a measure of the age of the artifact; so obsidian is sometimes called "the dating stone." Most investigators [I. Friedman and F. W. Trembour, *Amer. Sci.*, **66**, 44 (1978)] have reported that the depth of hydration is proportional to the square root of time. Show that this is consistent with diffusion of water through the hydrate followed by very fast heterogenous chemical reaction at the hydrated–unhydrated interface. In this, assume that the humidity outside the obsidian is constant but that at the interface it is essentially zero.

11.   Ancient air is trapped in bubbles deep within polar ice. The bubbles get fewer and smaller as the depth of ice gets deeper. These decreases are the result of the formation of an air hydrate phase. However, both bubbles and air hydrate coexist, even when only the hydrate phase is expected to form. P. B. Price [*Science*, **267**, 1802 (1995)] explains this by postulating a layer of air hydrate around each bubble. To form more

hydrate, water must diffuse through the hydrate to the inner surface of the bubble; the diffusion coefficient for this process is

$$D = 2100 \frac{\text{cm}^2}{\text{sec}} \exp^{-1.5 \, \cdot \, 10^{-19} \, \text{J} / k_B T}$$

As the water reaches the inner surface of the hydrate, it reacts to form more hydrate, so the resistance to water diffusion increases with time. Estimate the time to react all the air at $-46\,°C$ (the temperature of South Pole ice) in a 0.01-cm bubble of air at 1 atm. Repeat the estimate for a 0.10-cm bubble.

Further Reading

Aris, R. (1975). *The Mathematical Theory of Diffusion and Reaction in Permeable Catalysts.* Oxford: Clarendon Press.

Bard, A.J. and Faulkner, L.R. (2001). *Electrochemical Methods.* New York: Wiley.

Bradley, J.N. (1975). *Fast Reactions.* Oxford: Clarendon Press.

Fogler, H.S. (1998). *Elements of Chemical Reaction Engineering*, 3rd ed. New York: Wiley.

Laidler, K.J. (1997). *Chemical Kinetics*, 3rd ed. London: Prentice Hall.

Levenspiel, O. (1998). *Chemical Reaction Engineering*, 3rd ed. New York: Wiley.

Schmidt, L.D. (2004). *The Engineering of Chemical Reactions*, 2nd ed. London: Oxford.

Wright, M. (2004). *Introduction to Chemical Kinetics.* New York: Wiley.

CHAPTER 17

Homogeneous Chemical Reactions

Diffusion rates can be tremendously altered by chemical reactions. Indeed, these alterations are among the largest effects discussed in this book, routinely changing the mass fluxes by orders of magnitude. The effects of a chemical reaction depend on whether the reaction is homogeneous or heterogeneous. This question can be difficult to answer. In well-mixed systems, the reaction is heterogeneous if it takes place at an interface and homogeneous if it takes place in solution. In systems that are not well mixed, diffusion clouds this simple distinction, as detailed in Section 16.1.

The effects of chemical reactions are exemplified by the data for ammonia adsorption in water summarized in Fig. 17.0-1. The overall mass transfer coefficient, in cm/sec, is based on a liquid side driving force given in mol/cm^3. The specific values shown are for a hollow-fiber membrane contactor, though similar values would be obtained in a packed tower or other more conventional apparatus.

The different mass transfer coefficients shown in Fig. 17.0-1 represent different forms of ammonia and different rate-controlling steps for mass transfer. At pH above 5, the mass transfer coefficient is small, somewhat less than typical values for liquids. Below pH 4, the mass transfer coefficient rises. This rise, which is linear in the concentration of acid, occurs because the NH_3 that is transferred is being converted into NH_4^+. Below pH 1, the mass transfer coefficient again approaches an asymptote. This asymptote occurs because mass transfer in the liquid has been accelerated so much that the overall coefficient is now limited by diffusion in the gas.

The overall rate of a homogeneous reaction like that of ammonia is determined by a nonlinear combination of effects of diffusion and chemical reaction. The effects of such a reaction on the rates of mass transfer are analyzed in the first two sections of this chapter. In Section 17.1, we describe the simplest case, that of a first-order irreversible chemical reaction. We also summarize extensions of this case, extensions that produce significant gains only at the cost of major effort. In Section 17.2, we describe some results for second-order reactions. In Section 17.3, we apply these ideas to a specific case, that of H_2S scrubbing with amines.

The last two sections in this chapter are concerned with reactions commonly described as "fast" or "diffusion controlled." In Section 17.4, we discuss chemical reactions whose rates are controlled not by chemical kinetics but by Brownian motion of the reagents. These reactions are studied by suddenly changing the temperature or pressure and measuring the decay of the resulting perturbation. In Section 17.5, we investigate the speed of second-order reactions in turbulent flow. If these reactions are fast, their speed depends on mixing, not on chemistry. Thus their reaction rates are determined not by chemical kinetics but by the turbulent dispersion summarized in Chapter 4. These reactions can be described with mathematics like that for diffusion, and so are best treated here.

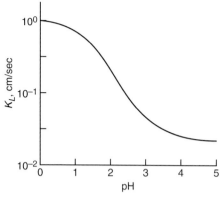

Fig. 17.0-1.   The overall mass transfer coefficient K_L for ammonia absorption into water. At high pH, the mass transfer is small, unaffected by ionization of ammonia. At lower pH, it is increased by instantaneous reaction until it is limited by the gas phase resistance.

17.1 Mass Transfer with First-Order Chemical Reactions

Chemical reaction increases the rate of interfacial mass transfer. The reaction reduces the reagent's local concentration, thus increasing its concentration gradient and its flux. Because chemical reaction rates can be very fast, the increase in mass transfer can be large.

In this section we want to calculate the increased mass transfer caused by a first-order, irreversible chemical reaction. This special case is the limit with which more elaborate calculations are compared. As a result, we shall go over the calculation in considerable detail so that its nuances are explicitly stated.

One might wonder why we make such a fuss over first-order reactions. After all, these reactions are uncommon. Real chemical reactions involve two reagents, like sodium hydroxide plus hydrochloric acid or methane plus oxygen. This focus on first-order reactions may seem a scientific ploy, emphasizing problems we can solve rather than problems that are important.

This skepticism has some justification, for there certainly are important reactions that are not first order. However, in many cases, all but one of the reagents will be present in excess; in stoichiometric terms, only one of the reagents is limiting. In this case, we can accurately approximate the reactions as first order. For the examples given earlier, we might have

$$\left(\begin{array}{c}\text{reaction}\\\text{rate}\end{array}\right) = [\kappa_1 c_{HCl}] c_{NaOH} \qquad (17.1\text{-}1)$$

for excess hydrogen chloride, or

$$\left(\begin{array}{c}\text{reaction}\\\text{rate}\end{array}\right) = [\kappa_1 c_{NaOH}] c_{HCl} \qquad (17.1\text{-}2)$$

for excess sodium hydroxide, or

$$\left(\begin{array}{c}\text{reaction}\\\text{rate}\end{array}\right) = [\kappa_1 c_{O_2}] c_{CH_4} \qquad (17.1\text{-}3)$$

for excess oxygen. In each case, the quantity in square brackets is a pseudo-first-order reaction-rate constant, with, of course, a different numerical value in each case. Under the circumstances given, each of these reactions can be treated as if it were first order.

We are interested in how a first-order chemical reaction alters the mass transfer in industrial equipment. For example, imagine that we are scrubbing ammonia out of air with water, using equipment like that shown in Fig. 10.2-1. To increase our equipment's capacity, we are considering adding small amounts of hydrogen chloride to the water. We want to predict the effect of this acid. However, the a-priori prediction of mass transfer in a scrubber is a tremendously difficult problem, requiring expensive numerical calculation.

As a result, we are much better off to use existing experimental correlations for mass transfer without reaction and to calculate a correction factor for the chemical reaction. Calculating this correction turns out to be easy for a first-order system. Moreover, we make good use of the 50 years of empirical correlations carefully obtained for industrial equipment. We next detail how this is achieved.

17.1.1 Irreversible Reactions

To calculate the correction to mass transfer due to reaction, we again adopt the simple film model shown in Fig. 17.1-1. In this model, a liquid is in contact with a well-mixed gas containing the material to be absorbed. The liquid is not well mixed. Near its surface, there is a thin film across which the absorbing species 1 is diffusing steadily. At the gas–liquid interface, this solute species is in equilibrium with the gas; at the other side of the film, its concentration is zero.

We can easily write a mass balance on this film. If there is no chemical reaction, this is

$$0 = -\frac{d}{dz}n_1 \doteq -\frac{dj_1}{dz} = D\frac{d^2c_1}{dz^2} \qquad (17.1\text{-}4)$$

This is subject to the boundary conditions

$$z = 0, \quad c_1 = c_{1i} \qquad (17.1\text{-}5)$$

$$z = l, \quad c_1 = 0 \qquad (17.1\text{-}6)$$

Integration and evaluation of the diffusion flux is easy, just as it was in Sections 2.2 and 9.1:

$$c_1 = c_{1i}\left(1 - \frac{z}{l}\right) \qquad (17.1\text{-}7)$$

as shown by the dotted line in Fig. 17.1-1. The flux is

$$j_1 = \frac{D}{l}(c_{1i} - 0) \qquad (17.1\text{-}8)$$

The mass transfer coefficient is the same old friendly value:

$$k^0 = \frac{D}{l} \qquad (17.1\text{-}9)$$

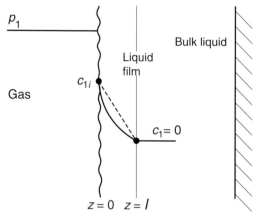

Fig. 17.1-1. Mass transfer with first-order chemical reaction. Species 1 is being absorbed from a gas into a liquid. If this species reacts in the liquid, its concentration profile changes from the dashed line to the solid line. Such a reaction increases the rate of mass transfer. The picture shown here implies the film theory.

where the superscript 0 indicates no chemical reaction. Note that we have again implicitly made the familiar assumption of dilute solution.

However, if there is a homogeneous, first-order chemical reaction, the mass balance becomes

$$0 = D\frac{d^2 c_1}{dz^2} - \kappa_1 c_1 \tag{17.1-10}$$

Integration of this equation gives

$$c_1 = a e^{\sqrt{\kappa_1/D}z} + b e^{-\sqrt{\kappa_1/D}z} \tag{17.1-11}$$

Evaluation of the integration constants a and b using the boundary conditions in Eqs. 17.1-5 and 17.1-6 gives

$$\frac{c_1}{c_{1i}} = \frac{\sinh\left[\sqrt{\kappa_1/D}(l-z)\right]}{\sinh\left[\sqrt{\kappa_1/Dl}\right]} \tag{17.1-12}$$

This concentration profile is curved like the solid line in Fig. 17.1-1. Note that

$$\lim_{\kappa_1 \to 0} \frac{c_1}{c_{1i}} = \frac{\sqrt{\kappa_1/D}(l-z) + \dots}{\sqrt{\kappa_1/D} + \dots}$$

$$= 1 - \frac{z}{l} \tag{17.1-13}$$

As the reaction gets slow, the concentration profile approaches the usual film result in Eq. 17.1-7.

The dimensionless group $(\kappa_1 l^2/D)$ which first appears in these equations is important, a key to many problems involving diffusion and chemical reaction. In physical terms, it can be rewritten as

$$\frac{\text{diffusion time}}{\text{reaction time}} = \frac{\kappa_1 l^2}{D} = \left[\frac{l^2/D}{1/\kappa_1}\right] \tag{17.1-14}$$

In discussions of mass transfer, this ratio is called the second Damköhler number and given the symbol Dm. In discussions of catalysis, it is called the square of the Thiele modulus and given the symbol ϕ^2. In either case, it is central to the description of coupled diffusion and homogeneous chemical reaction.

The flux in the presence of reaction is found by combining this concentration profile with Fick's law:

$$j_1 = -D\frac{dc_1}{dz}$$

$$= \sqrt{D\kappa_1}c_{1i}\left(\frac{\cosh\left[\sqrt{\kappa_1/D}(l-z)\right]}{\sinh\left[\sqrt{\kappa_1/D}l\right]}\right) \tag{17.1-15}$$

At the interface, where $z = 0$, this is

$$j_1 = \left[\sqrt{D\kappa_1}\coth\left(\sqrt{\frac{\kappa_1}{D}}l\right)\right]c_{1i} \tag{17.1-16}$$

Thus, in the case of chemical reaction, the mass transfer coefficient is

$$k = \sqrt{D\kappa_1}\coth\left(\sqrt{\frac{\kappa_1}{D}}l\right) = \sqrt{D\kappa_1}\coth\left(\sqrt{\frac{\kappa_1 D}{(k^0)^2}}\right) \tag{17.1-17}$$

This result reduces to Eq. 17.1-9 as κ_1 becomes small.

This important result is now used in two very different applications. First, it is used to calculate the *enhanced* mass transfer caused by the reaction. Second, it is used to calculate the *reduced* reaction which takes place in catalysis. I know it sounds strange that in one case the rate is said to be enhanced, and in the other it is felt to be reduced. As you will see, it is the same effect but with a different standard of comparison.

17.1.2 Mass Transfer: Enhancement Factors

We first consider the amount that the mass transfer is increased by chemical reaction. In other words, we are interested in the enhancement factor ε of the mass transfer with reaction k compared to that without k^0:

$$\varepsilon = \frac{k}{k^0} \tag{17.1-18}$$

By combining Eqs. 17.1-9, 17.1-18 and 17.1-17, we find

$$\varepsilon = \sqrt{\frac{D\kappa_1}{\left(k^0\right)^2}} \coth \sqrt{\frac{D\kappa_1}{\left(k^0\right)^2}} \tag{17.1-19}$$

Note that we can calculate this enhancement factor using information that we already have. The diffusion coefficient D and the reaction rate κ_1 will normally be known from experiment. The mass transfer coefficient without reaction k^0 can be estimated from correlations like those in Chapter 8.

Two limits of Eq. 17.1-19 are instructive. First, when the reaction is slow, κ_1 is small, ε is one, and

$$k = k^0 \tag{17.1-20}$$

Second, when the chemical reaction is fast, κ_1 is large, the hyperbolic cotangent equals one and

$$k = \sqrt{D\kappa_1} \tag{17.1-21}$$

The mass transfer coefficient now has nothing to do with k^0, but is simply the square root of the diffusion coefficient times the rate constant. I find this one of the most charming results in engineering.

Of course, these results did assume the film model. Because this model is an unsatisfying method for calculating mass transfer coefficients, we might expect it to be inaccurate. We can, with considerable effort, show that predicted corrections for chemical reaction are all nearly the same, independent of the specific model chosen. Some of these are shown in Fig. 17.1-2. Differences are minor, so the use of the film theory is justified.

This coupling between diffusion and reaction means that the mass transfer coefficient in a rapidly reacting system can vary sharply with temperature. If the reaction rate doubles every $10\,^\circ\mathrm{C}$, then the mass transfer will double every $20\,^\circ\mathrm{C}$. In contrast, doubling the mass transfer coefficient when no reaction is present usually requires increasing the temperature about $50\,^\circ\mathrm{C}$.

17.1.3 Catalysis: Effectiveness Factors

The second application of this result is the estimation of the extent that a catalytic reaction is compromised by diffusion. In other words, we are interested in the ratio of reaction with mass transfer to that without any limitations of mass transfer. This ratio is called an effectiveness factor η:

$$\eta = \frac{(\text{flux into catalyst})}{(\text{reaction per volume})}\frac{(\text{area }A)}{\text{volume}} = \frac{(j_1|_{z=0})\,A}{((\kappa_1 c_1))\,Al} \tag{17.1-22}$$

By inserting the flux found from Eq. 17.1-16, we find

$$\eta = \sqrt{\frac{D}{\kappa_1 l^2}} \coth \sqrt{\frac{\kappa_1 l^2}{D}} \tag{17.1-23}$$

This equation gives the reduction in reaction rate due to diffusion.

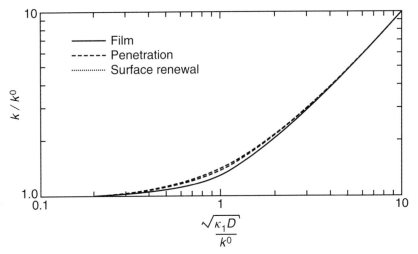

Fig. 17.1-2. Mass transfer corrected for first-order reaction. For slow reaction, the mass transfer coefficient is unchanged; for fast reaction, it equals $\sqrt{\kappa_1 D}$. Note that all theories give very similar results. (Redrawn from Sherwood *et al.*, 1975, with permission.)

Like the expressions for the enhancement factor, Eq. 17.1-22 for the effectiveness factor turns out to be more valuable than expected. It is derived for a first-order, irreversible chemical reaction occurring in a flat microporous catalyst pellet. However, making the catalyst pellet cylindrical or spherical doesn't matter much, as shown in Fig. 17.1-3. Changing the reaction order has a larger effect but even that is not that dramatic. In particular, for a reaction which is of order m in the reagent "1", the effectiveness factor does not change much, as shown in Fig. 17.1-4. Effectiveness factors are thus reliably estimated.

To me, the intriguing aspect is the difference between the enhancement factor and the effectiveness factor. While both are based on the same derivation, the enhancement

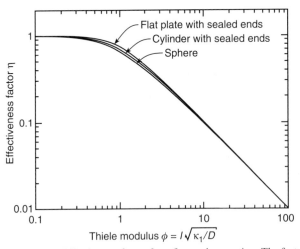

Fig. 17.1-3. Effectiveness factor for a first-order reaction. The factor gives the amount that the reactivity is compromised by diffusion within a porous catalyst particle. It varies little with particle shape. (Redrawn from Aris, 1975, with permission.)

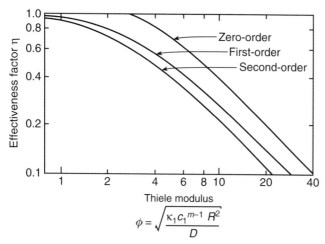

$$\phi = \sqrt{\frac{\kappa_1 c_1^{m-1} R^2}{D}}$$

Fig. 17.1-4. Effectiveness factor for different reaction orders. The Thiele modulus in this case varies with the surface concentration c_1. (Redrawn from Aris, 1975, with permission.)

factor increases and the effectiveness factor decreases as the reaction gets faster. The reason is not the way diffusion and reaction interact but the different standard of definition. The enhancement factor, calculated relative to diffusion without reaction, increases as the reaction rate increases. The effectiveness factor, calculated relative to reaction unlimited by diffusion, decreases as the reaction rate increases. The reason is not a different coupling between diffusion and reaction, but the different basis. These ideas are illustrated by the examples that follow.

Example 17.1-1: Mass transfer required for kinetic studies We are planning a series of experiments of the reactions of methyl iodide with pyridine and similar compounds:

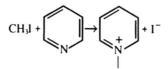

We are going to contact these reagents in a small laboratory reactor by bubbling methyl iodide vapor diluted with nitrogen through benzene solutions containing around 0.1 mol/l pyridine. We expect the rate constant of this and similar reactions to be about $1.46 \cdot 10^{-4}$ l/mol sec at 60 °C (Grimm *et al.*, 1931). How large must the mass transfer coefficient be to make sure we are studying the chemical kinetics?

Solution If the methyl iodide vapor is present at much lower concentrations than the pyridine, we can approximate this as a first-order reaction whose rate r_1 is

$$r_1 = -\left[\kappa' c_{\text{pyridine}}\right] c_{\text{CH}_3\text{I}}$$

$$= -\left[(1.46 \cdot 10^{-4} \text{l/mol sec})(0.1 \text{mol/l})\right] c_{\text{CH}_3\text{I}}$$

$$= -\frac{1.46 \cdot 10^{-5}}{\text{sec}} c_{\text{CH}_3\text{I}}$$

From Eq. 17.1-17 or Fig. 17.1-2, we see that the transition from usual mass transfer to the reaction-limited case occurs when

$$\frac{D\kappa_1}{(k^0)^2} = 1$$

The diffusion coefficient of methyl iodide in benzene is about $2.0 \cdot 10^{-5} \text{ cm}^2/\text{sec}$. Thus

$$\frac{(1.46 \cdot 10^{-5}/\text{sec})(2.0 \cdot 10^{-5} \text{cm}^2/\text{sec})}{(k^0)^2} = 1$$

$$k^0 = 1.7 \cdot 10^{-5} \text{cm}/\text{sec}$$

Because mass transfer coefficients are usually around 10^{-3} cm/sec, we can saturate the liquid with the methyl iodide. We can successfully study the chemical kinetics in this way.

Example 17.1-2: The reaction rate in a large catalyst pellet   We want to set up a packed-bed laboratory reactor to study a first-order reaction for which the rate constant is 18.6 sec^{-1}. We plan to use 0.6 cm spheres of a porous catalyst for this gas-phase reaction. The diffusion coefficient of reagents in these particles is about 0.027 cm^2/sec.

How much will diffusion reduce the speed of reaction in these spheres?

   Solution   We first need to calculate a characteristic size for these spheres, which is

$$l = \frac{\frac{4}{3}\pi R^3}{4\pi R^2} = \frac{R}{3} = \frac{(0.6/2)\,\text{cm}}{3} = 0.1 \text{ cm}$$

The Thiele modulus is

$$\phi^2 = \frac{\kappa_1 l^2}{D} = \frac{18.6/\text{sec}\,(0.1\text{ cm})^2}{0.027 \text{ cm}^2/\text{sec}} = 6.9$$

From Figure 17.1-3, we see that

$$\eta = 0.3$$

This diffusion will reduce the speed of the reaction by roughly a factor of three.

Example 17.1-3: Finding the reaction-rate constant from mass transfer data   In studies with a wetted-wall absorption column, we find that the mass transfer coefficient for chlorine into water is $16 \cdot 10^{-3}$ cm/sec. The chlorine presumably is irreversibly reacting with the water:

$$Cl_2 + H_2O \rightarrow Cl^- + H^+ + HOCl$$

From similar experiments with nonreacting systems, we expect that the mass transfer coefficient without reaction is around $1 \cdot 10^{-3}$ cm/sec. What is the rate constant for this reaction?

Solution To solve this problem, we must make two assumptions. First, we assume that the reaction kinetics are first order and irreversible. In other words, we linearize the reaction:

$$r_{Cl} = -[\kappa' c_{H_2O}]c_{Cl_2} = -[\kappa_1]c_{Cl_2}$$

We identify the quantity in brackets with a first-order rate constant κ_1, thus assuming that the water concentration changes little. Second, we assume that because the coefficient with reaction is higher than expected, mass transfer is influenced by reaction. Because the diffusion coefficient of Cl_2 in water is $1.25 \cdot 10^{-5}$ cm^2/sec, we find

$$k = \sqrt{\kappa_1 D}$$

$$16 \cdot 10^{-3} \text{cm/sec} = \sqrt{\kappa_1(1.25 \cdot 10^{-5} \text{cm}^2/\text{sec})}$$

$$\kappa_1 = 20 \text{ sec}^{-1}$$

The value obtained from a more complete study of mass transfer is 14 sec^{-1}; that found from fast-reaction techniques is 25 sec^{-1}.

Example 17.1-4: Variation of mass transfer with fluid flow Imagine a spinning disc of reagent 1 immersed in a dilute solution containing reagent 2. We plan to measure reagent 1 lost from the disc as a function of the rotation speed of the disc. How will this rate vary if the reagent dissolves and then irreversibly reacts, that is, if the reaction is homogeneous? How will it vary if the reaction is heterogeneous?

Solution The answer in this case depends on how the mass transfer coefficient varies with fluid flow, or, in more general terms, on how the Sherwood number varies with the Reynolds number. This variation depends on the specific experimental situation. For the spinning disc described in Section 3.4,

$$\frac{k^0 d}{D} = b \left(\frac{dv}{\nu}\right)^{1/2}$$

The quantity b includes variables like the Schmidt number.

For a first-order irreversible homogeneous reaction, this flow dependence can be combined with Eq. 17.1-17 to give

$$k = \sqrt{D\kappa_1} \coth\left[\frac{1}{b}\left(\frac{\kappa_1 dv}{Dv}\right)^{1/2}\right]$$

At low flow, the hyperbolic cotangent is a constant, and thus k is also constant. At high flow, the hyperbolic cotangent approaches the reciprocal of its argument, and k varies with the square root of flow. This behavior is shown in Fig. 17.1-5.

The behavior for a heterogeneous reaction is completely different, a special case of Eq. 16.3-7:

$$\frac{1}{k} = \frac{1}{\kappa_2} + \frac{1}{k_3}$$

$$= \frac{1}{\kappa_2} + \frac{1}{a}\left(\frac{dv}{D^2\nu}\right)^{1/2}$$

where a is the interfacial area per volume. Note that the same symbol κ_2 is used here

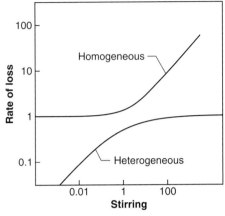

Fig. 17.1-5. Mass transfer for different types of reactions. At low flow, the mass transfer co-efficient is dominated by chemical kinetics if the reaction is homogeneous, but it is independent of kinetics if the reaction is heterogeneous. At high flow, the reverse is true.

to represent a heterogeneous rate constant. The results in this case are also given in Fig. 17.1-5.

The completely different variation with flow that results provides an easy way to distinguish between heterogeneous and homogeneous reactions. I have found it especially useful in biochemical systems, where ambiguity between the two types of reactions is frequent.

17.2 Mass Transfer with Second-Order Chemical Reactions

Like first-order reactions, second-order reactions can enhance interfacial mass transfer. Unlike the situation with first-order reactions, this enhancement cannot be easily calculated. Because second-order reactions are common and important, we resort to a variety of limiting cases to predict mass transfer coefficients in these situations.

The reason that predictions are difficult for second-order reactions is again best illustrated by the film theory, as shown in Fig. 17.2-1(a). The mass balances in this film are

$$0 = D_1 \frac{d^2 c_1}{dz^2} - \kappa_1 c_1 c_2 \tag{17.2-1}$$

and

$$0 = D_2 \frac{d^2 c_2}{dz^2} - \kappa_1 c_1 c_2 \tag{17.2-2}$$

The boundary conditions are typically

$$z = 0, \quad c_1 = c_{1i}, \quad \frac{dc_2}{dz} = 0 \tag{17.2-3}$$

$$z = l, \quad c_1 = 0, \quad c_2 = c_{2l} \tag{17.2-4}$$

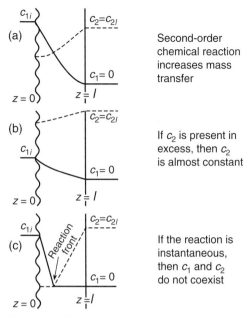

Fig. 17.2-1. Mass transfer with second-order chemical reaction. The mass transfer coefficients for the general case (a) cannot be easily calculated. They can be found for the special cases (b) and (c).

where c_{1i} and c_{2l} are the appropriate concentrations at the interface and in the bulk, respectively. Solving these equations is difficult because of the nonlinear reaction term. Various numerical solutions are available, but I never really understand what they are saying.

A more satisfying strategy is to consider three limiting cases. The most obvious limit, shown in Fig. 17.2-1(b), occurs when reagent 2 is present in excess, so that the second-order reaction is equivalent to a first-order reaction. This limit was discussed in the previous section.

A second, more interesting limit occurs when the reaction is very fast and irreversible. Here, finite concentrations of the two reagents cannot coexist, but simultaneously disappear at the reaction front, shown schematically in Fig. 17.2-1(c). The result is like two film theories, slapped one on top of the other.

Finding the mass transfer in this case is easy. For example, for the reaction

$$\left(\begin{array}{c} \text{species} \\ 1 \end{array} \right) + \nu \left(\begin{array}{c} \text{species} \\ 2 \end{array} \right) \rightarrow (\text{products}) \tag{17.2-5}$$

(where ν is a stoichiometric coefficient) we have

$$n_1 = \frac{D_1}{z_c}(c_{1i}) \tag{17.2-6}$$

$$n_2 = -\frac{D_2}{l - z_c}(c_{2l}) \tag{17.2-7}$$

and

$$\nu n_1 + n_2 = 0 \tag{17.2-8}$$

The distance z_c is the location of the reaction front. Combining these results to eliminate z_c, we find

$$n_1 = \left[\frac{D_1}{l} \left(1 + \frac{D_2 c_{2l}}{\nu D_1 c_{1i}} \right) \right] c_{1i} \tag{17.2-9}$$

The quantity in the square brackets corresponds to the mass transfer coefficient with chemical reaction. Remembering that the mass transfer coefficient without reaction k^0 equals D_1/l, we have

$$\frac{k}{k^0} = 1 + \frac{D_2 c_{2l}}{\nu D_1 c_{1i}} \tag{17.2-10}$$

which is the desired result. Again, we can extend this result to, for example, penetration and surface-renewal theories of mass transfer (Astarita et al., 1983). I believe that these extensions rarely produce significant improvements.

The third limit occurs when the second-order reaction is very fast and reversible, so that it essentially reaches equilibrium. The exact form of the result depends on the stoichiometry. As an example, consider the reaction

$$(\text{species } 1) + (\text{species } 2) \rightleftharpoons (\text{species } 3) \tag{17.2-11}$$

so

$$c_3 = K c_1 c_2 \tag{17.2-12}$$

where K is the equilibrium constant of the reaction. If species 2 and 3 are nonvolatile, their fluxes are zero at the gas–liquid interface. The calculation of the mass transfer in this case is similar to that for facilitated diffusion, given in detail in Section 18.5. Accordingly, only the result is given here:

$$\frac{k}{k^0} = 1 + \frac{D_3}{D_1} \left(\frac{K c_{2l}}{1 + K(D_3/D_2) c_{1i}} \right) \tag{17.2-13}$$

where k/k^0 represents the correction to the mass transfer coefficient caused by this kind of chemical reaction.

Example 17.2-1: Oxygen uptake by a synthetic blood   Oxygen uptake by blood is faster than oxygen uptake by water because of the reaction of oxygen and hemoglobin. Many chemists have dreamed of inventing a new compound capable of fast, selective reaction with oxygen. Aqueous solutions of this compound, of molecular weight around 500, could then be used as the basis of a process for oxygen separation from air. Such a compound would complex with oxygen at low temperatures, but would give up the oxygen at high temperatures.

How concentrated would this solution have to be to increase the oxygen concentration in water 50 times? How much faster would oxygen mass transfer into this solution be? The diffusion coefficient of oxygen in water is $2.1 \cdot 10^{-5}$ cm^2/sec; that of the new compound would be about $5 \cdot 10^{-6}$ cm^2/sec.

Solution The solubility of oxygen in water is $3 \cdot 10^{-7}$ mol/cm^3 so we want a solubility of $1.5 \cdot 10^{-5}$ mol/cm^3. This implies approximately a 0.7 wt% solution of our new compound. If the stoichiometric coefficient ν for this compound is one, the rate of mass transfer can now be estimated from Eq. 17.2-10:

$$\frac{k}{k^0} = 1 + \frac{(5 \cdot 10^{-6} \text{cm}^2/\text{sec})(1.5 \cdot 10^{-5} \text{mol/cm}^3)}{(2.1 \cdot 10^{-5} \text{cm}^2/\text{sec})(3 \cdot 10^{-7} \text{mol/cm}^3)}$$

$$= 13$$

There is about a 13-fold increase in rate.

Example 17.2-2: Sulfur dioxide absorption in a packed tower We are using an absorption tower 12 m high and 2 m in diameter to remove sulfur dioxide from a process gas. From previous experiments on nonreacting systems, we know that when the tower uses 2 °C water at the desired rate, the gas-side mass transfer is characterized by a $k_G^0 a$ of 1.7 sec^{-1} and the liquid-side mass transfer by a $k_L^0 a$ of $3.8 \cdot 10^{-3}$ sec^{-1}. We also know that under the current process conditions, sulfur dioxide is present at a partial pressure of around 10 mm Hg, producing a solution that contains 0.1 wt% at equilibrium.

We want to know how much the mass transfer in this column will be improved if we replace water with dilute solutions of sodium hydroxide. For this case, we expect

$$\text{SO}_2 + 2\text{NaOH} \rightarrow \text{Na}_2\text{SO}_3$$

Estimate the size of these improvements as a function of NaOH concentration.

Solution Because acid–base reactions like this are essentially instantaneous, we can estimate the improved mass transfer in the liquid from Eq. 17.2-10. Because c_{1i} in this equation is the interfacial value, we are forced to parallel the derivation in Section 8.5.

The flux across the interface must be

$$j_1 = k_p(p_1 - p_{1i})$$
$$= k_L(c_{1i} - 0)$$
$$= k_L^0\left(c_{1i} + \frac{D_2 c_{2i}}{\nu D_1}\right)$$

where the subscripts 1 and 2 refer to SO$_2$ and NaOH, respectively. We know that the gas and liquid concentrations at the interface are in equilibrium:

$$p_{1i} = Hc_{1i}$$

We now can find the interfacial concentration:

$$c_{1i} = \frac{k_p^0 p_1 - k_L(D_2 c_{2i}/\nu D_1)}{k_p H + k_L^0}$$

Note that this interfacial concentration is zero when

$$c_{2l} \geq \left(\frac{k_p \nu D_1}{k_L^0 D_2} \right) p_1$$

When it is greater than zero, the flux is

$$j_1 = K_L(p_1/H)$$

where

$$K_L a = \frac{1 + D_2 c_{2l} H/\nu D_1 p_1}{1/k_p a H + 1/k_L^0 a}$$

From the problem statement, we know that $k_L^0 a$ is $3.8 \cdot 10^{-3}$ sec^{-1}. We can find $k_p a$ by converting units:

$$k_p a = \frac{1.7/\text{sec}}{(82 \text{ atm cm}^3/\text{mol}K)(293K)}$$
$$= 7.1 \cdot 10^{-5} \text{mol/cm}^3 \text{ sec atm}$$

The Henry's law constant involves other unit conversions:

$$p_1 = Hc_1$$

$$(10 \text{ mm Hg}) \left(\frac{\text{atm}}{760 \text{ mm Hg}} \right) = H \left(\frac{0.001 \text{ g}}{\text{cm}^3} \right) \left(\frac{\text{mol}}{64 \text{ g}} \right)$$

$$H = 840 \text{ cm}^3 \text{ atm/mol}$$

The diffusion coefficient for NaOH is $2.1 \cdot 10^{-5}$ cm^2/sec; that for SO$_2$ is about $1.9 \cdot 10^{-5}$ cm^2/sec. To produce sulfite, $\nu = 2$.

Using these values, we find the c_{1i} equals zero when c_{2l} is 0.44 mol/l. At smaller hydroxide concentrations, the change in the overall mass transfer coefficient is

$$\frac{K_L a}{K_L^0 a} = 1 + \frac{D_2 c_{2l} H}{\nu D_1 p_1}$$

$$= 1 + \frac{(2.1 \cdot 10^{-5} \text{cm}^2/\text{sec}) c_{2l}(840 \text{cm}^3 \text{ atm/mol})(1/1000 \text{ cm}^3)}{2(1.9 \cdot 10^{-5} \text{cm}^2/\text{sec})(10/760 \text{ atm})}$$

$$= 1 + (35 \, l/\text{m}) c_{2l}$$

The results are shown in Table 17.2-1. At low hydroxide concentrations, the overall coefficient approaches the limit of no reaction; after an increase of a factor of 16, it becomes limited by mass transfer in the gas phase.

17.3 Industrial Gas Treating

In the previous two sections, we showed how mass transfer can be accelerated by a chemical reaction. We showed that for a first-order irreversible reaction, the mass transfer

Table 17.2-1. *Increases of sulfur dioxide (1) mass transfer effected with sodium hydroxide (2)*

c_2 (mol/liter)	$K_L a / K_L^0 a$
0.0	1
0.01	1.4
0.1	4.5
0.5	16
2.0	16

coefficient can become independent of the mass transfer coefficient without reaction, varying instead with the square root of the rate constant of the chemical reaction. We also showed that when the reaction became so fast that the various reagents cannot coexist, the mass transfer is proportional to the value without reaction times a correction factor involving ratios of concentrations in solution. These two limits describe much of the effect of chemical reaction on mass transfer, so most books stop at this point.

I want to go beyond this point because I believe there is still a large step between these idealized models and what actually happens in industrial gas treatment. In doing so, I am trying to avoid the usual chemical reaction of

$$A + B \rightarrow C \tag{17.3-1}$$

which some have suggested represents argon reacting with boron to produce carbon. Instead, I want to talk about what actual gas mixtures are treated and how chemical reactions can improve the treatment. In so doing, I am temporarily breaking the mold of most of this book, which is phrased in abstract terms. I am trying instead to talk about how the mathematical abstractions can be connected with real chemistry.

To make this connection, we first need to ask what gas mixtures we want to treat. There are three common targets. First, we frequently have chemical process gas streams containing 5 to 50% carbon dioxide and somewhat smaller amounts of hydrogen sulfide. We want to reduce the carbon dioxide concentration to around 0.1%. We want to cut the hydrogen sulfide concentration even further, to below 0.01% or 100 ppm. A second target is flue-gas desulfurization. Flue gas is produced by burning a hydrocarbon fuel that contains sulfur. As a result, the gas that goes up the stack contains typically 0.25% sulfur dioxide. We want to remove 90% of this sulfur dioxide. A third common target is the removal of organic sulfur compounds, like carbonyl sulfide (COS) or mercaptans, which are organic materials with an SH group instead of an OH group. Because these compounds have noxious odors, we generally want to reduce their concentration in any effluent to less than 1 ppm.

Gas mixtures like these occur widely. They occur in hydrogen manufacture, petroleum desulfurization and coal liquifaction. They occur in the manufacture of ammonia.

They occur frequently in natural-gas purification, both in the upgrading of pipeline gas and in the purification of liquid natural gas (LNG). Gas streams like these occur in the manufacture of such commodity organic chemicals like ethylene and ethyl acetate.

The current treatment of these materials depends on the concentration of the undesired gas to be removed in the mixed feed. When this undesired gas concentration is high, above perhaps 20%, the gas to be removed can often be absorbed in nonreactive liquids. Such nonreactive liquids are called "physical solvents." When the concentration in the feed is smaller, the free energy required for any separation will be larger. In this case, the solubility in nonreactive liquids is frequently insufficient to achieve the desired separation in reasonably sized equipment. Instead, we must absorb the target species in reactive liquids, which are called "chemical solvents."

We should stress that in many cases, other separation processes will be competitive to gas absorption, even gas absorption with highly reactive chemical solvents. Gases can frequently be separated using adsorbents, along the lines outlined in Chapter 15. Pressure-swing adsorption (PSA) can produce highly purified gases. Alternatively, where high purity is less important, the inexpensive membrane separations described in Chapter 18 may be appropriate. This is especially true in the upgrading of natural gas.

In this section, however, we want to focus on the treatment of gas mixtures by absorption in chemical solvents. To make this discussion easier to understand, we will focus on the particular example of the absorption of hydrogen sulfide into aqueous solutions of amines. Gas streams containing significant amounts of hydrogen sulfide are produced from sour crude oil by hydrodesulfurization. In most cases, we want to reduce the hydrogen sulfide concentration in the gas streams to less than 4 ppm.

As suggested above, the particular solvents in which we choose to carry out this gas absorption depend on the concentration in the feed stream. If the hydrogen sulfide concentration is greater than 20%, we will depend purely on solubility and choose physical solvents. If the hydrogen sulfide concentration is between 5 and 20%, we will choose aqueous amines. These are the focus of the discussion below. If the hydrogen sulfide concentration is less than about 5%, we may use solutions or slurries of carbonates, especially potassium carbonate. Finally, when we are anxious to remove the final traces of hydrogen sulfide, we will choose to use strong bases like sodium hydroxide, even though these cannot be easily regenerated, and so represent a solid waste stream.

A dramatically simplified diagram of one overall process for removing hydrogen sulfide with amines is shown in Fig. 17.3-1. In this process, the mixed gas feed enters the bottom of the gas absorber on the left-hand side of the figure. The hydrogen sulfide reacts with the amine solution that trickles down within this packed absorption tower. The sulfide–amine mixture passes out of the bottom of the tower into a second, warmer tower, where the hydrogen sulfide is driven off by heat. The hydrogen sulfide removed in this warmer tower goes to a Claus plant containing two reactors. Part of this hydrogen sulfide enters one reactor where it is burned to sulfur dioxide:

$$2H_2S + 3O_2 \rightarrow 2SO_2 + 2H_2O \tag{17.3-2}$$

The gas exiting from this reactor is mixed with more hydrogen sulfide in the second reactor to make elemental sulfur:

$$2H_2S + SO_2 \rightarrow 3S + 2H_2O \tag{17.3-3}$$

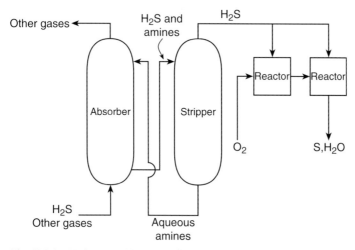

Fig. 17.3-1. Hydrogen sulfide removal with aqueous amines. This schematic process makes elemental sulfur in a Claus plant shown as the two reactors.

Thus the overall process essentially adds oxygen to hydrogen sulfide to make sulfur:

$$2H_2S + O_2 \rightarrow 2S + 2H_2O \tag{17.3-4}$$

While this Claus process is tangential to our interest in gas scrubbing, it represents established, reliable technology.

The design of a gas absorber like that in Fig. 17.3-1 is straightforward. It depends on the same three key equations that we used in our discussion of gas absorption in Chapter 10: a rate equation, an operating line, an equilibrium line, and a value of the mass transfer coefficient. The rate equation giving the height of the tower l is

$$l = \left(\frac{G}{K_y a}\right) \int_{y_l}^{y_0} \frac{dy}{y - y^*} \tag{17.3-5}$$

As in Section 10.3, the quantity in the parentheses corresponds to the height of a transfer unit and the integral corresponds to the number of transfer units. The operating line is just a mass balance given by Eq. 10.3-5 and will not be discussed further. The equilibrium line can be determined completely by experiment, but often is facilitated by considering the chemistry along the lines given below. The overall mass transfer coefficient, which has units of moles per area per time, is based on a gas-side mole-fraction driving force. It is related to the individual mass transfer coefficients based on concentration differences as described in Section 8.5. The liquid-side individual coefficient, which is affected by the chemical reaction, will be discussed after we explore the chemistry of the equilibrium line.

To find the equilibrium line, we must decide which forms of hydrogen sulfide are most important. To do so, we first consider the ionization of hydrogen sulfide in solution. This ionization involves two steps

$$H_2S \rightleftharpoons H^+ + HS^- \tag{17.3-6}$$

$$HS^- \rightleftharpoons H^+ + S^{2-} \tag{17.3-7}$$

We can decide which of these steps is more important by looking at the association constants of these different ions, and, more specifically, at their pK_a's. The association constant K_a for the first reaction is defined as

$$K_a = \frac{[\text{H}^+][\text{HS}^-]}{[\text{H}_2\text{S}]} \tag{17.3-8}$$

In this expression, we have used square brackets to indicate the concentration of the various species; thus $[\text{HS}^-]$ indicates the concentration of bisulfide in the system. The associated pK_a is defined by

$$pK_a = -\log_{10} K_a \tag{17.3-9}$$

Notice that the logarithm in this case is base 10 and is not the natural logarithm. In this sense, the pK_a echos the definition of the pH.

The pK_a is a convenient measure of how easily a particular species like H_2S ionizes. A large pK_a implies a small association constant K_a and hence a small degree of ionization. Phrased in other terms, a large pK_a implies a weak acid. One way to see this is to imagine that the pH is equals to the pK_a. This implies that in the case given above, the concentration of bisulfide equals the concentration of hydrogen sulfide. In other words, hydrogen sulfide is about half ionized. Thus, when the pH is much less than the pK_a, very little of the acid will be ionized. When the pH is much greater than the pK_a, almost all of the acid will be ionized.

In this particular case, the pK_a of hydrogen sulfide (H_2S) is about 7, and the pK_a of the bisulfide (HS^-) is about 12. Thus hydrogen sulfide is the main species below pH 7, bisulfide is central between pH 7 and pH 12, and sulfide is key above pH 12. Since most amine solutions are around pH 10, bisulfide is the chief form here.

We now return to the estimation of the equilibrium line. We are trying to decide how the concentration of H_2S in the vapor will vary with the total concentration of sulfides. We expect the fundamental reaction with the amine (B) will be of the form

$$\text{H}_2\text{S} + \text{B} \overset{K}{\rightleftharpoons} \text{BH}^+ + \text{HS}^- \tag{17.3-10}$$

where the equilibrium constant K is defined by

$$K = \frac{[\text{BH}^+][\text{HS}^-]}{[\text{H}_2\text{S}][\text{B}]} \tag{17.3-11}$$

We expect that the total concentration of amine $[\overline{\text{B}}]$ will be the sum of the free amine $[\text{B}]$ and the protonated amine $[\text{BH}^+]$. The amount of protonated amine will equal the amount of bisulfide in the system. In addition, we expect that the concentration of hydrogen sulfide in solution will be proportional to the concentration in the gas:

$$y^*_{\text{H}_2\text{S}} = m x_{\text{H}_2\text{S}}$$

$$= \left(\frac{m}{c}\right)[\text{H}_2\text{S}] \tag{17.3-12}$$

in which m is a Henry's law constant and c is the total molar concentration in the liquid. Combining Eq. 17.3-11 with the various mass balances, we find the equilibrium line

$$y^*_{H_2S} = \left(\frac{m}{K}\right) \frac{x^2_{HS}}{\left([\bar{B}]/c - x_{HS}\right)} \qquad (17.3\text{-}13)$$

where x_{HS} is the mole fraction of bisulfide (HS$^-$) in the liquid. This relation says that the equilibrium concentration in the gas phase be quadratic at low concentrations of bisulfide and vary still more strongly with bisulfide concentration at higher concentration. This prediction gives a reasonably good fit for the data, which always surprises me a little.

We now turn to the mass transfer coefficients themselves and the degree to which they are affected by chemical reaction. The experimental values of these coefficients for the reaction of H$_2$S with aqueous amines are given in Fig. 17.3-2. These figures show that the mass transfer coefficient varies linearly with the concentration of amine $[\bar{B}]$. We want to explain this result.

There are two possibilities, suggested by the results of Sections 17.1 and 17.2. Because in the liquid the amine is present in excess, the reaction is essentially first-order in H$_2$S, with an apparent first-order rate constant equal to the rate constant for the second-order reaction times the amine concentration. If this reaction is fast, the mass transfer coefficient can be found from an extension of Eq. 17.1-21.

$$k = \sqrt{D\kappa_1 [\bar{B}]} \qquad (17.3\text{-}14)$$

where D is the diffusion coefficient of the H$_2$S in the aqueous scrubbing solution, and κ_1 is the second-order rate constant for the reaction between the H$_2$S and the amine. Note that the second-order rate constant has units of reciprocal concentration per time, so its product with the concentration $[\bar{B}]$ has dimensions of reciprocal time and the mass transfer coefficient has dimensions of length per time. Alternatively, the analysis in Section 17.2 showed for an instantaneous, irreversible, second-order reaction, the mass transfer coefficient should be:

$$k = k^0 \left(1 + \frac{D_{BH}[\bar{B}]}{D_{H_2S}[H_2S]}\right) \qquad (17.3\text{-}15)$$

where k^0 is the mass transfer coefficient in the absence of chemical reaction. The data in Fig. 17.3-2 are inconsistent with the first of these two relations but consistent with the second of these two relations. The experimental data in the figure show that the mass transfer coefficient varies linearly with the amine concentration, as predicted by Eq. 17.3-15.

This qualitative agreement is difficult to make more quantitative. To attempt this quantification, we could look at the slope of the data shown in Fig. 17.3-2. This slope, which should correspond to the ratio of the diffusion coefficients, is equal to about 1.8. If we use the result in Eq. 17.3-15, which is based on the film theory, we find that the ratio of the diffusion coefficients of amine to hydrogen sulfide is 1.5 (cf. Chapter 5), less than the experimental result. The ratio of the diffusion coefficients of hydroxide to hydrogen sulfide is 3.7, greater than the experimental result. Alternatively, if we use the penetration theory rather than the film theory to derive a relationship equivalent to Eq. 17.3-15, we predict that the ratio of the diffusion coefficients should be replaced by the same ratio raised to the 1/2 power. For the amine and hydrogen sulfide, this implies a slope of the

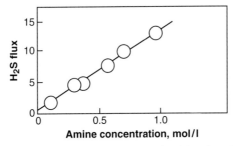

Fig. 17.3-2. Hydrogen sulfide uptake with amines. The data are consistent with an instantaneous reaction but not with a fast reaction.

data in Fig. 17.3-2 of 1.2, less than experiment. If we assume that important species are hydroxide and hydrogen sulfide, we predict that this ratio should be 1.9. This result is closest to the experimental value. Still, so many alternative predictions make everything seem like guessing.

My own conclusion is to accept the predictions of Sections 17.1 and 17.2 as qualitative guides that are not quantitatively exact. In other words, I use them more to make initial approximations and to organize experimental results than I use them to make a-priori predictions. This is especially true because the kinetics in many cases may not fall cleanly into the limit of a fast reaction or an instantaneous one. Here, the half life of the reaction is believed to be about 10^{-9} seconds, so that this reaction is effectively instantaneous. In contrast, the reaction of carbon dioxide is less easy to characterize. Even with sodium hydroxide, this reaction can fall in the transition between the fast and the instantaneous limits. Some of the difficulties with a carbon dioxide system are explored in the example that follows.

Example 17.3-1: Carbon dioxide absorption with amines   In many industrial processes, carbon dioxide must be removed from a gas mixture. This removal is frequently accomplished by scrubbing with aqueous solutions of pH 8 to 10 containing compounds like monoethanol amine:

$$NH_2CH_2CH_2OH$$

Carbon dioxide can react with both hydroxyl and the amine groups. However, the pK_a of the hydroxyl group is about 11, so that this reaction will be important only at a pH above 11. Such extremely basic conditions occur infrequently.

Reaction with the amine group involves a plethora of possibilities:

$$CO_2 + H_2O \rightleftharpoons H_2CO_3 \qquad\qquad K = 1 \cdot 10^3$$

$$H_2CO_3 \rightleftharpoons H^+ + HCO_3^- \qquad\qquad K = 4 \cdot 10^{-7}\ mol/l$$

$$HCO_3^- \rightleftharpoons H^+ + CO_3^{2-} \qquad\qquad K = 4 \cdot 10^{-11}\ mol/l$$

$$RNH_2 + H^+ \rightleftharpoons RNH_3^+ \qquad\qquad K = 3 \cdot 10^9\ l/mol$$

$$RNH_2 + HCO_3^- \rightleftharpoons RNH_3^+ + CO_3^{2-} \qquad K = 8 \cdot 10^{-2}\ l/mol$$

$$RNH_2 + HCO_3^- \rightleftharpoons RNHCOO^- + H_2O, \quad K = 50\ l/mol$$

In these equations, RNH_2 represents the monoethanol amine, and the K's are the equilibrium constants of the various reactions. For example, the last one is

$$[RNHCOO^-] = 50[RNH_2][HCO_3^-]$$

where the square brackets signify concentrations. Note that water does not appear in these equilibria because it is assumed to be present in excess.

How will the rate of mass transfer in these systems vary with the total amine concentration?

Solution The key aspect in this problem is making some sense out of all the possible chemical alternatives. First, we need to know in which forms the carbon dioxide occurs. For example,

$$[H^+][HCO_3^-] = 4 \cdot 10^{-7}[H_2CO_3]$$

Over the pH range to be used, $[H^+]$ will be 10^{-8} to 10^{-10} mol/l; thus $[HCO_3^-]$ will far exceed $[H_2CO_3]$. Similar reasoning shows that $[HCO_3^-]$ will exceed $[CO_3^{2-}]$. Turning to the reactions with amine, we expect $[RNH_2]$ to be relatively large, certainly greater than 10^{-2} mol/l. Thus $[RNH_3^+]$ and $[RNHCOO^-]$ are the chief reaction products, and the overall reaction involved here is

$$CO_2 + 2RNH_2 \rightleftharpoons RNH_3^+ + RNHCOO^-$$

This reaction is restricted to the pH range studied.

We next need to approximate the kinetics involved in this reaction. Because it is an acid–base reaction, we expect the reaction to be instantaneous, described by Eq. 17.2-10:

$$\frac{k}{k^0} = 1 + \frac{D_2 c_2}{\nu D_1 c_{1i}}$$

where 1 and 2 refer to carbon dioxide and amine at the boundaries of the interfacial reaction region. In a given experiment, we expect c_{1i} to be fixed and

$$c_2 = \bar{c}_2(1 - \theta)$$

where $\bar{c}_2$ is the total amine concentration and θ is the fraction of the amine already combined with carbon dioxide. Thus we expect

$$\frac{k}{k^0} = 1 + \frac{D_2 \bar{c}_2}{2 D_1 c_{1i}}(1 - \theta)$$

This prediction is verified for industrial absorption towers, for which

$$\frac{k}{k^0} = 1 + \frac{5.561}{9\,\text{mol}}\bar{c}_2(1 - \theta)$$

This holds over the pH range of 8 to 10; outside of this range, other chemical reactions may be more important. Note that the difficulty in this problem, and many like it, is not in the mathematics but in the chemistry.

17.4 Diffusion-Controlled Fast Reactions

In this section, we turn to chemical reactions that take place faster than a few milliseconds, often faster than microseconds. The most common example is the reaction of acid and base; other good examples are the combustion of methane and the action of the enzyme urease. These reactions occur so quickly that in industrial situations they are always diffusion controlled. However, they are of scientific interest.

To see how these fast reactions can be studied, we turn to the specific example of the temperature-jump apparatus shown in Fig. 17.4-1. In this apparatus, a cell containing perhaps $0.3\,\text{cm}^3$ of conducting solution is suddenly heated by discharging a capacitor through the solution. This heating, typically about $10\,°C$, shifts any reactions in the cell away from equilibrium. These reactions then move to a new equilibrium at the new, higher temperature. The speed with which they reach this new state is measured with a spectrophotometer attached to an oscilloscope.

One reaction studied with this temperature-jump apparatus is

$$H^+ + OH^- \rightleftharpoons H_2O \tag{17.4-1}$$

(a) Apparatus

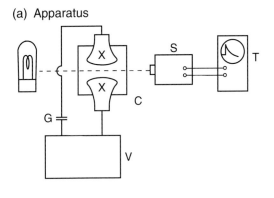

(b) Typical output

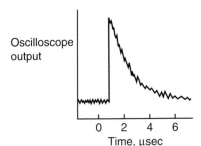

Fig. 17.4-1. The temperature-jump apparatus. Two large electrodes (X) are placed in a small cell (C) containing the solution of interest. The electrodes are charged with a power supply (V) until a spark flashes across the gap (G). The sudden current rapidly increases the cell's temperature, perturbing any reaction equilibria. These perturbations are recorded with the spectrophotometer (S) and displayed on an oscilloscope as a trace (T). A typical trace is also shown.

To study this reaction, the cell might be filled with an aqueous solution of KCl and a pH indicator. The KCl makes the solution conducting, so that capacitors can discharge; the indicator must be chosen so that its color change is still more rapid than the acid–base reaction. When the temperature is suddenly increased, the ionization of water is slightly increased, so that the indicator's color will slightly change. This change, monitored by the oscilloscope, can be used to find the rate constant κ_1.

In the remainder of this section, we shall describe how the oscilloscope trace can be related to the rate constant and how the rate constant varies with the diffusion coefficient.

17.4.1 Finding the Rate Constant

The experimental signal found on the oscilloscope is rarely a smooth, exactly defined curve. Far more frequently it contains considerable noise, more than that shown in Fig. 17.4-1(b). One method of reducing this noise is to repeat the experiment and to average electronically the various signals obtained. Even so, the signal remains inexactly known.

As a result, the signals obtained in this sort of experiment are almost always analyzed as if they are a first-order exponential decay. In other words, the logarithm of the signal's intensity is plotted versus time; the slope of this plot is the measured parameter. This slope, which has the units of reciprocal time, is a pseudo-first-order rate constant. These slopes are rarely reported; instead, their reciprocals, called relaxation times τ, are the data given.

We want to relate these relaxation times to kinetic rate constants. Such relations depend on the particular stoichiometry involved. As an illustration, we consider the reaction

$$\text{(species 1)} + \text{(species 2)} \underset{\kappa_{-1}}{\overset{\kappa_1}{\rightleftharpoons}} \text{(species 3)} \tag{17.4-2}$$

For this case, a mass balance on the cell gives

$$\frac{dc_1}{dt} = -\kappa_1 c_1 c_2 + \kappa_{-1} c_3 \tag{17.4-3}$$

We now let c_1' be the small perturbation in concentration caused by the change in temperature:

$$c_1 = \bar{c}_1 + c_1' \tag{17.4-4}$$

$$c_2 = \bar{c}_2 + c_1' \tag{17.4-5}$$

$$c_3 = \bar{c}_3 - c_1' \tag{17.4-6}$$

where the $\bar{c}_i$ are the equilibrium concentrations at the new, higher temperature. Combining these relations,

$$\frac{dc_1'}{dt} = -\kappa_1(\bar{c}_1 + c_1')(\bar{c}_2 + c_1') + \kappa_{-1}(\bar{c}_3 - c_1') \tag{17.4-7}$$

However, at equilibrium,

$$0 = \kappa_1 \bar{c}_1 \bar{c}_2 - \kappa_{-1} \bar{c}_3 \tag{17.4-8}$$

We now subtract these equations and, recognizing that c'_1 is small, neglect any terms in $(c'_1)^2$:

$$\frac{dc'_1}{dt} = -[\kappa_1(\bar{c}_1 + \bar{c}_2) + \kappa_{-1}]c'_1 \tag{17.4-9}$$

Integrating, we see that

$$\ln c'_1 = (\text{constant}) - [\kappa_1(\bar{c}_1 + \bar{c}_2) + \kappa_1]t \tag{17.4-10}$$

Thus the relaxation time τ is

$$\frac{1}{\tau} = \kappa_1(\bar{c}_1 + \bar{c}_2) + \kappa_{-1} \tag{17.4-11}$$

In cases like this, we also know the equilibrium constant K ($= \kappa_1/\kappa_{-1}$). Thus, from measurements of τ, we can find both the forward rate constant κ_1 and the reverse rate constant κ_{-1}. Note also that for this stoichiometry, τ^{-1} varies linearly with $(\bar{c}_1 + \bar{c}_2)$, permitting another experimental check of this argument.

17.4.2 Relating the Relaxation Time and the Diffusion Coefficient

At this point, we have described the temperature-jump method, a fast-reaction technique, and we have found relations between the relaxation times found from this method and the rate constants implied by the reaction's stoichiometry. In general, these rate constants will depend on factors like electronic structure. However, if the two reagents are very reactive, then they will react whenever they collide. In this case, their reaction rate depends not on electronic structure but on how often they collide (i.e., on their diffusion).

To explore this dependence of reaction rate on diffusion, we consider the forward reaction in Eq. 17.4-2. Imagine that species 2 is dilute and stationary. We then choose as a system a volume $1/c_2 \tilde{N}$ having a molecule of species 2 at its center. We assume that whenever a molecule of species 1 reaches this center, it reacts. We then write a mass balance on species 1 in this system:

$$\begin{pmatrix} \text{reaction} \\ \text{volume} \end{pmatrix} \begin{pmatrix} \text{system} \\ \text{volume} \end{pmatrix} = \begin{pmatrix} \text{diffusion} \\ \text{flux into} \\ \text{system} \end{pmatrix}$$

$$(r_1)\left(\frac{1}{c_2 \tilde{N}}\right) = 4\pi r^2 j_1 \tag{17.4-12}$$

where r is the approximate radius of the system. The flux of species 1 for such a system has already been calculated (see Eq. 2.4-24):

$$j_1 = -\left(\frac{D_1 \sigma_{12}}{r^2}\right) c_1 \tag{17.4-13}$$

in which σ_{12} is the collision distance between the two species and c_1 is the concentration in the bulk, far away from species 2. Thus

$$r_1 = -4\pi D_1 \sigma_{12} \tilde{N} c_1 c_2 \qquad (17.4\text{-}14)$$

However, species 2 is not actually stationary; if it too is allowed to move, this relation must be modified:

$$r_1 = -4\pi (D_1 + D_2) \sigma_{12} \tilde{N} c_1 c_2 \qquad (17.4\text{-}15)$$

Now the forward reaction rate is

$$r_1 = -\kappa_1 c_1 c_2 \qquad (17.4\text{-}16)$$

Thus the forward rate constant for this type of diffusion-controlled reaction is

$$\kappa_1 = 4\pi (D_1 + D_2) \sigma_{12} \tilde{N} \qquad (17.4\text{-}17)$$

This is the desired result.

The foregoing derivation – due to Smoluchowski (1917); see also Debye (1942) – is approximate. Most obviously, it neglects any potential surrounding the two species, a potential that might greatly accelerate or inhibit their interaction. It also ignores molecular shape. Because differently shaped reagents might need to rotate before they react with each other, the diffusion coefficients in Eq. 17.4-17 might include rotational contributions. Still, this derivation provides a first approximation for diffusion-controlled fast reactions.

Example 17.4-1: Diffusion-controlled reactions   Which of the following reactions is diffusion controlled?

(a)   Reaction of a proton and a hydroxyl:

$$H^+ + OH^- \rightarrow H_2O, \qquad \kappa_1 = 1.4 \cdot 10^{11} \, l/mol\,sec$$

(b)   Reaction of ethylenediaminetetraacetic acid:

$$OH^- + EDTA^{3-} \rightarrow EDTA^{4-}, \qquad \kappa_1 = 3.8 \cdot 10^7 \, l/mol\,sec$$

Solution   As a rule of thumb, reactions are diffusion controlled if their rate constants are faster than 10^9 l/mol sec. To make this more quantitative, we can insert into Eq. 17.4-17 estimates of diffusion coefficients from Chapters 5 and 6 and sizes from solid-state radii and thus estimate the rate constants if the reactions are diffusion controlled. For reaction (a),

$$\kappa_1 = 4\pi (D_1 + D_2) \sigma_{12} \tilde{N}$$

$$= 4\pi (9.3 + 5.3) \left(10^{-5} \frac{cm^2}{sec} \right) (2.8 \cdot 10^{-8} cm) \left(\frac{1}{10^3 cm^3} \right) \left(\frac{6.02 \cdot 10^{23}}{mol} \right)$$

$$= 3 \cdot 10^{10} \, l/mol\,sec$$

The experimental value is still more rapid, apparently because of the electrostatic interaction. Thus reaction (a) is diffusion controlled. For reaction (b),

$$\kappa_1 = 4\pi(5.3 + 0.8)\left(10^{-5}\frac{cm^2}{sec}\right)(5 \cdot 10^{-8}cm)\left(\frac{1}{10^3 cm^3}\right)\left(\frac{6.02 \cdot 10^{23}}{mol}\right)$$

$$= 2 \cdot 10^{10}\, 1/mol\, sec$$

This estimate, based on the assumption that the reaction is diffusion controlled, is much faster than the experimentally observed rate constant. Thus the kinetics of reaction (b) is not controlled by diffusion, but by chemical factors.

17.5 Dispersion-Controlled Fast Reactions

As the final topic in this chapter, we consider the rates of chemical reaction in turbulent flow. Such flow produces rapid mixing, so that the fluid appears homogeneous. Such mixing turns out to be only macroscopic. In other words, if we take 10 samples, each of 1 cm^3, we find that the average concentrations of the samples differ by only a few tenths of a percent. However, if we take ten samples of 10^{-12} cm^3, we find that their concentrations vary widely. For example, if we are mixing acid and base, we might find that some samples contain 10^{-2} mol/l H$^+$, and other have 10^{-12} mol/l H$^+$.

Such microscopic heterogeneity can sharply reduce the rate of a chemical reaction. Such reductions are not automatically bad. For example, in an automobile engine, we may wish to slow the combustion in order to reduce the maximum temperature and hence retard the formation of nitrogen oxide pollutants. It is important to know how this altered reaction rate can be estimated. Such estimates are the subject of this section.

To begin these estimates, imagine that we continuously inject a small amount of a concentrated solution of dye into a rapidly flowing solvent. We measure the dye concentration downstream of this injection. Far downstream, the dye's concentration will be a uniform value of $\bar{c}_1$, as shown in Fig. 17.5-1. Closer to the injection point, the concentration c_1 will fluctuate, both in position and time. The concentration fluctuations $(c_1 - \bar{c}_1)$ will sometimes be positive and sometimes negative; their average over time will be zero. However, the squares of these fluctuations will always be positive. We can characterize the size of an average fluctuation as a root mean square

$$\overline{(c_1 - \bar{c}_1)^2} = \frac{1}{\tau}\int_o^\tau (c_1 - \bar{c}_1)^2\, dt \tag{17.5-1}$$

Measurements of these fluctuations are reported as fluctuation sizes divided by the average concentration

$$\theta = \frac{\sqrt{\overline{(c_1 - \bar{c}_1)^2}}}{\bar{c}_1} \tag{17.5-2}$$

The relative fluctuation size θ provides a measure of the effectiveness of the turbulent mixing in our particular experiment.

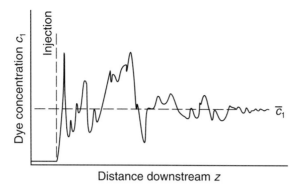

Fig. 17.5-1. Concentration fluctuations in turbulent flow. The concentration of continuously injected dye first varies wildly but reaches an average value after mixing.

We can plot our measurements of dimensionless fluctuations θ vs. downstream position z, divided by the fluid velocity v. Typically, we get a plot like that shown in Fig. 17.5-2, which often has the form of an exponential decay. In other words, a plot of log θ varies linearly with (z/v). Such a plot implies a characteristic decay time τ. In other words, we find from our mixing experiment that

$$\theta = \theta_0 e^{-\frac{z}{v\tau}} \tag{17.5-3}$$

where θ_0 is the fluctuation size where mixing starts. The experimentally determined relaxation time τ is a measure of how good our mixing really is. It shows that no matter how hard we try, we cannot get instantaneous mixing; we will always need some time. This time τ is then one easy way to compare mixers. For example, we can compare static mixers, which are vanes set in the flow to promote mixing. Vanes which give smaller values of τ are providing better mixing.

Now imagine a different experiment in which we inject acid into a rapidly flowing solution of base. As before, we measure concentration (as pH); as before, we measure the size of fluctuations (as θ). As before, we find values of θ drop with z/v. But we also find that the fluctuations of the acid θ_{acid} are exactly the same as the fluctuations caused by mixing alone

$$\theta_{acid} = \theta \tag{17.5-4}$$

In other words, fluctuations in the reacting acid–base system decay at the same rate as fluctuations in dye concentration. Indeed, if we injected hot water into cold water, we would find that the temperature fluctuations decayed in very much the same way, with the same time constant τ.

At first glance, it may seem startling that fluctuations in a nonreacting system, in a reacting system, and of heat, all fit the same equation. On reflection, this becomes less surprising. The reaction is so fast that acid and base never coexist, but at least one is destroyed as soon as contact is made. Thus how fast acid and base react does not depend on reaction kinetics, but only on turbulent mixing. How fast they react is a function of physics, but not of chemistry.

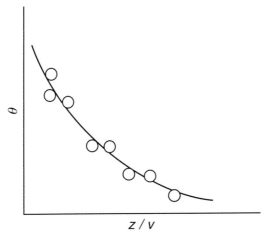

Fig. 17.5-2. Average fluctuation size vs. residence. Fluctuations often decay exponentially as turbulent mixing becomes more complete.

We now want to turn to the estimation of the relaxation time τ. Normally, we will want to measure τ by experiment because it will be peculiar to the specific turbulent flow which is involved. However, we can anticipate that this relaxation will involve two mutually dependent mechanisms. The first is the time τ' to cause the formation of small packets of fluid, estimated by

$$\tau' = \frac{d^2}{4E} \tag{17.5-5}$$

where d is a characteristic of the reactor size, typically the reactor diameter; and E is the dispersion coefficient in the flow. The second mechanism involves the time τ'' to diffuse into these packets

$$\tau'' = \frac{l^2}{4D} \tag{17.5-6}$$

where l is the packet size, normally around 30 µm; and D is the normal diffusion coefficient, about 10^{-5} cm^2/sec in normal liquids. Because these two mechanisms occur sequentially, the total relaxation time for the process τ is just the sum of the steps:

$$\tau = \tau' + \tau'' \tag{17.5-7}$$

Often, one of these relaxation times will be longer than the other, and hence control the process. These ideas are illustrated in the example which follows.

Example 17.5-1: Acid and base mixing   A small amount of aqueous 1-M nitric acid is injected into excess aqueous 0.1-M sodium hydroxide, flowing at 2 m/sec in a 5 cm pipe. Estimate the relaxation time τ in this case.

Solution   We first calculate the Reynolds number Re

$$Re = \frac{dv}{\nu} = \frac{5 \text{ cm } (200 \text{ cm/sec})}{0.01 \text{ cm}^2/\text{sec}} = 10^5$$

The flow is highly turbulent. Under this situation, we find from Eq. 4.2-2 that

$$\frac{dv}{E} = 2$$

$$\frac{5 \text{ cm } (200 \text{ cm/sec})}{E} = 2$$

$$E = 500 \text{ cm}^2/\text{sec}$$

We can find the time τ' from Eq. 17.5-5:

$$\tau' = \frac{d^2}{4E} = \frac{(5 \text{ cm})^2}{4\left(500 \text{ cm}^2/\text{sec}\right)}$$

$$= 0.013 \text{ sec}$$

The diffusion coefficient of nitric acid found from Section 6.1 is $3.2 \cdot 10^{-5}$ cm^2/sec, so

$$\tau'' = \frac{l^2}{4D} = \frac{\left(30 \cdot 10^{-4} \text{ cm}\right)^2}{4\left(3.2 \cdot 10^{-5} \text{ cm}^2/\text{sec}\right)}$$

$$= 0.070 \text{ sec}$$

Thus τ from Equation 17.5-7 is about 0.08 sec and is largely documented by diffusion within the small packets generated by the turbulent flow.

17.6 Conclusions

The material in this chapter describes how homogeneous chemical kinetics can alter rates of mass transfer and how mass transfer can alter rates of chemical reaction. The first three sections of the chapter emphasize how these altered rates affect the production of industrial chemicals. These sections distinguish between a "fast reaction" and an "instantaneous reaction." At first such a distinction may seem silly: after all, if a reaction is instantaneous, it must be fast, right? But what is meant by this distinction is more subtle and merits thought.

In this chapter, a "fast reaction" is one whose chemical kinetics is the same speed or faster than diffusion, but one where the reagents can coexist. In this case, the actual rate is a function both of diffusion coefficients and of reaction rate constants, as detailed in Section 17.1. This interaction leads both to the enhancement factors used in reactive gas treating and to the effectiveness factors important for porous catalysts.

In contrast, an "instantaneous reaction" has chemical kinetics that is so fast that the reagents do not coexist. It is the ultimate chemical civil war: as soon as the reagents are together, at least one is completely exterminated. In this case, the actual reaction rate depends on diffusion coefficients but not on reaction rate constants, as described in Section 17.2. The way in which these two limits interact in one practical situation is covered in Section 17.3.

These sections go beyond industrial chemical production to raise more fundamental questions about whether a reaction is heterogeneous or homogeneous. An instantaneous reaction can certainly involve reagents in solution without any physical interface between phases. However, such an instantaneous reaction occurs only at a "reaction front," like that shown in Figure 17.2-1(c). If it takes place only at such a front, it is really behaving as if it is heterogeneous. These fronts may be created as perturbations, as they are in the case of diffusion-controlled fast reaction. They can be generated as the result of turbulent mixing. In each case, they push the limits of what is meant by "diffusion controlled." That is why the material in this chapter is so interesting.

Questions for Discussion

1. What is the difference between a fast reaction and an instantaneous one?
2. What is the difference between a second Damköhler number and a Thiele modulus?
3. If the rate constant of a fast reaction is increased 10 times, how much does the mass transfer coefficient increase?
4. If the rate constant of an instantaneous reaction is increased 10 times, how much does the mass transfer coefficient increase?
5. What is an enhancement factor?
6. What is the effectiveness factor?
7. Why do different models of mass transfer give such similar results for the mass transfer coefficient with fast reaction?
8. Why do differently shaped catalyst particles give similar results for fast chemical reaction?
9. Sketch how the mass transfer coefficient of carbon dioxide in flue glass will vary with the kinetics of a reactive liquid solution.
10. Why do so many acid–base reactions have similar rate constants?
11. If you wanted to study acid–base kinetics by mixing aqueous acid with aqueous base, what reactor design would you use?
12. How can experimental measurements mixing hot and cold water be used to predict the kinetics of mixing acid and base?

Problems

1. You have been recovering an antibiotic from a fermentation broth by adsorption on activated carbon and have found that the mass transfer coefficient for this adsorption is $6.1 \cdot 10^{-4}$ cm/sec. In an effort to accelerate this adsorption, you switch to a cation-exchange resin of the same size beads and keep all details of your experiment the same. You find that the coefficient is now $1.03 \cdot 10^{-2}$ cm/sec. Because this coefficient is highly temperature-dependent, you suspect that it is influenced by chemical reaction. The

antibiotic has a diffusion coefficient of about $9.4 \cdot 10^{-7}$ cm^2/sec. What is the rate constant for the reaction? (E. Frieden)

2. (a) Estimate the change in mass transfer coefficient k/k^0 for 1.2-atm partial pressure H$_2$S being quickly absorbed by large quantities of 0.1-M monoethanolamine in water instead of by pure water. The chief reaction

$$H_2S + RNH_2 \rightleftharpoons HS^- + RNH_3^+$$

has an equilibrium constant of 275; the Henry's law constant for H$_2$S in water is 545. (H. Beesley) (b) How large an error is made if the reaction is assumed to be irreversible? (S. Gehrke)

3. The hydrocracking of a heavy oil at 1,080 °C and 1 atm uses a porous catalyst and excess hydrogen. Batch experiments with finely divided catalyst show a half life for the oil of 0.082 sec. Moreover, the oil has a diffusion coefficient of 0.014 cm^2/sec in the catalyst. Estimate the apparent rate constant in 0.1-cm spherical catalyst pellets.

4. The reaction

$$CO + 3H_2 \rightarrow CH_4 + H_2O$$

has a rate on a single nickel crystal given by [D. W. Goodman, R. D. Kelley, T. E. Madey, and J. T. Yates, Jr., *J. Catal.*, **63**, 226 (1980)]

$$\text{rate}\left([=]\frac{\text{molecules}}{\text{sec site}}\right) = 2.2 \cdot 10^8 e^{(-24.7\,\text{kcal/mol})/RT} p_{H_2}^{0.77} p_{CO}^{-0.31}$$

where the pressures are in atmospheres. This rate is known to be the same as that for a supported nickel catalyst [M. A. Vannice, *J. Catal.*, **37**, 449 (1975)]. With this in mind, estimate the reaction rate at 430 °C for a feed containing 1 atm each of CO and H$_2$ and a catalyst containing $1.1 \cdot 10^{14}$ sites/cm^3 in 1.6-cm spherical pellets.

5. The plasma membranes of cells in a suspension 0.7 cm deep are stained with an organic dye by layering the dye in a tetradecane solution on top of an agar-stabilized suspension in a 10-cm Petri dish. The dye has a diffusion coefficient of $2.8 \cdot 10^{-6}$ cm^2/sec. Its concentration in the tetradecane solution is 0.1 M, and the partition coefficient between the tetradecane and the suspending medium is 13. The cells are roughly spherical, approximately $2.1 \cdot 10^{-6}$ meters in diameter and present at a volume fraction of 0.05. In the range of dye concentrations being considered, the concentration of dye on the cells c_2 is proportional to the dye concentration in the suspending medium c_1:

$$c_2 = 3.4 \cdot 10^3 c_1$$

The dye cannot penetrate the cell membrane. How much dye will have entered the solution after 30 min? (K. H. Keller)

6. The data in the figure below give the moisture content of rough brown rice as a function of time [Bakshi and Singh, *J. Food Sci.*, **45**, 1387 (1980)]. Estimate the mass transfer coefficient versus temperature in the rice. Compare this temperature dependence with that expected if the water did not react with starch in the rice.

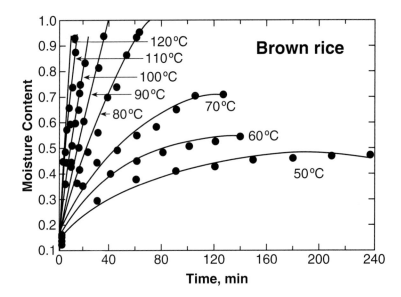

7. Imagine mass transfer across a thin film like that shown in Fig. 17.1-1. A zero-order reaction is taking place with the film. (a) What is the mass transfer coefficient in the absence of chemical reaction? (b) What is the differential equation and boundary conditions for the concentration in the film? The reaction is

$$r_1 = -\kappa \quad \text{when} \quad c_1 > 0$$
$$r_1 = 0 \quad \text{when} \quad c_1 = 0$$

(c) What is the mass transfer coefficient in this second situation?

8. Hydrogen sulfide is being removed from a hydrocarbon stream by washing with excess reactive liquid amine. The reaction is irreversible and fast but not instantaneous; its approximate form is

$$H_2S + RNH_2 \rightarrow RNH_3^+ + HS^-$$

Sketch the logarithm of the mass transfer coefficient in this liquid versus the logarithm of the reaction rate constant, and versus the logarithm of the velocity. Discuss different regions of these sketches.

9. Ammonia in dilute solution is often air-stripped in packed towers as a method of water pollution control. The stripping is a strong function of pH because of the equilibrium

$$NH_4^+ \rightleftharpoons NH_3 + H^+$$

Use the film theory to derive an expression giving the overall mass transfer coefficient as a function of pH, the total ammonia concentration c, and the mass transfer coefficients in the liquid and the gas. (a) Write continuity equations in terms of c_1 ($=$ $[NH_3]$) and ($= [NH_4^+]$). Integrate these to find the total flux J in terms of concentrations on either side of the film. (You may assume all species have equal diffusion coefficients.) (b) Use boundary conditions at $x = l$ within the liquid to rewrite the flux equation in terms of known bulk concentrations. (c) Write boundary conditions at $z = 0$, but don't use them. Instead, assume the acid concentration c_3 is in excess. As a result

$$c_{1i}c_3 = Kc_{2i}$$

Use this to find J in terms of c_{1i}, c and $k_0 (= D/l)$. (d) Remove c_{1i} by including a mass transfer resistance in the gas k_G to find the overall coefficient $K_L(k_L, k_G, c, pH)$.

10. The absorption of CO_2 from gas into mixed concentrated aqueous carbonates has the stoichiometry

$$CO_2 + H_2O + CO_3^{2-} \rightarrow 2HCO_3^-$$

This reaction can be inhibited by slow reaction kinetics; this inhibition can be reduced using hypochlorite ions (ClO^-) as a catalyst. The resulting reaction rate is

$$r = \left\{ \left(1.5 + \frac{2700 \, 1}{mol} [ClO^-] \right) sec^{-1} \right\} [CO_2]$$

where the concentration $[CO_2]$ is in the liquid, not in the gas.

Estimate the change in mass transfer caused by this chemical reaction using the penetration theory. In particular, (a) write an appropriate mass balance and boundary conditions for CO_2. (b) One integrated form of this result is

$$j_1|_{z=0} = \sqrt{Dk}[CO_2]\left\{ erf(\kappa t)^{1/2} + \frac{e^{-\kappa t}}{\sqrt{\pi \kappa t}} \right\}$$

Find the limits of this result for fast and slow reactions.

11. Under one set of experimental conditions, oxygen diffuses to the surface of carbon particles and reacts to form carbon monoxide:

$$O_2 + 2C \, (solid) \xrightarrow{\kappa_2} 2CO$$

As the carbon monoxide diffuses away, it rapidly reacts to form carbon dioxide:

$$O_2 + 2CO \xrightarrow{\kappa_3} 2CO_2$$

In the region where this second reaction occurs, carbon monoxide has a much higher concentration than oxygen. Use the film theory to estimate the flux of oxygen as a function of both rate constants.

12. In the text, we used the film theory to show that for a fast irreversible second-order reaction,

$$\frac{k}{k^0} = 1 + \frac{D_2 c_{20}}{\nu D_1 c_{10}}$$

This derivation assumed implicitly that the system was always dilute, so that there was no diffusion-engendered convection (i.e., $n_1 = j_1$). Imagine, instead, that the solution is concentrated, diluted only with product c_3. The stoichiometry is simple:

$$(species \, 1) + (species \, 2) \rightarrow (species \, 3)$$

Find the mass transfer coefficient in this case, assuming that diffusion obeys binary equations, with diffusion coefficients that are the same for all species.

13. In 1874, Louis Pasteur was commissioned by Napoleon III to determine why opening a bottle of full-bodied red wine hours before it is to be consumed gives the wine a more "mature" flavor than the same wine when drunk immediately after opening. Pasteur reported that the aging process is associated with a reaction consuming oxygen. More recently, some investigators have argued that this process involved oxygen diffusion into the wine. They used the oxygen concentration profile for unsteady diffusion without reaction in the wine and claimed justification of Pasteur's conclusions. (a) Write the differential equation and boundary conditions for this problem without chemical reaction. (b) Solve this equation, or determine the key dimensionless variables. (c) Use typical values for diffusion coefficients and wine bottles to discuss this result. (d) Qualitatively describe the effect of chemical reaction.

Further Reading

Aris, R. (1975). *The Mathematical Theory of Diffusion and Reaction in Permeable Catalysts*. Oxford: Clarendon Press.

Astarita, G. (1966). *Mass Transfer and Chemical Reaction*. Amsterdam: Elsevier.

Astarita, G., Savage, D.W., and Bisio, A. (1983). *Gas Treating with Chemical Solvents*. New York: Wiley.

Damköhler, G. (1937). *Deutsch Chemie Ingenieur*, **3**, 430.

Debye, P. (1942). *Transactions of the Electrochemical Society*, **82**, 265.

Grimm, H. G., Reif, H., and Wolff, H. (1931). *Zeitschrift für Physikalische Chemie*, **B13**, 301.

Hatta, S. (1928). *Tohoku Imperial University Technical Reports*, **8**, 1; (1932) **10**, 119.

Levenspiel, O. (1998). *Chemical Reaction Engineering*, 3rd ed. New York: Wiley.

Sherwood, T.K., Pigford, R.L., and Wilke, C.R. (1975). *Mass Transfer*. New York: McGraw-Hill.

Smoluchowski, M.V. (1917). *Zeitschrift für Physikalische Chemie*, **92**, 129.

Thiele, E.W. (1939). *Industrial and Engineering Chemistry*, **31**, 916.

CHAPTER 18

Membranes

Diffusion across thin membranes can sometimes produce chemical and physical separations at low cost. These low costs have spurred rapid development of membrane separations, especially during the last 20 years. This rapid development has sought both high fluxes and high selectivities. It has included the separation of gases, of sea water, and of azeotropic mixtures. It has used hollow fibers and spiral wound modules; it has centered on asymmetric membranes with selective layers as thin as 10 nm. This rapid development is a sharp contrast to other diffusion-based separations like absorption, where the basic ideas have been well established for 50 years.

We will describe membrane separations in this chapter. In this description, we must recognize that membrane separations usually employ a somewhat different vocabulary than that used on other separations. The stream that flows into any membrane module is sensibly called the feed. That part of the feed retained by the membrane is called the retentate rather than the raffinate. That part of the feed that crosses the membrane is called the permeate. Any stream added to improve permeate flow is usually called a sweep.

Our description of membrane separations also involves two overlapping traditions: filtration and diffusion. These traditions use a vocabulary that is confusing. The biggest confusion comes from the term permeability. In ultrafiltration, the solvent's permeability relates the solvent flow through the membrane's pores to the pressure drop across the membrane. As such, it is like a Darcy's law permeability for flow through porous media. In contrast, in gas separations, the permeability results from a gas flux across a non-porous membrane. This permeability is the product of the diffusion coefficient and a partition coefficient. In some cases, those using the term permeability do not know which effect they are discussing. Not surprisingly, this makes learning about these effects more difficult.

With this warning, we now bravely embark on our description. In Section 18.1, we discuss membrane construction and the membrane modules used for the separations. In Section 18.2, we describe gas separations. These separations use a gas feed at high concentration diffusing across a membrane into a gas permeate at low concentration. The only unusual idea is that the high concentration is expressed as a high partial pressure, and the low concentration as a low partial pressure.

In Section 18.3, we discuss reverse osmosis and ultrafiltration, two separations that are accomplished in similar ways. In doing so, we invite the confusion mentioned above: reverse osmosis is commonly believed to depend on pore-free diffusion, but ultrafiltration is a form of filtration with small pores. We will stress how these processes are similar and how they differ. In Section 18.4, we describe pervaporation, in which a liquid feed yields a vapor permeate. This method contrasts with distillation because the less volatile species may be in higher concentration in the permeate vapor.

Finally, we turn to membranes capable of both diffusion and a variety of chemical reactions. Such membranes, analyzed in Section 18.5, can aid – or "facilitate" – the transport of specific solutes. All these membrane processes imply new ways to effect separations.

18.1 Physical Factors in Membranes

In this section, we want to describe the fundamentals of diffusion across membranes and the actual physical construction of the membrane. We will extend these basic ideas to specific types of separations in latter sections. The fundamentals of diffusion across membranes include the effects of partition coefficients, concentration units, and resistances in series. The physical construction of membranes includes both the membranes themselves and the modules in which the membranes are used. The membranes themselves may be symmetric or asymmetric; the modules include hollow fibers, spiral-wound elements, and plate-and-frame assemblies.

18.1.1 Fundamentals of Diffusion

The basic flux across the membrane is exactly that for transport across a thin film. The flux j_1 is proportional to the concentration difference

$$j_1 = \frac{D}{l}(c_{10} - c_{1l}) \tag{18.1-1}$$

in which D is the diffusion coefficient, l is the membrane's thickness, c_{10} is the concentration in the membrane on the feed side of the membrane, and c_{1l} is the concentration within the membrane on the permeate side of the membrane. The quantity D/l obviously corresponds to a mass transfer coefficient. If we were to double the diffusion coefficient in the membrane, the flux would double. If we were to double the thickness of the membrane, the flux would be cut in half. If we were to double the concentration difference across the membrane, the flux would double.

There are four key points about the basic equation that are relevant for membrane separations. First, the separations are inherently based on rate of transport. They depend on diffusion. In this sense, they are like absorption and adsorption. They are much less like the analysis of staged distillation, which depends largely on thermodynamics.

The second key point about this membrane separation is that it is strongly influenced by the partition of the solute between the membrane and the adjacent solution. This means that Eq. 18.1-2 must be replaced by

$$j_1 = \frac{DH}{l}(C_{10} - C_{1l}) \tag{18.1-2}$$

where H is the partition coefficient or "solubility" between the membrane and the adjacent solution, defined by

$$c_1 = HC_1 \tag{18.1-3}$$

In these equations, the concentration C_{10} is the concentration in the fluid adjacent to the feed surface of the membrane. Similarly, the concentration C_{1l} is the concentration in the fluid adjacent to the permeate surface of the membrane. These partition coefficients mean that the concentration within the membrane may be higher or lower than the concentration in the adjacent solution. As such, this focus on partition is an echo of Example 2.2-1.

The third key point in the analysis of membrane behavior is that the simple Eqs. 18.1-1 and 18.1-2 are often obscured by the use of different concentration units. For example, for gas separations, concentrations are not ordinarily expressed in terms of moles per volume, but in terms of partial pressures because partial pressures are convenient. The gas separation parallel to Eq. 18.1-2 still retains the basic phenomenology characteristic of diffusion. But while the phenomenology may be the same, the units of the coefficients involved are now very different. Previously, the coefficient was D/l, which was equivalent to a mass transfer coefficient. The mass transfer coefficient had the dimensions of velocity. Now, the coefficient is the quantity $[DH/lRT]$ which is sometimes called the permeance. The dimensions of the permeance are no longer those of velocity, but must be modified to reflect the expression of the concentration as a partial pressure.

The fourth key point is that the membrane itself is always one of several resistances in series, as suggested in Fig. 18.1-1. In the past, the membrane has been the principal mass of the transfer resistance, dominating the resistances in the feed and in the permeate. However, as those working in the area have become more skilled, they have reduced the membrane resistance, sometimes by increasing the partition coefficient, but most frequently by dramatically reducing the membrane thickness. Twenty years ago, membrane thicknesses were as much as 100 μm. Now, they are often less than 1 μm. Thus the overall flux will be

$$ j_1 = \frac{C_{10} - C_{1l}}{\dfrac{1}{k_1} + \dfrac{l}{DH} + \dfrac{1}{k_3}} \tag{18.1-4} $$

where k_1 and k_3 are the mass transfer coefficients in the feed and permeate, respectively. These four key points about membrane diffusion will occur again and again in the more specific membrane descriptions in subsequent sections of this chapter.

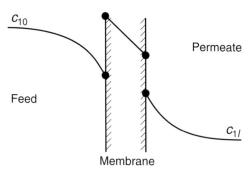

Fig. 18.1-1. The concentration profile. The membrane resistance, the dominant element in the past, is increasingly similar in size to the resistances in the feed and in the permeate.

18.1.2 Membrane Architecture

We now turn to the membranes themselves. These are most easily subdivided into symmetric and asymmetric structures. The symmetric membranes, which are less important, may be nonporous or porous. Nonporous symmetric membranes are typically made by spreading a polymer solution on a glass plate and allowing the solvent slowly to evaporate. Typically, 10 percent polymer dissolved in a volatile solvent like chloroform works well.

Porous symmetric membranes can be made by radiation, by stretching, and by leaching. To make a membrane by radiation, a homogeneous polymer film is exposed to a source of alpha particles. After exposure, the membrane is treated with chemicals like hydrofluoric acid to leach away the polymer structure damaged by the radiation. The result is a membrane with very homogeneous pores, like those shown in Fig. 18.1-2(a). Membranes with higher concentrations of more regularly packed mondisperse pores can also be made by self-assembly of some block copolymers, as shown in Fig. 18.1-2(b). The practical value of these membranes is uncertain. In still another case, polymers like polypropylene can be stretched at 143 °C to form tiny tears characteristically about 30 nanometers across. After the membrane is torn, it is then exposed to a second heat treatment to allow the polymer chains to relax in this slightly torn form. The result is a membrane like that shown in Fig. 18.1-2(c). Alternatively, suspended solids in polymer films can be removed by chemical leaching, which generates a structure similar to the membranes made by stretching.

Asymmetric membranes, which are more important commercially, come in a wide variety of types. I will discuss three types here. The first type of asymmetric membrane is that made by phase inversion. This process, originally developed by Leob and Sourirajan, involves spreading a polymer solution on a moving web. The polymer solution typically contains 20% polymer in a volatile solvent like acetone. Some of the solvent evaporates to form a thin skin of polymer. The moving web then dips into a nonsolvent like water

(a)   **(b)**   **(c)**

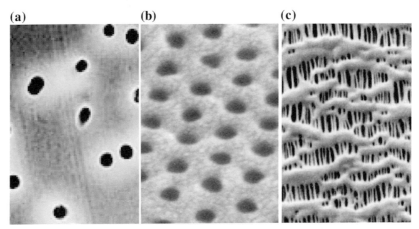

Fig. 18.1-2. Three porous membranes. The pores are formed (a) by radiation followed by etching; (b) by self-assembly followed by leaching; or (c) by stretching followed by heat treatments. These pores all have about 10 nm diameters.

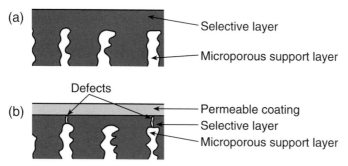

Fig. 18.1-3. Asymmetric membranes made by phase inversion. The membrane in (a) has a thin selective layer mechanically supported by a thicker spongy layer. Sometimes, a thin coating on top of this structure can repair membrane defects, as shown in (b).

and the remaining polymer is precipitated. The resulting membrane, shown schematically in Fig. 18.1-3(a), is idealized as two layers. The selective layer, perhaps one μm thick, frequently consists of glassy polymer. The remaining part of the membrane, typically 30 μm thick, is a spongy support that is usually assumed to have no resistance to mass transfer but that mechanically stabilizes the membrane. The thin, selective skin on membranes made in this way often contains defects. One can sometimes repair these defects by coating the membrane with a second permeable polymer layer, especially one of polydimethylsiloxane. This technique, used in some gas separation membranes, is suggested schematically in Fig. 18.1-3(b).

A second important type of asymmetric membrane is made by interfacial polymerization. In one example of this method, an asymmetric microporous support is impregnated with a polyamine. This polyamine is then exposed to a diisocyanate to form a cross-linked polymer film. This cross-linked layer can be extremely thin, perhaps around 0.1 μm. It is responsible for the membrane's selectivity. The microporous support, again typically 30 μm, provides mechanical stability.

The third method of making an asymmetric membrane is the most exotic. It consists of spreading a dilute polymer solution on a liquid like water. The resulting film can be extremely thin, perhaps 10 nm. As such, it is close to a monolayer. Such monolayers, sometimes called Langmuir–Blodgett films, are a popular research topic because they represent the thinnest possible membranes and because they are believed to be a model for some biological membranes. Fragile membranes like this have been produced commercially for one device that produces oxygen-enriched air for patients with emphysema. The unstated disadvantage of these ultrathin membranes is that their selectivity is often compromised by boundary layers adjacent to the membrane.

18.1.3 Membrane Modules

All the different types of membranes described above can be used in the three kinds of membrane modules shown in Fig. 18.1-4. One common type is based on hollow fibers. Hollow fibers are essentially very small pipes, typically 300 μm in diameter with

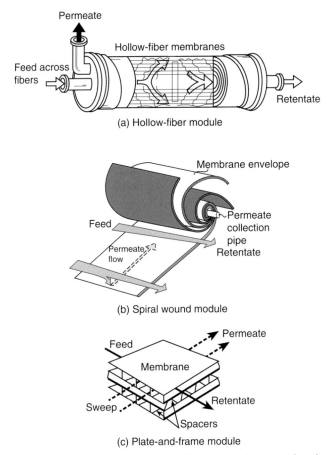

Fig. 18.1-4. Membrane modules. The three principal types are based on hollow fibers, spiral-wound elements, or plate-and-frame assemblies.

a 30 μm wall. They can be melt-spun, wet-spun or formed by interfacial polymerization. Often they are coated with a selective layer. Hollow-fiber membrane modules offer the greatest surface area per volume and hence the most efficient type of membrane separation. They are widely used for gas separations, although they plug easily and must be discarded if only a few of the fibers rupture.

The second type of membrane module, which is the hardest to visualize, is the spiral-wound element. This module essentially consists of a large membrane envelope loosely rolled like a jelly roll. The feed stays outside the envelope and products are harvested from the inside via a central tube. In some more sophisticated designs, many envelopes may come out from the central tube, so that a cross-section of the module would look like a daisy with petals twisted in a circular direction. This type of module has become the dominant geometry for reverse osmosis. While it has less membrane area per volume than a hollow-fiber module, it plugs less easily. However, even if only part of the membrane fails, the entire module must be discarded.

Table 18.1-1 *Characteristics of membrane module designs*

	Hollow-fiber	Spiral-wound	Plate-and-frame
Manufacturing cost	Moderate	High	High
Packing density	High	Moderate	Low
Resistance to fouling	Poor	Moderate	Good
High pressure operation	Yes	Yes	Difficult
Limited to specific membranes	Yes	No	No

The third type of membrane module is the plate-and-frame assembly, which closely parallels that used in filtration. Such a module has a high capital cost but is resistant to fouling. Moreover, if the membrane in one of the plates fails, it can be individually replaced; the entire module does not have to be discarded. This geometry is often used for ultrafiltration and has been used for pervaporation. I would expect its use for physical separations to be sustained but its use for diffusion-based separations to wane.

These module characteristics are summarized in Table 18.1-1. Manufacturing costs are lowest for hollow fibers, but operating costs are usually lowest for plate-and-frame units. In the past, new membrane technologies have tended to begin with the plate-and-frame geometry, then move to the spiral-wound geometry, and finally evolve to the hollow-fiber geometry. After that evolution, they return to the geometry that is cost effective for the specific application. I expect this trend to continue.

Example 18.1-1: Aeration of a fermentation broth You are currently using a sparger in a small fermenter to carry out an aerobic fermentation. With no microorganisms present, you measure the oxygen concentration in this fermenter and obtain

$$[O_2] = [O_2]_{sat}(1 - e^{-kat})$$

where ka equals 0.09 sec^{-1}. As an alternative, you want to aerate the fermentation with a hollow-fiber module. The microporous hollow fibers that you plan to use have no membrane resistance; the only resistance is in the boundary layers adjacent to the hollow fibers. You decide to feed the beer normal to the fibers and to blow excess air through the bore of each fiber. In this case, the only significant mass transfer resistance is in the broth, where the mass transfer coefficient is given by

$$\frac{kd}{D} = 0.8 \left(\frac{dv}{\nu}\right)^{0.47} \left(\frac{\nu}{D}\right)^{0.33}$$

You plan to use a velocity past the fibers of 10 cm/sec; and you expect the diffusion coefficient of oxygen in your beer to be the same as that of water, 2.1×10^{-5} cm^2/sec. Each fiber diameter is 240 μm, and the void fraction in the module is 0.5.

How much more rapidly can you aerate the fermentation beer with the hollow-fiber modules?

Solution The mass transfer coefficient in the beer is:

$$\frac{k(240 \cdot 10^{-4}\text{cm})}{2.1 \cdot 10^{-5}\text{cm}^2/\text{sec}} = 0.8 \left(\frac{240 \cdot 10^{-4}\text{cm}(10 \text{ cm/sec})}{0.01 \text{ cm}^2/\text{sec}}\right)^{0.47}$$

$$\times \left(\frac{0.01 \text{ cm}^2/\text{sec}}{2.1 \cdot 10^{-5}\text{cm}^2/\text{sec}}\right)^{0.33}$$

$$k = 0.024 \text{ cm/sec}$$

The surface area per volume a of module is

$$a = \frac{(\pi dl)\,(1-\varepsilon)}{\frac{\pi}{4}d^2 l} = \frac{4(1-\varepsilon)}{d}$$

where ε is the void fraction in the module. Thus

$$a = \frac{4(0.5)}{240 \cdot 10^{-4}\text{ cm}} = 83 \text{ cm}^{-1}$$

Thus the product ka is 2.0 sec^{-1}, roughly 20 times that of the sparger. This is a typical result. The improvement in the membrane module comes not from an improved mass transfer coefficient, which would be the same as or smaller than that in conventional equipment. The improvement comes from the greater surface area per volume. That this area is much larger than those offered conventionally can be easily seen by comparison with the areas for conventional packing like those listed in Table 10.2-1.

18.2 Gas Separations

In this section, we apply the general ideas of diffusion presented in the previous section to the separation of gaseous mixtures. In principle, these separations are straightforward: one simply pumps a high-pressure mixture down one side of a membrane. The membrane is commonly a hollow fiber, though it can equally be a spiral-wound element or plate-and-frame array. Some components in the gas mixture dissolve in the membrane and diffuse across it faster than other components. The more permeable species can be collected in the permeate; the less permeable species are concentrated in the retentate.

To put these ideas in a more quantitative basis, we return to the basic flux equation given in Eq. 18.1-2, which can be rewritten as

$$j_1 = \frac{DH}{l}(C_{10} - C_{1l})$$

$$= \frac{P}{l}(p_{10} - p_{1l}) \tag{18.2-1}$$

where p_1 is the partial pressure of gas "1," equal to RTC_1; and P is the permeability, equal to (DH/RT). In some cases the membrane's thickness is known, so the permeability can be calculated from measurements of flux and partial pressure.

Table 18.2-1 *Nitrogen and carbon dioxide permeabilities in various polymers at 30°C*

Film	$P_{CO_2} \times 10^{11}$	$P_{N_2} \times 10^{11}$	Selectivity
Saran	0.29	0.009	31
Mylar	1.5	0.05	31
Nylon	1.6	0.010	16
Neoprene	250	12	21
Polyethylene	350	19	19
Natural rubber	1310	80	16

Note: The permeabilities are given in barrers (see Eq. 18.2-2).

Typical permeabilities for carbon dioxide and nitrogen are given for some common polymers in Table 18.2-1. While the permeability is a simple quantity, its units are not. These units are most commonly given in terms of barrers. A barrer is defined as:

$$1 \text{ barrer} = \frac{10^{-10} \text{ cm}^3 \text{ gas(STP) (cm thickness)}}{(\text{cm}^2 \text{ membrane area}) (\text{cm Hg pressure})\text{sec}} \qquad (18.2\text{-}2)$$

Such a unit was originally defined for convenience because many polymer membranes had permeabilities around 1 barrer. In the cases where the membrane thickness is not known, the permeance (P/l) becomes the operating parameter. In its simplest form, the permeance is just the mass transfer coefficient for the membrane, with dimensions of velocity. Like the permeability, however, more common dimensions are gas permeation units (GPU), defined as

$$1 \text{ GPU} = \frac{10^{-6} \text{ cm}^3 \text{ (STP)}}{(\text{cm}^2 \text{ membrane area})\text{sec}(\text{cm Hg pressure})} \qquad (18.2\text{-}3)$$

Thus a membrane with a permeability of 1 barrer and a thickness of 1 μm has a permeance of 1 GPU.

18.2.1 Target Separations

Four groups of gas mixtures have most often been targets of membrane separation. The first group involves the separation of hydrogen from nitrogen, methane, and carbon monoxide. These separations have been successful, especially using hollow fibers. Their success depends on the high permeability of hydrogen, which is a consequence of hydrogen's low molecular weight and high mobility. Applications exist in ammonia purge gas, in oil refining, and in synthesis-gas manufacture. These separations seem mature, important because they were the earliest examples commercialized.

The second group of gas separations concerns the removal of carbon dioxide and hydrogen sulfide from low-grade natural gas. This includes gas produced by drilling land fills to collect the methane produced by the long-term degradation of garbage. Applications using spiral-wound modules have been compromised by the membrane selectivity. More selective membranes seem feasible, so this area should grow.

A third important area for gas separation is the removal of water vapor from air or from hydrocarbons. This separation is easily accomplished using adsorption, but the adsorbent beds require periodic regeneration. Membranes, which do not require this regeneration, show selectivities over air and hydrocarbons of thousand to one. Hydrocarbon losses, which are significant now, should improve as the membranes evolve.

The largest potential market for gas membrane separations is the separation of air. At present, 95% nitrogen can be effectively produced from a variety of commercially available membrane units. This particular separation has been used for producing inert atmospheres used, for example, to fill grain elevators. Doing so stops rats from living in the grain elevators and prevents the spontaneous dust explosions which could occur in the past. The production of oxygen with membranes has proved more difficult. Small units producing oxygen-enriched air are commercially available, but membrane separations for 90% oxygen remain elusive. Doing so requires membranes with an oxygen/nitrogen selectivity around 50, well beyond that available now.

18.2.2 Rubbery Polymer Membranes

Four kinds of membranes are used to effect gas separations: rubbery polymers, glassy polymers, porous membranes, and membranes capable of capillary condensation. Rubbery polymer membranes can be single polymers or copolymers. Their behavior is essentially that of a viscous liquid. Gas molecules diffusing in these membranes tend to have diffusion coefficients of the order of 10^{-7} cm^2/sec. Any selectivity in these membranes tends to come from differences in the partition coefficients.

Some of the rubbery membranes are chemically simple, but others are considerably more exotic. For example, membranes of polyvinylammonium thiocyanate

$$\left[\begin{array}{c} CH \\ | \\ NH_3^+ \ SCN^- \end{array} \quad CH_2 \right]_n$$

are highly selective for ammonia over nitrogen and hydrogen. While the exact selectivity is not known, it is probably greater than a thousand to one. As a second example, copolymers of

$$HO\left(C_2H_4 \ O\overset{O}{\underset{\|}{C}} \ C_4 \ H_8 \ \overset{O}{\underset{\|}{C}} \ O \right) COOH$$

and

$$O=C=N-\!\!\!\left\langle \bigcirc \right\rangle\!\!\!-CH_2-\!\!\!\left\langle \bigcirc \right\rangle\!\!\!-N=C=O$$

are selective for aromatics. The selectivity of toluene over isooctane is about seven. Again, the difference in the selectivity of rubbery membranes is dominated by differences in the partition coefficient, and not by differences in the diffusion coefficient.

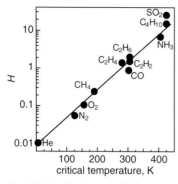

Fig. 18.2-1. Partition coefficients vs. critical temperature for various diffusing gases. This correlation means that major permeability differences must often come from differences in diffusion coefficient.

18.2.3 Glassy Polymer Membranes

Glassy polymer membranes are a contrast to rubbery ones. These consist of ultrathin polymer layers that are supported on inert microporous layers. The ultrathin layer, which is responsible for the membrane's selectivity, contains no large crystals, but smaller structures sometimes described as fringed micelles. In these membranes, the partition coefficient H correlates closely with the critical temperature of the diffusing gas, as shown in Fig. 18.2-1. Thus much of the selectivity between species with similar critical temperatures must be dominated by differences in diffusion. The diffusion coefficients in glassy polymers are smaller than in rubbery polymers, characteristically around 10^{-10} cm^2/sec. These small diffusion coefficients are of course why these membranes must be used as ultrathin layers.

The design of membrane structures capable of selective diffusion has been a research focus over the last decade. There seem to be three keys in this development:

(1) Inhibiting chain packing and tortional mobility around flexible linkages tends to increase both the permeability and the selectivity.

(2) Reducing the concentration of the mobile linkages in the polymer backbone tends to sustain the permeability and to increase the selectivity.

(3) Exposing the membrane to high gas concentrations can temporarily alter membrane selectivity. This "membrane conditioning" disappears with time.

These heuristics are empirical, the best that one can currently offer. They may not prove infallible in the future.

18.2.4 Microporous Membranes

Microporous membranes for gas separations are often very fragile inorganics, intended for use at higher temperatures. The pores in these membranes are often small. When they are smaller than the mean free path in the feed gas, their selectivity is modest, the result of Knudsen diffusion.

Table 18.2-2 *Sizes of typical gas molecules*

Solute	Molecular diameter/Å
H_2	2.89
CO_2	3.30
O_2	3.46
N_2	3.64
CH_4	3.80

Note: When these sizes are similar to the pore sizes in porous membranes, the membranes may be highly selective.

However, when their pores are of molecular dimensions, their selectivity can be much higher. Typical molecular sizes for gases that are common separation targets are given in Table 18.2-2. Membranes with pores of these sizes are potentially extremely selective. For example, one report of a zeolite membrane suggests a selectivity of seventeen between normal pentane and isopentane. Such promised selectivity has made these membranes an active area for research, even though their fragility remains a concern.

18.2.5 Capillary Condensation

Another separation mechanism in microporous membranes depends on the fact that the vapor pressure of a liquid in a pore p is different than the vapor pressure of the bulk liquid p_0.

$$p = p_0 \exp^{-2\gamma \cos \theta / C_L RTr} \tag{18.2-4}$$

where γ is the liquid's surface tension, c_L is its molar concentration, θ is the contact angle of the liquid in the pore, and r is the pore radius. As an example, imagine water vapor in contact with a hydrophilic membrane with pores 4 nm in diameter. Because the water is hydrophilic, the contact angle θ is zero; water's surface tension at room temperature is around 70 dyne/cm. Thus the saturation vapor pressure p is around half of the equilibrium vapor pressure. We can use this effect to remove condensable components at partial pressures below their equilibrium vapor pressure. We explore separations like these in the examples which follow.

Example 18.2-1: Alternative units of membrane permeability   The permeability of a specific membrane is given as one barrer at 25 °C. What would this permeability be in square centimeters per second?

    Solution   Because the partial pressure p_1 equals RTC_1, we see from Eq. 18.2-1 that

$$DH = PRT$$

From the definition of a barrer,

$$
DH = \frac{10^{-10} \text{cm}^3 (STP) \text{cm}}{\text{cm}^2 \sec \text{cm Hg}} \left[\frac{\text{mol}}{22.4 \cdot 10^3 \text{cm}^3} \right] \left[\frac{6240 \text{ cm Hg cm}^3}{\text{mol K}} \right] 298 \text{ K}
$$

$$
= 8.3 \cdot 10^{-9} \text{cm}^2/\sec
$$

If the partition coefficient H were also one, a permeability of one barrer would imply a diffusion coefficient typical of a rubbery polymer.

Example 18.2-2: Ammonia recovery A stream containing 4 mol% NH_3, 72 mol% H_2, and 24 mol% of N_2 at 135 atm and 35°C is being recycled to an ammonia synthesis reactor. You want to explore passing this stream through an ammonia-selective hollow-fiber module to recover 90 mol% of the ammonia. The module has the following properties:

fiber diameter d	320 μm
module void fraction ε	0.50
membrane permeability P	$1 \cdot 10^{-4} \text{cm}^2/\sec$ (12,000 barrers)
membrane thickness l	2.3 μm

How long should the gas spend in the module?

Solution We assume that the ammonia pressure in the permeate is much less than that in the feed. We then write a mass balance on the hollow-fiber module:

$$
0 = -v \frac{dC_1}{dz} - \left(\frac{P}{l} \right) a C_1
$$

where v is the feed velocity, C_1 is the concentration in the high-pressure stream, and a is the surface area per module volume, equal to $4(1 - \varepsilon)/d$. This is subject to the initial condition

$$
z = 0', \quad C_1 = C_{10}
$$

where C_{10} is the concentration entering the module.
Integrating, we find

$$
\ln \frac{C_{10}}{C_1} = \left(\frac{P}{l} \right) a \left(\frac{z}{v} \right)
$$

To find the residence time (z/v), we remember that we want to recover 90% of ammonia

$$
\ln \left[\frac{1}{0.1} \right] = \left(\frac{1 \cdot 10^{-4} \text{ cm}^2/\sec}{2.3 \cdot 10^{-4} \text{ cm}} \right) \frac{4(0.5)}{320 \cdot 10^{-4} \text{ cm}} \left(\frac{z}{v} \right)
$$

Thus

$$
\frac{z}{v} = 0.08 \sec
$$

The gas is effectively separated in this module.

18.3 Reverse Osmosis and Ultrafiltration

Reverse osmosis and ultrafiltration are processes for separating liquid solutions. The driving force for these separations comes from a pressure difference. This pressure difference does increase the chemical potentials on the feed side, but it does not dramatically alter the concentrations on this side. As a result, these separations are different than the gas separations described in the previous section, where the feed concentrations are directly proportional to the feed pressure.

Traditionally, the difference between ultrafiltration and reverse osmosis was defined in terms of solute size. More recently, reverse osmosis tends to be restricted to separations that occur by a diffusion–solubility mechanism like that for gas separations. Ultrafiltration implies transport through actual pores within the membrane and hence is related more to fluid flow than to diffusion. With this distinction in mind, we suggest approximate size ranges for ultrafiltration and reverse osmosis shown in Fig. 18.3-1. These are not universally accepted and so should be used with caution.

18.3.1 Target Separations

Reverse osmosis and ultrafiltration are already well-developed separation processes with defined market niches. They are expected to show continued growth, but they seem past their rapid expansion. However, this may change if the world's crisis in water supplies develops more quickly than expected.

Reverse osmosis continues to be dominated by desalination, that is, by removing salt from water. The substantial range of applications includes seawater, brackish water, and reclamation of municipal waste water. Similar membranes can be used to pretreat boiler feed water, desirable because pretreated water forms scale in boilers more slowly. There

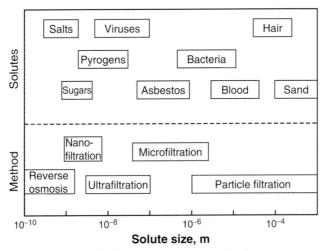

Fig. 18.3-1. Solute size for reverse osmosis and ultrafiltration. These processes use a high flow across the membrane surface to avoid the equivalent of the filter cake in conventional filtration. (Redrawn from Baker, 2004.)

is an emerging market for ultrapure water, both for semiconductors and for injectable pharmaceuticals. Other markets include water for domestic use, sweetener concentration, fruit-juice concentration, and fermentation product recovery. In the short term, these markets will grow more rapidly than conventional large-scale desalination.

Ultrafiltration is currently dominated by two large applications, the recovery of electropaint waste water and the recovery of proteins from dairy wastes. The former application results from the use of solvent-free paints, especially for automobiles. These paints are electrostatically applied. Any wash water is then processed to recover suspended pigments and other colloidal material. In the dairy industry, cheese whey can be concentrated and purified. In some cases, ultrafiltration concentrates valuable albumins that are lost in conventional processes.

Other applications of ultrafiltration are scattered. Not surprisingly there is a growing market in microelectronics, where smaller and smaller integrated circuits require purer and purer water. There is also a market in pyrogen removal for pharmaceuticals. Ultrafiltration is used to sterilize beer without heating and hence to produce draft beer in cans. Enzyme recovery, gelatin concentration, and juice clarification are possible, but their success depends on specific situations.

18.3.2 Three Basic Effects

The analysis of reverse osmosis and ultrafiltration depends on three phenomena that are tangential in other processes but are central here. Many descriptions of these membrane processes presume that everyone understands these phenomena. When I began research in this area, I did not understand them. Because I suspect others may be in my original position, I review them here.

The first of these phenomena is osmotic pressure. This pressure occurs whenever a membrane separates a solvent from a solution, as illustrated schematically in Fig. 18.3-2. For simplicity, we regard the membrane as permeable to solvent but completely impermeable to solute. In this case, the pure solvent has a higher free energy than that in the solution, and so it will tend to flow from left to right. It will continue to flow until the pressure in the solution rises enough to hold back this flow. When the flow ceases, the system is in equilibrium, with the osmotic pressure difference giving a measure of the concentration.

We want to know the relation between osmotic pressure and concentration more explicitly. To find this, we recognize that at equilibrium, the solvent's chemical potential must be constant:

$$\mu_2(T,p) = \mu_2(T,p+\Delta\Pi) \tag{18.3-1}$$

where $\Delta\Pi$ is the osmotic pressure. The chemical potential of the pure solvent is, of course, already the standard state, but that of the solution must be corrected for both solute concentration and pressure:

$$\mu_2^0(T,p) = \mu_2^0(T,p+\Delta\Pi) + RT\ln x_2$$

$$= \mu_2^0(T,p) + \bar{V}_2\Delta\Pi + RT\ln(1-x_1) \tag{18.3-2}$$

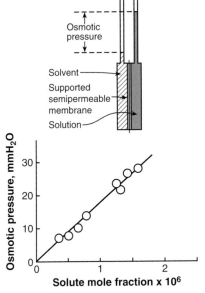

Fig. 18.3-2. Osmotic pressure. At equilibrium, the pressure on the solution must be higher than that on the solvent. This increased pressure, a measure of solution free energy, is often simply related to solute concentration. The specific data shown are for a polyethylene glycol (molecular weight *ca.* 20,000) dissolved in water.

where the superscript 0 signifies the standard state, $\bar{V}_2$ is the partial molar volume of solvent, and x_2 and x_1 are the mole fractions of solvent and solute, respectively. Note that Eq. 18.3-2 implies an ideal solution. Thus osmotic pressure and solute concentration are related by

$$\bar{V}_2 \Delta \Pi = -RT \ln(1 - x_1) \simeq RT x_1 + \cdots \tag{18.3-3}$$

For a dilute solution, the partial molar volume of water $\bar{V}_2$ equals the reciprocal of the total molar concentration c. Thus the mole fraction of solute x_1 divided by this partial molar volume equals the solute concentration C_1 and

$$\Delta \Pi = RT C_1 + \cdots \tag{18.3-4}$$

This relation, restricted to dilute ideal solutions, is called van't Hoff's law. When it was first proposed, it excited enormous interest, for it looked similar to the ideal-gas law. As its restrictions to dilute ideal solutions were realized, it has been deemphasized. For our purposes, it remains a central precept.

The second key experimental observation is for solute diffusion across a thin membrane at constant temperature and pressure. The flux j_1 due to this diffusion is like that calculated in Section 18.1:

$$j_1 = \frac{DH}{l}(C_{10} - 0) \tag{18.3-5}$$

in which D is the diffusion coefficient, H is the partition coefficient between membrane and adjacent solution, and l is the membrane thickness. For reverse osmosis, it is convenient to write this in a different form:

$$j_1 = \omega \Delta \Pi \tag{18.3-6}$$

where the osmotic pressure difference now equals $RT(C_{10} - 0)$, and the solute permeability ω is given by

$$\omega = \frac{DH}{lRT} \tag{18.3-7}$$

This solute permeability typically has dimensions of moles per area per time per pressure, but it is another form of mass transfer coefficient like those in Chapter 8. However, ω is the money of biophysics, and mass transfer coefficients are the coin of engineering, and these currencies are almost never exchanged.

The third basic effect concerns not solute diffusion but solvent transport. Imagine we were using a pressure difference to force pure water across the membrane. The flux of water could be described by

$$j_v = \frac{\text{volume water/time}}{\text{area membrane}} = v_2 \tag{18.3-8}$$

where v_2 is the velocity of water. Thus j_v is a volumetric flux, in units of volume per area per time, as opposed to the molar and mass fluxes used elsewhere in this book. Reverse osmosis is one of the few cases where this type of flux is used.

The water flux j_v can occur either by diffusion or by flow through pores in the membrane. If it occurs by diffusion, we expect the water flux in moles per area per time to be given by

$$j_2 = C_2 j_v = \frac{D}{l} \Delta c_2 \tag{18.3-9}$$

where C_2 is the water concentration outside of the membrane and Δc_2 is the water concentration difference within the membrane. Calculating this difference is tricky. We begin by assuming equilibrium across each membrane interface, so that the pressure and the water's chemical potential are each constant:

$$\mu_2^0 + \bar{V}_2^0 p + RT \ln C_2 = \mu_2^* + \bar{V}_2^* p + RT \ln c_2 \tag{18.3-10}$$

where the superscripts 0 and * refer to the solvent and the membrane phases, respectively. This condition can be rearranged to give

$$c_2 = \left[He^{(\bar{V}_2^0 - \bar{V}_2^*)(p - \bar{p})/RT} \right] C_2 \tag{18.3-11}$$

where H is the partition coefficient at some average reference pressure $\bar{p}$

$$H = \exp^{[\mu_2^0 - \mu_2^* + (\bar{V}_2^0 - \bar{V}_2^*)\bar{p}]/RT} \tag{18.3-12}$$

We can expand these relations as a Taylor series in pressure to find

$$c_2 = H\left[1 + \frac{\bar{V}_2^0 - \bar{V}_2^*}{RT}(p - \bar{p})\right]C_2 \tag{18.3-13}$$

$$\Delta c_2 = \left[\frac{HC_2(\bar{V}_2^0 - \bar{V}_2^*)}{RT}\right]\Delta p \tag{18.3-14}$$

where Δp is the pressure difference across the membrane. When we combine this with Eq. 18.3-9, we obtain

$$j_v = \left[\frac{DH(\bar{V}_2^0 - \bar{V}_2^*)}{RTl}\right]\Delta p$$

$$= L_p\Delta p \tag{18.3-15}$$

where L_p is called the coefficient of solvent permeability.

Equation 18.3-15 describes the solvent transport across the membrane by means of pressure-driven diffusion. However, we can also describe this solvent flow in a very different way if we assume the membrane contains small cylindrical pores – tubes – of diameter d. In this case, the velocity across the membrane would be given by the Hagen–Pouiseuille law:

$$v_2 = \varepsilon\left(\frac{\Delta p d^2}{32\mu l}\right) \tag{18.3-16}$$

where ε is the void fraction occupied by the tubes and μ is the fluid's viscosity. This result presumes laminar flow within the membrane's pores, which, because of their small size, is accurate. When we combine this relationship with the definition in Eq. 18.3-8, we obtain

$$j_v = \left[\frac{\varepsilon d^2}{32\mu l}\right]\Delta p$$

$$= L_p\Delta p \tag{18.3-17}$$

The parameters that contribute to this new solvent permeability are a dramatic contrast to those that contribute to the diffusion-based value in Eq. 18.3-15.

We might feel that it should be easy to distinguish between the diffusion-based and the flow-based explanations for solvent permeability. In fact, it often is difficult to do so. The reason is that the partition coefficient or the diameter of any pores in a thin selective layer is difficult to measure. Even the membrane thickness is frequently not exactly known. As a result, there has been considerable controversy over which of these explanations is correct. For reverse osmosis, the consensus is strongly in favor of the diffusion-solubility mechanism of Eq. 18.3-15. For ultrafiltration, there is an overwhelming belief that the membrane does contain pores and hence should be described with Eq. 18.3-17, modified by appropriate tortuosity factors. Fortunately, this controversy does not affect the development of flux equations, which is our next objective.

18.3.2 Flux Equations

The flux equations for membrane transport can now be developed by combining the effects of osmotic pressure, solute diffusion, and solvent transport. To do this, we consider the situation shown schematically in Fig. 18.3-3. In this situation, a concentrated solution at high pressure is being forced across a membrane into a dilute solution at lower pressure. The membrane is more permeable to solvent than to solute, and so a concentration difference develops. This concentration difference in turn produces an osmotic pressure opposing the flow.

We want flux equations describing the total flux across the membrane and the flux of solute alone. The solvent flux j_v must be that caused by the applied pressure minus that caused in the opposite direction by the osmotic pressure:

$$j_v = L_p(\Delta p - \sigma \Delta \Pi) \tag{18.3-18}$$

where σ is a new "reflection coefficient" characteristic of the membrane. If the membrane is permeable to solvent but completely impermeable to solute, σ equals unity. If the membrane is equally permeable to both solute and solvent, σ equals zero. The solute flux j_1, defined relative to the membrane, is again viewed as that due to diffusion plus that caused by convection:

$$j_1 = \omega \Delta \Pi + (1 - \sigma')\bar{C}_1 j_v \tag{18.3-19}$$

in which $\bar{C}_1$ is the average solute concentration $(C_{10} + C_{1l})/2$, and σ' is a second reflection coefficient to be discussed later.

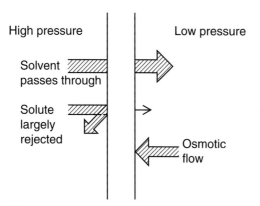

Fig. 18.3-3. Transport across a semipermeable membrane. Here, the solvent flux can be reduced by osmotic pressure and the solute flux can be altered by convection. Describing this transport process requires at least three independent coefficients.

These two new flux equations are strongly parallel to the common result for binary diffusion:

$$n_1 = -D\nabla c_1 + c_1 v^0 \tag{18.3-20}$$

Both Eq. 18.3-19 and Eq. 18.3-20 describe the solute flux, and ∇c_1 and $\Delta\Pi$ both reflect changes in solute concentration. The quantities v^0 and j_v represent the amount of convection. At the same time, there are differences between Eqs. 18.3-18 and 18.3-19 for reverse osmosis and Eq. 18.3-20 for binary diffusion. These differences are reflected by the number of transport coefficients involved. For binary diffusion, there is one: D. For reverse osmosis, there are four: L_p, σ, ω, and $1 - \sigma'$. This is because there is only one force, the concentration gradient, responsible for binary diffusion. For reverse osmosis, there are two independent forces, the concentration difference and the pressure difference. Reverse osmosis is more like ternary diffusion, where the three components are the solute, the solvent, and the membrane.

There is an alternative way of formulating these flux equations that is parallel to ternary diffusion and based on irreversible thermodynamics. The basic strategy is identical to that given in Section 7.2, so details are not given here. The key results are the water flux

$$j_v = (L_p) \Delta p + (-L_p\sigma)\Delta\Pi \tag{18.3-21}$$

and a solute flux j_D relative to the water flux

$$j_D = \frac{j_1}{\bar{C}_1} - j_v$$

$$= (-L_p\sigma')\Delta p + \left(\frac{\omega}{\bar{C}_1} + L_p\sigma\sigma'\right)\Delta\Pi \tag{18.3-22}$$

Although these equations are inconvenient in practice, they have interesting theoretical implications. These center on the relation between the coefficients σ and σ'. If the Onsager reciprocal relations are valid across membranes, then these two are equal. In this case, the number of coefficients necessary to describe membrane diffusion will drop from four to three. However, some studies suggest that the Onsager reciprocal relations are not always valid across membranes. As a result, I use Eqs. 18.3-18 and 18.3-19 assuming that σ and σ' are not equal, unless there is experimental evidence that they are. We next explore these ideas by means of examples.

Example 18.3-1: Fruit-juice concentration   Some farmers produce a variety of fruit juices which they wish to dehydrate to prolong shelf life and facilitate transportation. One dehydration method is to put the juice in a plastic bag and drop the bag into brine at $10\,°C$. If the bag is permeable to water, but not to salt or juice components, then osmotic flow will concentrate the juice.

Is the osmotic pressure generated in this way significant? Assume that the juice contains solids equivalent to 1 wt% sucrose and that the brine contains 35 g of sodium chloride per 100 g of water.

Solution For simplicity, assume that the juice and brine are ideal solutions. Then the osmotic pressure difference is

$$\Delta\Pi = \frac{RT}{\bar{V}_2}\ln\left(\frac{(1-x_1)_{juice}}{(1-x_1)_{brine}}\right)$$

$$= \frac{(0.082\ 1\ atm/mol\ K)(283\ K)}{0.018\ 1/mol}$$

$$\times \ln\left(\frac{1-\dfrac{(0.01\ g)/(342\ g/mol)}{(0.01\ g/(342\ g/mol)) + (0.99\ g)/(18\ g/mol)}}{1-\dfrac{2[(35g)/(58.5g/mol)]}{2[(35g)/(58.5g/mol)] + (100\ g)/(18\ g/mol)}}\right) = 250\ atm$$

This large pressure can cause rapid dehydration.

Example 18.3-2: Finding membrane coefficients for ultrafiltration To study the transport properties of glucose and water across an ultrafiltration membrane, we clamp a piece of the membrane across one end of a tube 0.86 cm in diameter and immerse the tube 2.59 cm into a large beaker of buffered water. We then fill the tube with a 0.03 mol/l glucose solution. The result is somewhat like the diaphragm-cell apparatus described in Sections 2.2 and 5.6-1.

We make two experiments, both at 25 °C. In the first, we adjust the tube so that the solution and solvent levels are initially the same and we leave both solution and solvent open to the atmosphere. We find that the solute concentration drops 0.4% in 1.62 h, and the total volume of solution increases 0.35%. In the second experiment, we initially adjust the solution level to be the same as the solvent level and leave the solution open to the atmosphere, but pull a vacuum of 733 mm Hg on the solvent. In this case, after 0.49 h, the solute concentration decreases 0.125%, and the solution volume also decreases 0.5%. Find the transport coefficients across this membrane.

Solution These experiments have been made over such short times than neither volume nor concentration has changed much. As a result, we can use the flux equations directly, without integration. We first consider the total flow. In the first experiment, there is no applied pressure difference, so from Eqs. 18.3-4 and 18.3-18

$$j_v = -(0.0035)\left(\frac{\pi(0.43\ cm)^2(2.59\ cm)}{\pi(0.43\ cm)^2(1.62\ hr)(3,600\ sec/hr)}\right)$$

$$= -L_p\sigma(C_1 RT)$$

$$= -L_p\sigma[(0.03\ mol/l)(0.082\ 1\ atm/mol\ K)(298\ K)]$$

$$L_p\sigma = 2.1\cdot10^{-6}\ cm/sec\ atm$$

In the second experiment, there are both pressure and osmotic differences; so

$$j_v = (0.005)\left(\frac{2.59\ cm}{(0.490\ hr)(3,600\ sec/hr)}\right)$$

$$= L_p\Delta p_1 - L_p\sigma\Delta\Pi$$

$$= L_p\left(\frac{733\ mm\ Hg}{760\ mm\ Hg/atm}\right) - \left(\frac{2.1\cdot10^{-6}\ cm}{sec\ atm}\right)\left[\left(\frac{0.03\ mol}{1}\right)\left(\frac{0.0821\ atm}{mol\ K}\right)(298\ K)\right]$$

Thus

$$L_p = 2.4 \cdot 10^{-6} \, \text{cm/atm sec}$$

and

$$\sigma = 0.90$$

About 90% of the glucose is rejected by this membrane.

Next, we turn to the glucose flux. From Eq. 18.3-19, we have

$$j_1 = \omega \, \Delta \Pi + (1 - \sigma') \bar{C}_1 j_v$$

Using the values of j_v from the foregoing, we obtain

$$0.004 \left(\frac{(2.59 \, \text{cm})(3 \cdot 10^{-5} \, \text{mol/cm}^3)}{1.62 \, \text{hr} \, (3600 \, \text{sec/hr})} \right)$$

$$= \omega[(0.03 \, \text{mol/l})(0.082 \, 1 \, \text{atm/mol K})(298 \, \text{K})]$$
$$+ (1 - \sigma')(0.015 \, \text{mol}/1000 \, \text{cm}^3)(-1.55 \cdot 10^{-6} \text{cm/sec})$$

for the first experiment, and

$$0.00125 \left(\frac{(2.59 \, \text{cm})(3 \cdot 10^{-5} \, \text{mol/cm}^3)}{0.490 \, \text{hr}(3600 \, \text{sec/hr})} \right)$$

$$= [(0.03 \, \text{mol/l})(0.082 \, 1 \, \text{atm/mol K})(298 \, \text{K})]$$
$$+ (1 - \sigma')(0.015 \, \text{mol}/1000 \, \text{cm}^3)(-0.73 \cdot 10^{-6} \text{cm/sec})$$

for the second. Solving these,

$$\omega = 7.4 \cdot 10^{-11} \, \text{mol/cm}^2 \, \text{sec atm}$$
$$\sigma' = 0.95$$

Whether or not σ' equals σ depends in this case on the experimental errors involved.

18.4 Pervaporation

We now turn to pervaporation, another membrane separation. In pervaporation, a liquid mixture fed on one side in the membrane produces a vapor permeate on the other side of the membrane. The basic pervaporation process is shown schematically in Fig. 18.4-1. In this figure, a liquid feed enters the membrane module, shown schematically as a box with a diagonal line through it. The diagonal line represents the membrane. The liquid that does not pass through this membrane – the retentate – exits from the module. The liquid that passes through the membrane evaporates to exit as a vapor permeate. This vapor permeate is at least partially condensed to produce a liquid product. Any uncondensable vapor is purged from the system. Because the amount of uncondensable vapor is usually small, the pumping required for this method is frequently minor.

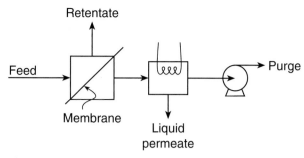

Fig. 18.4-1. A schematic representation of pervaporation. The membrane, represented by the diagonal, separates a liquid feed into a vapor permeate and a liquid retentate.

This method is a complement to the gas separations and reverse osmosis discussed in Sections 18.2 and 18.3, respectively. In the gas separations, a gas feed produces a gas permeate. In reverse osmosis, a liquid feed yields a liquid permeate. In pervaporation, a liquid feed produces a vapor permeate.

Pervaporation may initially seem rather like a single-stage distillation and hence incapable of providing much selectivity. This is incorrect: the selectivity possible with the one stage distillation may be dramatically improved by a selective membrane. The classic example is the separation of ethanol–water mixtures across a polyvinyl alcohol pervaporation membrane, which was the original application of pervaporation. The effect of this pervaporation process is contrasted with that of distillation in Fig. 18.4-2. In this figure, the ethanol concentration in the vapor is plotted versus the ethanol concentration in the liquid. The dotted line in the figure gives the conventional vapor–liquid equilibrium. For example, if the feed contains 20 mol% ethanol, the vapor in a single-stage distillation contains about 65 mol% ethanol. At a feed-liquid concentration of around 85 mol%, the vapor liquid equilibrium shows an azeotrope. At liquid concentrations below this azeotrope, the vapor is enriched in ethanol. At liquid compositions above this azeotrope, the vapor is depleted in ethanol. Azeotropes like this can make distillation difficult.

The behavior of the pervaporation process, shown as the solid line in Fig. 18.4-2, is a vivid contrast. Now, the vapor phase is always depleted in ethanol. For example, if the liquid feed contains 20 mol% ethanol, the vapor permeate will contain about 17 mol% ethanol. If the liquid contains 85 mol% ethanol, the vapor permeate contains less than 5 mol% ethanol. There is no azeotrope: The difficulties of distillation have been circumvented. Moreover, the less volatile component can be the one enriched in the permeate. This means that pervaporation will be especially appropriate for removing small traces of high-boiling impurities from large amounts of water.

18.4.1 Pervaporation Targets

Industrial pervaporation is currently dominated by separations of alcohol and water. Initially, this may be surprising because alcohol and water are relatively completely separated by simple distillation. However, they do form an azeotrope that can be avoided by pervaporation. Other systems where pervaporation is attractive are diverse.

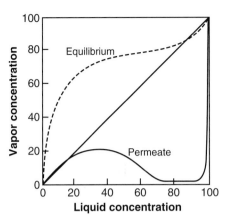

Fig. 18.4-2. Distillation versus pervaporation. The vapor–liquid equilibrium is given by the
dotted line and the pervaporation into vacuum is given by the solid line. The data are for
ethanol–water using a polyvinyl alcohol membrane. Note there is no azeotrope for pervaporation.

Major oil companies have looked at separating aliphatic and aromatic hydrocarbons
because this could offer a major energy saving. Others have looked at the separation of
ketones from olefins, of acidic acid from water, and of cumene from phenol. Still, the
future scale of pervaporation is uncertain.

18.4.2 Basic Equations

The basic analysis of pervaporation is complicated by the existence of two kinds
of partition coefficients. First, there are partition coefficients from the liquid feed into
the membrane. Second, there are partition coefficients from the membrane into the gas.
We begin by defining the selectivity β of the pervaporation membrane

$$\beta = \frac{j_1/j_2}{C_{10}/C_{20}} \tag{18.4-1}$$

where j_i is the flux of species i and C_{i0} is its concentration in the liquid feed. The flux of
each species will be proportional to the concentration of that species times the convective
velocity, both on the permeate side

$$j_i = C_{il}v \tag{18.4-2}$$

Moreover, because the permeate is a vapor at low pressure, its concentration C_{il} is
proportional to its partial pressure:

$$C_{il} = \frac{p_{il}}{RT} \tag{18.4-3}$$

Thus, the selectivity of this pervaporation process is

$$\beta = \frac{p_{1l}/p_{2l}}{C_{10}/C_{20}} \tag{18.4-4}$$

We are concerned with two major limits of this selectivity: the case where the total permeate pressure is close to the equilibrium vapor pressure and the case where the total permeate pressure is near zero. We consider these two cases sequentially.

When the total permeate pressure is close to the equilibrium vapor pressure, the membrane has no effect. In this case, the partial pressures in the vapor are in equilibrium with the partial pressures in the liquid:

$$p_{1l} = H_1' C_{10} \tag{18.4-5}$$

$$p_{2l} = H_2' C_{20} \tag{18.4-6}$$

where the H_i' is the volatility of species i. The selectivity of the pervaporation is now

$$\beta = \frac{H_1'}{H_2'} \tag{18.4-7}$$

This selectivity is the same as that for a conventional single-stage distillation. When the permeate pressure approaches the vapor pressure, there is absolutely no reason to use a membrane at all.

The second case, when the total permeate pressure is close to zero, is more interesting. This case is far from equilibrium and is influenced by the membrane permeability. In particular, the flux of species 1 is given by

$$j_1 = \frac{D_1}{l}(c_1|_{z=0} - c_1|_{z=l}) \tag{18.4-8}$$

where the concentrations $c_1|_z$ are those within the membrane at one of the two interfaces. We want to write out what these concentrations are in terms of the actual liquid and vapor concentrations in the column. Writing this out is tricky. At the liquid side of the membrane, where the coordinate z begins

$$z = 0, \quad c_1|_{z=0} = H_1 p_{10}$$
$$= H_1 H_1' C_{10} \tag{18.4-9}$$

where H_1 is the partition coefficient relating the concentration within the membrane to the partial pressure p_{10} outside of the membrane. This partial pressure is hypothetical; it does not exist at the membrane interface. However, this hypothetical partial pressure is related to the actual liquid concentration C_{10} by the volatility H_1'. Thus the concentration inside the membrane $c_1|_{z=0}$ is related to the real liquid concentration C_{10} by the expression given in Eq. 18.4-9.

Similarly, at the permeate side of the interface we may write

$$z=l, \quad c_1|_{z=l} = H_1 p_{1l}$$
$$= H_1 H_1' C_{1l} \tag{18.4-10}$$

This expression relates the concentration inside the membrane at the permeate interface to the partial pressure p_{1l}, which actually exists adjacent to the membrane. By analogy with Eq. 18.4-9 we can use the volatility H_1' to relate this concentration in the membrane to a hypothetical concentration C_{1l}, the concentration of the liquid that would be in equilibrium with the actual partial pressure of species 1 on the permeate side of the membrane.

After we write parallel equations for species 2, we combine these to find the relative flux across the membrane

$$\frac{j_1}{j_2} = \left[\frac{D_1 H_1}{D_2 H_2}\right] \left\{\frac{H_1'}{H_2'}\right\} \left(\frac{C_{10} - p_{1l}/H_1'}{C_{20} - p_{2l}/H_2'}\right) \qquad (18.4\text{-}11)$$

The permeate partial pressures p_{1l} and p_{2l} are both small because we are pulling a vacuum on the permeate side. Hence we can combine this relationship with Eq. 18.4-1 to find the selectivity of the pervaporation

$$\beta = \left[\frac{D_1 H_1}{D_2 H_2}\right] \left\{\frac{H_1'}{H_2'}\right\} \qquad (18.4\text{-}12)$$

This shows that pervaporation depends on two sources of selectivity. The first source, given in the square brackets, is the relative permeability across the membrane. As expected, this relative permeability is the product of diffusion and partition coefficients. Second, the selectivity of the pervaporation is influenced by the relative volatility given in the braces. This volatility, a thermodynamic factor, is independent of any dynamic concerns. This combination of dynamic and equilibrium factors explains why the less volatile species may be concentrated in the permeate stream of a pervaporation process.

Example 18.4-1: Volatile organic carbon (VOC) recovery   You want to reduce trichloroethylene concentration from 0.1% to 0.001% in a stream of water flowing at 1 l/sec. To do so, you are testing a pervaporation membrane which is 1.3 μm thick (excluding the support). The membrane has the following properties

Solvent	Trichloroethane	Water
D, cm^2/sec	$3.0 \cdot 10^{-7}$	$1.4 \cdot 10^{-6}$
H, mol/cm^3 atm	0.18	$2.2 \cdot 10^{-3}$
H' atm cm^3/mol	13	0.6

What is the membrane's selectivity? What membrane process might be effective?
   Solution   The membrane selectivity is easily found:

$$\beta = \left[\frac{D_1 H_1}{D_2 H_2}\right] \left\{\frac{H_1'}{H_2'}\right\}$$

$$= \left[\frac{3.0 \cdot 10^{-7} \text{cm}^2/\text{sec}\,(0.18\,\text{mol/cm}^3\,\text{atm})}{1.4 \cdot 10^{-6} \text{cm}^2/\text{sec}\,(2.2 \cdot 10^{-3}\,\text{mol/cm}^3\,\text{atm})}\right] \left\{\frac{13\,\text{atm cm}^3/\text{mol}}{0.6\,\text{atm cm}^2/\text{mol}}\right\}$$

$$= 380$$

The selectivity favors trichloroethylene 20 times more than volatility alone. The selectivity is augmented by the membrane even though the trichloroethylene has a smaller diffusion coefficient in the membrane. One possible process for this separation is a hollow-fiber module, in which there is a liquid feed down the hollow-fiber bore, and a vapor permeate on the shell side of the hollow-fiber module.

18.5 Facilitated Diffusion

Some membranes are much more selective than would be expected. For example, membranes containing tertiary amines can be much more selective for copper than for nickel and other metal ions. Ion-exchange membranes can be much more selective for olefins than for alkanes. Solid membranes containing synthetic porphyrins are sometimes reported to be highly selective for oxygen over nitrogen. In some cases, the membranes actually move specific solutes from a region of low concentration to a region of high concentration.

Many of these highly selective membranes are believed to operate by mechanisms different than more common membranes. As detailed above, more common membranes often function by a diffusion–solubility mechanism. In such a mechanism, the flux is proportional to the product of the diffusion coefficient and a partition coefficient, sometimes called the solubility. In contrast, these highly selective membranes sometimes function not only as a consequence of diffusion and solubility but also as a consequence of chemical reaction. Transport by combined diffusion and reaction is called facilitated diffusion.

These membranes often function by means of a "mobile carrier," a reactive species imprisoned by solubility within the membrane. The simplest way in which one of these chemically well-defined mobile carriers operates is shown schematically in Fig. 18.5-1. The two vertical lines bounding the cross-hatched area represent the membrane, which separates a concentrated solution on the left-hand side from a dilute solution on the right-hand side. Almost all the solute transported across the membrane is transported via complex formation with the mobile carrier. This mechanism qualitatively explains why the fluxes are much more selective than expected, and why they reach a constant value at high concentration differences.

In this section, we first describe where facilitated diffusion might be used for commercially attractive separations. We then develop equations for the mobile-carrier mechanism shown in Fig. 18.5-1. Finally, we describe other characteristics of this type of transport.

18.5.1 Target Separations

Facilitated transport does not have any significant commercial applications. It has failed commercially because the membranes used do not have the three-year lifetime normally demanded for practical processes. Even under ideal circumstances, membrane lifetimes are less than a month. In most cases, they are much shorter.

Still, the high selectivity promised by these membranes continues to encourage research. The targets are those where other types of membrane separations have been found to be limiting. Among commodity chemicals, these include the separation of air and of olefins and alkanes. The recovery of carbon dioxide from flue gas will receive

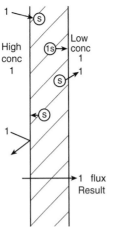

Step 1 Carrier s reacts with
solute 1.

Step 2 The complexed carrier
diffuses across the membrane.

Step 3 Because the adjacent
solution is dilute, the solute–
carrier reaction is reversed,
releasing solute 1.

Step 4 The carrier returns
across the membrane.

Step 5 Uncomplexed solute can
not diffuse across the membrane
because of low solubility.

Result The reaction with the mobile
carrier enhances or "facilitates"
the flux of solute.

Fig. 18.5-1. The simple mobile-carrier mechanism. This scheme can explain why facilitated diffusion is unexpectedly rapid, unusually selective, and nonlinear in the concentration difference across the membrane.

renewed attention as global warming continues. The recovery and purification of metals, including copper, nickel, and gold, may be revisited if the price of these compounds makes new efforts attractive.

Other applications of facilitated transport at a smaller scale may also be attractive. One possibility is controlled release because facilitated transport can give a steady flux, independent of the concentration difference. Another possibility is chemical signal amplification because a target solute can be concentrated by this mechanism. These small-scale applications may require a long shelf life but not a membrane stable under-operating conditions for three years. Nonetheless, until membrane stability is improved, facilitated transport will not be widely used commercially.

18.5.2 Basic Equations

As before, we consider the simplest case for detailed analysis, and later explore extensions to more complex situations. We assume that solute and carrier are nonionic and constantly reacting within the membrane:

$$(\text{solute } 1) + (\text{carrier } s) \rightleftharpoons (\text{complex } 1s) \tag{18.5-1}$$

These three components must satisfy the continuity equations for one-dimensional steady-state transport across the membrane:

$$0 = D\frac{d^2 c_1}{dz^2} - r_{1s} \tag{18.5-2}$$

$$0 = D\frac{d^2 c_s}{dz^2} - r_{1s} \tag{18.5-3}$$

$$0 = D\frac{d^2 c_{1s}}{dz^2} + r_{1s} \tag{18.5-4}$$

where r_{1s} is the rate of formation of the complex within the membrane. Note that we have assumed that the diffusion coefficients of all species in the membrane are equal. This assumption is not necessary, but it greatly simplifies the analysis and focuses attention on the effects of the solute–carrier reaction.

These continuity equations are subject to the restraints that

$$z = 0, \qquad c_1 = HC_{10} \tag{18.5-5}$$

$$z = l, \qquad c_1 = 0 \tag{18.5-6}$$

where 0 and l denote the two sides of the membrane. Thus C_{10} is the concentration outside of the membrane at the upstream side, where z is zero. From a mass balance on all forms of carrier,

$$\frac{1}{l} \int_0^l (c_s + c_{1s}) dz = \bar{c} \tag{18.5-7}$$

where $\bar{c}$ is the average carrier concentration in the membrane.

For a complete solution, three more restraints are needed. To simplify the mathematics, we also assume that the rate of complex formation is so fast that

$$c_{1s} = K c_1 c_s \tag{18.5-8}$$

where K is the equilibrium constant of the reaction in Eq. 18.5-1. This means that

$$z = 0, l, j_s + j_{1s} = 0$$

$$-D\left(\frac{dc_s}{dz} + \frac{dc_{1s}}{dz}\right) = 0 \tag{18.5-9}$$

at the membrane boundaries. Equations 18.5-8 and 18.5-9 are missing restraints.

Because these assumptions seem reasonable, we should briefly discuss why they cannot be exact. Equation 18.5-9 is the key. This relation implies that uncomplexed mobile-carrier molecules diffuse right up to the membrane wall, instantaneously react, and move away at the same rate but in the opposite direction. The molecules never have zero velocity; they instantaneously change from positive to negative velocity. This will be approximately true only when the second Damköhler number is large:

$$\frac{l^2}{Dt_{1/2}} \gg 1 \tag{18.5-10}$$

where $t_{1/2}$ is the half-time of either forward or reverse mobile-carrier reaction. When l is several micrometers, this condition is easily satisfied. For a membrane 10 nm thick, this condition is stringent: for example, if the reaction half-time is 10^{-7} seconds, the derivation presented here is in error by about 10%.

With these approximations, the flux across the membrane can be found in a straightforward fashion. Equations 18.5-3 and 18.5-4 are added, integrated, combined with Eq. 18.5-9, integrated again, and combined with Eq. 18.5-7 to give

$$c_s + c_{1s} = \bar{c} \tag{18.5-11}$$

everywhere throughout the membrane. This result and Eq. 18.5-8 are then combined with the sum of Eqs. 18.5-2 and 18.5-3 and integrated once:

$$- (j_1 + j_{1s}) = D \left[\frac{dc_1}{dz} + \frac{d}{dz} \left(\frac{K\bar{c}c_1}{1 + Kc_1} \right) \right] \tag{18.5-12}$$

where the integration constant $j_1 + j_{1s}$ represents the total flux of solute 1. When this result is integrated from $z = 0$ to $z = l$,

$$j_1 + j_{1s} = \frac{DH}{l} C_{10} + \frac{DH}{l} \left[\frac{K\bar{c}C_{10}}{(1 + HKC_{10})} \right] \tag{18.5-13}$$

The first term on the right-hand side represents the flux due to uncomplexed solute, and the second is the flux caused by carrier-assisted diffusion.

This result explains the experimental characteristics for facilitated diffusion listed at the start of this section. Frequently the diffusion of uncomplexed solute is much less than that of complexed solute, so that only the second term on the right-hand side of Eq. 18.5-13 is experimentally significant. When C_{10} is small, the flux is

$$j_1 + j_{1s} \doteq j_{1s} = \left(\frac{DHK\bar{c}}{l} \right) C_{10} \tag{18.5-14}$$

This flux is larger and more selective because it is proportional to $K\bar{c}$ and thus is altered by the chemical reaction. If C_{10} becomes large, the flux reaches a constant value;

$$j_1 + j_{1s} \doteq j_{1s} = \frac{D\bar{c}}{l} \tag{18.5-15}$$

which again is consistent with experiment. If the solute–carrier reaction is irreversible, K becomes infinite, and

$$j_1 + j_{1s} \doteq j_1 = \frac{DH}{l} (C_{10} - C_{1l}) \tag{18.5-16}$$

The carrier-assisted flux is effectively poisoned by an irreversible reaction. However, when the reacton is fast, strong, but reversible, the possibility of much more selective membranes remains undimmed.

18.5.3 Special Cases

As explained above, the enormous attraction of facilitated diffusion is the chance that chemical reaction can be used to enhance membrane selectivity. This has sparked immense theoretical and experimental research on this topic. In these paragraphs, I want to review some of the results of this research.

One result is an explanation for permeabilities that decrease with concentration, which are reported frequently. In these reports, the permeability is defined as

$$P = j_1 / (C_{10} - 0) \tag{18.5-17}$$

At low concentrations, P is observed to be constant, independent of C_{10}; but at high C_{10}, P drops. Because this is consistent with Eq. 18.5-13, some experimentalists conclude that they are observing a mobile-carrier mechanism. They may, but often they cannot, identify possible chemical species. In such cases, their invocation of this mechanism is usually just curve fitting.

A second result is the presence of facilitated transport in solid, reactive membranes. If such solid membranes can be made, they could circumvent the main failure of current commercial applications of facilitated tranport where membranes are unstable. Over the last 20 years, there have been over 200 claims of mobile carriers in solid membranes. So far, none have withstood experimental scrutiny in laboratories where they were not invented. Moreover, there are major theoretical objections asserting that such membranes are unlikely to exist. Nonetheless, these efforts continue to show small glitters of promise.

A third result is the restraint offered by the assumption of fast reaction made for the simple case above in Eq. 18.5-8. The antithesis of this assumption is the opposite limit, that fast diffusion mixes all species in the membrane and that the reaction-rate kinetics limits transport. In this case, the continuity equation for solute 1 becomes

$$0 = D\frac{d^2 c_1}{dz^2} - \kappa_1 c_1 \bar{c}_s + \kappa_{-1} \bar{c}_{1s} \tag{18.5-18}$$

where $\bar{c}_s$ and $\bar{c}_{1s}$ are the constant concentrations of carrier and complex in the membrane and κ_1 and κ_{-1} are the forward and reverse rate constants for the reaction in Eq. 18.5-1. Because c_1 is the only variable, this equation can be integrated directly. The total flux of solute 1 is then found by differentiating the concentration profile:

$$-(j_1 + j_{1s}) = H\left(\sqrt{\kappa_1 \bar{C}_s D}\right)\left(\frac{1 + \cosh(\kappa_1 \bar{C}_2 l^2/D)^{1/2}}{2 \sinh(\kappa_1 \bar{C}_2 l^2/D)^{1/2}}\right)(C_{10} - 0) \tag{18.5-19}$$

where

$$\bar{c}_s = \frac{\kappa_{-1}}{\kappa_1 \bar{c}_1 + \kappa_{-1}} \bar{c} \tag{18.5-20}$$

$$\bar{c}_{1s} = \frac{\kappa_1 \bar{c}_1}{\kappa_1 \bar{c}_1 + \kappa_{-1}} \bar{c} \tag{18.5-21}$$

and

$$\bar{c}_1 = \frac{H}{2}(C_{10} + C_{1l}) \tag{18.5-22}$$

As before, $\bar{c}$ represents the average total carrier concentration. This situation, which also implies that the flux does not always vary linearly with the concentration difference, occurs much less frequently than the fast-reaction case discussed earlier.

Table 18.5-1 *Glucose uptake in a human erythrocyte at 37 °C*

Glucose concentration ($\times 10^{-3}$ M)	Glucose flux ($\times 10^{-3}$ mol/min)
1.0	0.09
1.5	0.12
2.0	0.14
3.0	0.20
4.3	0.25
5.0	0.28

Note: This uptake is consistent with facilitated diffusion.

Example 18.5-1: Analyzing data for facilitated diffusion   The data shown in Table 18.5-1 were obtained for glucose transfer across a human erythrocyte. How can these data be analyzed to determine characteristic coefficients of facilitated diffusion?

 Solution   Presumably, glucose concentration on one side of the membranes is small, and facilitated diffusion is the dominant transport mechanism. In this case, we can use Eq. 18.5-13 to find

$$\frac{1}{(j_1 + j_{1s})} = \frac{l}{D\bar{c}} + \left(\frac{1}{DK\bar{c}}\right)\frac{1}{C_{10}}$$

or in notation more commonly used in biology

$$\frac{1}{(j_1 + j_{1s})} = \frac{1}{v_{max}} + \left(\frac{K_m}{v_{max}}\right)\frac{1}{C_{10}}$$

Where v_{max} and K_m are diffusion-based analogues of the maximum rate and the Michaelis constant in enzyme kinetics. We should plot the reciprocal of the total flux versus the reciprocal of the concentration. The plot, which is roughly linear, gives values of 0.56 M/min for $l/DA\bar{c}$ or v_{max} and 190 M^{-1} for KA where A is the erythrocyte area. These values are consistent with a large number of experiments.

Example 18.5-2: The flux in facilitated diffusion   Lithium, sodium, or potassium chloride is diffusing from a 0.1-M aqueous solution across a 32 μm organic membrane into pure water. The membrane is largely made of liquid chlorinated hydrocarbons, but it also contains as a mobile carrier $6.8 \cdot 10^{-3}$ M of the macrocyclic carrier dibenzo-18-crown-6. This carrier selectively complexes alkalai metals. For lithium chloride, the association constant is 260 1/mol; for sodium chloride, it is $1.3 \cdot 10^4$ 1/mol; for potassium chloride, it is $4.7 \cdot 10^6$ 1/mol. The partition coefficients of the various salts are $4.5 \cdot 10^{-4}$, $3.4 \cdot 10^{-4}$, and $3.8 \cdot 10^{-4}$, respectively. Assume that all salts and complexes have diffusion coefficients of $2 \cdot 10^{-5}$ cm^2/sec. Find the total flux for each of these salts.

Solution   These fluxes are easily found from Eq. 18.5-13, which for these cases reduces to

$$j_1 + j_{1s} = \frac{DHC_{10}}{l} + \frac{DHK\bar{c}C_{10}}{l(1 + HKC_{10})}$$

For lithium chloride, this is

$$j_1 + j_{1s} = \frac{2 \cdot 10^{-5}\, \text{cm}^2/\text{sec}}{32 \cdot 10^{-4}\, \text{cm}} (4.5 \cdot 10^{-4})(10^{-4}\text{mol/cm}^3)$$

$$+ \frac{\left(\dfrac{2 \cdot 10^{-5}\, \text{cm}^2}{\text{sec}}\right)(4.5 \cdot 10^{-4})\left(\dfrac{2.6 \cdot 10^5\, \text{cm}^3}{\text{mol}}\right)\left(\dfrac{6.8 \cdot 10^{-6}\, \text{mol}}{\text{cm}^3}\right)\left(\dfrac{10^{-4}\, \text{mol}}{\text{cm}^3}\right)}{(32 \cdot 10^{-4}\, \text{cm})\left[1 + (4.5 \cdot 10^{-4})\left(\dfrac{2.6 \cdot 10^5\text{cm}^3}{\text{mol}}\right)\left(\dfrac{10^{-4}\, \text{mol}}{\text{cm}^3}\right)\right]}$$

$$= (2.8 \cdot 10^{-10} + 4.9 \cdot 10^{-10})\,\text{mol/cm}^2\,\text{sec}$$

$$= 7.7 \cdot 10^{-10}\,\text{mol/cm}^2\,\text{sec}$$

Thus ordinary diffusion is responsible for about one-third of lithium transport. For sodium chloride, the total flux is $1.32 \cdot 10^{-8}$ mol/cm^2 sec, and for potassium chloride, it is $4.25 \cdot 10^{-8}$ mol/cm^2 sec. For both sodium and potassium chloride, facilitated transport is dominant.

18.6 Conclusions

Diffusion across membranes is an alternative route to separations where transport rates can enhance thermodynamic differences. In most cases, membrane diffusion depends on the permeability, which is the product of a diffusion coefficient and a partition coefficient, often called a solubility. Such membranes, coated onto porous support to provide mechanical strength, may be used as hollow fibers or in spiral-wound modules to increase the area per volume and hence increase the flux per volume.

Membrane separations vary with the mixture to be separated. Gases fed at high pressure are usually separated through thin glassy polymer films. Reverse osmosis of liquid mixtures also depends on pressure to overcome osmotic pressure. Pervaporation uses a warm liquid feed to give a cooler vapor permeate. Facilitated diffusion couples diffusion and chemical reaction to enhance membrane selectivity. This catalog of membrane separations both reviews the ideas of mass transfer and suggests new commercial opportunities.

Questions for Discussion

1.   What are the dimensions in mass M, length L, and time t of a permeability?
2.   What are they of a permeance?
3.   Membranes often have a thin, selective skin over a porous, mechanically strong substrate. Describe how such a structure can be made.

4.   What is an advantage and a disadvantage of a spiral-wound membrane module?
5.   Describe how to make a hollow fiber.
6.   If a gas separation membrane is said to have a permeance of 68 GPU, what is its mass transfer coefficient in cm/sec?
7.   Will a glassy membrane normally be better for separating air than a rubbery membrane?
8.   Membranes for separating air are normally more permeable to oxygen than nitrogen (molecular diameters 3.46 Å and 3.64 Å, respectively). Discuss the mechanism for separation, remembering that these membranes are nonporous.
9.   If you blow up a balloon with oxygen and then let it sit in air, it will get smaller with time. If you blow it up with nitrogen, it can get bigger with time. Why?
10. The sea contains about 0.5 M NaCl. Estimate its osmotic pressure.
11. Plot solvent flow through a reverse osmosis membrane vs. the pressure difference across the membrane.
12. Phenol is more permeable through reverse osmosis membranes than sodium chloride even though phenol is a larger molecule. How can this be?
13. What is responsible for the selectivity of pervaporation?
14. Facilitated diffusion is often said to describe how oxygen and hemoglobin interact in blood. Discuss whether this is true.

Problems

1.   You are studying a membrane said to be selective for oxygen, whose oxygen permeability is given as 63 barrers at 40°C. From measurements of gas absorption, you determine that the Henry's law constant is 3600 atm. What is the diffusion coefficient in this membrane?

2.   You are using a hollow-fiber membrane module made of the membranes in the previous problem. The 1.8 m module contains 30,000 hollow fibers 340 μm in diameter, packed with 41% voids. Each membrane has an active coating on the wall of 7.2 μm (i.e., the selective part of the membrane is 7.2 μm thick). This gives a membrane permeance or overall mass transfer coefficient of 7.6. 10^{-3} cm/sec, based on the fiber area per module volume. The membrane's selectivity for oxygen over nitrogen is said to be 8, the feed pressure is 120 psi, the permeate pressure is small, and the shell side is under vacuum. How much gas containing 95 mol% nitrogen can this module make per hour?

3.   To get better product purity, an engineer at the drug company, Pfizer, was recently carrying out a reaction in a glove bag. This is a large, clear, plastic bag with gloves put into the wall so that an engineer can work in the bag in an oxygen-free environment. To get this environment, the engineer continuously and steadily had nitrogen flowing in and out of the bag.

It didn't work; the oxygen concentration was still too high. He checked carefully for leaks; it still didn't help. He bought super-pure nitrogen, but the nitrogen concentration in the outlet didn't change.

After three months of frustration, he finally wondered if oxygen was diffusing across the wall of the bag. (a) If this is true, sketch how you expect the oxygen concentration in the bag will change with the flow of nitrogen. (b) Derive a differential equation for the oxygen concentration in the bag of volume V, and area A, and thickness l as a function of nitrogen flow Q, the permeability (DH), and the time t.

4. A stream at 220 atm and 100 °C containing 27.2% NH_3, 54.5% H_2, and 18.2% NH_3 is currently being recycled to an ammonia synthesis reactor. You want to feed it through a hollow-fiber module with a fiber volume fraction of 0.5 to recover 90% of the ammonia. The module's membranes are 240 μm in diameter, have a permeability P of $4.0 \cdot 10^{-5}$ cm^2/sec, and a selective layer thickness l equal to 35 μm. How long should gas spend in this module?

5. Imagine a thin hydrogel membrane consisting largely of water separating two well-stirred organic solutions. One of the solutions contains acetic acid, which diffuses across the membrane. (a) If the acetic acid concentration is very high, it will largely be un-ionized everywhere. What is the flux across the membrane in terms of the concentrations in the organic phases? (b) If the acetic acid concentration is moderate, it will be un-ionized outside the membrane, but ionized in it. Again, what is the flux? (c) If the acetic acid concentration is very low, it will be ionized everywhere. What is the flux now? (d) Use the results of (a)–(c) to plot log (flux) versus log (concentration difference). What is the slope on this graph?

6. One commercially available ultrafiltration membrane is claimed to have a permeability of 0.62 m^3/m^2 day under a pressure difference of 3.4 atm. This membrane initially rejects 96% of a 3 wt% suspension of partially hydrolyzed starch (molecular weight 17,000). However, if 4.2 cm^2 of membrane separates 65 cm^3 of a starch solution from the same volume of pure water, the volumetric flow is zero, and the osmotic difference is 85% of the original value in one week. Assuming the temperature is 25 °C, find the permeability L_p, the solute permeabilty ω, and the reflection coefficient σ. *Answer:* $L_p = 0.2$ m/day atm, $\omega = 0.4$ mol/m^2 day atm, $\sigma = 0.985$.

7. You are using a 64 m^2 spiral-wound module to remove 90% of the water in 6.8 m^3 of a feed containing of 0.082% polyethylene glycol of 18,000 molecular weight. For pure water, the membrane has a flux of 0.6 m^3/m^2 day at a pressure drop of 5500 kPa. How long will it take to remove 90% of the water?

8. A water-cooling process evaporates a small amount of water, which is replenished from a river. As a result, dissolved species in the river water slowly accumulate in the recycled cooling water stream until they precipitate in the heat exchangers and slow the heat transfer.

    To avoid this fouling, the process water must be periodically removed from the cooling-water system and treated to remove dissolved impurities. Such "cooling-tower blowdown" can be treated by a variety of methods. For an unsaturated feed at 0.2–0.4 m^3/min, the recommended method is reverse osmosis (RO) plus crystallization. In your case, you have a volume of 0.26 m^3 in 10 min at a temperature of 20 °C, initially containing solutes at 0.16 N, with a solubility limit of 10.92 N. You plan to concentrate this feed until the concentration in the water is 0.60 N by rapidly recycling it through a reverse osmosis unit using a pressure drop of 2400 kPa and a membrane which rejects nearly all the salt. Because the salt concentration will not be constant, you may assume that net osmotic pressure is the average at the start and end of the run. If the membrane's permeance is $1.3 \cdot 10^{-6}$ m/kPa min, how large a membrane area is needed?

9. In the text, we calculate the flux in facilitated diffusion by homogeneous chemical reaction with a mobile carrier. Rework this analysis, assuming that the carrier reaction is very fast but heterogeneous, occurring only at the membrane boundaries (J. Zasadzinzki).

10. Imagine a thin liquid film bounded by porous electrodes. The film contains Fe^{2+} and Fe^{3+}. At one electrode, there is the reaction

$$Fe^{3+} + e^- \xrightarrow{\text{fast}} Fe^{2+}$$

At the other electrode this reaction is quickly reversed. Thus the electrodes engender fluxes of Fe^{2+} and Fe^{3+} within the film. Both sides of the film are exposed to equal pressures of nitric oxide (NO) gas. The following reversible reaction occurs in the liquid:

$$NO + Fe^{2+} \rightleftharpoons FeNO^{2+}$$

The following reaction *does not* occur:

$$NO + Fe^{3+} \rightleftharpoons FeNO^{3+}$$

The net result will be a flux of NO, even though there is no NO concentration difference between the gases. Find the size of this flux and how it is related to the electrode current.

Further Reading

Baker, R. W. (2004). *Membrane Technology and Applications,* 2nd ed. New York: Wiley.

Baker, R. W., Cussler, E. L., Eykamp, W., *et al.* (1990). *Membrane Separation Systems.* (Department of Energy ER/30133-H1) Plainfield, NJ: Noyes Data Corporation.

Ho, W. S. and Sirkar K. K., eds. (1992). *Membrane Handbook.* New York: Van Nostrand-Reinhold.

Katchalsky, A. and Curran, P. F. (1967). *Non-Equilibrium Thermodynamics in Biophysics.* Cambridge, MA: Harvard University Press.

Leob, S. and Sourirajan, S. (1962). Sea water demineralization by means of an osmotic membrane. *Advances in Chemistry*, Series 38, 117.

Noble R. D. and Stern S. A., eds. (1967). *Handbook of Membrane Separations.* New York: Marcel Dekker.

Shao, P. and Huang, R. Y. M. (2007). "Polymeric membrane pervaporation" *J. Membrane Science,* **287**, 162.

CHAPTER 19

Controlled Release and Related Phenomena

Controlled-release technologies are used to supply compounds like drugs, pesticides, or fragrances at prescribed rates. The prescribed rates offer improved efficacy, safety, and convenience. The most commonly cited example is that of a drug dosed either by periodic pills or by a controlled-release technology. The concentration of the drug in the blood is shown schematically in Fig. 19.0-1. When the drug is given in a pill form, its concentration rises abruptly right after the pill is taken. This rise can carry the drug concentration past the effective level and briefly above the toxic level. The concentration then drops below the effective level. In contrast, when the drug is delivered by controlled release, its concentration rises above the level required to be effective and stays there, without sudden excursions to toxic or ineffective levels. Such delivery is often called zero-order release.

Typical products using controlled release are listed in Table 19.0-1. In the case of drugs, we normally want to release a single solid species, typically with a molecular weight greater than 600 daltons. The water solubility of these molecules is often strongly pH dependent because of pendant carboxylic acid or amino groups. The molecules normally will have several chiral centers. While this species is normally nonvolatile, it is usually unstable if it is warmed. It is often crystalline but may be a polymorph. As the previous paragraph suggests, we most often will seek to release the drug at a constant rate, although in some cases, we may want a periodic discharge.

The second case, agrochemicals, includes fertilizers, pesticides and herbicides. The fertilizers are mostly low molecular weight organics and inorganics, which may have a significant solubility in water. The insecticides and herbicides are often halogenated organics. These compounds may be either solids or liquids. The form of controlled-release sought may be zero order but may also be a burst, triggered by monsoon rains or by the first day of spring. Unlike the case of drugs, the cost of any controlled-release system for agrochemicals is important.

The third group of compounds targeted for controlled release is flavors. These are normally hydrophobic organic liquids of molecular weights less than 250. Here, our biggest concern is protecting these compounds from oxidation or retarding their evaporation. If this is achieved, we may seek release as a burst rather than as zero order. These release systems must be made at modest cost and of food-grade ingredients.

The actual release tends to occur from two types of systems. The first is a microcapsule. In its simplest form, this is just a small bubble with a solid, usually glassy, shell and a fluid or gel core. Because diffusion of the active species is much slower in the shell than in the core, release is often controlled by diffusion of the active species through the shell. In the second type of system, the active species is distributed throughout solid particles. Release is controlled either by diffusion out of the active species or because of diffusion in a solvent, commonly water, which releases the active species.

In this chapter, we have divided our discussion as cases of diffusion out of the active species, and of diffusion in solvent. More specifically, in Section 19.1, we describe cases

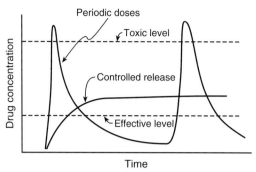

Fig. 19.0-1. Drug concentrations vs. time. Periodic dosages, like pills, can cause the concentration to go above the toxic level and below the effective level. Controlled release mitigates this burst effect.

regulated by diffusion of solutes like drugs. In Section 19.2, we analyze cases controlled by diffusion of solvent, most commonly water. The first section is the simplest, a direct extension of material in earlier chapters. The second section involves more invention, and includes a discussion of osmotic pumps. Both sections include situations used commercially.

We should stress that some cases of controlled release are not regulated by diffusion and are not detailed in this chapter. Some involve release controlled by chemical kinetics. For example, imagine a drug suspended in a water-insoluble polymer. The polymer slowly reacts with water and then dissolves. As the polymer dissolves, the drug will be released at a rate controlled by the polymer's hydrolysis. Alternatively, imagine an implantable pump that releases a drug in response to the patient's demand. While these cases involve controlled release, neither is controlled by diffusion, so neither is discussed here.

Other applications of diffusion are also discussed in this chapter. In Section 19.3, we discuss diffusion barriers. These are the antithesis of controlled release because our objective is to retard diffusion, rather than regulate it. The most obvious method of retardation is to choose the barrier so that it is impermeable to any solute. For example,

Table 19.0-1 *Targets for controlled release*

	Drugs	Agrochemicals	Flavors
Target species	Single compound	Single compound	Mixtures
Typical properties	Organic solids; $\tilde{M} > 600$	Organic and inorganic solids and liquids; $\tilde{M} < 200$	Hydrophobic, organic liquids; $\tilde{M} < 250$
Desired release	Zero order	Zero order or burst	Burst in mouth or food matrix
Desired protection	Water, oxygen	Water, light	Oxygen, evaporation
Major constraint	Safety	Cost	Food-grade ingredients

we could try to make a crystalline paint to keep out water, and hence retard corrosion. However, such a paint would tend to crack and so be easily breeched. How we can achieve flexible barriers is discussed in this section.

Finally, in Section 19.4 we describe problems involving diffusion and phase change. In many of these systems, diffusion causes solid dissolution. In others, diffusion can produce precipitation or spontaneous emulsification, even when the initial and final solutions are only a single phase. These unusual situations extend our understanding of diffusion.

19.1 Controlled Release by Solute Diffusion

In most cases, we want the release rate of a solute, like a drug, to be constant with time. This constant release rate will often give a constant concentration when in use, as suggested by Fig. 19.0-1. Often, we will test for this type of release rate by placing this solute, or a device containing solute, in a beaker of stirred solvent. We will measure the solute concentration versus time. We hope for a linear variation, for this would mean a constant release rate. Such a variation is called "zero-order release," as shown in Fig. 19.1-1.

We usually do not get this constant release rate without careful work. If we directly place a drug in water, its dissolution rate will often be mass-transfer controlled. As a result, the amount released M will vary with time t according to the relation

$$M = M(\text{sat}) \left(1 - e^{-\frac{kAt}{V}} \right) \tag{19.1-1}$$

where $M(\text{sat})$ is the drug dissolved at saturation, k is its mass transfer coefficient, A is its total solid surface area, and V is the volume of water (cf. Section 8.1). This type of behavior, where the logarithm of $(M(\text{sat}) - M)$ varies linearly with time, is often called first-order release, an echo of a first-order chemical reaction. This dissolution is fastest at short times and then slowly approaches zero. It is far from what we want partly because the release of the active will normally be too fast.

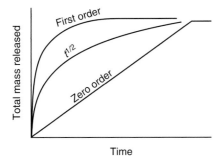

Fig. 19.1-1. Mass of drug released by different mechanisms. The steady zero-order release is the common target.

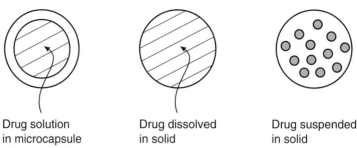

Drug solution   Drug dissolved   Drug suspended
in microcapsule   in solid   in solid

Fig. 19.1-2.   Three devices not giving zero-order release. These methods all depend on solute diffusion.

We would prefer a system where we could alter the release rate more systematically. Three systems where we can systematically alter drug dissolution are shown schematically in Fig. 19.1-2. In the first example, we could make small microcapsules containing a solution of the drug. We would find that the drug's release followed the same mathematical form as Eq. 19.1-1. Before, however, the mass transfer coefficient k would refer to the drug itself in the stirred solvent. Now, the mass transfer coefficient would be given by

$$k = \frac{P}{l} = \frac{DH}{l} \tag{19.1-2}$$

where P is the drug's permeability in the microcapsule's wall, D is its diffusion coefficient in the wall, H is the drug's partition coefficient between the solvent and the wall, and l is the wall's thickness. Because by varying the wall's material we can vary D, H, and l, we now can control the release. However, this release will still be first order and not the zero order which we seek.

As an alternative, we could disperse the drug evenly in a slab of water-swollen polymer. When this slab is dropped into water, the drug will diffuse out through the polymer. If the drug is initially dissolved at concentration c_1, as shown in the center of Fig.19.1-2, the amount released M will change at small times according to

$$M = \sqrt{\frac{4Dt}{\pi}}(Ac_1) \tag{19.1-3}$$

where A is now the surface area of the polymer slab and D is the drug's diffusion coefficient (cf. Section 2.3). If the drug is initially present as small particles, as shown at the right of Fig. 19.1-2, the amount released at small times will be approximately

$$M = A\sqrt{2Dtc_{10}c_1(\text{sat})} \tag{19.1-4}$$

where c_{10} and $c_1(\text{sat})$ are the total initial drug concentration and the saturation concentration, respectively. Because in these two cases M is proportional to the square root of time, this release is sometimes called $t^{1/2}$ kinetics. While this release may be slower than that of the pure drug, it will still be very fast at the beginning and get slower

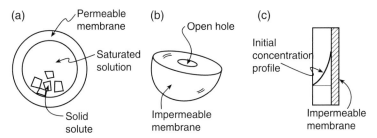

Fig. 19.1-3. Three devices giving near zero-order release. These methods are often superior to those in the previous figure.

and slower. At small times it will show what in French is called "le burst effect" and be far from the zero-order release which we seek. This type of release is not what we want either.

The contrast between the zero-order release which we seek and the other orders which we usually get is vividly shown in Fig. 19.1-1. We clearly need other ways to control release. The first-order release rate, which depends on the drug's solubility and mass transfer, is often too fast and decreases with time. The $t^{1/2}$ release rate can be slower, but is still time dependent, difficult to change.

However, there are many ways in which we can both control drug release and get the zero-order behavior. Three common ways are a reservoir system, an altered device geometry, and an altered initial concentration profile. These three, shown schematically in Figure 19.1-3, are analyzed in the paragraphs that follow.

19.1.1 Reservoir Systems

The simplest system, shown in Fig. 19.1-3(a), consists of a reservoir of saturated drug solution that contains extra solid drug. The drug is released from the reservoir by diffusion across the surrounding permeable membrane into a large volume with near-zero drug concentration. The solution inside remains saturated because the solid keeps quickly dissolving; thus the drug's flux is a constant, and the amount released is given by

$$M = \left[\frac{DH}{l} c_1(\text{sat}) \right] At \tag{19.1-5}$$

where (DH) and l are the membrane's permeability and thickness, respectively. We now have the adjustable, constant release rate that we seek.

Reservoir systems have a major advantage and a major disadvantage. The advantage is the enormous flexibility allowed by the choice of membrane properties. This allows the constant drug release rates to be adjusted over a wide range. The disadvantage is the risk of rupture. If this rupture occurs, all the drug may be quickly released, which can be dangerous. While there are ways to guard against the quick release, the risk has often meant reservoirs are not chosen.

19.1.2 Coated Hemispheres

The second system, shown in Fig. 19.1-3(b), consists of a hemisphere of insoluble polymer containing a drug solution. Dissolved drug can diffuse easily in the polymer, which often is slightly swollen with water. However, all of the outer surface of the polymer hemisphere – except a small central hole – is coated with an impermeable barrier. Dissolved drug diffusing within the polymer can only diffuse out of the hemisphere through this small central hole.

Now imagine what happens when this device is placed in water. Often, the drug concentration in this surrounding water will be near zero, at least compared with that in the polymer. As a result, the drug diffuses radially inward, at first with a fast flux, but later with a slower and slower flux. However, while the flux per area is dropping, the area supplying the drug is at the same time increasing. The total drug released M is approximately given by

$$M = [4\pi D R_0 c_1] t \tag{19.1-6}$$

where R_0 is the radius of the inner hole, which is much smaller than the radius of the device; and c_1 is the initial uniform drug concentration. Note that this device is less prone to problems caused by accidental rupture: if the hemisphere breaks open, release can be faster, but not as fast as in a reservoir device.

19.1.3 Uneven Initial Profiles

The third system for achieving steady drug release, shown in Fig. 19.1-3(c), again uses a polymer slab with dissolved drug. Now, however, the drug is not initially present uniformly, but is present at varying low concentrations within the slab. During storage, the slab is kept dry, so that its concentration profile decays only very slowly, perhaps over years. Once the slab is wet, it quickly hydrates uniformly, and the drug starts diffusing out. That near the surface has only a small driving force, that is, a small concentration gradient; but it doesn't have far to diffuse to get out of the slab. The drug deep within the slab has a larger driving force and a larger distance to diffuse. Thus the amount released can be nearly constant with time.

This system also has significant advantages and disadvantages. It is cheap to manufacture. It is little affected by accidental rupture. It may be hard to store: After all, those initial concentration profiles will be decaying, leading to a uniform concentration profile and hence non-zero order release.

Example 19.1-1: Cattle ear tags   Ear tags for protecting cattle from horn flies can reduce the annoyance from the flies and hence enhance weight gain by the cattle. The ear tag consists of a reservoir of mixed permethrin (molecular weight 391 daltons) and piperonyl butoxide, surrounded by 12 cm^2 of a 63-μm-thick membrane. The permethrin is the actual insecticide, whose release is enhanced by the presence of the second compound. The permethrin's release is a constant 19 μg per day; its vapor pressure is estimated to be 0.045 pascal at 25 °C.

What is the permeability of the membrane (in m^2/sec)?

Solution This solution is a routine application of Eq. 19.1-5; the only problem is the units:

$$\frac{dM_1}{dt} = \left[\frac{P}{l}c_1(\text{sat})\right]A\frac{19 \cdot 10^{-6}\text{g}}{24(3600)\text{sec }391\frac{\text{g}}{\text{mol}}}$$

$$= \left[\frac{P}{63 \cdot 10^{-6}\text{m}} \frac{0.045 \cdot 10^{-3}\text{kg/m sec}^2}{8.31\frac{\text{kg}}{\text{sec}^2}\frac{\text{m}^2}{\text{mol }K}298\ K}\right]12 \cdot 10^{-4}\text{m}^2$$

$$P = 1.6 \cdot 10^{-6}\text{m}^2/\text{sec}.$$

The large value of P results from a very large partition coefficient.

Example 19.1-2: Bollworm pheromones You have been using 1.0-mm-diameter micro-capsules filled with a saturated solution of the pheromone gossyplure, for the pink bollworm. When the capsules contain excess solid pheromone, they attract bollworm moths into insecticide-loaded traps for 27 days.

You are considering trying to increase the catch by using the same mass of pheromone in an equal mass of 50-μm microcapsules that have the same thin wall thickness (i.e., the same P/l). How long will the smaller microcapsules last?

Solution Because the wall of the microcapsules is thin and all other properties are unchanged, the gossyplure flux per area out of the capsules is constant. However, the surface area per mass of the 50 μm capsules is larger. In particular, if the microcapsules have about the same density ρ_M, independent of their size, then

$$\frac{\text{surface area}}{\text{mass}} = \frac{1}{\rho_M}\left[\frac{\pi d^2}{\frac{4}{3}\pi\left(\frac{d}{2}\right)^3}\right] = \frac{6}{d\rho_M}$$

where d is the diameter of the capsules. Since the 50-μm particles are 20 times smaller than the 1-mm particles, they will only last 5% as long, or about half a day.

19.2 Controlled Release by Solvent Diffusion

We can also regulate the release of a solute, again like a drug, by diffusion of solvent. As in the previous section, we will usually want the solute's release rate to be nearly constant over time. Unlike the cases in the previous section, we want to use solvent or polymer diffusion, not solute diffusion, to control the solute's release.

19.2.1 Swelling-Controlled Systems

The two common ways to achieve this result are swelling-controlled systems and so-called osmotic pumps. Swelling-controlled systems are simpler. The most com-mon case consists of a slab of a dry, water-soluble polymer containing a dispersion of small particles of drug. Before the slab is wet, drug diffusion is slow. When this slab is dropped in water, the polymer starts to dissolve and the drug is released.

How this release occurs depends critically on the particular polymer–water interactions. For most polymers (but not for most used to control release), the polymer will swell in water even while its edges dissolve in water. In this case, the drug will usually diffuse through the ever-increasing layer of water-swollen gel. The release rate will not be constant; it will often decrease with the square root of time. If this were the only behavior observed, swelling of a drug-containing polymer slab would be a poor route to controlled release.

Interestingly, polymer dissolution does not always involve polymer swelling. In some cases, the polymer must react before it will dissolve. This reaction may be a solvation; in other cases, it may involve hydrolysis of the polymer chain to produce smaller, more mobile segments. Often, the polymer reacts with dilute acid or base in the water.

The polymer dissolution now involves two sequential steps, much like those described for mass transfer and heterogeneous reactions in Section 16.3. In the first step, the reacted, water-soluble polymer must disentangle and detach from other polymer molecules at the surface. In some cases, this detachment may include aspects of non-Fickian diffusion, i.e., type II transport. In the second step, the dissolved polymer must diffuse away from the surface into the surrounding bulk. The polymer's dissolution rate will depend on the overall resistance to mass transfer of these two steps.

In this situation, the drug follows the behavior of the polymer. It remains trapped within the polymer and away from the solution until the polymer reacts; then it is released at the rate of polymer dissolution. This release will normally be at zero order. Moreover, because the drug's release is controlled by the properties of the polymer and not of the drug, this swelling system can be successful for many drugs. This is an advantage compared with the reservoir systems discussed above.

19.2.2 Osmotic Pumps

The second method of obtaining constant drug release with solvent diffusion is the osmotic pump, shown schematically in Fig. 19.2-1. In its simplest form, this device consists of a rigid housing, one end of which is covered by a semipermeable membrane. The other end of the housing is open only through a small hole, which leads into a bag filled with a solution or a dispersion of the drug. The elastic bag does not completely fill the housing; the remainder is filled with salt crystals suspended in saturated brine.

The osmotic pump works as follows. The hole attached to the drug reservoir is so small that little drug diffuses out of it. Salt can't diffuse out of the semipermeable

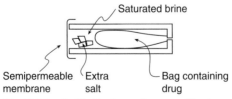

Fig. 19.2-1. An idealized osmotic pump. The drug release does not depend on drug properties, but on those of the semipermeable membrane.

membrane, either. However, water can diffuse across this membrane into the pump. If the membrane is truly semipermeable and at uniform pressure, then the volumetric water flux j_v is:

$$j_v = L_p \Delta \Pi \qquad\qquad (19.2\text{-}1)$$

where L_p is the solvent's permeability and $\Delta \Pi$ is the osmotic pressure. However $\Delta \Pi$ is proportional to the salt concentration (cf. Eq. 18.3-4), and this concentration is always saturated because the solution contains suspended salt crystals. As more water enters, more salt dissolves; because water diffusion is slower than the salt dissolution, the osmotic pressure remains constant. Thus j_v is constant, independent of time.

The constant water flux j_v now squeezes the drug-containing bag, ejecting drug solution at a constant rate. This ejection is much faster than any drug diffusion. More importantly, the drug ejection is not only constant, but independent of the properties of the drug. It depends only on the rate of osmosis, on the rate of water uptake by the salt solution. Like the dissolving polymer, this form of controlled release is generic because all kinds of drugs can be released at the same fixed constant rate. This fixed rate depends on properties like the geometry and water permeability of the semipermeable membrane. These properties can be adjusted independently of the drug properties.

The osmotic pump is an important method of controlled release. Its disadvantages are similar to those of other reservoir systems. The chief one is again the risk of rupture, which might accidentally release all of the drug in the device at once. This release is potentially dangerous and can be a major deterrent to the use of this technology.

Example 19.2-1: Steady delivery of a water-insoluble drug We want to develop a device to deliver nifedipine, a calcium channel blocker with extremely low water solubility (<10 ppm). We need to deliver the drug as a 0.006 g/cm^3 suspension at a rate of 25 µg/hr using an osmotic pump functioning like that shown in Fig. 19.2-1. The osmotic pump uses a brine whose saturated concentration is 10^{-3} M. If the semipermeable membrane that you plan to use has permeance (P/l) of $1.4 \cdot 10^{-4}$ cm/sec, what membrane area will we need?

Solution The volume of drug solution per time is just the drug delivery rate divided by the concentration:

$$\frac{25 \cdot 10^{-6}\,\text{g}}{\text{hr}} \left(\frac{\text{hr}}{3600\,\text{sec}}\right) \frac{\text{cm}^3}{0.006\,\text{g}} = 1.16 \cdot 10^{-6}\,\frac{\text{cm}^3}{\text{sec}}$$

This flow of drug suspension equals the osmotic flow of water:

$$1.16 \cdot 10^{-6}\,\frac{\text{cm}^3}{\text{sec}} \left[\frac{\text{mol}}{18\,\text{cm}^3}\right] = \frac{6.43 \cdot 10^{-8}\,\text{mol}}{\text{sec}} = \left(\frac{P}{l}\right) A \Delta c_1$$

$$= \left(\frac{1.4 \cdot 10^{-4}\,\text{cm}}{\text{sec}}\right) A \left(\frac{0.006\,\text{equivalent}}{\text{cm}^3}\right)$$

$$A = 0.075\,\text{cm}^2$$

You need a membrane that if square is about 3 mm on a side.

19.3 Barriers

The other sections of this chapter, indeed the rest of this book, are largely concerned with making diffusion as rapid as possible. This is because diffusion coefficients are commonly small, and so limit the rate at which the process occurs. Such slow speeds are a major limitation for chemical separations. Diffusion limits the rate of acid gas treating and the rate of leaching of soybeans. Diffusion causes nonideal breakthrough curves in adsorption and makes distillation inefficient. Thus most of this book is concerned with ways to increase diffusion, for example, by stirring or by chemical reaction.

In contrast, this section aims to decrease diffusion, often by as much as possible. This decrease is often important to retain flavors in foods, or to ensure chemical stability of herbicides or pesticides. Such decreases, a key part of the packaging business, are most obviously identified by considering the flux j_1 across a thin film

$$j_1 = \frac{DH}{l} \Delta c_1 \tag{19.3-1}$$

where D is the solute's diffusion coefficient in the film; H is its partition coefficient; l is the film's thickness; and Δc_1 is the concentration difference across the film, often fixed by the particular situation. Obviously, we can get a smaller flux by using a thicker film, or by choosing a film material with a small H. These obvious first steps are the sensible way to start.

Sometimes, however, we want a smaller flux than changes in l and H can provide. To do so, we must reduce the effective value of D, at least at short times. To see how we can do so, we consider the system shown in Fig. 3.5-1. We imagine that the upper volume in this figure is the interior of our package, and that the lower volume is the surroundings. The volume of the package will, of course, be much smaller than the volume of the surroundings, but that does not affect the arguments below. We see by analogy with Example 3.5-1 that the concentration in the package is given by

$$c_1 = \left[\frac{A\,c_{10}}{V}\right] \frac{DH}{l} (t - t_{\text{lag}}) \tag{19.3-2}$$

The quantity in square brackets is a geometrical parameter, characteristic of this experiment. Thus data like those in Figure 3.5-1 can be described with two parameters: a leak rate (DH/l) and a large time t_{lag}.

We can retard diffusion either by decreasing the leak rate or by increasing the lag time. In the paragraphs below, we discuss three specific cases. First, we look at ways to choose materials with smaller values of the diffusion coefficient D. Second, we look at decreasing D – and increasing the lag t_{lag} – by using composite films. Third, we consider how films with sacrificial reagents can increase t_{lag}.

19.3.1 Smaller Diffusion Coefficients

The most obvious way to reduce diffusion in packaging is to use a metal foil. Such a foil, made of the solid metal, has a diffusion coefficient millions of times smaller than that in a polymer film. However, such films are expensive and opaque. For most consumer packaging we strongly prefer transparent films. We want not metals but high polymers.

We can reduce diffusion by working with polymers with a high glass transition temperature T_g. Above the T_g, the polymer behaves as a viscous liquid, with diffusion coefficients typically around 10^{-7} cm^2/sec. Below the T_g, the polymer forms a glass willing to form small crystals whose growth is inhibited by the polymer chains running between crystals. Diffusion in such crystals is small, like that in metals; diffusion between crystals includes Knudsen transport and diffusion in a viscous liquid.

We will normally choose materials which are below their T_g to be diffusion barriers. For flavors, this means choosing amorphous carbohydrates, especially mallodextrins. We will choose high molecular weight materials to get robust mechanical properties but lower molecular weight materials to have fewer intercrystalline flaws. Fewer flaws mean lower permeabilities for solutes like oxygen. We will often manufacture these systems by spray drying to produce 100-μm capsules containing 1-μm dispersed droplets of the flavor. While I know this is only an approximation, I think of these systems as hard candy with dispersed flavor.

19.3.2 Composite Films

The second route for a diffusion barrier is to add an impermeable filler to a polymer particle. Fillers of sand or clay are good choices if their particle size is below the wavelength of visible light. Under these conditions, the film will remain transparent. If the volume fraction of the fillers is high, the mechanical properties of the film, like Young's modulus, can be compromised. This should not be crippling because we can, in principle, reduce the permeability and match the Young's modulus by starting with a polymer with a smaller Young's modulus.

The problem with composites is that the continuous phase dominates the diffusion. For example, (impermeable) spherical particles of sand have little effect on diffusion, as shown in Section 6.4.4. The largest exception, summarized by Eq. 6.4-26, is for films with aligned impermeable flakes. There, we can achieve reductions in diffusion of 10–50 times as shown in Fig. 19.3-1.

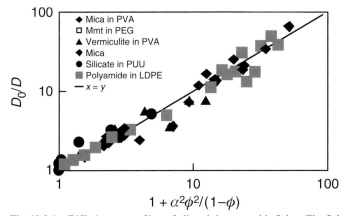

Fig. 19.3-1.   Diffusion across films of aligned, impermeable flakes. The flakes increase the lag before permeation and reduce permeation once it occurs. The permeation reductions shown vary with the amount and shape of the flakes. (Data from De Rocher *et al.*, 2004.)

19.3.3 Sacrificial Reagents

A third way to build a diffusion barrier is to include a sacrificial reagent within the film. For example, we can use a polymer with unsaturated double bonds to consume diffusing oxygen or a film with suspended particles of an inorganic base to capture and consume acids. If the reaction is rapid and irreversible, it will delay the penetration of the oxygen or the acid across the film. It will increase the lag, though it will not change the steady-state leak rate.

To estimate the size of this effect, we assume that the barrier includes a diffusion zone of thickness l' within a film of thickness l. The immobile sacrificial reagent is at a constant concentration c_{20} throughout the zone $(l-l')$. At the position $z = l'$, the reaction occurs

$$\text{species "1"} + \text{species "2"} \quad \rightarrow \quad \text{products} \tag{19.3-3}$$

In the diffusion zone $z < l'$, the concentration c_2 is zero, all consumed by the solute c_1. This unwanted solute, which we are trying to exclude, is diffusing through this zone to the reaction front where $z = l'$.

We can now write a mass balance on the sacrificial reagent to find how these zones change with time

$$\begin{bmatrix} \text{amount of} \\ \text{reagent "2"} \\ \text{consumed} \end{bmatrix} = \begin{bmatrix} \text{diffusion} \\ \text{flux of} \\ \text{solute "1"} \end{bmatrix} \tag{19.3-4}$$

$$\frac{d}{dt}\left[(Al')\, c_{20}\right] = \frac{DHA}{l'}\,(c_{10} - 0)$$

and A is the cross-sectional area. This is subject to the initial condition:

$$t = 0, \quad l' = 0 \tag{19.3-5}$$

Integrating, we find

$$l' = \left[\frac{2DHc_{10}t}{c_{20}}\right]^{\frac{1}{2}} \tag{19.3-6}$$

When the sacrificial reagent is completely consumed, l' equals the total thickness l, and the film will start to leak. The time when the leak begins is

$$t_{lag} = \frac{l^2 c_{20}}{2DHc_{10}} \tag{19.3-7}$$

Estimates of this lag time are in close agreement with those measured experimentally, as shown in Fig. 19.3-2.

The case of an immobile sacrificial reagent, which is useful for barriers, stands in sharp contrast to films with mobile reagents, as summarized in Table 19.3-1. These results are most easily discussed in terms of the leak rate and the lag time. If the reagent is immobile, the leak rate is not altered from the case of simple diffusion, as shown in the second column of the table. The leak rate is the same as a nonreactive film. However, the

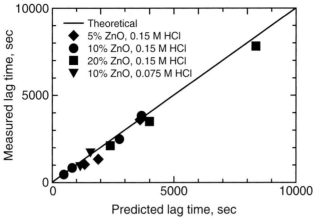

Fig. 19.3-2. Diffusion of hydrochloric acid across films with sacrificial zinc oxide. The graph plots vs. time the acid concentration downstream, which is initially acid free. The sacrificial zinc oxide greatly increases the lag before acid permeation occurs, but has much less effect on the permeation after the oxide is exhausted. (Data from Yang, et al., 2001.)

lag time does increase far beyond that expected for a reagent-free film. This is how an immobile reagent can improve a barrier.

The results for a reversible reaction with an immobile species are similar, as also shown in Table 19.3-1. Again, the steady-state leak rate is the same as the nonreactive case. The lag can increase dramatically, as discussed in Example 2.3-2. Again, with an immobile reagent, the lag can increase, but the leak is unchanged.

Table 19.3-1 *Barrier properties of different reactive films.*

System	Leak rate	Lag time
Simple diffusion	$j_1 = \dfrac{DHC_{10}}{l}$	$\dfrac{l^2}{6D}$
Diffusion and irreversible reaction with immobile species	$j_1 = \dfrac{DHC_{10}}{l}$	$\dfrac{l^2}{6D}\left(1 + \dfrac{3c_{20}}{HC_{10}}\right)$
Diffusion and reversible reaction with immobile species	$j_1 = \dfrac{DHC_{10}}{l}$	$\dfrac{l^2(1 + K)}{6D}$
Diffusion and reversible reaction with mobile species	$j_1 = \dfrac{Dc_{20}}{l}$	$\dfrac{l^2}{6D(1 + KC_{10})}$

In this table, the diffusion coefficient is assumed to have the same value D for all mobile species.

In contrast, a mobile reactant which binds reversibly increases the leak rate and decreases the lag time. The limits shown in Table 19.3-1, for a strongly reactive mobile species, are an example of facilitated diffusion, detailed in Section 18.4. Now the leak rate depends on the concentration of mobile reagent c_{20} and becomes independent of the concentration of diffusing species. The lag is shorter, not longer, in the presence of mobile reactant because the process now occurs as a form of facilitated diffusion. Mobile reagents make barriers worse, not better.

Example 19.3-1: Cesium barriers   Cesium-137 is one of the most dangerous species in nuclear waste not because it is especially radioactive, but because it is water-soluble. One way to capture this species is by selective adsorption on crystalline silicon titanate (CST). If we add 10 wt% CST to a geotextile used to retard accidental cesium releases, by how much will the lag time be increased?

Solution   From Table 19.3-1, we see that

$$\frac{\text{lag with CST}}{\text{lag without CST}} = 1 + K$$

With a feed concentration of 10^{-3} M $CsNO_3$, we expect K will be about 20. Thus the ratio of lags should also be about 20. By experiment, we find the ratio is about 30. The performance is better than predicted, possibly because of the nonlinear adsorption isotherm for cesium on CST.

19.4 Diffusion and Phase Equilibrium

Thus far in this chapter, we have described diffusion as a means of controlling or retarding release of solutes like drugs or flavors. These solutes may initially be present as a solution or a suspension. If the suspension dissolves quickly, diffusion is key. We now turn to other cases of diffusion in two-phase systems. Like controlled release, these cases can show dramatic and unexpected behavior. Like controlled release, this behavior is the consequence of rapid chemical reactions.

The effects we want to discuss are exemplified by the dissolution of slaked lime, $Ca(OH)_2$, in aqueous solutions of a strong acid like HCl:

$$Ca(OH)_2 + 2H^+ \rightleftharpoons Ca^{2+} + 2H_2O \tag{19.4-1}$$

How this dissolution proceeds depends on the relative speed of diffusion and reaction. When the bulk of the solution next to the solid is rapidly stirred, the acid can diffuse to the solid's surface very quickly. It then reacts with the solid's surface. If the solid is essentially impermeable, containing a very few pores, then any ions produced by the dissolution are quickly swept back into the bulk solution. Because diffusion and chemical reaction occur sequentially, the overall dissolution rate is like that of a heterogeneous reaction, depending on the sum of the resistances of diffusion and of reaction (see Chapter 16). Such a process represents an important limit of corrosion and is that usually studied.

Alternatively, the solution next to the solid may not be well stirred, and the solid may be highly porous, as shown schematically by Fig. 19.4-1. In this case, the acid

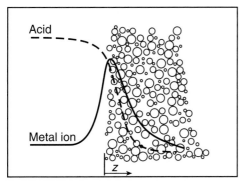

Fig. 19.4-1. Dissolution of a porous solid. In this schematic representation, acid diffusing from left to right is consumed by chemical reaction with the solid. The metal ions produced by this reaction can, under some conditions, diffuse into the pores and precipitate as more solid. (Data from Cussler and Featherstone, 1981, with permission.)

concentration will drop as it approaches the solid's surface and continue to drop within the solid's pores. The ions produced as the result of the acid–solid reaction will be present in highest concentration near the solid's surface. From this maximum, they can diffuse out into the bulk solution or further into the porous solid. Within the solid, diffusion and reaction occur simultaneously, so that the overall dissolution rate is like that of a homogeneous reaction, not a simple sum of the resistances of diffusion and reaction (see Chapter 17).

To calculate the dissolution rate of the porous solid r_1, we write continuity equations for calcium ions (species 1) and protons (species 2):

$$\frac{\partial c_1}{\partial t} = D\frac{\partial^2 c_1}{\partial z^2} + r_1 \tag{19.4-2}$$

$$\frac{\partial c_2}{\partial t} = D\frac{\partial^2 c_2}{\partial z^2} - 2r_1 \tag{19.4-3}$$

where, as before, we assume that the diffusion coefficients are equal. We also assume that the reaction in Eq. 19.4-1 is so fast that it reaches equilibrium:

$$[Ca^{2+}] = \left\{\frac{K'[Ca(OH)_2]}{[H_2O]^2}\right\}[H^+]^2 \tag{19.4-4}$$

where K' is the equilibrium constant of the dissolution. The $Ca(OH)_2$ is solid and of unit activity; the water is present in excess; so everything in the braces is a new equilibrium constant K. Thus

$$c_1 = Kc_2^2 \tag{19.4-5}$$

$$\frac{\partial c_1}{\partial t} = 2Kc_2\frac{\partial c_2}{\partial t} \tag{19.4-6}$$

and

$$\frac{\partial^2 c_1}{\partial z^2} = 2K\left[\left(\frac{\partial c_2}{\partial z}\right)^2 + c_2\frac{\partial^2 c_2}{\partial z^2}\right] \tag{19.4-7}$$

Inserting these into Eq. 19.4-2 and then combining with Eq. 19.4-3, we find the dissolution rate:

$$r_1 = -\left[\frac{2Dc_1(\partial \ln c_2/\partial z)^2}{1+4(c_1/c_2)}\right] \tag{19.4-8}$$

The exact values depend on the particular boundary conditions involved.

The remarkable feature about Eq. 19.4-8 is that it predicts that the dissolution rate within the pores is negative, because all terms in the brackets in Eq. 19.4-8 are positive. In physical terms, this means that $Ca(OH)_2$ will precipitate in front of the acid wave shown in Fig. 19.4-1. This predication is verified experimentally: dissolution can produce precipitation.

To explain the origin of these effects in more physical terms, we imagine that the boundary conditions in the porous solid are those of free diffusion (see Section 2.3):

$$t < 0, \quad \text{all } z, \quad c_1 = 0, \quad c_2 = 0 \tag{19.4-9}$$

$$t > 0, \quad z = 0, \quad c_1 = c_{10}, \quad c_2 = c_{20} \tag{19.4-10}$$

$$z = \infty, \quad c_1 = 0, \quad c_2 = 0 \tag{19.4-11}$$

For the moment, we pretend that there is no reaction, so $r_1 = 0$. Then, if both species have the same diffusion coefficient D,

$$\frac{c_1}{c_{10}} = \frac{c_2}{c_{20}} = 1 - \text{erf}\frac{z}{\sqrt{4Dt}} \tag{19.4-12}$$

We thus know c_1 as a function of z and t. We also know c_1 as a function of c_2, and we can graph this variation as the diffusion path in Fig. 19.4-2.

However, when there is a reaction, the concentrations in the system must follow the equilibrium in Eq. 19.4-5, producing a path that is also shown in Fig. 19.4-2. This equilibrium line is essentially a phase diagram. If at fixed acid concentration c_2, the dissolved calcium concentration c_1 is below this equilibrium line, then solid will dissolve to produce more c_1. If at fixed c_2, c_1 is above this line, then solid will precipitate to reduce c_2.

When we compare the equilibrium and the diffusion paths, we see that diffusion of H^+ into porous $Ca(OH)_2$ will tend to carry us into the two-phase region, and so produce precipitation in front of the acid wave. In contrast, for the reaction

$$CaCO_3 + H^+ \rightleftharpoons Ca^{2+} + HCO_3^- \tag{19.4-13}$$

we might write

$$c_1^2 = Kc_2 \tag{19.4-14}$$

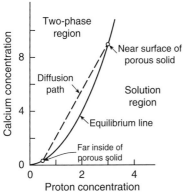

Fig. 19.4-2. How dissolution can cause precipitation. When no solid is present, the calcium and proton concentrations are approximately linearly related, as shown by the dashed diffusion path. When solid is present, they are related by the equilibrium solubility, although diffusion constantly tries to push the system into the two-phase region.

where 1 is Ca^{2+} or HCO_3^- and 2 is H^+. Here, the equilibrium curve would show the opposite curvature, and we would expect dissolution both at the solid's surface and within the pores.

This type of effect, which is intuitively surprising, is also responsible for spontaneous emulsification in liquids (Ruschak and Miller, 1972), for some phase separations in metals (Kirkaldy and Brown, 1963), and for the formation of fogs in gases (Toor, 1971). These problems are conceptually parallel, but their details are more complex. This complexity comes largely from the replacement of simple stoichiometric relations like Eqs. 19.4-1 and 19.4-13 by more complicated phase diagrams. In many interesting cases, these phase diagrams do not have a single species present in excess, and so they require representation on triangular coordinates. The phase diagrams may also imply additional interfaces. This area contains unsolved problems of practical significance.

19.5 Conclusions

This short chapter suggests how diffusion and phase changes can sometimes interact. For example, when drug release is controlled by slow drug diffusion across a membrane, the rate of release can be constant if the drug concentration on one side of the membrane is kept saturated, refreshed by the solid drug quickly dissolving. When acid diffuses into a porous solid, the acid can dissolve the solid's surface while it effects precipitation of the same solid below the surface. While the examples discussed here are not exhaustive, they illustrate the more basic ideas in the book that find broad application.

Questions for Discussion

1. What is zero-order release? Why is it desirable?
2. Briefly explain why the coated hemisphere shown in Fig. 19.1-3(b) can give approximately zero-order release.

3. How does an osmotic pump work?
4. Write equations describing flavor release from a spherical microcapsule.
5. Estimate flavor release from a microcapsule which is flavor impermeable but which is burst by osmotic flow.
6. Write equations describing release from a polymer–drug matrix whose dissolution is controlled by polymer hydrolysis.
7. How much will 10 vol% spheres reduce the permeability?
8. How much will 10 vol% flakes of aspect ratio 30 reduce the permeability?
9. Imagine you have an activated carbon which can absorb 1% of its weight in polychlorinated biphenyls (PCBs). How much should such a carbon retard PCB release across a 0.2-cm sheet of polyethylene?
10. Imagine a polymer layer containing 1% suspended particles of NaOH. If such a film is challenged with acid, will extra NaOH precipitate in advance of the reaction front?

Problems

1. Drugs are sometimes administered as solid particles suspended in water and trapped in microcapsules like those shown at the left of Fig. 19.1-3. Because the drug concentration in the surrounding solution is near zero, the flux of drug is constant with time. In some cases, this allows the drug dosage to be dramatically reduced. (a) Assuming the capsule wall is thin, find the total steady-state flux in moles per time out of the microcapsule. You may neglect the osmotic flow and diffusion of water. (b) Sketch this flux versus the wall thickness $(R_0 - R_i)$. (c) Explain the previous result in physical terms. (d) Now consider diffusion as a ternary system of drug, water, and wall. How could the flux change?

2. Microcapsules are small polymer bubbles in which a thin polymer wall surrounds a core of active solute. These microcapsules are often used to deliver special ingredients as part of a product. For example, in a hand soap, microcapsules could release perfume while one is washing their hands.
    The trouble with these capsules is that the thin polymer wall of thickness l is often very slightly porous with porosity ε. In this case, perfume diffuses through the non-porous part of the wall, which has a diffusion coefficient D and a partition coefficient H. It also diffuses through the pores with a different diffusion coefficient D'. (a) If the perfume concentration inside the capsules is a (roughly) constant c_{10}, what is the total flux j across the wall? (b) What is the total resistance across the wall?

Further Reading

Baker, R. W. (1987). *Controlled Release of Biologically Active Agents*. New York: Wiley.

Cussler, E. L. and Featherstone, J. D. (1981). *Science*, **213**, 1018.

De Rocher, J. P., Gettlefinger, B. T., Wang, J., Nuxoll, E. E., and Cussler, E. L. (2005). Barrier membranes with different sizes of aligned flakes. *Journal of Membrane Science*, **254**, 21.

Kirkaldy, J. S. and Brown, L. C. (1963). *Canadian Metallurgical Quarterly*, **2**, 89.

Rathbone, J. J., Hadgraft, J., and Roberts, M. S. (2002). *Modified-Release Drug Delivery*. New York: Informa Healthcare.

Ruschak, K. J. and Miller, C. A. (1972). *IEC Fundamentals*, **11**, 534.

Smith, K. L. and Herbig, S. M. (1992). 'Controlled Release'. *In Membrane Handbook*, eds. Ho, W.S. and K.K. Sirkar. New York: Van Nostrand Reinhold.

Toor, H. L. (1971). *American Institute of Chemical Engineers Journal,* **17**, 5.

Ubink, J. and Schoonman, A. (2003). '*Flavor Delivery and Systems*'. *In Kirk–Othmer Encyclopedia of Chemical Technology.* New York: Wiley.

Wise, D. L. (2000). *Handbook of Pharmaceutical Controlled Release Technology.* Boca Raton, FL: CRC.

Yang, C. F., Nuxoll, E. E., and Cussler, E. L. (2001). Reactive barrier films. *AIChE J.,* **47**, 295.

CHAPTER 20

Heat Transfer

In this chapter, we briefly describe fundamental concepts of heat transfer. We begin in Section 20.1 with a description of heat conduction. We base this description on three key points: Fourier's law for conduction, energy transport through a thin film, and energy transport in a semi-infinite slab. In Section 20.2, we discuss energy conservation equations that are general forms of the first law of thermodynamics. In Section 20.3, we analyze interfacial heat transfer in terms of heat transfer coefficients, and in Section 20.4, we discuss numerical values of thermal conductivities, thermal diffusivities, and heat transfer coefficients.

This material is closely parallel to the ideas about diffusion presented in the rest of this book. This parallelism is not unexpected, for heat transfer and mass transfer are described with equations that are very similar mathematically. The material in Section 20.1 is like that in Chapter 2, and the general equations in Section 20.2 are conceptually similar to those in Chapter 3. The material on heat transfer coefficients in Section 20.3 closely resembles the mass transfer material in Chapters 9 through 15, and the numerical values in Section 20.4 are parallel to those in Chapters 5 and 8.

Thus we are abstracting ideas of heat transfer in a few sections, whereas we detailed similar ideas of mass transfer over many chapters. This represents a tremendous abridgment. As those skilled in heat transfer recognize, the heat transfer literature is immense, of far greater size than the mass transfer literature. To be sure, this book is about diffusion, and so an emphasis on mass transfer is appropriate. But if the description of heat transfer is to be so terse, why include it at all?

I have included the description of heat transfer because I want to discuss simultaneous heat and mass transfer in the next chapter. This simultaneous transport process is important practically and is interesting intellectually, with implications ranging beyond the particular problems presented. However, to discuss this simultaneous process, we need to assure a background in heat transfer. I expect that many who read this book will not have such a background, for this topic is usually buried well inside the engineering curriculum. Accordingly, this chapter is a synopsis to provide this background.

20.1 Fundamentals of Heat Conduction

The fundamental understanding of heat conduction rests on the work of Jean Joseph Fourier, who was born March 21, 1768. Orphaned before 10 years of age, Fourier got an education by joining the church. He started teaching school but then advanced rapidly through the government bureaucracy as the French Revolution eliminated those above him. It was a risky business, and Fourier spent some time in prison. Nonetheless, under Napoleon he became prefect of the department of Isère. He did his work on heat conduction while holding that position. It was as if the governor of Minnesota was doing first-rate mathematical physics in his spare time, during evenings and on weekends.

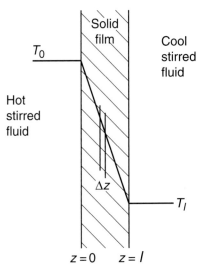

Fig. 20.1-1. Steady heat conduction across a thin film. Heat conduction across a thin film is like diffusion across a membrane (see Section 2.2). The resulting temperature profile is linear, and the flux is constant and inversely proportional to the film thickness *l*.

In 1807, Fourier presented his work to the most qualified doctoral committee in history, including Laplace, Lagrange, and Monge. Poisson was also involved. Lagrange was critical; so Fourier's degree was delayed.

In his 1807 paper, Fourier used the experiments of Biot to argue that the heat flux *q* should be proportional to the temperature gradient ∇T:

$$q = -k_T \nabla T \tag{20.1-1}$$

where the proportionality constant k_T is the thermal conductivity. Note that the dimensions of k_T are not simple, but are commonly energy per length temperature time. This equation is a close parallel to Fick's law; indeed, as explained in Section 2.1, Fick developed the diffusion law by analogy with Fourier's work.

To calculate heat fluxes or temperature profiles, we make energy balances and then combine these with Fourier's law. The ways in which this is done are best seen in terms of two examples: heat conduction across a thin film and into a semi-infinite slab. The choice of these two examples is not casual. As for diffusion, they bracket most of the other problems, and so provide limits for conduction.

20.1.1 Steady Heat Conduction Across a Thin Film

As a first example, consider a thin solid membrane separating two well-stirred fluids, as shown schematically in Fig. 20.1-1. Because one fluid is hotter than the other, energy will be conducted from left to right across the thin film. To find the amount of conduction, we make a steady-state energy balance on a thin layer located between z and $z + \Delta z$:

$$\begin{pmatrix} \text{energy} \\ \text{accumulation} \end{pmatrix} = \begin{pmatrix} \text{energy} \\ \text{conducted in} \end{pmatrix} - \begin{pmatrix} \text{energy} \\ \text{conducted out} \end{pmatrix} \tag{20.1-2}$$

At steady state, this is

$$0 = Aq|_z - Aq|_{z + \Delta z} \qquad (20.1\text{-}3)$$

where A is the cross-sectional area and q is the heat flux in the z direction. Dividing by the layer's volume $A\Delta z$ and taking the limit as this volume goes to zero, we find

$$0 = -\frac{dq}{dz} \qquad (20.1\text{-}4)$$

Combining with Fourier's law for conduction in the z direction, we find

$$0 = k_T \frac{d^2 T}{dz^2} \qquad (20.1\text{-}5)$$

This differential equation is subject to the boundary conditions

$$z = 0, \quad T = T_0 \qquad (20.1\text{-}6)$$

$$z = l, \quad T = T_l \qquad (20.1\text{-}7)$$

Integration to find the temperature profile is simple:

$$T = T_0 + (T_l - T_0)\frac{z}{l} \qquad (20.1\text{-}8)$$

Note that this profile, shown in Fig. 20.1-1, does not depend on the thermal conductivity. Finding the heat flux is also easy:

$$q = -k_T \frac{dT}{dz}$$

$$= \frac{k_T}{l}(T_0 - T_1) \qquad (20.1\text{-}9)$$

This is a complete parallel to Eq. 2.2-10.

The results in Eqs. 20.1-8 and 20.1-9 are extraordinarily useful, the basis of much thinking about heat transfer. Still, you may have trouble taking them seriously because you are not mathematically intimidated by the derivation. To test your understanding, try to answer the following questions.

(1) *How is the temperature profile changed if the fluid at $z = 0$ and T_0 is replaced by a different liquid that is at the same temperature?* There is no change as long as the interfacial temperature is constant.

(2) *What will the temperature profile look like across two thin slabs of different materials that are clamped together?* In steady state, the heat flux is constant. Thus the temperature drop across the poorly conducting slab will be larger than that across the better conductor.

(3) *Imagine that for the system in Fig. 20.1-1 the fluid at $z = l$ has a small volume, V, but the fluid at $z = 0$ has a very large volume. How will T_l change with time?* To answer this, we write an energy balance on the fluid at $z = l$:

$$\frac{d}{dt}\left(\rho V \hat{C}_p T_l\right) = Aq|_{z = l}$$

$$= A\frac{k_T}{l}(T_0 - T_l) \qquad (20.1\text{-}10)$$

in which ρV is the mass of fluid located at $z = l$, $\hat{C}_p$ is the specific heat capacity of this fluid, and A is the area available for heat transfer. Initially, the temperatures are known:

$$t = 0, \quad T_l = T_l(t = 0) \tag{20.1-11}$$

Integrating,

$$T_l = T_l(t = 0) + [T_0 - T_l(t = 0)](1 - e^{-(k_T A/l\rho V\hat{C}_p)t}) \tag{20.1-12}$$

The temperature rises to a limit of T_0. Note that in this analysis, we use the steady-state result for a thin film in conjunction with an unsteady energy balance on the fluid. The justification for this is that the film volume is much less than the fluid volume. The same justification was used for the diaphragm-cell method of measuring diffusion coefficients (see Example 2.2-4).

20.1.2 Unsteady Heat Conduction into a Thick Slab

Our second example involves thermal conduction into the large solid slab shown in Fig. 20.1-2. This slab is the antithesis of the thin film discussed earlier. To be sure, both the slab and the film are in contact at $z = 0$ with hot fluid at T_0; but here, the slab has no other boundary. Instead, far within the slab the temperature remains equal to the initial value T_∞.

To solve this problem, we again make an energy balance on a differential layer located between z and $z + \Delta z$:

$$\begin{pmatrix} \text{accumulation} \\ \text{of energy} \end{pmatrix} = \begin{pmatrix} \text{energy} \\ \text{conduction in} \end{pmatrix} - \begin{pmatrix} \text{energy} \\ \text{conduction out} \end{pmatrix} \tag{20.1-13}$$

When I derive these equations, I find it easiest to build up the terms I want:

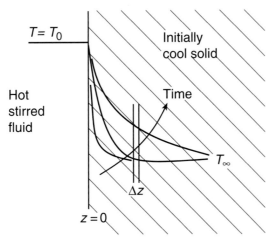

Fig. 20.1-2. Unsteady heat conduction into a semi-infinite slab. The temperature profile in this case is an error function, just like the concentration profile in Section 2.3. This profile depends on the variable $z/\sqrt{4\alpha t}$, where $\alpha \, (= k_T/\rho\hat{C}_p)$ is the thermal diffusivity.

$A\Delta z$ is the volume of the differential layer,
$\rho A\Delta z$ is the mass,
$\rho A\Delta z \hat{C}_v$ is its energy per temperature,
$\rho A\Delta z \hat{C}_v T$ is its energy, and
$\partial/\partial t(\rho A\Delta z \hat{C}_v T)$ is the energy accumulation.
Thus

$$\frac{\partial}{\partial t}\left(\rho A\Delta z \hat{C}_v T\right) = Aq|_z - Aq|_{z + \Delta z} \tag{20.1-14}$$

Dividing by $\rho A\Delta z \hat{C}_v$ and taking the limit as Δz becomes small,

$$\frac{\partial T}{\partial t} = -\frac{1}{\rho \hat{C}_v}\frac{\partial q}{\partial z} \tag{20.1-15}$$

Combining with Fourier's law and making the accurate assumption that in a solid the heat capacities at constant volume and at constant pressure are the same, i.e. $C_v = C_p$,

$$\frac{\partial T}{\partial t} = \alpha \frac{\partial^2 T}{\partial z^2} \tag{20.1-16}$$

where $\alpha\ (= k_T/\rho \hat{C}_p)$ is the thermal diffusivity, with dimensions of length squared per time. This equation occurs so frequently that it is sometimes called "the heat conduction equation," as if there were no other forms.

For the specific case of interest here, Eq. 20.1-16 is subject to the initial condition

$$t = 0, \quad \text{all } z, \quad T = T_\infty \tag{20.1-17}$$

and to the boundary conditions

$$t > 0, \quad z = 0, \quad T = T_0 \tag{20.1-18}$$

$$z = \infty, \quad T = T_\infty \tag{20.1-19}$$

For these boundary conditions, the solution to Eq. 20.1-16 is easily obtained by combination of variables, just as was discussed in Section 2.3. The results are

$$\frac{T - T_0}{T_\infty - T_0} = \text{erf}\frac{z}{\sqrt{4\alpha t}} \tag{20.1-20}$$

$$q|_{z=0} = -k_T \frac{\partial T}{\partial z}\bigg|_{z = 0}$$

$$= \sqrt{k_T\rho\hat{C}_p/\pi t}\ (T_0 - T_\infty) \tag{20.1-21}$$

The temperature profile is the same as that for diffusion but with the thermal diffusivity replacing the diffusion coefficient.

As in the case of the thin film, this result may be so familiar that it is difficult to think about carefully. As before, you can test your understanding by trying to answer these three questions:

(1)   To what depth does the temperature change penetrate in, for example, a steel slab? To a first approximation, the temperature changes occur to a depth where $z^2/4\alpha t$

equals unity. For steel, α equals about $0.1 \text{ cm}^2/\text{sec}$; if the steel is heated for 10 min, the temperature penetrates about 15 cm. This result is independent of heating or cooling, and it is like similar arguments for diffusion (see Section 2.6).

(2) *How does the flux vary with physical properties for the thick slab as compared with the thin film?* Doubling the temperature difference doubles the heat flux in both cases. Doubling the thermal conductivity increases the flux by $\sqrt{2}$ for the thick slab and by 2 for the thin film. Doubling the heat capacity increases the flux by $\sqrt{2}$ for the thick slab, but has no effect for the steady-state conduction across a thin film.

(3) *How much heat is transferred over a time t_0?* To find this, we integrate Eq. 20.1–21 over time:

$$\begin{pmatrix} \text{total heat} \\ \text{transferred} \\ \text{per area} \end{pmatrix} = \int_0^{t_0} q|_{z\,=\,0}\,dt$$

$$= 2\sqrt{k_T \rho \hat{C}_p t_0/\pi}\,(T_0 - T_\infty) \qquad (20.1\text{-}22)$$

To double the heat transferred in t_0, we need to wait four times as long.

The limits of heat conduction across a thin film and into a thick slab are the two most important cases of a rich variety of examples. This variety largely consists of solutions of Eq. 20.1-16 for different geometries and boundary conditions. The geometries include slabs, spheres, and cylinders, as well as more exotic shapes like cones. The boundary conditions are diverse. For example, they include boundary temperatures that vary periodically because this is important for diurnal temperature variations of the earth. They include boundary conditions in which the heat flux at the surface is related to the temperature of the surroundings, T_{surr}; for example,

$$-k_T \frac{\partial T}{\partial z}\bigg|_{z\,=\,0} = h\left(T|_{z\,=\,0} - T_{\text{surr}}\right) \qquad (20.1\text{-}23)$$

where h is an individual transfer coefficient, a rough analogue to an individual mass transfer coefficient. This type of constraint is sometimes called a radiation condition.

At the same time, Eq. 20.1-16 is only one of a wide variety of energy balances that are useful. These more general balances are the subject of the next section, which comes after some simple examples.

Example 20.1-1: Determining thermal diffusivity A thick slab of a polymer composite at $40\,^\circ\text{C}$ is immersed in a large stirred oil bath kept at $4\,^\circ\text{C}$. A thermocouple 1.3 cm below the slab's surface reads $26.2\,^\circ\text{C}$ after 3 min. What is the thermal diffusivity of the slab?

Solution Because the slab is thick, we can use Eq. 20.1-20:

$$\frac{T - T_0}{T_\infty - T_0} = \text{erf}\frac{z}{\sqrt{4\alpha t}}$$

$$\frac{26.2 - 4}{40 - 4} = \text{erf}\frac{1.3\text{ cm}}{\sqrt{4\alpha(180\text{ sec})}}$$

Thus, $\alpha = 3.1 \cdot 10^{-3}$ cm^2/sec, a typical value for this type of material.

Example 20.1-2: Heat loss from a well-insulated pipe   Imagine a well-insulated pipe used to transport saturated steam (Fig. 20.1-3). How much will the heat loss through the pipe's walls be reduced if the insulation thickness is doubled? Assume that the thermal conductivity of the pipe's walls is much higher than that of the insulation.

    Solution   We begin this problem by making a steady-state energy balance on a cylindrical shell of insulation of volume $2\pi r \Delta r L$:

$$\begin{pmatrix} \text{energy} \\ \text{accumulation} \end{pmatrix} = \begin{pmatrix} \text{energy in minus energy out} \\ \text{by conduction} \end{pmatrix}$$

$$0 = (2\pi r L q)_r - (2\pi r L q)_{r+\Delta r}$$

where r is the radial distance measured from the centre of the shell.

    Dividing by the shell's volume and taking the limit as Δr goes to zero,

$$0 = -\frac{d}{dr}(rq)$$

Integration gives

$$rq = R_0 q_0$$

where q_0, an integration constant, is equal to the heat flux at the pipe's outer surface. Combining this with Fourier's law, we find

$$-rk_T \frac{dT}{dr} = R_0 q_0$$

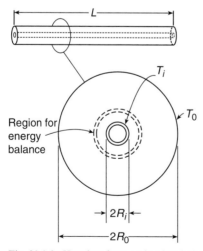

Fig. 20.1-3. Heat loss from an insulated pipe. This problem illustrates the extension of ideas of heat transfer to systems with cylindrical symmetry. When the pipe's diameter is large, the results approach the limit of the thin film shown in Fig. 20.1-1.

or, after rearrangement,

$$q_0 = -\frac{k_T \int_{T_i}^{T_0} dT}{R_0 \int_{R_i}^{R_0} \frac{dr}{r}}$$

The limits of the integrals are those shown in Fig. 20.1-3. Thus

$$q_0 = \frac{k_T(T_i - T_0)}{R_0 \ln(R_0/R_i)}$$

But we want to know the effect of doubling the insulation thickness. In other words, we want to double δ ($= R_0/R_i - 1$). We can rearrange the foregoing equation to show

$$\left(\frac{\text{flux with double insulation}}{\text{flux with single insulation}}\right) = \frac{(1 + \delta)\ln(1 + \delta)}{(1 + 2\delta)\ln(1 + 2\delta)}$$

which is the desired result. A good exercise is to show that this ratio approaches 0.5 as the pipe diameter R_i becomes large (i.e., as δ goes to zero).

20.2 General Energy Balances

Energy balances can be more complicated and more difficult to understand than mass balances. Mass balances are easy because chemical compounds are uniquely defined. For example, glucose not only has a particular ratio of atoms of carbon, hydrogen, and oxygen but also has a specific structure of these atoms. Making a mass balance on glucose is straightforward, and any appearance or disappearance of glucose is described by a chemical reaction.

Energy balances can be more difficult because energy and work can take so many different forms. Internal, kinetic, potential, chemical, and surface energies are all important. Work can involve forces of pressure, gravity, and electrical potential. As a result, a truly general energy balance is extraordinarily complicated, so much so that it is difficult to use.

As a result, most people do not try to use truly general energy balances, but use simplified versions appropriate for special problems. I find this specialization reminiscent of a Tibetan painting that hangs on the wall of my study. The top of the painting shows the Buddha of the Yellow Cap, sitting like a star on a Christmas tree. Other manifestations of the god are spread out below him like the tree's ornaments. These demigods are vividly represented. Some are ferocious, some look ineffective, others seem kind and approachable.

These different forms of the Buddha are like the different forms of the energy equation. All forms of this equation are derived from the same difficult and complex spirit. These derived forms can look very different. Some are much more tractable than others and are more useful for solving specific problems.

In this section, we shall focus our discussion on energy balances for a single pure component, possibly a fluid, that has internal and kinetic energy. The use of a pure fluid is equivalent to our earlier assumption of dilute solution, for the physical properties will be those of the one component. The discussion hinges on three forms of the energy balance. The most general form is

$$\frac{\partial}{\partial t}\rho\left(\hat{U}+\frac{1}{2}v^2\right) = -\nabla \cdot \rho v\left(\hat{U}+\frac{1}{2}v^2\right) - (\nabla \cdot q) + \rho(v \cdot g)$$

$$\begin{pmatrix} \text{energy} \\ \text{accumulation} \end{pmatrix} = \begin{pmatrix} \text{energy convection} \\ \text{in minus that out} \end{pmatrix} + (\text{conduction}) - \begin{pmatrix} \text{work by} \\ \text{gravity} \end{pmatrix}$$

$$-\nabla \cdot \rho v\left(\frac{p}{\rho}\right) - (\nabla \cdot [\tau \cdot v])$$

$$-\begin{pmatrix} \text{work by} \\ \text{pressure} \\ \text{forces} \end{pmatrix} - \begin{pmatrix} \text{work by} \\ \text{viscous} \\ \text{forces} \end{pmatrix}$$

$$(20.2\text{-}1)$$

This scalar equation can be simplified by subtracting the mechanical energy balance and thus removing the kinetic terms:

$$\frac{\partial(\rho\hat{U})}{\partial t} = -(\nabla \cdot \rho v\hat{U}) - (\nabla \cdot q)$$

$$\begin{pmatrix} \text{energy} \\ \text{accumulation} \end{pmatrix} = \begin{pmatrix} \text{energy} \\ \text{convection} \\ \text{in minus that out} \end{pmatrix} + (\text{conduction})$$

$$-p(\nabla \cdot v) - (\tau : \nabla v)$$

$$-\begin{pmatrix} \text{reversible} \\ \text{work} \end{pmatrix} - \begin{pmatrix} \text{irreversible} \\ \text{work} \end{pmatrix} \qquad (20.2\text{-}2)$$

A third useful form can be derived by combining the energy convection and the reversible work as the enthalpy:

$$\frac{\partial(\hat{U})}{\partial t} = -(\nabla \cdot \rho v\hat{H}) - (\nabla \cdot q)$$

$$\begin{pmatrix} \text{energy} \\ \text{accumulation} \end{pmatrix} = \begin{pmatrix} \text{enthalpy} \\ \text{convection} \\ \text{in minus that out} \end{pmatrix} + (\text{conduction})$$

$$+ (v \cdot \nabla p) - (\tau : \nabla v)$$

$$-\begin{pmatrix} \text{part of the} \\ \text{enthalpy} \\ \text{definition} \end{pmatrix} - \begin{pmatrix} \text{irreversible} \\ \text{work} \end{pmatrix} \qquad (20.2\text{-}3)$$

in which $\hat{H}(=\hat{U}+p/\rho)$ is the specific enthalpy.

These equations are formidable, and so may best be understood by comparing them with more easily remembered results. For example, consider the form of the first law most commonly remembered by scientists:

$$\Delta U = Q + W \qquad (20.2\text{-}4)$$

This equation is for a batch system with no flow in or out, and it is restricted to changes in internal energy only. If we simplify Eq. 20.2-2 for these conditions, we have

$$\rho \frac{\partial \hat{U}}{\partial t} = -(\mathbf{V} \cdot \mathbf{q}) - [p(\mathbf{V} \cdot \mathbf{v}) + \boldsymbol{\tau} : \mathbf{V}\mathbf{v}] \tag{20.2-5}$$

The term ΔU on the left-hand side of Eq. 20.2-4 corresponds to the accumulation in Eq. 20.2-5; the term Q in Eq. 20.2-4 refers to the conduction $(-\mathbf{V} \cdot \mathbf{q})$ in Eq. 20.2-5; the work $-W$ is represented by the quantity in brackets in Eq. 20.2-5. Equation 20.2-4 is extensive, referring to the total system. In contrast, Eq. 20.2-5 is intensive, written on a small differential volume within the system.

Another familiar form of the first law used largely by engineers is that for a steady-state open system of fixed volume

$$\Delta H = Q + W_s \tag{20.2-6}$$

where ΔH is the enthalpy change from inlet to outlet and W_s is the shaft work. For such a system, Eq. 20.2-3 becomes

$$0 = -(\mathbf{V} \cdot \rho \mathbf{v}\hat{H}) - (\mathbf{V} \cdot \mathbf{q}) - (\boldsymbol{\tau} : \mathbf{V}\mathbf{v}) \tag{20.2-7}$$

The enthalpy and conduction terms in this equation are analogous, but the work terms here include subtle differences.

At this point, you may justifiably wonder why the more complex equations have been introduced at all. Such wonder is legitimate because these equations are rarely as useful as the simpler energy balances used for conduction problems in the previous section. They are much less useful than the corresponding equations used for fluid mechanics. Still, I find these equations a reliable way to check my derivation of energy balances, especially in cases of simultaneous conduction and flow. Some of these are illustrated in the examples that follow.

Example 20.2-1: Conduction in a thin film and a thick slab Derive Eq. 20.1-5 for a thin film and Eq. 20.1-16 for a thick slab from the generalized energy balance in Eq. 20.2-2.

Solution For the thin film, the conduction is in steady state; so accumulation is zero. The film is solid, so there is no energy convection or work. Thus

$$0 = -\mathbf{V} \cdot \mathbf{q}$$

Because transport is one dimensional,

$$0 = -\frac{d}{dz}q$$

When we combine with Fourier's law,

$$0 = k_T \frac{d^2}{dz^2} T$$

This is the desired result.

For a solid slab of constant density, again there is no convection, so Eq. 20.2-2 becomes

$$\rho \frac{\partial \hat{U}}{\partial t} = -\mathbf{V} \cdot \mathbf{q}$$

Transport is still one dimensional, and q can be restated in terms of Fourier's law. In addition, for a solid,

$$\hat{U} = \hat{C}_v[T - (\text{some reference temperature})]$$

For solids, $\hat{C}_v$ equals $\hat{C}_p$ and is almost a constant, so

$$\rho \hat{C}_p \frac{\partial T}{\partial t} = k_T \frac{d^2 T}{dz^2}$$

which is the same as Eq. 20.1-16. In both cases, deriving the differential equation is easy; solving it for particular boundary conditions may be difficult.

Example 20.2-2: Heating a flowing solution   A viscous solution in laminar flow is flowing steadily through a narrow pipe. At a known distance along the pipe, the pipe's wall is heated with condensing steam. Find a differential equation from which the temperature distribution in the pipe can be calculated.

Solution   Because this is a flow system, we decide to begin with the general energy balance in Eq. 20.2-3. The system is in steady state, so the accumulation is zero. We usually can anticipate that heating due to viscous dissipation is small, and so take $(\tau : \mathbf{V}v)$ as nearly zero. Equation 20.2-3 then becomes

$$0 = -(\nabla \cdot \rho v \hat{H}) - (\nabla \cdot \mathbf{q})$$

To solve this problem, we often assume that the energy transfer along the pipe axis is largely by convection and that energy transport in the radial direction is by conduction:

$$0 = -\frac{\partial}{\partial z} \rho v \hat{H} - \frac{1}{r} \frac{\partial}{\partial r} rq$$

We expect that

$$\hat{H} = \hat{C}_p[T - (\text{some reference temperature})]$$

and that ρ, v, and $\hat{C}_p$ are constants. With these simplifications, we combine with Fourier's law to find

$$\frac{\partial}{\partial z} T = \left(\frac{k_T}{\rho \hat{C}_p v} \right) \frac{1}{r} \frac{\partial}{\partial r} r \frac{\partial T}{\partial r}$$

Solutions of a somewhat similar diffusion problem were discussed in Section 9.4.

20.3 Heat Transfer Coefficients

The material presented in the first two sections focused on Fourier's law of heat conduction. This law, which is especially useful for heat conduction in solids, allows calculation of the temperature and heat flux at any position and time. It has been tremendously useful, a pillar of scientific thought. However, it can be difficult to use in fluid systems, especially when heat is transferred across phase boundaries.

Instead, we often use a different model for heat transfer, one better suited to approximate calculations of the heat transferred across interfaces. In this model, the separate phases are imagined to be well mixed, and hence isothermal. The only temperature gradients are close to the interface, in some vaguely defined interfacial region. The heat flux in this model is assumed to be

$$q = U\Delta T \tag{20.3-1}$$

where the heat flux q is taken as normal to the interface, the temperature difference ΔT is from one bulk phase across the interface into a second bulk phase, and the proportionality constant U is called the overall heat transfer coefficient.

This new model is similar to that using mass transfer coefficients (see Chapter 8). Like the mass transfer model, it is comprised of a variety of definitions using different temperature differences. One common choice of temperature difference is based on that at some particular position z:

$$q(z) = U\Delta T(z) \tag{20.3-2}$$

This local definition is that applied in this book except where explicitly stated otherwise. Another common choice is an arithmetic average temperature difference:

$$q = U'\frac{\Delta T(\text{inlet}) + \Delta T(\text{outlet})}{2} \tag{20.3-3}$$

where U' is a different overall heat transfer coefficient, related to U but not normally equal to it. This definition is sometimes used for correlations of data on full-size industrial equipment.

The overall heat transfer coefficient across an interface is often the average of several sequential steps. One classic example of this averaging is shown in Fig. 20.3-1. In this example, energy is transferred from a hot fluid to a solid wall, is conducted across the wall, and then is transferred into a cooler fluid. The heat flux in this case is

$$\begin{aligned} q &= h_1(T_1 - T_{1i}) \\ &= h_2(T_{1i} - T_{3i}) \\ &= h_3(T_{3i} - T_3) \end{aligned} \tag{20.3-4}$$

where the various temperatures and heat transfer coefficients are defined in the figure. The various h_i, called individual heat transfer coefficients, are characteristics of the fluids near the wall and of the wall itself.

We now want to calculate the overall heat transfer coefficient U as a function of these individual coefficients. We do this in two steps. First, we compare the second line of Eq. 20.3-4 with Eq. 20.1-9 to discover that

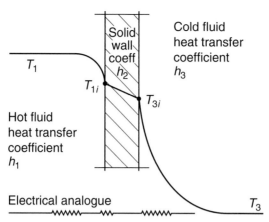

Fig. 20.3-1. Heat transfer across an interface. The overall heat transfer coefficient is a harmonic average of the individual heat transfer coefficients for the hot fluid, the wall, and the cold fluid. This averaging, which corresponds to the electrical problem of several resistances in series, is simpler than the corresponding mass transfer problem examined in Section 8.5.

$$h_2 = \frac{k_{T_2}}{l_2} \tag{20.3-5}$$

where k_{T_2} is the thermal conductivity and l_2 is the thickness of the solid wall. Second, we can combine the various equalities in Eq. 20.3-4 and eliminate the interfacial temperatures T_{1i} and T_{3i}. The final result is (after some algebra)

$$q = \frac{T_1 - T_3}{1/h_1 + l_2/k_{T_2} + 1/h_3} \tag{20.3-6}$$

By comparison with Eq. 20.3-2, we see that

$$U = \frac{1}{1/h_1 + l_2/k_{T_2} + 1/h_3} \tag{20.3-7}$$

The overall coefficient is a harmonic average of the individual coefficients.

Two features of this result are noteworthy. First, this heat transfer problem is a complete analogue to the electrical problem of three resistances in series. The heat flux q corresponds to the current, and the temperature $\Delta T (= T_1 - T_3)$ is like the voltage. The reciprocal of the overall heat transfer coefficient is analogous to the overall resistance. This overall resistance equals the sum of the resistances of the individual steps $1/h_1$, l_2/k_{T_2}, and $1/h_3$. This parallel is schematically suggested in Fig. 20.3-1.

At the same time, the overall heat transfer coefficient is simpler than the overall mass transfer coefficient developed in Section 8.5. Both coefficients are related to a sum of resistances, but the mass transfer case also involves weighting factors that are often confusing. These factors relate the concentrations on different sides of the interface. In the heat transfer case, the interfacial temperature in, for example, the hot fluid at the wall equals the interfacial temperature of the solid wall in contact with the hot fluid. This equality means no weighting factors and a simpler mathematical form.

We now illustrate the use of these coefficients by means of several simple examples.

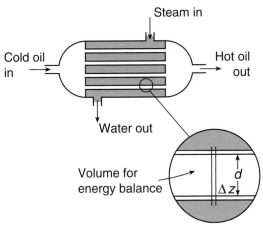

Fig. 20.3-2. A heat exchanger for crude oil. The oil flows through the tubes of the exchanger and is heated with condensing steam. The problem is to calculate the overall heat transfer coefficient from data given in the text.

Example 20.3-1: Finding the overall heat transfer coefficient A total of 18 cm³/hr crude oil flows in a heat exchanger with forty tubes 0.05 m in diameter and 2.8 m long (Fig. 20.3-2). The oil, which has a heat capacity of 2 J/kg K and a specific gravity of 900 kg/m³, is heated with 240 °C steam from 20 °C to 140 °C. The steam is condensed at 240 °C but is not cooled much below that temperature. What is the overall heat transfer coefficient based on the local temperature difference? What is it when based on the average temperature difference?

Solution We begin with an energy balance on the volume $(\pi/4)d^2\Delta z$, as shown in Fig. 20.3-2:

$$\begin{pmatrix} \text{energy} \\ \text{accumulation} \end{pmatrix} = \begin{pmatrix} \text{energy in minus energy out} \\ \text{by convection} \end{pmatrix} + \begin{pmatrix} \text{energy transferred} \\ \text{through tube walls} \end{pmatrix}$$

At steady state, this is

$$0 = \left(\frac{\pi}{4}d^2\right)\left(v\rho\hat{C}_p T|_z - v\rho\hat{C}_p T|_{z+\Delta z}\right) + (\pi d\Delta z)U(240 - T)$$

where U is the value defined by Eq. 20.3-2. The velocity in these equations is the average value in one of the tubes. Dividing by the volume $(\pi/4)d^2\Delta z$, taking the limit as Δz goes to zero, and rearranging the result,

$$\frac{dT}{dz} = \left(\frac{4U}{v\rho\hat{C}_p d}\right)(240 - T)$$

This is subject to the condition that

$$z = 0, \quad T = 20\,°C$$

Integrating,

$$\ln\left(\frac{240-20}{240-T}\right) = \frac{4Uz}{v\rho\hat{C}_p d}$$

All data are given except v, which is easily found from the total flow:

$$v = \frac{18 \text{ m}^3}{3600 \text{ sec }\left[40\frac{\pi}{4}(0.05\,\text{m})^2\right]}$$

$$= 0.064 \text{ cm}/\text{sec}$$

Inserting this and the other values given,

$$U = \frac{v\rho\hat{C}_p d}{4z}\ln\left(\frac{240-20}{240-T}\right)$$

$$= \frac{(0.064\,\dfrac{\text{m}}{\text{sec}}\,\dfrac{900\,\text{kg}}{\text{m}^3}(2\,\dfrac{\text{J}}{\text{kgK}})(0.05\,\text{m})}{4(2.80\,\text{m})}\ln\left(\frac{240-20}{240-140}\right)$$

$$= 0.41 \text{ W}/\text{m}^2 \text{ K}$$

This value is based on the local temperature difference.

Alternatively, we might base the heat transfer coefficient on the average temperature difference (see Eq. 20.3-3):

$$\Delta T = \frac{1}{2}(\Delta T_{\text{in}} + \Delta T_{\text{out}})$$

$$= \frac{1}{2}(220\,^{\circ}\text{C} + 100\,^{\circ}\text{C})$$

$$= 160\,^{\circ}\text{C}$$

The heat flux is easily found:

$$q = \frac{\dfrac{18\,\text{m}^3}{3600\,\text{sec}}\left(\dfrac{900\,\text{kg}}{\text{m}^3}\right)\dfrac{2\,\text{J}}{\text{kg K}}(140\,^{\circ}\text{C} - 20\,^{\circ}\text{C})}{40\pi\,(0.05\,\text{ m})(2.8\,\text{m})}$$

$$= 61 \,\frac{\text{W}}{\text{m}^2}$$

Thus the overall heat transfer coefficient is

$$U' = 0.38 \text{ W}/\text{m}^2 \text{ K}$$

The difference between this value and that found earlier illustrates the importance of making sure which definition we are using.

Example 20.3-2: The time for tank cooling A 100-gallon tank filled with water initially at 80 °F sits outside in air at 10 °F. The overall heat transfer coefficient for heat lost from the water-containing tank is 3.6 Btu/hr ft^2 °F, and the tank's area is 27 ft^2. How long can we wait before the water in the tank starts to freeze?

Solution As before, we begin with an energy balance on the tank:

$$\left(\begin{array}{c} \text{energy} \\ \text{accumulation} \end{array}\right) = \left(\begin{array}{c} \text{heat loss} \\ \text{from tank} \end{array}\right)$$

$$\frac{d}{dt}(\rho V \hat{C}_v T) = -UA(T - 10\,°F)$$

where ρ and C_v are the density and heat capacity of the water, V and A are the volume and area of the tank, and T is the water's temperature. This equation is subject to the initial condition

$$t = 0, \quad T = 80\,°F$$

We can use this condition in integrating the previous equation to find

$$\frac{T - 10\,°F}{80\,°F - 10\,°F} = e^{-(UA/\rho V \hat{C}_v)t}$$

We want the time at which freezing will begin:

$$\frac{32\,°F - 10\,°F}{80\,°F - 10\,°F} = \exp\left[-\left(\frac{(3.6\,\text{Btu/hr ft}^2\,°F)(27\,\text{ft}^2)}{(8.31\,\text{lb/gal})(100\,\text{gal})(1\,\text{Btu/lb}\,°F)}\right)t\right]$$

$$t = 10\,\text{hr}$$

The tank will probably be safe overnight, but not much longer. Note that we implicitly assume that the tank's contents are isothermal and hence well mixed.

Example 20.3-3: The effect of insulation Insulation advertisements claim that, in Minnesota, we can save 40% on our heating bills by installing 10 in of glass wool as insulation. The glass wool has a thermal conductivity of about 0.03 Btu/hr ft °F; the average winter temperature in Minnesota is 15 °F, and the house temperature is 68 °F. If the advertisements are true, and if heat loss from doors and windows is minor, how much can we save with 2 ft of insulation?

Solution Imagine that the heat loss in our current home is

$$q = h\Delta T$$

By adding 10 in of glass wool, we have, from Eq. 20.3-6,

$$0.6q = \frac{1}{(1/h) + (10/12\,\text{ft})/(0.03\,\text{Btu/hr ft}\,°F)}\Delta T$$

Dividing these equations, we find that

$$h = 0.024\,\text{Btu/hr ft}^2\,°F$$

With the thicker insulation, we have a heat loss q' of

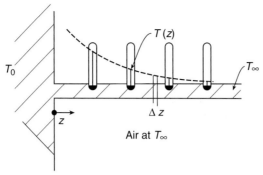

Fig. 20.3-3. Heat loss from a heated bar. The long bar shown is heated by contact with the large body at the left, which is at the high temperature T_0. The temperature in the bar drops exponentially along the bar. This situation is important historically, for it gave Fourier a major clue in developing his law for heat conduction.

$$q' = \frac{1}{(\text{hr ft °F}/0.024 \text{ Btu}) + (2 \text{ ft})/(0.03 \text{ Btu/hr ft °F})} \Delta T$$

$$= (0.0092 \text{ Btu/hr ft}^2 \text{°F}) \Delta T$$

This represents a saving of about 36% over the house with 10 in of insulation. Obviously, the additional gain should be balanced against the insulation's cost.

Example 20.3-4: Heat loss from a bar   M. P. Crosland wrote in 1970 that "in 1804 Biot carried out an experimental investigation of the conductivity of metal bars by maintaining one end at a high known temperature and taking readings of thermometers placed in holes along the bar. . . [He found] that the steady state temperature decreased exponentially along the bar." Biot could not explain this; Fourier could. Can you?

Solution   To solve this problem, we make an energy balance on a differential length Δz of the bar shown in Fig. 20.3-3:

$$\begin{pmatrix} \text{energy} \\ \text{accumulation} \end{pmatrix} = \begin{pmatrix} \text{energy in minus energy out} \\ \text{by conduction} \end{pmatrix} - \begin{pmatrix} \text{energy lost to} \\ \text{the surroundings} \end{pmatrix}$$

Because we are at steady state, the accumulation is zero, and

$$0 = Wlq|_z - Wlq|_{z+\Delta z} - 2(W+l)\Delta z h(T - T_\infty)$$

where W is the bar width, l is its vertical height, and h is the heat transfer coefficient between the bar and the surroundings. We now divide by the volume $Wl\Delta z$, take the limit as Δz goes to zero, and combine the result with Fourier's law:

$$0 = k_T \frac{d^2 T}{dz^2} - \frac{h}{L}(T - T_\infty)$$

where L equals $Wl/(2W + 2l)$. This equation is easily integrated to give

$$T - T_\infty = a e^{\sqrt{\frac{h}{k_T L}}\, z} + b e^{-\sqrt{\frac{h}{k_T L}}\, z}$$

where a and b are integration constants. These constants can be found from the boundary conditions

$$z = 0, \quad T = T_0$$
$$z = \infty, \quad T = T_\infty$$

We can show that a is zero by use of the second condition, equivalent to the assumption of a long bar. From the first condition, we find that

$$\frac{T - T_\infty}{T_0 - T_\infty} = e^{-\sqrt{\frac{h}{k_T L}} z}$$

Thus the temperature drops off exponentially as we get farther and farther away from the bar's base. This is what Biot observed experimentally.

This problem is difficult because of the boundary condition that the bar is very long and because of the term in the energy balance for heat loss into the air. This second aspect gave Fourier himself a lot of trouble, so if it is not clear the first time, try again.

20.4 Rate Constants for Heat Transfer

Up to this point, we have treated the thermal conductivity k_T, the thermal diffusivity α, and the heat transfer coefficient h as unknowns, adjustable parameters in any calculation. In fact, we often want to use previously measured values of these quantities to make predictions about new situations. Values for gases can be predicted from kinetic theory, and values for liquids and solids are best found by experiment. In this section, we report a few selected values of these quantities.

Estimates of thermal conductivities of gases depend on the following result of kinetic theory:

$$k_T = \frac{1.99 \cdot 10^{-4} \sqrt{T/\tilde{M}}}{\sigma^2 \Omega_k} \tag{20.4-1}$$

in which the thermal conductivity k_T is in cal/cm sec K, σ is the collision diameter in Å, T is the temperature in Kelvin, and $\tilde{M}$ is the molecular weight. The dimensionless quantity Ω_k is of order 1 and a weak function of $k_B T/\varepsilon$ where ε is an energy of interaction. Values of σ and ε are given in Table 5.1-2; some selected values of Ω_k are given in Table 20.4-1. These calculations are straightforward, completely parallel to those in Section 5.1.

Typical values of thermal conductivities and thermal diffusivities in gases, liquids, and solids are given in Table 20.4-2. Some of the values are expected from experience; for example, the thermal conductivities of metals are much higher than those of liquids or gases. Less obviously, the thermal diffusivities of nonmetallic solids and liquids are more nearly the same, indicating that unsteady heat transfer proceeds at more similar rates in these materials.

The effective thermal conductivity of composite materials tends to be dominated by the continuous phase, just as in the case of diffusion (see Section 6.4.4). Composite materials that can partially melt show anomalous thermal diffusivities; examples include some hydrated salts and foods like ice cream. Still, thermal conductivities like those in Table 20.4-2 represent a norm from which there are few departures.

Table 20.4-1 *Selected values of the collision integral Ω_k for use in Eq. 20.4-1*

$k_B T/\varepsilon$	Ω_k
0.4	2.49
0.6	2.07
0.8	1.78
1.0	1.59
1.2	1.45
1.4	1.35
1.6	1.28
2.0	1.18
3.0	1.04
4.0	0.97
6.0	0.90
10.0	0.82

Source: Hirschfelder *et al.* (1954).

Table 20.4-2 *Thermal conductivities and thermal diffusivities of various materials*

	T (K)	k_t (10^{-4} cal/cm sec K)	α (cm^2/sec)
Gases			
Hydrogen	273	4.03	1.55
Nitrogen	273	0.57	0.22
Oxygen	273	0.58	0.22
Carbon dioxide	273	0.35	0.11
Liquids			
Water	293	14.3	0.0014
Ethanol	293	4.4	0.00093
Hexane	293	2.9	0.0011
Octane	293	3.5	0.0010
Toluene	293	3.6	0.0010
Mercury	293	210	0.46
Solids			
Steel, 1% carbon	293	1000	0.12
Copper	293	9500	1.17
Silver	293	10,200	1.71
Silver bromide	273	25	0.0055
Sodium chloride	273	88	0.020
Brick (masonry)	293	16	0.0046
Concrete (dry)	293	3.1	0.0049
Glass wool ($\rho = 200$ kg/m^3)	293	1.0	0.0028
Glass (window)	293	19	0.0034

Source: Data from Handbook of Chemistry and Physics (2008), and International Critical Tables (1933).

Some common correlations of heat transfer coefficients are reported in Table 20.4-3. These all refer to heat transfer across a solid–fluid interface because other situations either are rare or are described in different terms. Like the mass transfer correlations in Section 8.3, these are best presented in terms of dimensionless groups. The two most

Table 20.4-3 *A few correlations of heat transfer coefficients*

Physical situation	Correlation[a]	Specific definitions	Remarks
Turbulent flow in pipes	$\dfrac{hd}{k_T} = 0.027 \left(\dfrac{dv\rho}{\mu}\right)^{0.8} \left(\dfrac{\mu\hat{C}_p}{k_T}\right)^{0.33}$	d: tube diameter, v: velocity in tube	Widely quoted, often with slightly different constants or with small correction factors
Turbulent flow over banks of tubes	$\dfrac{hd}{k_T} = 0.33 \left(\dfrac{dv\rho}{\mu}\right)^{0.6} \left(\dfrac{\mu\hat{C}_p}{k_T}\right)^{0.33}$	d: tube diameter, v: velocity at minimum-flow area	Valid when $dv\rho/\mu > 6{,}000$, and for 10 or more rows of tubes
Flow around a solid sphere	$\dfrac{hd}{k_T} = 2.0 + 0.6 \left(\dfrac{dv\rho}{\mu}\right)^{0.5} \left(\dfrac{\mu\hat{C}_p}{k_T}\right)^{0.33}$	d: sphere diameter, v: sphere velocity	Valid only when free convection is absent
Free convection between vertical plates	$\dfrac{hl}{k_T} = \alpha \left(\dfrac{l^3 g\rho\Delta\rho}{\mu^2}\right)^{\beta} \left(\dfrac{l}{L}\right)^{1/9}$	l: distances between plates, L: length of plates, $\Delta\rho$: density change caused by temperature change	When $2 \cdot 10^3 < (l^3 g\rho\Delta\rho/\mu^2) < 2 \cdot 10^4$, $\alpha = 0.18$, and $\beta = 0.25$; when $2 \cdot 10^4 < (l^3 g\rho\Delta\rho/\mu^2) < 2 \cdot 10^5$, $\alpha = 0.065$, and $\beta = 0.33$

Note: $^a hd/k_T$ is the Nusselt number; $\mu\hat{C}_p/k_T$ is the Prandtl number; $dv\rho/\mu$ is the Reynolds number; and $l^3 g\rho\Delta\rho/\mu^2$ is the Grashof number.
Source: Adapted from Kreith and Bohn (2000).

important new groups are the Nusselt number and Prandtl number. The Nusselt number explicitly contains the heat transfer coefficient:

$$\left(\begin{array}{c} \text{Nusselt} \\ \text{number} \end{array}\right) = \frac{hl}{k_T} \tag{20.4-2}$$

where l is some characteristic length. The Prandtl number is more complex:

$$\left(\begin{array}{c} \text{Prandtl} \\ \text{number} \end{array}\right) = \frac{\mu \hat{C}_p}{k_T} = \frac{\nu}{\alpha} \tag{20.4-3}$$

It essentially represents the relative importance of viscosity and thermal conductivity. These groups are closely parallel to the Sherwood and Schmidt numbers for mass transfer, a point detailed in Section 21.1. Now we turn to illustrations of heat transfer using these numerical values.

Example 20.4-1: The overall heat transfer coefficient of a heat exchanger   As part of a chemical process, we plan to use a shell-tube heat exchanger of twenty banks of 0.05 m outside-diameter steel tubes with 3 mm walls. Outside the tubes, we plan to use 400 °C flue gas; inside, we expect to be heating aromatics like benzene and toluene fed at around 30 °C. The gas flow will be 17 m/sec, and the liquid flow will be 2.7 m/sec. What overall heat transfer coefficient can we expect in this exchanger?
      Solution   From Eq. 20.3-7, we see that

$$U = \frac{1}{1/h_1 + l_2/k_{T_2} + 1/h_3}$$

where h_1 is the coefficient in the hot flue gas, k_{T_2}/l_2 refers to the steel wall, and h_3 is the coefficient in the liquid.
      These coefficients are easily calculated. For h_1, we assume that the flue gas has the properties of nitrogen, so, from Eq. 20.4-1,

$$\begin{aligned} k_T &= \frac{1.99 \cdot 10^{-4}\sqrt{T/\tilde{M}}}{\sigma^2 \Omega_k} \\ &= \frac{1.99 \cdot 10^{-4}\sqrt{673\,°\text{K}/28}}{(3.80\,\text{Å})^2(0.87)} \\ &= 0.77 \cdot 10^{-4}\,\text{cal/cm sec K} = 0.032\,\text{W/m K} \end{aligned}$$

We then use the correlation for flow over tube banks in Table 20.4-3:

$$\begin{aligned} h_1 &= 0.33\left(\frac{k_T}{d}\right)\left(\frac{dv\rho}{\mu}\right)^{0.6}\left(\frac{\mu \hat{C}_p}{k}\right)^{0.3} \\ &= 0.33\left(\frac{0.032\,\text{W/m K}}{0.05\,\text{m}}\right) \\ &\quad \cdot \left(\frac{(0.05\,\text{m})(17\text{m/sec})(051\,\text{kg/m}^3)}{3.3 \cdot 10^{-5}\,\text{kg/cm sec}}\right)^{0.6} \\ &\quad \cdot \left(\frac{(3.3 \cdot 10^{-5}\,\text{kg/m sec})\left(1100\,\frac{\text{J}}{\text{kg K}}\right)}{.032\,\frac{\text{W}}{\text{m K}}}\right)^{0.33} \\ &= 65\,\text{W/m}^2\,\text{K} \end{aligned}$$

The value for the steel wall, h_2, is easily found using data from Table 20.4-2:

$$h_2 = \frac{k_{T_2}}{l_2} = \frac{42 \frac{W}{m\,K}}{3 \cdot 10^{-3}\,m}$$

$$= 14,000\ W/m^2\,K$$

The value for the liquid inside the tube comes again from Table 20.4-3:

$$h_3 = 0.027 \left(\frac{k_T}{d}\right) \left(\frac{dv\rho}{\mu}\right)^{0.8} \left(\frac{\mu \hat{C}_p}{k}\right)^{0.33}$$

$$= 0.027 \left(\frac{0.15 \frac{W}{m\,K}}{0.044\ m}\right)$$

$$\cdot \left(\frac{0.044\ m(2.7\ m/sec)870\ kg/cm^3}{5.3 \cdot 10^{-4}\ kg/m\ sec}\right)^{0.8}$$

$$\cdot \left(\frac{5.3 \cdot 10^{-4}\ \frac{kg}{m\ sec}\ \frac{1700\ J}{kg\ K}}{\frac{0.15\ J}{m\ sec\ K}}\right)^{0.33}$$

$$= 1,100\ W/m^2\,K$$

Combining,

$$U = \frac{1}{\frac{1}{65} + \frac{1}{14,000} + \frac{1}{11,000}}$$

$$= 64\ W/m^2\,K$$

Note that the gas-side heat transfer coefficient is less than that on the liquid side, which in turn is less than that of the wall. This is the usual sequence; heat transfer in gases tends to limit the overall process. In contrast, mass transfer in liquids tends to be the rate-limiting step.

Example 20.4-2: The design of storm windows Rising energy prices have led to a renaissance in the insulation of houses. Advertisements state that the cost of home heating can be substantially reduced by using two sets of storm windows. In other words, these advertisements urge the use of windows with three layers of glass in existing window frames, which are about 3 cm deep and 1 m long. Use your knowledge of heat transfer to decide whether or not this is a good idea. Assume that the outside temperature is $-10\,°C$ and the room temperature is $+20\,°C$.

Solution The physical situation in this problem is illustrated schematically in Fig. 20.4-1. In the simplest case, the window consists of a single pane of glass. Heat loss through this window depends mostly on the thermal conductivity of this pane, for the adjacent air tends to be stirred by free convection.

The heat loss might be substantially reduced using storm windows of two or more panes of glass. Ideally, we would hope that this loss would now be governed by heat conduction across the gap between the panes. In fact, free convection stirs the air in this gap, so that the heat loss is much greater than that due to conduction. Still, this new resistance to heat transfer sharply reduces heat loss.

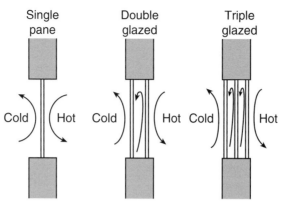

Fig. 20.4-1. Heat loss through storm windows. The heat loss through one pane of glass is much greater than the heat loss through two. Interestingly, the heat loss through two panes is greater than that through three, even though the thermal conductivity of air is less than that of the third pane of glass. This result illustrates the importance of free convection.

We want to know if the heat loss is further reduced by an additional pane of glass added between those in a conventional two-pane storm window. To answer this question, we first find the heat transfer coefficient for a two-pane storm window, using correlations for free convection given in Table 20.4-3. These correlations involve the Grashöf number, which for air is

$$\frac{l^3 g \rho \Delta p}{\mu^2} = \frac{l^3 g (\Delta \rho / \rho)}{\nu^2} = \frac{l^3 g \Delta (p\tilde{M}/RT)}{\nu^2} \frac{1}{p\tilde{M}/RT} = \frac{l^3 g}{\nu^2} \frac{\Delta T}{T}$$

Thus, for a window with two panes 3 cm apart with a 30 °C temperature drop and filled with air of kinematic viscosity ν equal to 0.14 cm^2/sec

$$h\left(\begin{array}{c} \text{two} \\ \text{pane} \end{array}\right) = 0.065 \left(\frac{k_T}{l}\right)\left(\frac{l^3 g}{\nu^2}\frac{\Delta T}{T}\right)^{1/3}\left(\frac{l}{L}\right)^{1/9}$$

$$= 0.065 \left(\frac{0.57 \cdot 10^{-4}\,\text{cal/cm sec K}}{3\,\text{cm}}\right)$$

$$\cdot \left(\frac{(3\,\text{cm})^3 (980\,\text{cm/sec}^2)}{(0.14\,\text{cm}^2/\text{sec})^2}\left(\frac{30}{278}\right)\right)^{1/3}\left(\frac{3\,\text{cm}}{100\,\text{cm}}\right)^{1/9}$$

$$= 0.44 \cdot 10^{-4}\,\text{cal/cm}^2\,\text{sec K}$$

In contrast, the gaps in a three-pane window will be 1.5 cm. For each gap, the temperature drop will be about 15 °C and the heat transfer coefficient will be

$$h\left(\begin{array}{c} \text{one gap of} \\ \text{three pane} \end{array}\right) = 0.065 \left(\frac{0.57 \cdot 10^{-4}\,\text{cal/cm sec K}}{1.5\,\text{cm}}\right)$$

$$\cdot \left(\frac{(1.5\,\text{cm})^3 (980\,\text{cm/sec}^2)}{(0.14\,\text{cm}^2/\,\text{sec})^2}\left(\frac{15}{278}\right)\right)^{1/3}\left(\frac{1.5\,\text{cm}}{100\,\text{cm}}\right)^{1/9}$$

$$= 0.32 \cdot 10^{-4}\,\text{cal/cm}^2\,\text{sec K}$$

Because there are two such gaps, the overall heat transfer coefficient is

$$U\left(\frac{\text{three}}{\text{pane}}\right) = 0.16 \cdot 10^{-4} \, \text{cal/cm}^2 \, \text{sec K}$$

Having an additional pane of glass cuts the heat loss through the windows by over half.

20.5 Conclusions

This chapter contains a synopsis of heat transfer. Although the presentation is brief, the chapter can supply a review or a summary for those already skilled in diffusion. The sections here are like those earlier in the book: a basic law, differential equations leading to fluxes, approximate models of interfacial transport, and values of the various coefficients.

The analysis of heat transfer is parallel to that for diffusion because heat transfer and diffusion are described with the same mathematical equations. Indeed, many experts argue that because of this mathematical identity, the two are identical. I do not agree with this view because the two processes are so different physically. For example, heat conduction is faster in liquids than in gases, but diffusion is faster in gases than in liquids. This relation between the mathematical similarity and the physical difference affects the problems involving both heat and mass transfer, problems central to the next chapter.

Questions for Discussion

1. Compare Fourier's law of heat conduction with Fick's law of diffusion.
2. What are the dimensions in mass M, length L, time t, and temperature T of the thermal conductivity k_T, the thermal diffusivity α, and the heat transfer coefficient h?
3. Will the thermal conductivity k_T be changed by stirring? Will the overall heat transfer coefficient U?
4. How could you measure a thermal conductivity?
5. How could you measure a heat transfer coefficient?
6. If you double the thermal conductivity, how much will the heat flux across a thin film change?
7. How much will it change for conduction into a thick slab?
8. Which has the highest thermal conductivity, air, water, or steel? Which has the smallest thermal diffusivity?
9. Compare an overall heat transfer coefficient with an overall mass transfer coefficient.
10. What are the more common dimensionless groups used in heat transfer correlations?

Problems

1. Find the heat lost per external area from a house at 18 °C on a winter day at –14 °C. The house is insulated with the equivalent of 8 cm of glass wool. The resistance to heat

transfer at the inner walls is negligible, but that at the outer walls is controlled by an air layer equivalent to 0.2 cm.

2. Find the heat loss per external area from a water pipe containing water at 18 °C on a winter day at –14 °C. The pipe has a diameter of 5 cm and is insulated with an 8-cm layer of polymer foam. The resistances to heat transfer inside the pipe and of the pipe wall are negligible. That outside the pipe is controlled by an air layer equivalent to 0.2 cm.

3. The values given below are abstracted from [*Wind Chill: Equivalent Temperatures*. Washington, DC: NOAA (1974)] of windchill versus temperature.

Windchill

Wind velocity (mph)	Dry-bulb temperature			(°F)
	20	0	−20	−40
4	20	0	−20	−40
10	3	−22	−46	−71
20	−10	−39	−67	−95
30	−18	−49	−79	−109
40	−21	−53	−84	−115

Note: The perceived temperature is given as a function of the wind velocity and the actual "dry-bulb" temperature.

Windchill is popularly interpreted as how cold the weather "feels" at the true temperature and the given wind. In fact, these values are based on the time to freeze water in a sausage casing hung over a Quonset hut in Antarctica. Since these results were published, several have asserted that they are equivalent to the heat loss from a cylinder, and so can be predicted from standard engineering correlations. Test this assertion using the values in the table.

4. The energy balance on a differential volume can be written in a variety of ways, including

$$\frac{\partial}{\partial t}\rho\left(\hat{U}+\tfrac{1}{2}v^2\right) = -\nabla\cdot\rho v\left(\hat{U}+\tfrac{1}{2}v^2\right) - (\nabla\cdot q) + \rho(v\cdot g) - \nabla\cdot pv - (\nabla\cdot[\tau\cdot v])$$

$$\rho\frac{\partial\hat{U}}{\partial t} = \rho\left(\frac{\partial\hat{U}}{\partial t}+v\cdot\nabla\hat{U}\right) = -(\nabla\cdot q) - p(\nabla\cdot v) - (\tau:\nabla v)$$

$$\frac{\partial}{\partial t}\rho\hat{U} = -(\nabla\cdot\rho v\hat{H}) - (\nabla\cdot q) - (\tau:\nabla v) + v\cdot\nabla p$$

Prove that these equations are equivalent.

5. Assume that a straight wire in a large volume of fluid is suddenly connected to an electrical power supply that puts a constant wattage through the wire. The resistance of the wire is then measured as a function of time. Because this resistance is a function of temperature and the wire's temperature depends on the thermal conductivity, this measurement of resistance provides a way of determining the thermal conductivity. Derive an equation that allows calculation of thermal conductivity from this resistance

[as references, see H. Ziebland. in: *Thermal Conductivity*, ed. R. P. Tye. London: Academic Press (1969); J. K. Horrocks and E. McLaughlin (1963), *Proc. Roy. Soc.*, **A273**, 259].

6. Polymer fibers are often melt-spun by forcing a polymer melt through small holes into cold air. The specific polymer is first a rubber and then becomes a glass at T_g; this transition involves a negligible enthalpy of fusion, but does result in altering thermal diffusivity from α_1 to α_2. At the same time, the polymer surface quickly reaches the temperature of the surrounding air. Find the radius where the transition occurs as a function of time.

8. One possible automotive improvement would be a radiator based on hollow fibers. Such a device offers a huge surface area per volume, and hence a smaller, lighter radiator for the same job. However, because of the small diameter of the hollow fibers, coolant flow will be laminar. Use your knowledge of interfacial transport to suggest appropriate correlations for the overall heat transfer coefficient in such a device.

Further Reading

Carslaw, H. S., and Jaeger, J. C. (1959). *Operational Methods in Applied Mathematics*. Oxford: Clarendon Press.

Crosland, M. P. (1970). Jean-Baptiste Biot. In *Dictionary of Scientific Biography*, ed. C. C. Gillispie. New York: Scribner.

Handbook of Chemistry and Physics (2008) ed. R. C. Weast. Boca Raton, FL: Chemical Rubber Publishing Co.

Herivel, J. (1975). *Joseph Fourier*. Oxford: Clarendon Press.

Hirschfelder, J. O., Curtiss, C. F., and Bird, R. B. (1954). *Molecular Theory of Gases and Liquids*. New York: Wiley.

Incropera, F. P., DeWitt, D. P., Bergman, T. L., and Lavine, A. S. (2006). *Introduction to Heat Transfer*, 5th ed. Wiley: New York.

International Critical Tables (1933). New York: McGraw-Hill.

Kreith, F., and Bohn, M. S. (2000). *Principles of Heat Transfer*, 6th ed. New York: Cengage-Engineering.

Poling, B. E., Prausnitz, J. M., and O'Connell, J. P., (2000). *The Properties of Gases and Liquids*, 5th ed. New York: McGraw-Hill.

CHAPTER 21

Simultaneous Heat and Mass Transfer

Processes involving coupled heat and mass transfer occur frequently in nature. They are central to the formation of fog, to cooling towers, and to the wet-bulb thermometer. They are important in the separation of uranium isotopes and in the respiration of water lilies. This chapter analyzes a few of these processes. Not unexpectedly, such processes are complex, for they involve equations for both diffusion and heat conduction. These equations are coupled, often in a nonlinear way. As a result, our descriptions will contain approximations to reduce the complexities involved.

We begin this chapter with a comparison of the mechanisms responsible for mass and heat transfer. The mathematical similarities suggested by these mechanisms are discussed in Section 21.1, and the physical parallels are explored in Section 21.2. The similar mechanisms of mass and heat transfer are the basis for the analysis of drying, both of solids and of sprayed suspensions. However, the detailed models differ, as shown by the examples in Section 21.3. In Section 21.4, we outline cooling-tower design as an example based on mass and heat transfer coefficients. Finally, in Section 21.5, we describe thermal diffusion and effusion.

21.1 Mathematical Analogies Among Mass, Heat, and Momentum Transfer

Analogies among mass, heat, and momentum transfer have their origin either in the mathematical description of the effects or in the physical parameters used for quantitative description. The mathematically based analogies are useful for two reasons. First, they can save mathematical work; if the solution to a heat conduction problem is known, the solution to the corresponding diffusion problem is also known. We have already discussed this type of analogy in Section 3.5. Second, mathematical analogies often suggest dimensionless groups that are helpful in correlating the results of physical experiments. It is this second use that has broad scope and that is of interest in this section.

To explore these analogies, we remember that the diffusion of mass and the conduction of heat obey very similar equations. In particular, diffusion in one dimension is described by the following form of Fick's law:

$$-j_1 = D\frac{dc_1}{dz} \tag{21.1-1}$$

where D is the diffusion coefficient. If this diffusion takes place into a semi-infinite slab, as shown in Fig. 21.1-1, the concentration profile can be shown to be (see Section 2.3)

$$\frac{c_1 - c_{10}}{c_{1\infty} - c_{10}} = \mathrm{erf}\frac{z}{\sqrt{4Dt}} \tag{21.1-2}$$

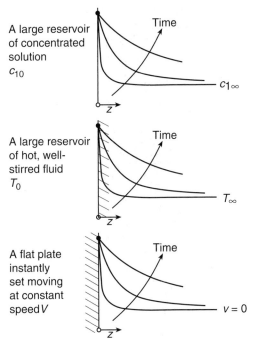

Fig. 21.1-1. Profiles for concentration, temperature, and fluid velocity. The diffusion of mass, the conduction of heat, and the laminar flow of fluids all obey laws of the same mathematical form. Accordingly, for mathematically identical boundary conditions, like those shown for a semi-infinite system, the profiles of concentration, temperature, and velocity are the same.

where c_{10} and $c_{1\infty}$ are the concentrations at the slab's surface and far within the slab, respectively. Similarly, heat conduction is described by Fourier's law:

$$- q = k_T \frac{dT}{dz} \tag{21.1-3}$$

where k_T is the thermal conductivity. If heat conduction takes place into the semi-infinite slab in Fig. 21.1-1, the temperature profile is (see Section 20.1)

$$\frac{T - T_0}{T_\infty - T_0} = \mathrm{erf}\,\frac{z}{\sqrt{4\alpha t}} \tag{21.1-4}$$

where $\alpha\,(= k_T/\rho\hat{C}_\mathrm{p})$ is the thermal diffusivity and T_0 and T_∞ are the temperatures of the surface of the slab and far within the slab, respectively.

Although we have not discussed momentum transport in this book, we should mention that this process is also described within the same framework. The basic law is due to Newton:

$$- \tau = \mu \frac{dv}{dz} \tag{21.1-5}$$

where τ is the momentum flux or the shear stress and μ is the viscosity. If a flat plate is suddenly moved in an initially stagnant fluid, the velocity v of the fluid is

$$\frac{v - V}{0 - V} = \mathrm{erf}\frac{z}{\sqrt{4\nu t}} \tag{21.1-6}$$

where the plate's velocity is V, the fluid's velocity far from the plate is zero, and the fluid's kinematic viscosity is ν.

At this point it has become conventional to draw an analogy among mass, heat, and momentum transfer. Each process uses a simple law combined with a mass or energy or momentum balance. If each process is described by the same mathematical equations and is subject to the mathematically equivalent boundary conditions, then each leads to results of the same mathematical form. Many believe it is more elegant to say that each process depends on combining a linear constitutive equation and a conservation relation to yield mathematically congruent results. The phenomenological coefficients of diffusion D, of thermal conductivity k_T and of viscosity μ are thus analogous.

As a student, I found this conventional analogy confusing. Sure, Eqs. 21.1-1, 21.1-3, and 20.1-5 all say that a flux varies with a first derivative. Sure, Eqs. 21.1-2, 21.1-4, and 21.1-6 all have an error function in them. But D, k_T, and μ do not have the same physical dimensions. Moreover, D appears in both Eq. 21.1-1 and Eq. 21.1-2. In contrast, k_T appears in Eq. 21.1-3, but it must be replaced by the thermal diffusivity α in Eq. 21.1-4. The viscosity μ in Eq. 21.1-5 is replaced by the kinematic viscosity ν in Eq. 21.1-6. These changes confused me, and initially they undercut any value that these analogies might have.

The source of my confusion stemmed from the ways in which the basic laws are written. In Fick's law (Eq. 21.1-1), the mass flux is proportional to the gradient of mass per volume, or the molar flux varies with the gradient in moles per volume. To be analogous, the energy flux q should be proportional to the gradient of the energy per volume $(\rho\hat{C}_p T)$. In other words, Eq. 21.1-3 should be rewritten as

$$-q = \frac{k_T}{\rho\hat{C}_p}\frac{d}{dz}(\rho\hat{C}_p T)$$

$$= \alpha\frac{d}{dz}(\rho\hat{C}_p T) \tag{21.1-7}$$

(In suggesting this alternative form, we imply that $\hat{C}_p$ equals $\hat{C}_v$, which is nearly true for liquid and solids.) If we use Eq. 21.1-7 instead of Eq. 21.1-3, then mass flux and heat conduction are truly analogous. Just as Eq. 21.1-2 follows from Eq. 21.1-1, so Eq. 20.1-4 follows from Eq. 21.1-7.

Newton's law for momentum transport can also be rewritten so that the momentum flux is proportional to the gradient of the momentum per volume (ρv), that is, by replacing Eq. 21.1-5 with

$$-\tau = \frac{\mu}{\rho}\frac{d}{dz}(\rho v)$$

$$= \nu\frac{d}{dz}(\rho v) \tag{21.1-8}$$

where ν is the kinematic viscosity. This new form, which implies a constant density, leads directly to Eq. 21.1-6.

Just as the fundamental laws for mass, heat, and momentum transfer can be made more nearly parallel, so can expressions for mass transfer coefficients and heat transfer coefficients. The interfacial mass flux already varies with the difference in mass per volume:

$$N_1 = n_1|_{z=0} = k(c_{10} - c_{1i}) \tag{21.1-9}$$

The interfacial heat flux must be modified so that the energy flux varies with the energy difference per volume:

$$q|_{z=0} = h\Delta T$$
$$= \frac{h}{\rho\hat{C}_p}\Delta(\rho\hat{C}_p T) \tag{21.1-10}$$

Thus the mass transfer coefficient k corresponds less directly to the heat transfer coefficient h than to the quantity $h/\rho\hat{C}_p$. The appropriate parallel for momentum transfer is the dimensionless friction factor f, defined as

$$\tau|_{z=0} = f\left(\frac{1}{2}\rho v^2\right)$$
$$= \left(\frac{fv}{2}\right)(\rho v - 0) \tag{21.1-11}$$

Thus $fv/2$ is like k and $h/\rho\hat{C}_p$.

When these equations are written in these parallel forms, they automatically suggest the most common dimensionless groups. For example, the ratio of the coefficient in Eq. 21.1-8 to that in Eq. 21.1-1 is ν/D, the Schmidt number. The ratio of the coefficient in Eq. 21.1-9 to that in Eq. 21.1-11 is $[k/v\,(2/f)]$. Because $2/f$ is itself dimensionless, this is equivalent to (k/v), the Stanton number.

These and other dimensionless groups formed in this way are shown in Table 21.1-1. Some of these analogies used to be surprising to me as a student; I never understood the assertion that the Prandtl number $\mu\hat{C}_p/k_T$ is analogous to the Schmidt number $\mu/\rho D$, although I learned to give that answer on exams. When I look at Eqs. 21.1-9, 21.1-7 and 20.1-1, I see that the Prandtl number ν/α and the Schmidt number ν/D are simply ratios of the coefficients of these equations. By similar arguments, the two Stanton numbers in the table represent the same kinds of ratios.

Thus the parallels in the descriptions of these processes suggest not only ways to save mathematical work but also parallels between different kinds of measurement (e.g., between heat transfer coefficients and mass transfer coefficients). These similarities suggest that the numerical values of these different kinds of coefficients are also similar. This more powerful quantitative analogy is the subject of the following section.

Table 21.1-1 *Dimensionless analogies between heat and mass transfer*

Key properties	Heat transfer		Mass transfer	
	Common forms	Fundamental forms	Common forms	Fundamental forms
Variable	Temperature T	Energy per volume $\rho \hat{C}_p T$	Concentration c_1	Concentration c_1
Property	Thermal conductivity k_T	Thermal diffusivity α	Diffusion coefficient D	Diffusion coefficient D
Coefficient	Heat transfer coefficient h	$\dfrac{h}{\rho \hat{C}_p}$	Mass transfer coefficient k	Mass transfer coefficient k
Dimensionless groups often used as dependent variables[a]	Nusselt number $\dfrac{hl}{k_T}$	Stanton number $\dfrac{h}{\rho \hat{C}_p v}$	Sherwood number $\dfrac{kl}{D}$	Stanton number $\dfrac{k}{v}$
Dimensionless groups often used as independent variables[a]	Prandtl number $\dfrac{\mu \hat{C}_p}{k_T}$	Prandtl number $\dfrac{\nu}{\alpha}$	Schmidt number $\dfrac{\mu}{\rho D}$	Schmidt number $\dfrac{\nu}{D}$
	Reynolds number $\dfrac{lv\rho}{\mu}$	Reynolds number $\dfrac{lv}{\nu}$	Reynolds number $\dfrac{lv\rho}{\mu}$	Reynolds number $\dfrac{lv}{\nu}$
	Grashof number[b] $\dfrac{l^3 \rho g \Delta\rho}{\mu^2}$	Grashof number[b] $\dfrac{l^3 g \Delta\rho/\rho}{\nu^2}$	Grashof number[c] $\dfrac{l^3 \rho g \Delta\rho}{\mu^2}$	Grashof number[c] $\dfrac{l^3 g \Delta\rho/\rho}{\nu^2}$

Notes: [a]Remember that the characteristic length l is different in different physical situations. [b]The density change $\Delta\rho$ is caused by a temperature difference here. [c]The density change $\Delta\rho$ is caused by a concentration difference here.

Example 21.1-1: Cooling metal spheres We want to quench a liquid metal quickly to make fine powder. We plan to do this by spraying metal drops into an oil bath. How can we estimate the cooling speed of the drops?

Solution No heat transfer correlation for this situation is given in Table 20.4-3. However, several mass transfer correlations for drops are given in Table 8.3-2. For example, for large drops without stirring,

$$\frac{kd}{D} = 0.31 \left(\frac{d^3 \Delta \rho g}{\rho \nu^2}\right)^{1/3} \left(\frac{\nu}{D}\right)^{1/2}$$

where d and ν are the drop's diameter and the fluid's kinematic viscosity. From Table 21.1-1, we see that the Sherwood number kd/D is equivalent to the Nusselt number hd/k_T and that the Schmidt number ν/D is analogous to the Prandtl number ν/α or $\mu \hat{C}_p/k_T$. Thus we expect as a heat transfer correlation

$$\frac{hd}{k} = 0.31 \left(\frac{d^3 \Delta \rho g}{\rho \nu^2}\right)^{1/3} \left(\frac{\mu \hat{C}_p}{k_T}\right)^{1/2}$$

This correlation will be reliable only if the Grashof number for the cooling falls in the same range as that used to develop the mass transfer correlation.

Example 21.1-2: Heat transfer from a spinning disc Imagine that a spinning metal disc electrically heated to 30 °C is immersed in 1,000 cm³ of an emulsion at 18 °C. The disc is 3 cm in diameter and is turning at 10 rpm. The emulsion's kinematic viscosity is $0.082 \text{ cm}^2/\text{sec}$. After an hour, the emulsion is at 21 °C. What is its thermal diffusivity?

Solution We begin with an energy balance on the emulsion:

$$\left(\begin{array}{c} \text{energy} \\ \text{accumulation} \end{array}\right) = \left(\begin{array}{c} \text{energy gained} \\ \text{from disc} \end{array}\right)$$

$$(\rho \hat{C}_p V)\frac{dT}{dt} = (\pi R_0^2)q$$

$$= (\pi R_0^2)h(T_{\text{disc}} - T)$$

where V is the emulsion volume, T and T_{disc} are the emulsion and disc temperatures, respectively, R_0 is the disc radius, and h is the heat transfer coefficient. We have also assumed that $\hat{C}_v$ equals $\hat{C}_p$. This equation is subject to the initial condition

$$t = 0, \quad T = T_0$$

Integrating,

$$\frac{T_{\text{disc}} - T}{T_{\text{disc}} - T_0} = e^{-(h/\rho \hat{C}_p)(\pi R_0^2/V)t}$$

Inserting the numbers given,

$$\frac{30 - 21}{30 - 18} = e^{-(h/\rho \hat{C}_p)[\pi(1.5 \text{ cm})^2/(1, 000 \text{ cm}^3)](3, 600 \text{ sec})}$$

Thus

$$\frac{h}{\rho \hat{C}_p} = 0.011 \, \text{cm/sec}$$

We now need to relate this quantity to the thermal diffusivity. We have no direct way to do so. We do have a similar correlation for mass transfer away from a spinning disc (see Table 8.3-3):

$$\frac{kd}{D} = 0.62 \left(\frac{d^2 \omega}{\nu}\right)^{1/2} \left(\frac{\nu}{D}\right)^{1/3}$$

Using Table 21.1-1, we see that the corresponding correlation must be

$$\frac{(h/\rho \hat{C}_p)d}{\alpha} = 0.62 \left(\frac{d^2 \omega}{\nu}\right)^{1/2} \left(\frac{\nu}{\alpha}\right)^{1/3}$$

or

$$\alpha = \left[\frac{1}{0.62}\left(\frac{h}{\rho \hat{C}_p}\right)\nu^{1/6}\omega^{-1/2}\right]^{3/2}$$

Inserting the numerical values,

$$\alpha = \left[\frac{1}{0.62}\left(0.011 \, \frac{\text{cm}}{\text{sec}}\right)\left(0.082 \, \frac{\text{cm}^2}{\text{sec}}\right)^{1/6}\left(\frac{2\pi(10)}{60 \, \text{sec}}\right)^{-1/2}\right]^{3/2}$$

$$= 1.2 \cdot 10^{-3} \, \text{cm}^2/\text{sec}$$

This value is comparable to those given in Table 20.4-2.

21.2 Physical Equalities Among Mass, Heat, and Momentum Transfer

In this section, we want to discuss situations in which mass transfer, heat transfer, and fluid flow occur at the the same rate. Such equivalence may be startling, for much of our earlier discussion emphasized differences between these processes. To be sure, the previous section described the parallel equations called Fick's law, Fourier's law, and Newton's law; but this parallelism was one of mathematics. The diffusion coefficient, the thermal conductivity or diffusivity, and the viscosity all had different numerical values, and so should give different rates.

21.2.1 The Reynolds Analogy

Nonetheless, the rates of mass, heat, and momentum transfer can be essentially the same for fluids in turbulent flow. This subject was first studied by the Englishman

Osborne Reynolds, who lived from 1842 to 1912. The descendant of generations of clergy who had served in the same Irish parish, Reynolds deliberately went to work in mechanical engineering before going to Cambridge University. Shortly after his graduation in 1867, Reynolds became professor of engineering at Owens College, Manchester, where he remained for his entire professional life.

Reynolds argued that mass or heat transport into a flowing fluid must involve two simultaneous processes: "1. the natural diffusion of the fluid when at rest [and] 2. the eddies caused by visible motion which mixes the fluid up and brings fresh particles into contact with the surface." He went on: "The first of these causes is independent of the velocity of the fluid [but] the second cause, the effect of eddies, arises entirely from the motion of the fluid." Note that Reynolds implies that any flowing fluid contains eddies. Nine years later, Reynolds discovered the distinction between laminar flow and turbulent flow, and that eddies occur only in the latter.

These arguments have considerable value even when restricted to turbulent flow. To see this, we write expressions for the various fluxes. For example, the mass flux should be

$$N_1 = k\Delta c_1 = [a + bv]\Delta c_1 \tag{21.2-1}$$

where the quantity in brackets is equivalent to the mass transfer coefficient k. This coefficient has two parts: a, which is due to diffusion, the "natural internal diffusion," and bv, which represents the effect of eddies "which mix the fluid up." This seems very simple but very sensible.

In a similar fashion, we can write other flux equations. For energy, we find

$$q = h\Delta T = [a' + b'v]\Delta(\rho \hat{C}_p T) \tag{21.2-2}$$

where the heat transfer coefficient h reflects heat conduction a' and the effect of eddies $b'v$. For momentum, we write

$$\tau = f\left(\tfrac{1}{2}\rho v^2\right) = \left[\frac{fv}{2}\right]\rho v$$

$$= [a'' + b''v]\rho v \tag{21.2-3}$$

where the friction factor f is made of a viscous contribution a'' and the effects of eddies $b''v$.

We now turn to the limit of rapid turbulent flow, where the effect of eddies will dominate any diffusion, conduction, or viscosity. In other words, a, a', and a'' have very little effect. However, if the eddies dominate, then all transport is due to that "mixing up" and is independent of any diffusion coefficient or thermal conductivity or viscosity. All transport is due to the same turbulent mechanism. In Reynolds's words, these "various considerations lead to the supposition that"

$$b = b' = b'' \tag{21.2-4}$$

Although this always seems to me a big intuitive leap, it does make more sense as I think about it.

This supposition provides a relation between the various transport coefficients. From Eq. 21.2-1, because a is relatively small,

$$k = bv \tag{21.2-5}$$

We can make similar arguments for the other coefficients:

$$\frac{h}{\rho \hat{C}_p} = b'v \tag{21.2-6}$$

$$\frac{fv}{2} = b''v \tag{21.2-7}$$

We then use Eq. 21.2-4 to combine the results:

$$\frac{k}{v} = \frac{h}{\rho \hat{C}_p v} = \frac{f}{2} \tag{21.2-8}$$

This result is called the "*Reynolds analogy*."

The Reynolds analogy is interesting because it suggests a simple relation between different transport phenomena. This relation should be accurate when transport occurs by means of turbulent eddies. In this situation, we can estimate mass transfer coefficients from heat transfer coefficients or from friction factors.

The Reynolds analogy is found by experiment to be accurate for gases, but not for liquids. We can rationalize this on the basis of the transport coefficients involved. We expect turbulent mixing to take place at two levels: a macroscopic level, where eddies are dominant, and a microscopic level, where diffusion, conduction, and viscosity are important. For gases, these microscopic processes are about the same because

$$D \doteq \alpha \doteq \nu \doteq 0.1 \, \text{cm}^2/\text{sec} \tag{21.2-9}$$

In the more dignified terms of dimensionless groups, the Schmidt and Prandtl numbers of gases are equal:

$$\frac{\nu}{D} \doteq \frac{\nu}{\alpha} \doteq 1 \tag{21.2-10}$$

However, for liquids, these groups are significantly different; the Schmidt number is about 1,000, but the Prandtl number is around 10. Thus the "mixing up" of turbulence may be nearly the same for gases, but it will not be for liquids.

21.2.2 The Chilton–Colburn Analogy

Because the Reynolds analogy was practically useful, many authors have tried to extend it to liquids. These extensions often included elaborate theoretical rationalizations. However, the most useful extension is the simple empiricism suggested by Chilton and Colburn.

Chilton and Colburn recognized that the Reynolds analogy worked well for gases but not for liquids. They also believed that the changes in liquids could best be represented as

Prandtl and Schmidt numbers. By an analysis of experimental data, they showed that Eq. 21.2-6 was better replaced by

$$b' = \frac{h}{\rho \hat{C}_p v} \left(\frac{\nu}{\alpha}\right)^{2/3}$$
(21.2-11)

By mathematical analogy, they extended this to mass transfer by replacing Eq. 21.2-5 by

$$b = \frac{k}{v} \left(\frac{\nu}{D}\right)^{2/3}$$
(21.2-12)

Thus from Eq. 21.2-4 and Eq. 21.2-7

$$\frac{k}{v} \left(\frac{\nu}{D}\right)^{2/3} = \frac{h}{\rho \hat{C}_p v} \left(\frac{\nu}{\alpha}\right)^{2/3} = \frac{f}{2}$$
(21.2-13)

This "Chilton–Colburn analogy" reduces to the Reynolds analogy (Eq. 21.2-8) for gases whose Schmidt and Prandtl numbers equal unity.

Because the Chilton–Colburn analogy turned out to be successful experimentally, we sometimes forget the frailty of its original basis. It was justified by available data for both fluid flow and heat transfer at solid walls. It was much more of a guess for mass transfer, where the important cases involved transfer across the fluid–fluid interfaces common to absorption and extraction. The mass transfer correlation at fluid–solid walls was based on just five data points. Clearly, Chilton and Colburn made an inspired guess.

Two other historical asides about this result are interesting. First, the dimensionless quantities b and b' suggested by Reynolds were renamed j-factors by Chilton and Colburn. These factors are common in the older literature, especially as j_D and j_H. Second, the exponent of $\frac{2}{3}$ on the Schmidt and Prandtl number is frequently subjected to theoretical rationalization, especially using boundary-layer theory. Chilton is said to have cheerfully conceded that the value of $\frac{2}{3}$ was not even equal to the best fit of the data, but was chosen because the slide rules in those days had square-root and cube-root scales, but no other easy way to take exponents.

21.2.3 The Wet-Bulb Thermometer

The best example of simultaneous heat and mass transfer using these analogies is the analysis of the wet-bulb thermometer. This convenient device for measuring relative humidity of air consists of two conventional thermometers, one of which is clad in a cloth wick wet with water. The unclad dry-bulb thermometer measures the air's temperature. The clad wet-bulb thermometer measures the colder temperature caused by evaporation of the water. This colder temperature is like that you feel by licking your finger and waving it about.

We want to use this measured temperature difference to calculate the relative humidity in air. This relative humidity is defined as the amount of water actually in the air divided by the amount at saturation at the dry-bulb temperature. To find this humidity, we first write equations for the mass and energy fluxes:

$$N_1 = k(c_{1i} - c_1) = kc(y_{1i} - y_1)$$
(21.2-14)

$$q = h(T_i - T)$$
(21.2-15)

where c_{1i} and c_1 are the concentrations of water vapor at the wet bulb's surface and in the bulk, y_{1i} and y_1 are the corresponding mole fractions, T_i is the wet-bulb temperature, and T is the bulk dry-bulb temperature. Note that y_{1i} is the value at saturation at T_i. The mass and energy fluxes are coupled:

$$N_1 \Delta \tilde{H}_{vap} = -q$$
(21.2-16)

where $\Delta \tilde{H}_{vap}$ is the heat of vaporization of the evaporating water. Thus

$$k \Delta \tilde{H}_{vap} c(y_{1i} - y_1) = h(T - T_i)$$
(21.2-17)

From Eq. 21.2-13, the Chilton–Colburn analogy,

$$k = \frac{h}{\rho \hat{C}_p} \left(\frac{D}{\nu}\right)^{2/3} = \frac{h}{c \tilde{C}_p} \left(\frac{D}{\nu}\right)^{2/3}$$
(21.2-18)

where ρ is the mass concentration, that is, the density; c is the molar concentration; $\hat{C}_p$ is the specific heat capacity, that is, per mass; and $\tilde{C}_p$ is the molar heat capacity, that is, per mole. For gases, the Lewis number α/D is about unity. Combining Eqs. 20.2-17 and 20.2-18 and rearranging, we find that

$$y_1 = y_{1i} - \left(\frac{\tilde{C}_p}{\Delta \tilde{H}_{vap}}\right)(T - T_i)$$
(21.2-19)

or

$$\left(\begin{array}{c} \text{relative} \\ \text{humidity} \end{array}\right) = \frac{p_1}{p_1(\text{sat at } T)} = \frac{y_1}{y_1(\text{sat at } T)}$$

$$= \frac{1}{y_1(\text{sat at } T)} \left[y_{1i}(\text{sat at } T_i) - \left(\frac{\tilde{C}_p}{\Delta \tilde{H}_{vap}}\right)(T - T_i)\right]$$
(20.2-20)

Thus the relative humidity should be independent of the flow past the thermometers and should vary with the temperature difference between the wet-bulb and dry-bulb readings.

21.3 Drying

Armed with the analogies above, we are now able to analyze many forms of drying. In this section, we give a synopsis of these analyses. We begin by discussing how

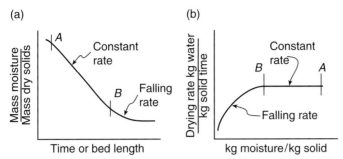

Fig. 21.3-1. Drying curves. The mass of a wet solid drops with drying time. Often, these data are replotted as drying rate vs. moisture concentration.

the mass of a wet solid changes as it dries, and from this infer the mechanism of drying. We show how this mechanism affects the results of various batch and continuous dryers. We then discuss the time required for batch drying and spray drying. The case of batch drying is really a generalization of the ideas of mass and heat transfer used in the wet-bulb thermometer example. The case of spray drying extends these ideas back into the concept of diffusion.

In many cases, the mass of a wet solid varies with time as shown in Fig. 21.3-1. Two points merit explanation. First, the plot vs. time is appropriate for a batch dryer. However, we could equally well plot mass vs. position divided by velocity, that is, vs. the residence time of a steady-state dryer. Second, while the liquid being evaporated can be any solvent, we will discuss only the most common case when that solvent is water.

The actual variation of the mass with time is on the left of the figure. After a brief induction time, the mass drops linearly with time, often until 90 percent or more of the water has evaporated. It then drops at a slower rate. This linear drop is the norm and is strikingly accurate: to test it for yourself, just hang a wet tee shirt on a lab balance and record the weight vs. time.

Ironically, the simple behavior on the left of Fig. 21.3-1 is often replotted in the more complicated way on the right of the figure. This right-hand way plots the rate of drying vs. the moisture concentration. In both cases, a constant drying region begins at point A and ends at point B, which is sometimes called the "critical drying concentration." In this region, the evaporating water is no different than any other bulk water. That is why it evaporates at a constant rate: it is the same pure water evaporating from the constant surface area of the solid. This period of constant drying rate of "free" water ends at point B in the figures.

At larger times, that is, at lower moistures, the drying slows. In this "falling rate" period, the water is more and more difficult to remove. It may be trapped within the solid, escaping only slowly through small pores; or it may be chemically bound to the solid, and require a greater enthalpy of vaporization to pry it loose. "Free" water may be drawn quickly to the surface by capillarity; but "bound" water is stuck in place, sometimes because it is part of a chemical compound. Evaporating the "free" water may cause 90 percent of the total change in mass, but evaporating the "bound" water may take 90 percent of the total drying time.

The evaporation of free and bound water affects the temperatures in most commercial dryers. During constant-rate drying, the solid temperature retains a constant wetbulb temperature, just as for the wet-bulb thermometer. The gas temperature often varies. To illustrate the ideas involved, we consider the cases of batch drying and spray drying in the next two subsections.

21.3.1 Batch Dryer Analysis

We first consider the easiest case of a batch of wet solid dried with excess hot air. In this case, the heat flux q and the mass flux N_1 are

$$q = h(T - T_i) \tag{21.3-1}$$

$$N_1 = k(c_{1i} - c_1) = k_y(y_{1i} - y_1) \tag{21.3-2}$$

where the subscript i indicates a temperature or concentration in the vapor but at the interface. As before, the mass transfer coefficients k and k_y are easily related

$$kc = k_y \tag{21.3-3}$$

where c is the total concentration in the vapor. When the dryer is adiabatic, the heat and mass fluxes are coupled by

$$q = N_1 \Delta \tilde{H}_{vap} \tag{21.3-4}$$

where $\Delta \tilde{H}_{vap}$ is the molar enthalpy of the evaporation of water. This coupling means that we can solve drying problems either in terms of heat transfer or in terms of mass transfer. Normally, temperature is easier to measure than concentration, and the mass flux (i.e., the moles of water lost per time) is easier to measure than heat flux. We will use this to make solving the problems easier, as the following example shows.

Example 21.3-1: Drying titania   Pans containing a 4-cm deep layer of titania particles are dried in an oven. The voids between the particles ($\varepsilon = 0.36$) contain pure water, but there is no water inside of the particles themselves. If the air flow is 3 m/sec, the heat transfer coefficient above the bed is about 30 W/m² °C. Because the pans have little contact with the racks in the dryer, the drying is essentially adiabatic. How long will it take to dry this titania using excess air at 75 °C and a humidity of 0.01 kg water/kg air?

Solution   Because only free water is evaporating, we will have constant rate drying, so that we can calculate the flux as the total amount of water removed per drying time. From the temperature and humidity given, the wet-bulb temperature is 31 °C. The drying time t can then be found from Eqs. 21.3-1 and 21.3-4,

$$N_1 = \frac{q}{\Delta \tilde{H}_{vap}} = \frac{h}{\Delta \tilde{H}_{vap}}(T - T_i)$$

$$\frac{0.04\,m(0.36)}{t}\frac{10^3\,kg}{m^3} = \frac{30\,W/m^2\,°C\,sec}{2300 \cdot 10^3\,J/kg}(75 - 31)\frac{3600\,sec}{hr}$$

$$t = 19\,hr$$

The drying takes the better part of a day.

21.3.2 Spray Dryer Analysis

The ideas used for the simple case of batch drying of wet solid can be extended to more complex situations like spray drying. We review one example of this intellectual extension in the following.

Spray drying involves pumping a slurry into hot air. The liquid trapped between particles in the slurry evaporates adiabatically, normally in a lot of air. Air flow can be concurrent, countercurrent, or a mixture of the two. The process is used to make dried milk, dried eggs, and dry grain. It is the basis of manufacturing instant coffee and laundry detergent.

The slurry particles which are dried are typically 30 to 100 µm. They can be produced either with nozzles or, more frequently, on the top of discs perhaps 0.3 m in diameter spinning at 5000 rpm. The gas used, which is typically 100 to 700 °C, must be cool enough not to compromise product stability. In many cases, the spray drying will produce both larger particles and a fine dust, and so the dryer must have some form of dust collection, often in a cyclone. Sometimes, when a fine powder is desired, all of the product will be captured in a cyclone.

We want to estimate the drying time for such a spray. Our objective is to dry the spray so quickly that a hard skin forms on the outside of each particle. This hard skin will often slow further removal of water, and so is something that we normally try to avoid. Here, however, we want to make the drying as fast as possible for two reasons. First, we want to retain product "quality." Sometimes, this means that we want to retain chemical integrity; at other times, we may want to inhibit flavor evaporation. Another common measure of quality is consistency, i.e., a uniform powder which is suitable for immediate packaging.

The second reason that we want the drying to be fast is to prevent agglomeration. Partially dried slurry particles are sticky, and tend to aggregate. They stick to each other, and they stick to the dryer walls. In extreme cases, they fuse together, filling the entire dryer with one huge chunk of porous solid. Obviously, fast drying which avoids this clogging is desirable.

To estimate the drying time, we assume that droplets in the spray are heated quickly, and dry more slowly. This implies that the thermal diffusivity is much greater than the diffusion coefficient. The concentration of water c_1 within one slurry droplet is described by

$$\frac{\partial c_1}{\partial t} = \frac{D}{r^2} \frac{\partial}{\partial r}\left(r^2 \frac{\partial c_1}{\partial r}\right) = D \frac{\partial^2 c_1}{\partial z^2} \tag{21.3-5}$$

where r is the distance out from the center of the droplet, and z is the distance from the surface of the particle towards the center. Note that the use of the z coordinate system neglects the droplet's curvature, which is appropriate because we are more interested in early stages of drying.

This equation is subject to the initial condition

$$t = 0, \quad \text{all } z, \quad c_1 = c_{1\infty} \tag{21.3-6}$$

where $c_{1\infty}$ is the initial concentration. It is also subject to the boundary conditions

$$t > 0, \quad z = \infty, \qquad c_1 = c_{1\infty} \tag{21.3-7}$$

$$z = 0, -D\frac{\partial c_1}{\partial z} = k\left(c_{10} - c_1^*\right) \tag{21.3-8}$$

where c_{10} is the surface concentration of water and c_1^* is the concentration that would be in the drop if it were in equilibrium with the surrounding air. The first of these boundary conditions repeats our focus on the initial stages of drying, because it says the concentration well below the surface doesn't change much. The second condition includes the effective diffusion coefficient D and the mass transfer coefficient k. This latter is for mass transfer in the surrounding gas, but is based on the liquid concentrations in the slurry particle

$$k = k_G H \tag{21.3-9}$$

where k_G is the more normal mass transfer coefficient in the gas, and H is the molar concentration of water at equilibrium in the gas divided by that in the particle. Because I think that H is best explained by an example, I ask that you accept k for the moment as a mass transfer coefficient.

The solution to Eqs. 21.3-5 to 21.3-8 is known but complex. The concentration at the surface is somewhat simpler

$$\frac{c_{10} - c_{1\infty}}{c_1^* - c_{1\infty}} = 1 - e^{k^2 t/D}\left(1 - \text{erf}\left[k\sqrt{\frac{t}{D}}\right]\right) \tag{21.3-10}$$

In many cases, we are interested in the time when c_1 is 10% of its initial value, because at this point, the product starts to behave as a solid powder. When

$$\frac{c_{10} - c_{1\infty}}{c_1^* - c_{1\infty}} = 0.1 \tag{21.3-11}$$

we find

$$\frac{k^2 t}{D} = 30 \tag{21.3-12}$$

If we know k and D, we can estimate the drying time t. This key result for spray drying is illustrated by the following example.

Example 21.3-2: Drying a particle of soap powder   A spray dryer injects 100 μm particles containing 30% water into air at 60 °C and a relative humidity of 70%. The diffusion coefficient within the particles is $3 \cdot 10^{-7}$ cm²/sec; that in the surrounding gas is 0.3 cm²/sec; and the gas flow is 10 m/sec. (a) What is the mass transfer coefficient k in this system? (b) How long will it take to dry the particle?

    Solution   Calculating the mass transfer coefficient requires a correlation for mass transfer from a sphere:

$$\frac{k_G d}{D} = 2 + 0.6\left(\frac{dv}{\nu}\right)^{1/2}\left(\frac{\nu}{D}\right)^{1/3}$$

For this case

$$\frac{k_G \ 100 \cdot 10^{-4} \, \text{cm}}{0.2 \, \text{cm}^2/\text{sec}} = 2 + 0.6 \left(\frac{100 \cdot 10^{-4} \, \text{cm} \ 10^{-3} \text{cm}^2/\text{sec}}{0.2 \, \text{cm}^2/\text{sec}} \right)^{1/2}$$

$$\cdot \left(\frac{0.2 \, \text{cm}^2/\text{sec}}{0.3 \, \text{cm}^2/\text{sec}} \right)^{1/3}$$

$$k_G = 170 \ \text{cm}/\text{sec}$$

We must now change this to a value based on a concentration difference in the wet particle, rather than in air

$$k = k_G H$$

where H is a partition coefficient, defined here as

$$H = \frac{\text{molar conc in gas}}{\text{molar conc in particle}}$$

At 60 °C, this is

$$H = \frac{\dfrac{150 \, \text{mm Hg}}{760 \, \text{mm Hg}} \left[\dfrac{1 \, \text{mol}}{22.4 \cdot 10^3 \, \text{cm}^3} \right] \dfrac{273}{333}}{\left(\dfrac{0.3 \, \text{g H}_2\text{O}}{\text{cm}^3 \, \text{particle}} \right) \left(\dfrac{\text{mol}}{18 \, \text{cm}^3} \right)}$$

$$= 4.3 \cdot 10^{-4}$$

Thus

$$k = 170 \ (4.3 \cdot 10^{-4}) = 0.074 \ \text{cm}/\text{sec}$$

From Eq. 21.3-12,

$$\frac{k^2 t}{D} = \frac{(0.074 \ \text{cm}/\text{sec})^2 \ t}{3 \cdot 10^{-7} \ \text{cm}^2/\text{sec}} = 30$$

$$t = 0.002 \ \text{sec}$$

The particle is dried quickly. This is an interesting case because fast mass transfer in the gas and slow diffusion within the particle mean that a dry skin on the particle is easily formed.

21.4 Design of Cooling Towers

The final practical problem in this chapter is the cooling of water. Chemical processes typically require large quantitites of cool water, which is why chemical plants are sited near rivers or lakes. Frequently, the river's water is too warm to cool the process efficiently, and cooling the water with refrigeration is prohibitably expensive. Cool water must be found in some other way.

Cooling towers like those shown in Fig. 21.4-1 are the cheapest way to cool large quantities of water. They are among the largest mass transfer devices in common use.

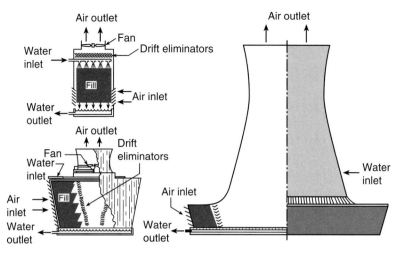

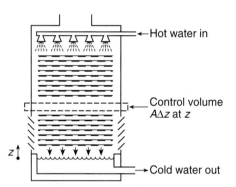

Fig. 21.4-1. Cooling towers. These devices, which are among the largest made for mass transfer, cool large quantities of water by evaporation of a small fraction of the water.

Fig. 21.4-2. Modeling a small cooling tower. We want to calculate the size of a cooling tower required to cool a given amount of water. We base this calculation on mass and energy balances on the small volume shown.

The basic operation of one common form of cooling tower is shown schematically in Fig. 21.4-2. The tower is packed with inert material, commonly with wooden slats. Hot water sprayed into the top of the tower trickles down through the wood, partially evaporating as it goes. Air enters the bottom of the tower and rises up through the packing. In smaller towers, the air can be pumped with a fan; in larger ones, it is often allowed to rise by natural convection.

   We want to calculate the size of a tower required to cool a given amount of water. This calculation is roughly parallel to that for gas absorption. That earlier problem involved three equations: a mass balance or operating line; an energy balance or equilibrium line; and a rate equation. However, for water cooling, these equations are written in terms of gas enthalpy and water temperature, as discussed below.

21.4.1 The Operating Line

To begin, we assume that the cooling tower operates adiabatically, so that any heat and mass exchange is between the liquid water and the wet air, and not with the outer walls of the tower. If this is true, we see that

$$
\begin{pmatrix} \text{energy} \\ \text{accumulation} \end{pmatrix} = \begin{pmatrix} \text{energy gained} \\ \text{by wet air} \end{pmatrix} + \begin{pmatrix} \text{energy lost} \\ \text{by liquid water} \end{pmatrix} \tag{21.4-1}
$$

$$
0 = -G\frac{d\tilde{H}}{dz} - L\tilde{C}_{p,L}\frac{dT_L}{dz} \tag{21.4-2}
$$

where G and L are the fluxes of dry air and liquid water, respectively; $\tilde{H}$ is the enthalpy of wet air per mole of dry air; $\tilde{C}_{p,L}$ is the molar enthalpy of liquid water; and T_L is the temperature of liquid water. Note that in this equation both G and L are taken to be constant. While the flux of dry air is nearly constant, the flux of liquid water certainly is not because some of the water is evaporating. After all, that is where the cooling comes from. However, because the heat of vaporization is large, the flow of liquid water doesn't change that much, and this assumption turns out to be reasonable.

We now can integrate this energy balance between the top of the column and some arbitrary position to find

$$
\tilde{H} = \left(\tilde{H}_0 - \frac{L}{G}\,\tilde{C}_{p,L}\,T_{L,0} \right) + \left(\frac{L}{G}\,\tilde{C}_{p,L} \right)T_L \tag{21.4-3}
$$

where $\tilde{H}_0$ and $T_{L,0}$ are, respectively, the molar gas enthalpy and liquid temperature, both at the top of the column where the liquid enters. A plot of $\tilde{H}$ vs. T_L is linear. For this operating line, the slope (L/G) is an echo of that for other separation processes. Note that $(\tilde{C}_{p,L}T)$ is the molar liquid enthalpy, and hence echoes our development of analogies in Section 21.2.

21.4.2 The Equilibrium Line

We next seek an equilibrium line which gives the molar enthalpy of the wet air in equilibrium with the liquid water, that is, at the liquid water temperature. To start to understand this, we need to examine the definition of the enthalpy of the wet air $\tilde{H}$ more carefully:

$$
\tilde{H} = \tilde{C}_p T + \tilde{H}_{vap}y \tag{21.4-4}
$$

The first term on the right-hand side is the enthalpy due to heating the air; the second represents the enthalpy increase caused by adding water vapor to the dry air. While the exact values of the enthalpy will, of course, depend on the reference values for zero enthalpy, we will ignore these because they do not affect the results of our calculations.

We are especially interested in the enthalpy of the wet air in equilibrium with the liquid water. Because the thermal conductivity of the liquid water is much greater than that of the air, the temperature of the air–water interface is almost that of the bulk liquid

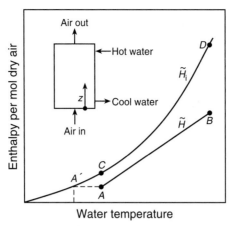

Fig. 21.4-3. Sizing a cooling tower. These calculations depend on evaluating the wet air's humidity vs. the water's temperature, shown as the line AB. They also involve the interfacial humidity vs. water temperature, shown as the line CD. The lines AB and CD are sometimes called the operating and equilibrium lines by analogy with gas absorption. Note that A' is the dew point of the entering air.

water. Then the concentration y must be in equilibrium with the saturation vapor pressure, which is in turn a function of T_L. Thus

$$\tilde{H}_i = \tilde{H}^* = \tilde{C}_p T_L + \Delta\tilde{H}_{vap} y(T_L) \tag{21.4-5}$$

A plot of $\tilde{H}_i$ vs. T_L curves upwards because the vapor pressure vs. temperature also curves upwards. This result is the equilibrium line which we seek.

The meaning of this equilibrium line and the operating line in Eq. 21.4-3 may be clearer if we plot both as shown in Fig. 21.4-3. The line $\overline{AB}$ is the operating line; the enthalpy $\tilde{H}$ shown by this line is that actually present at the liquid temperature T_L. (The point A' is the dew point of the water at the bottom of the tower.) The line $\overline{CD}$ is the equilibrium line, giving the enthalpy $\tilde{H}_i$ which would be present if air and liquid water were in equilibrium. The difference between these lines $(\tilde{H}_i - \tilde{H})$ is the enthalpy driving force responsible for water cooling.

21.4.3 The Rate Equation

The rate equation for water cooling is a combined mass and energy balance. To find this, we begin with a mass balance on the differential volume $A\Delta z$ shown schematically in Fig. 21.4-2. This volume, located at z, is filled with packing having a surface area per volume equal to a. We can make a mass balance on the water vapor in this volume as follows:

$$\begin{pmatrix} \text{water} \\ \text{accumulation} \end{pmatrix} = \begin{pmatrix} \text{water} \\ \text{convection in} \\ \text{minus that out} \end{pmatrix} + \begin{pmatrix} \text{water} \\ \text{added by} \\ \text{evaporation} \end{pmatrix} \tag{21.4-6}$$

or

$$0 = \left(Gy|_{z+\Delta z} - Gy|_z\right) + (\Delta za)k_y(y_i - y) \tag{21.4-7}$$

where y and y_i are the water vapor concentrations in the bulk and at the air–liquid interface, respectively; G is again the molar air flux; and k_y is the mass transfer coefficient for the water vapor. Strictly speaking, the concentration y is defined as moles of water per mole of dry air, and the flux G is the flux of dry air. However, because the water vapor is normally dilute, this detail is often not important. The mass transfer coefficient refers just to the gas phase and is based on a mole fraction driving force. It is related to other forms of mass transfer coefficients by the usual relations given in Section 8.2. We take the limit of Eq. 21.4-7 by dividing by Δz and letting this distance go to zero,

$$0 = \frac{d}{dz}(Gy) + k_y a(y_i - y) \tag{21.4-8}$$

This is the first part of our rate equation. This mass balance is basic to the calculations that follow.

We next make an energy balance on the wet air in the same differential volume $A\Delta z$:

$$\left(\begin{array}{c} \text{energy} \\ \text{accumulation} \end{array}\right) = \left(\begin{array}{c} \text{energy} \\ \text{convection in} \\ \text{minus that out} \end{array}\right) + \left(\begin{array}{c} \text{energy} \\ \text{added by} \\ \text{heat transfer} \end{array}\right) \tag{21.4-9}$$

$$0 = G\hat{C}_p \frac{dT}{dz} + ha(T_L - T) \tag{21.4-10}$$

where T and T_L are the temperatures of the bulk air and the air–water interface, $\hat{C}_p$ is the heat capacity of the air, and h is the heat transfer coefficient.

We now want to combine the mass and energy balances in a way that allows us to size the cooling tower. To do so, we multiply Eq. 21.4-8 by the heat of vaporization of water $\Delta \tilde{H}_{vap}$ and add the result to Eq. 21.4-10 to obtain

$$G\frac{d}{dz}(\hat{C}_p T + \Delta \tilde{H}_{vap} y) = ha(T_L - T) + k_y a \Delta \tilde{H}_{vap}(y_i - y) \tag{21.4-11}$$

The Chilton–Colburn analogy is

$$\frac{k}{v}\left(\frac{v}{D}\right)^{2/3} = \frac{h}{\rho \hat{C}_{p,\text{air}} v}\left(\frac{v}{\alpha}\right)^{2/3} \tag{21.4-12}$$

For gases, the Lewis number α/D is about unity, $\rho \hat{C}_p$ equals $c\tilde{C}_p$, and kc equals k_y. Thus

$$h = (\rho \hat{C}_p)k = (c\tilde{C}_p)k_y/c \tag{21.4-13}$$

Inserting this into the previous equation and rearranging, we find

$$G\frac{d}{dz}(\tilde{H}) = k_y a(\tilde{H}_i - \tilde{H}) \tag{21.4-14}$$

where

$$\tilde{H} = \tilde{C}_p T + \Delta \tilde{H}_{vap} y \tag{21.4-15}$$

$$\tilde{H}_i = \tilde{H}^* = \tilde{C}_p T_L + \Delta \tilde{H}_{vap} y_i \tag{21.4-16}$$

In physical terms, $\tilde{H}$ is the enthalpy of the wet air per mole of dry air. Correspondingly, the quantity $\tilde{H}_i$ is the enthalpy of wet air per mole dry air at the interface, where the air is saturated. Note that the mole fraction at this interface is the saturation value and hence is a function only of the water's temperature T_L.

Equation 21.4-14 can be numerically integrated to find the desired height of the cooling tower l:

$$l = \int_0^l dz = \left[\frac{G}{k_y a} \right] \int_{\tilde{H}_0}^{\tilde{H}_l} \frac{d\tilde{H}}{\tilde{H}_i - \tilde{H}} = [\text{HTU}] \int_{\tilde{H}_0}^{\tilde{H}_l} \frac{d\tilde{H}}{\tilde{H}^* - \tilde{H}} \tag{21.4-17}$$

This integration depends on the operating and equilibrium lines like those shown in Fig. 21.4-3. It is similar mathematically to the integration used in the gas absorption analysis in Chapter 10, except that the enthalpy replaces the gas concentration y, and the liquid temperature replaces the liquid concentration x. How the integation is accomplished is shown in the example below.

Example 21.4-1: Design of a cooling tower   A countercurrent cooling tower is needed to cool water flowing at 2,150 kg/min. The water enters at 60 °C and is to be cooled to 25 °C. The air is fed at 60 mol/m² sec, with a dry-bulb temperature of 30 °C and a dew point of 10 °C. The water flux should be 40% lower than the maximum allowed thermodynamically. The tower is packed with wood slats giving an HTU of 3 m. Find the height of tower required.

Solution   We begin by plotting the enthalpy of water-saturated air $\tilde{H}^*$ vs. temperature, as shown in Fig. 21.4-4. We know that $\tilde{H}$ for the entering air is based on a mole fraction of water equal to the saturated value at 10 °C, the wet-bulb temperature; we also know that the water temperature at this point is 25 °C. Thus we locate the point A. We can then solve for the specific values given.

We begin by remembering that the operating line is a plot of $\tilde{H}$ vs. liquid temperature T_L. The slope of this line is the liquid flux L times the liquid heat capacity $\tilde{C}_p$ divided by the gas flux G. If we increase the water flux, we increase the slope of this operating line. We can increase this water flux until the operating line is tangential to the equilibrium line, that is to the line $\overline{AB'}$ shown in Fig. 21.4-4. Any higher water flow would give an enthalpy of wet air higher than the equilibrium value. Thus from Fig. 21.4-4, we find

$$\left(\frac{\text{slope}}{AB'} \right) = \frac{\tilde{C}_{p,L} L}{G} = 230 \text{ J/g mol} \degree \text{C} \tag{21.4-18}$$

The flux G equals 60 mol/m² sec, and $\tilde{C}_{p,L}$ is 75 J/mol °C; so L is 180 mol H_2O/m² sec. The actual flow is to be 40% less:

$$L = 110 \text{ mol } H_2O/m^2 \text{ sec}$$

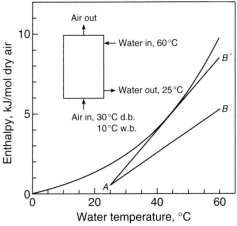

Fig. 21.4-4. Design of a cooling tower. The line AB' is used to find the minimum air flow. The line AB, which determines the actual enthalpy $\tilde{H}$, is key to finding the height of tower required.

Using this value, we can draw the actual operating line AB. The tower's height can then be found from Eq. 21.4-17:

$$l = [\mathrm{HTU}] \int_{\tilde{H},\mathrm{in}}^{\tilde{H},\mathrm{out}} \frac{d\tilde{H}}{\tilde{H}_i - \tilde{H}}$$

Values of $\tilde{H}_i$ and $\tilde{H}$ can easily be read off lines CD and AB in Fig. 21.4-3; the integral found by numerical integration of these data equals 2.7. Thus

$$l = (3\mathrm{m})(2.7)$$
$$= 8.1\,\mathrm{m}$$

This is the desired result.

21.5 Thermal Diffusion and Effusion

This section discusses several ways in which temperature gradients effect a solute flux. The phenomena involved occur in the absence of convection and are treated with models like those developed for diffusion in Chapters 2 and 3. Thus the approach is again based on a distributed-parameter model and is more fundamental than the approach based on mass transfer coefficients and used in earlier sections of this chapter.

The first effect, thermal diffusion, is exemplified by the two experiments shown in Fig. 21.5-1. In the first, a tall column of salt solution is heated at the top and cooled at the bottom. The salt's concentration is initially uniform, but later becomes more concentrated near the bottom of the tube. This experiment was originally made in 1856 by Fick's mentor, Carl Ludwig; more complete experiments were later made by Charles Soret, after whom this effect is named. A similar experiment, shown schematically in Fig.

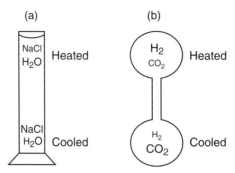

Fig. 21.5-1. Thermal diffusion. If a tall column of initially homogeneous salt solution is heated at the top and cooled at the bottom, the mole fraction of salt will become slightly greater at the bottom. If two bulbs connected with a capillary are filled with the same gas mixture and only one bulb is heated, the gas compositions will become unequal.

21.5-1(b), consists of two bulbs, both of which initially contain the same gaseous mixture. When one bulb is heated and the other is cooled, the gas no longer has the same mole fractions throughout. This experiment was originally made to check the theoretical prediction of its existence. It is a rare occurrence when the experiment follows the theory, rather than the other way around.

Results of experiments like these are often reported in slightly different ways. For liquids, the results are usually given in terms of the flux equation:

$$-j_1 = Dc(\nabla x_1 - \sigma x_1 x_2 \nabla T) \tag{21.5-1}$$

The Soret coefficient σ, which has the dimensions of reciprocal temperature, can be either positive or negative. For gases, the results are correlated using the equation

$$-j_1 = Dc\left(\nabla x_1 - \alpha x_1 x_2 \frac{\nabla T}{T}\right) \tag{21.5-2}$$

The dimensionless thermal diffusion factor α obviously equals the Soret coefficient times the temperature. Both σ and α are positive when species 1 concentrates in the hot region.

Two aspects of these equations are interesting. First, we are now writing the diffusion flux in terms of the gradient of mole fraction, not molar concentration. This is because we know that the molar concentration varies with temperature, but the mole fraction is much more nearly constant, independent of temperature. Such a flux equation implies a different reference frame than the volume average velocity emblazoned through this book. Second, we deliberately introduce a factor $x_1 x_2$ into the expression for thermal diffusion. This anticipates observations that the effect disappears rapidly for dilute solutions and is largest when solute and solvent concentrations are similar.

A few experimental values of α are shown in Table 21.5-1. The values of α are frequently small, especially in dilute solution. They are largest for solutes of very different molecular weights or for highly nonideal solutions. They are more nearly constant for near-ideal solutions and are concentration-dependent in nonideal liquid mixtures. In short, they behave much like the ternary diffusion coefficients discussed in Chapter 7. They are usually of minor practical importance, even though they can be used to effect surprisingly good separations.

Table 21.5-1 *Thermal diffusion coefficients*

Mixture[a]	Temperature (K)	α[b]
Gases		
(1) 50% H_2–(2) 50% D_2	290–370	+0.17
(1) 50% H_2–(2) 50% He	273–700	+0.15
(1) 50% H_2–(2) 50% CH_4	300–500	+0.29
(1) 50% H_2–(2) 50% O_2	293	+0.02
(1) 50% H_2–(2) 50% CO_2	290–400	+0.05
Liquids		
(1) 20% cyclohexane–(2) 80% CCl_4	313	+1.3
(1) 50% cyclohexane–(2) 50% CCl_4	313	+1.3
(1) 80% cyclohexane–(2) 20% CCl_4	313	+1.3
(1) 20% cyclohexane–(2) 80% benzene	313	−0.1
(1) 50% cyclohexane–(2) 50% benzene	313	−0.4
(1) 80% cyclohexane–(2) 20% benzene	313	−0.6
(1) 25% water–(2) 75% ethanol	298	−0.9
(1) 60% water–(2) 40% ethanol	298	−1.5
(1) 90% water–(2) 10% ethanol	298	+0.3
0.01–M KCl in (2) H_2O	303	−0.6
0.01–M NaCl in (2) H_2O	303	−0.9

Notes: [a]Concentrations in mole percent, except as noted.
[b]Taken as positive if the first species given concentrates in the hot region.
Source: Tyrrell (1961).

This table of values does not explain why the thermal diffusion occurs. This is not an easy explanation to give briefly, and so it is carefully avoided by authors whose knowledge of this field is much greater than mine. We can give a vague explanation by again referring to the gaseous experiment in Fig. 21.5-1(b). In steady state, the flux of molecules from left to right must equal that from right to left. These fluxes must have two parts: that due to thermal motion and that caused by the bulk flow necessary to maintain equal pressure. The thermal motion varies with the molecular weight of the particular species, but the second varies only with the average molecular weight. When the molecules interact as rigid spheres, the net flux is greater for the heavier molecules than for the lighter ones, so these heavier molecules usually will concentrate in the cooler region. Even this qualitative argument is compromised for more elaborate intermolecular potentials, for nonideal liquid solutions, and for solids. No simple, more general explanation seems possible.

Thermal diffusion, which has just been discussed, occurs in mixtures in which molecules of solute and solvent interact with each other. Thermal effusion, the effect discussed next, occurs when the molecules of a pure gas react largely with surroundings.

Thermal effusion is most clearly illustrated by the schematic drawings in Fig. 21.5-2. In these drawings, a pure gas is placed in a closed tube and separated into two volumes by a porous diaphragm. The gas on one side of the diaphragm is heated, so the gas pressure changes. After a while, the pressure reaches a constant value and can be measured.

Two distinctly different cases are observed. The first occurs when the gas pressure is high and the diaphragm has large pores. When the gas is heated, its pressure initially

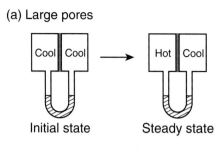

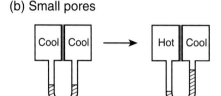

Fig. 21.5-2. Thermal effusion. When the gas on one side of a porous diaphragm is heated, its pressure may change to a new steady state value. If the diaphragm has large pores, this pressure on the hot side is the same as that on the colder side. If the diaphragm has small pores, the pressure on the hot side remains higher, an effect called thermal effusion.

increases, but this increase is quickly reduced by flow through the diaphragm. This flow can be described by Darcy's law, for the gas molecules collide largely with each other. At steady state, no pressure difference remains.

The second case involves a dilute gas and very small pores. Here, any gas molecules in the diaphragm collide mainly with the pore walls, not with other gas molecules. When the temperature is increased, the molecular velocity u in these collisions also increases; from kinetic theory, we can show that this velocity increases with temperature:

$$u \propto \sqrt{T/\tilde{M}} \qquad (21.5\text{-}3)$$

However, the concentration is decreased by temperature:

$$c = \frac{n}{V} = \frac{p}{RT} \qquad (21.5\text{-}4)$$

where n is the number of moles in the system. The flux out of the hot region is the product of the hot velocity and the hot concentration; that out of the cold side is analogous. Thus the total flux is

$$j = cu - c'u'$$

$$\propto \left(\frac{p}{RT} \sqrt{T/\tilde{M}} - \frac{p'}{RT'} \sqrt{T'/\tilde{M}} \right)$$

$$\propto \left(\frac{p}{\sqrt{T}} - \frac{p'}{\sqrt{T'}} \right) \qquad (21.5\text{-}5)$$

where the primed and unprimed value refer to the cold and hot regions, respectively. At steady state, this implies

$$\frac{p}{\sqrt{T}} = \frac{p'}{\sqrt{T'}}$$
(21.5-6)

The pressure on the hot side will be greater than the pressure on the cold side. The key to this process is that the holes in the diaphragm are very small. Thus thermal effusion is to Knudsen diffusion as thermal diffusion is to ordinary diffusion. Both thermal effusion and diffusion are illustrated in the following examples.

Example 21.5-1: The size of thermal diffusion Thermal diffusion is being studied in a two-bulb apparatus like that on the right of Fig. 21.5-1. Each bulb is 3 cm³ in volume; the capillary is 1 cm long and has an area of 0.01 cm². The left-hand bulb is heated to 50 °C, and the right-hand bulb is kept at 0 °C. The entire apparatus is initially filled with an equimolar mixture, either of hydrogen–methane or of ethanol–water. How much separation is achieved? About how long does this separation take?

Solution At steady state, the net flux must be zero. Thus we can rewrite Eq. 21.5-2 for a gas mixture to find

$$0 = Dc\left(\nabla y_1 - \alpha y_1 y_2 \frac{\nabla T}{T}\right)$$

Because we are in steady state, the temperature gradient can be replaced by the temperature difference divided by the tube length. This equation then simplifies to

$$y_1|_{\text{hot}} - y_1|_{\text{cold}} = \alpha y_1 y_2 \left(\frac{T_{\text{hot}} - T_{\text{cold}}}{T_{\text{avg}}}\right)$$

For the gas mixture, this is

$$y_1|_{\text{hot}} - y_1|_{\text{cold}} = (0.29)(0.50)(0.50)\left(\frac{50\,\text{K}}{298\,\text{K}}\right)$$

$$= 0.012$$

For the liquid system water–ethanol, we can use Eq. 21.5-2 as written, and find by interpolation that

$$x_1|_{\text{hot}} - x_1|_{\text{cold}} = (-1.3)(0.50)(0.50)\left(\frac{50\,\text{K}}{298\,\text{K}}\right)$$

$$= -0.05$$

The separations obtained are both small; that with liquids is slightly larger, but in the opposite direction.

We now turn to the time required for this separation to occur. To find this time, we parallel the analysis given for the diaphragm cell in Example 2.2-4. In this analysis, we assume that the temperature difference is suddenly applied at time zero. We also

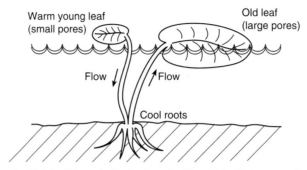

Fig. 21.5-3. Thermal effusion in water lilies. This effect apparently causes a flow of as much as 50 cm/min through the lily's hollow stems. This flow supplies the roots with oxygen.

assume that in spite of this difference, the total molar concentration c is a constant. Thus for the left-hand bulb of volume V_B, we find

$$V_B \frac{dx_{1B}}{dt} = \frac{A}{c} j_i = -\frac{AD}{l}[(x_{1B} - x_{1A}) + b]$$

where b $[= \alpha x_1 x_2 (T_{1B} - T_{1A})/T_{avg}]$ is the effect of thermal diffusion. For the right-hand bulb, of volume V_A

$$V_A \frac{dx_{1A}}{dt} = +\frac{AD}{l}[(x_{1B} - x_{1A}) + b]$$

Combining these equations,

$$\frac{d}{dt}(x_{1B} - x_{1A}) = -D\beta(x_{1B} - x_{1A} + b)$$

where β $[= (A/l)(1/V_B + 1/V_A)]$ was previously used as the calibration constant of the cell. Note that b depends only on average concentrations and constant temperatures, so this result can be integrated directly.

Alternatively, we note that $(D\beta)^{-1}$ is essentially the relaxation time of this cell. If the experiment takes less than this time, the steady state is still far away; if the experiment takes much longer than this time, then the steady state will be approached.

Thus, for gases $(\beta D)^{-1}$ is about 500 sec, and the steady state is reached in a few hours. For liquids, $(\beta D)^{-1}$ is half a year, and reaching the steady state requires a very long time. These slow rates and small separations mean that thermal diffusion usually is a bad route for separations.

Example 21.5-2: Flow in water lily stems   Some water lilies generate flow through their hollow stems in order to facilitate oxygen transfer to their roots. This flow represents an "internal wind" that can reach 50 cm/min. It is believed to occur because of differences in pore sizes between young and old lily leaves, as shown schematically in Fig. 21.5-3. Use kinetic theory to show how the warm young leaf can generate this pressure difference.

   Solution   The explanation of this effect (Dacey, 1980) asserts that the flow is caused by thermal effusion. Sun strikes the lily, warming both old and young leaves. The small pores in the young leaves produce a higher pressure, as suggested by Fig. 21.5-2 and Eq. 21.5-6. The larger pores in the older leaves do not produce any pressure change. Thus gas flows into the young leaves, down the stems, and out the older leaves. I find this delightful.

21.6 Conclusions

Simultaneous heat and mass transfer, the subject of this chapter, is a complicated process. Analyzing this process to find simple but useful results depends on making effective approximations. The approximations exploit both the similar mathematics used for the processes and the similar numerical values of the transport coefficients. This can be true for both distributed and lumped-parameter models. More specifically, for gases, D and α are nearly equal, and k and $h/\rho \hat{C}_p$ are very similar.

This strategy for analysis immediately raises the question of the similar effects possible in liquids and solids. Here, the mathematics remain similar, but the transport coefficients are very different: D is much less than α, and k is much smaller than $h/\rho \hat{C}_p$. What should you do now?

For liquids and solids, the heat transfer is much more rapid than the mass transfer, and so proceeds as if the mass transfer did not exist. In other words, for liquids and solids, the two processes are essentially uncoupled. As an example, imagine that both mass and energy are being transferred from a well-stirred reservoir into a thick solid slab. The energy will be transferred much more rapidly than the mass. In the region where the mass flux is large enough to be interesting, the temperature will be essentially constant, equal to the reservoir temperature.

This difference between gases and other phases illustrates the different analytical strategies possible. Your success in exploring problems like these rests on your ability to make effective approximations. Good luck in your efforts.

Questions for Discussion

1. What are Fick's law of diffusion, Fourier's law of heat conduction, and Newton's law of viscosity?
2. Mass transfer is described by the diffusion coefficient, and momentum transfer is described by the viscosity. Since these quantities have different dimensions, why are they regarded as parallel transport properties?
3. What quantity is the heat transfer parallel of the Sherwood number (kd/D)?
4. What quantity is the heat transfer parallel of the Schmidt number $(\mu/\rho D)$?
5. What quantity is the momentum transfer parallel of the Stanton number (k/v)?
6. Reynolds assumed that the "mixing up" of eddies was the same for mass, energy, and momentum transfer. Justify this assumption.
7. How would you plot mass and heat transfer coefficients to test whether they fit the same form of correlation?
8. Sketch a drying curve, i.e., wet solid mass vs. time. Describe the drying mechanism for each part of this curve.
9. Estimate how wet solid mass will vary with time during falling rate drying.

10. Water cooling is often described by enthalpy balances. Review the assumptions made in deriving these balances.
11. How is the HTU for mass transfer related to that for water cooling?
12. How would you analyze water cooling that was accomplished in stages?

Problems

1. You want to use a wet-bulb thermometer wet with carbon tetrachloride to determine the carbon tetrachloride concentration in air at 2 atm flowing at 62 °C. The wet bulb reads 23 °C. What is the carbon tetrachloride concentration?

2. Predict the mass transfer coefficient in cm/sec for liquid *n*-butyl alcohol vaporizing into air at 80 °F and 1 atm. You know that the heat transfer coefficient in the same system in 56 Btu/hr ft^2 °F.

3. A meteor is falling through the earth's atmosphere and burning as it falls. The burning can be approximated as a diffusion-controlled first-order chemical reaction oxidizing iron at the meteor's surface. Find the meteor's temperature in terms of only the heat of this reaction, the concentration of iron oxide vapor near the surface, and the properties of the air.

4. Antibiotic crystals in a filter cake are placed in trays at a depth of 2.0 cm and dried with excess solvent-free air at 40 °C (dry bulb). The crystals have a wet density of 0.77 g crystal/cm^3 cake and each tray originally contains 8.0 wt% acetone whose vapor concentration c in mol/cm^3 is approximated by the equation

$$c(\text{in mol/cm}^3) = 4.0 \cdot 10^{-6} + 0.24 \cdot 10^{-6} \, T \, (\text{in °C})$$

The heat of vaporization of acetone is about 7.8 kcal/mol; its molecular weight is 58; the heat capacity of air is about 5 cal/mol K. The final solvent concentration should be less than 0.6 wt%. Past experiments suggest that this drying should occur at constant rate, with a mass transfer coefficient (based on the vapor) equal to 1.8 cm/sec. How long does it take to dry these crystals?

5. Chemical plants will increasingly operate under the restraint of "zero discharge", so that they can no longer take processed water from the river next to the plant, use it to cool the heat exchangers, and dump it back into the river (cf. J.A. Dalan, *Chemical Engineering Progress*, November 2000, p. 71–76). Instead, the plants must install cooling towers, which cool the heat exchanger effluent and allow its recycle. This problem deals with such a water treatment system.

   Seven thousand kilograms of water per minute at 61 °C is to be cooled to 20 °C in a tower fed with dry air entering at 25 °C. The entering air flow is 50% greater than the minimum. (a) What is the total air flow in the cooling tower? (b) If the HTU in this tower is 2 m, estimate how tall it should be. In this estimate, you may assume that the equilibrium line is linear between $\tilde{H}_0^*$ and $\tilde{H}_l^*$.

6. You need to cool 50 gmol/sec of 60 °C water to 30 °C. To do so, you plan to build a cooling tower which has an HTU of 3 m, and operates at twice the minimum flow of air entering with an enthalpy of 1 kJ/mol. The equipment supplier suggests that the water flux in the tower should not exceed 80 gmol/m^2 sec, and that the average enthalpy difference can be approximated as half the average of the enthalpy differences at the tower's inlet and at the outlet. The heat capacity of liquid water can be

taken as 0.075 kJ/mol K. (a) What should the diameter of the tower be? (b) What should the total air flow be? (c) What should its height be?

7. Your plant has available a countercurrent cooling tower 10 m high and 6 m in diameter. The tower packing has a surface area per volume of about 63 m^{-1}. At present, it is effectively cooling water at 3200 kg/min from 66 °C to 20 °C, using air at 80 mol/m^2 sec at 18 °C and 20% relative humidity. (a) What is the mass transfer coefficient in this tower? (b) You need 1000 kg/min water at 15 °C or less to cool a new chemical reactor. By how much should you reduce the water flow to get this output?

8. Imagine that you fill the two-bulb capillary apparatus (see Fig. 3.1-2) with an equi-molar mixture of hydrogen and methane. Each bulb has a volume of 270 cm^3; the vertical capillary is 6 cm long. You place the lower bulb in ice water and heat the upper one with steam. (a) What is the maximum concentration difference due to thermal diffusion? *Answer:* $\Delta y_1 = 0.02$. (b) How many moles of hydrogen are there in the hot bulb? How many in the cold?

9. Track-etched membranes are made by exposing mica or polycarbonate sheets $15 \cdot 10^{-4}$ cm thick to an α-radiation source and then etching the sheet in hydrofluoric acid. The resulting membrane can have about 0.4% of its area pierced by 120-nm cylindrical pores. Imagine that you place a track-etched membrane across one end of a 2-cm-diameter glass pipe 36 cm long. You cover the other end with a filter that has a high Darcy's law permeability. If you set the pipe in the sun, air will flow into the pipe by thermal effusion through the track-etched membrane and out of the pipe by Darcy's law flow through the filter. How fast will the air flow if the air in the pipe is 47 °C and the surrounding air is 23 °C?

10. Extend the analysis for fast mass transfer given in Section 9.5 to include the effect of diffusion-engendered convection on heat transfer. Use the film theory in this extension. (a) Show that the energy equation in this situation is

$$0 = -\frac{d}{dz}(q + \rho \hat{C}_p T v)$$

subject to

$$z = 0, \quad T = T_0$$
$$z = l, \quad T = T_l$$

(b) Integrate this to find

$$q|_{z=0} = \rho \hat{C}_p v \left(\frac{T_l - T_0}{1 - e^{vl/\alpha}} \right)$$

(c) Defining

$$q|_{z=0} = h(T_0 - T_l) + \rho \hat{C}_p T_0 v$$

show that

$$h = \frac{\rho \hat{C}_p v}{e^{vl/\alpha} - 1}$$

11. The thermal conductivity of reacting gas mixtures is sometimes found to be larger than would be expected from molecular considerations. If a temperature gradient exists in a gas, then in different temperature regions, the concentrations of reactive species may be different; this concentration gradient can augment conduction because there is a transport of energy by molecular diffusion. A convenient system to study this phenomenon utilizes nitrogen dioxide. The reaction

$$2NO_2 \rightleftharpoons N_2O_4$$

is very rapid in both directions, and for most studies, the mixture may always be assumed to be in chemical equilibrium. (a) Assume two horizontal parallel plates separated by a distance of 0.16 cm. The gap is filled with a NO_2–N_2O_4 mixture. The lower plate is at 40 °C, so the mole fraction of $[NO_2]$ next to this plate is 0.48. The upper plate is at 80 °C; so the mole fraction of $[NO_2]$ adjacent to this plate is 0.85. If the diffusion coefficient of both species is 0.07 cm^2/sec, find the flux of $[NO_2]$ across the gap. (b) Find the molar average velocity across this gap. (c) Calculate the temperature profile, including that due to diffusion. (d) The thermal conductivity of these mixtures is about $4 \cdot 10^{-5}$ cal/cm sec K, and their heat capacity is about 7 cal/mol K. How much will the reaction in this system increase the heat flux?

12. Imagine a thin layer of gas of thickness l across which heat transfer occurs. The heat transfer is facilitated by a reactive gas 1 within the film. At the hot surface at T_0 and $z = 0$, the gas reacts catalytically, rapidly, and endothermically:

$$(2 \text{ moles of gas } 1) + (\text{heat}) \rightarrow (1 \text{ mol of gas } 2)$$

At the cold surface at T_l and $z = l$, the reaction is rapidly reversed:

$$(1 \text{ mol of gas } 2) \rightarrow (2 \text{ moles of gas } 1) + (\text{heat})$$

Heat conduction also occurs, but free convection does not. Only gas 1 and gas 2 are in the film, and the thermal conductivity is constant. Also assume that the thermal conductivity at the boundaries is much greater than in the bulk. Find the heat transfer coefficient across this thin film in three steps: (a) Find the concentration profiles in the film. (b) Find the temperature profile corrected for mass transfer. (c) Find the heat flux at the boundary $z = 0$.

Further Reading

American Institute of Chemical Engineers (1972). *Cooling Towers*. New York: American Institute of Engineers.

Chapman, S., and Dootson, F. W. (1917). *Philosophical Magazine*, **33**, 248.

Chilton, T. H., and Colburn, A. P. (1934). *Industrial and Engineering Chemistry*, **26**, 1183.

Dacey, J. W. H. (1980). *Science*, **210**, 1017.

Hoglund, R. L., Shacter, J., and von Halle, E. (1979). *Diffusion Separation Methods*. In *Encyclopedia of Chemical Technology*, ed. M. Grayson. New York: Wiley.

Ludwig, C. (1859). *Sitzungsberichte der Akademie der Wissenschaften-Wien*, **20**, 539.

Reynolds, O. (1874). *Proceedings of the Manchester Literary and Philosophical Society*, **14**, 7.

Reynolds, O. (1879). *Philosophical Transactions*, **170**, 727.

Sherwood, T. K., Pigford, R. L., and Wilke, C. R. (1975). *Mass Transfer*. New York: McGraw-Hill.

Soret, C. (1879). *Archives des Sciences Physiques et Naturelles*, **2** (3), 48; (1888). *Annales de Chimie et de Physique*, **22** (5), 293.

Tyrrell, H. J. V. (1961). *Diffusion and Heat Flow in Liquids*. London: Butterworth.

Index